Springer Series in Optical Sciences   Volume 3

Editor   David L. MacAdam

Springer Series in Optical Sciences

Volume 1 **Solid-State Laser Engineering**
By W. Koechner

Volume 2 **Table of Laser Lines in Gases and Vapors**
By R. Beck, W. Englisch, and K. Gürs

Volume 3 **Tunable Lasers and Applications**
Editors: A. Mooradian, T. Jaeger, and P. Stokseth

Volume 4 **Nonlinear Laser Spectroscopy**
By V. S. Letokhov and V. P. Chebotayev

Tunable Lasers and Applications

Proceedings of the Loen Conference, Norway, 1976

Editors
A. Mooradian   T. Jaeger   P. Stokseth

With 238 Figures

Springer-Verlag Berlin Heidelberg GmbH 1976

Dr. ARAM MOORADIAN
Leader of the Quantum Electronics Group,
MIT, Lincoln Laboratory,
P.O. Box 73, Lexington, Massachusetts 02173/USA

Dr. TYCHO JAEGER
Norwegian Defence Research Establishment,
Division for Electronics,
P.O. Box 25, Kjeller/Norway

Dr. PER STOKSETH
Electrooptics Group,
Norwegian Defence Research Establishment,
Division for Electronics,
P.O. Box 25, Kjeller/Norway

Dr. DAVID L. MACADAM
68 Hammond Street, Rochester, NY 14615/USA

This conference was sponsored by:

Norwegian Council of Industrial and
  Scientific Research
Norwegian Defence Research
  Establishment
Massachusetts Institute of Technology
Office of Naval Research

A/S SIMRAD
Kerr McGee Corporation
EXXON Corporation
Spectra Physics
Coherent Radiation
Laser Analytics

ISBN 978-3-662-13481-8   ISBN 978-3-540-37996-6 (eBook)
DOI 10.1007/978-3-540-37996-6

Preface

The Tunable Lasers and Applications Conference was held in Loen, Norway, on June 6-11, 1976. This conference dealt with the technology of tunable lasers from the vacuum ultraviolet to the far infrared and their application in the areas of photochemistry, chemical kinetics, isotope separation, atmospheric photochemistry and environmental studies, photobiology, and spectroscopy. The need for such a conference grew out of the rapidly expanding use of tunable lasers in a broad range of disciplines.

The conference was attended by 130 scientists representing Japan, Italy, West Germany, Canada, Israel, France, England, Norway, Sweden, Brazil, Denmark, Finland, the Netherlands, the Soviet Union, and the United States.

The location of the conference in Loen, Nordfjord, was chosen because of the magnificent beauty of its aqua-blue fjords surrounded by glacier-capped mountains and lush green hillsides. The Alexandra Hotel was a perfect host for such a conference with first class services, superb smorgesbord, and excellent audio-visual facilities. The atmosphere was free of distractions and provided for a relaxed interchange of ideas. An afternoon hike was arranged to the Briksdal glacier while the highlight of the outdoor activities was a bus-ship excursion to the magestically splendid Geiränger fjord. A sumptuous banquet was followed by an interesting and entertaining illustrated talk on high-speed and underwater photography by Professor Harold Edgerton from MIT.

Financial support from our sponsors is gratefully acknowledged in helping to make this conference a complete success.

Aram Mooradian
Tycho Jaeger
Per Stokseth

September 1976

Contents

I. Tunable and High Energy UV-Visible Lasers

    Tunable Coherent VUV Radiation. By B.P. Stoicheff and
    S.C. Wallace 1

    High Efficiency UV Lasers. By J.J. Ewing and C.A. Brau 21

    Tunable VUV Excimer Laser Systems
    By D.J. Bradley, M.H.R. Hutchinson and C.C. Ling 40

    Dye Laser Technology. By F.P. Schäfer 50

II. Tunable IR Laser Systems

    Scalable Tunable IR Lasers. By A. Mooradian 60

    Parametric Oscillators. By R.L. Byer 70

    Tunable Infrared Generation in Molecular Gases
    By J. Ducuing, R. Frey and F. Pradère 81

    Tunable High Power Raman Lasers and Their Applications
    By A.Z. Grasiuk and I.G. Zubarev 88

    Efficient High-Power 8.62 μm Infrared Radiation Source for
    Uranium Isotope Separation in UF_6. By R.L. Aggarwal, N. Lee
    and B. Lax 96

III. Isotope Separation and Laser Driven Chemical Reactions

    Laser Chemistry at Surfaces. By M.S. Djidjoev, R.V. Khokhlov,
    A.V. Kiselev, V.I. Lygin, V.A. Namiot, A.I. Osipov,
    V.I. Panchenko and B.I. Provotorov 100

    The Photophysics and Photochemistry of Formaldehyde
    By A.P. Baronavski, A. Cabello, J.H. Clark, Y. Haas,
    P.L. Houston, A.H. Kung, C.B. Moore, J. Reilly, J.C. Weisshaar
    and M.B. Zughul 108

    Future Applications of Selective Laser Photophysics and
    Photochemistry. By V.S. Letokhov 122

    Uranium Isotope Separation and its Demand on Laser Development
    By S. Rockwood 140

IV. Nonlinear Excitation of Molecules

Dissociation of Polyatomic Molecules by an Intense Infrared
Laser Field. By R.V. Ambartzumian 150

Collisionless Dissociation of Polyatomic Molecules by Multiphoton
Infrared Absorption. By N. Bloembergen, C.D. Cantrell and
D.M. Larsen 162

Laser Excitation of Molecules to High States of Vibration
By K.L. Kompa 177

Double Resonance and Energy Transfer in Sulfur Hexafluoride
By. J.I. Steinfeld and C.C. Jensen 190

V. Laser Photokinetics

Laser Induced Collisions. By S.E. Harris, R.W. Falcone,
W.R. Green, D.B. Lidow, J.C. White and J.F. Young 193

Application of Picosecond Laser Pulses to the Determination of
Vibrational Time Constants of Polyatomic Molecules in Liquids
By W. Kaiser and A. Laubereau 207

Optical Coherent Transients by Laser Frequency-Switching
By R.G. Brewer and A.Z. Genack 218

Relaxation in Macroscopic System: An Information Theoretic
Approach. By R.D. Levine 224

VI. Atmospheric Photochemistry and Diagnostics

Tropospheric Photochemical and Photophysical Processes
By J.N. Pitts, Jr. and B.J. Finlayson-Pitts 236

Photochemistry in the Stratosphere. By H.S. Johnston 259

Remote Sensing Using Tunable Lasers
By K.W. Rothe and H. Walter 279

VII. Photobiology

Resonance Raman Spectroscopy: Application of Tunable Lasers
to the Study of the Molecular Mechanisms and Dynamics of
Visual Excitation. By R. Mathies, A.R. Oseroff, T.B. Freedman
and L. Stryer 294

Laser-Induced Fluorescence of Biological Molecules
By A. Andreoni, A. Longoni, C.A. Sacchi, O. Svelto and
G. Bottiroli 303

Fluorescence Spectroscopy Applied to Dynamics and Structure
of Biopolymers. By M. Ehrenberg and R. Rigler 314

VIII. Spectroscopic Applications of Tunable Lasers

Applications of High Resolution Laser Spectroscopy
By T.W. Hänsch 326

Applications of Far Infrared Lasers. By B. Lax 340

Tunable Laser Spectroscopy in Mineral Prospecting
By S.T. Eng and E. Max 348

Study on Phase-Matching Characteristics of Optical Second Harmonic
Generation in Nonlinear Thin-Film Waveguides Using a Tunable
Parametric Oscillator. By H. Ito and H. Inaba 353

Control Techniques for CW Dye Lasers. By J.L. Hall and S.A. Lee ... 361

Optically Pumped Gas Lasers. By H. Kildal and T.F. Deutsch 367

Cars Techniques and Applications. By J.-P. Taran 378

Development of Cars for Measurement of Molecular Parameters
By S.A. Akhmanov, A.F. Bunkin, S.G. Ivanov, N.I. Koroteev,
A.I. Kovrigin and I.L. Shumay 389

List of Participants 398

I. Tunable and High Energy UV-Visible Lasers

TUNABLE COHERENT VUV RADIATION

B.P. Stoicheff and S.C. Wallace
Department of Physics, University of Toronto
Toronto, Canada M5S 1A7

I Introduction

The direct aproach to generation of coherent radiation in the vacuum ultraviolet (VUV) is, of course, through the development of suitable lasers. There has been considerable progress in this area of laser research, as shown in Table I. Molecular hydrogen provided the first medium for laser emission in the 1600 Å region in 1970[1], and later work has extended the emission to 1100 Å[2]. Such lasers have produced emission at several hundred different wavelengths spanning the region 1098 Å to 1644 Å. Carbon monoxide also provides a multi-line output, from 1811 to 1970 Å[3], and ionized carbon, CIV, produces emission in two resonance lines at 1548 and 1551 Å[4]. Molecular dissociation of Xe_2, Kr_2, and Ar_2 has provided high output powers in broad bands centred at 1726 Å, 1457 Å, 1261 Å, respectively[5]. These emissions should each be tunable over regions of 1000 to 1500 cm^{-1}, and some success has been achieved with Xe[6]. Recently ArF, with emission at 1940 Å [7] has been added to this family of excimer lasers. We look forward to learning the current status of these lasers in succeeding papers at this Conference.

An important and entirely new direction for the production of coherent radiation in the UV and VUV regions was opened with the observation by FRANKEN et al in 1961[8], of second harmonic radiation at 3472 Å generated from ruby laser light (6944 Å) incident on crystalline quartz. This was quickly followed in 1962 with the

classic theoretical paper on second- and third- order nonlinear
susceptibilities by ARMSTRONG et al[9], and with the beautiful
experiments on index matching by orienting birefringent crystals,
leading to high conversion efficiencies, carried out by GIORDMAINE,
and MAKER et al[10]. In 1963, third harmonic generation of ruby
light, at 2314 Å was demonstrated by MAKER et al[11] in crystals,
glasses and liquids, and other interesting third-order nonlinear
effects by MAKER and TERHUNE[12] in condensed matter. A major
problem to the generation of even shorter wavelengths was the
limited transparency of many nonlinear solids to the region above
~2000 Å.

The solution to this difficulty became evident with the generation of third harmonic radiation in a large number of atomic and
diatomic gases, by NEW and WARD in 1967[13]. Harris and his co-
workers proposed[14] and demonstrated[15] that high conversion
efficiency of third harmonic generation and sum frequency mixing
could be obtained by using phase-matched metal vapors as nonlinear
media. They succeeded in generating coherent radiation at 1182 Å
in Cd:Ar[16], at 886 Å in Ar[17], and tunable picosecond radiation
in the VUV by nonlinear mixing in Xe gas[18]. Detailed calculations of the third-order susceptibility by MILES and HARRIS[19],
showed that additional enhancement could be realized if the incident radiation frequency is tuned to one-, two-, or three-photon
allowed transitions in the metal vapors. Such resonant enhancement was observed by mixing visible and infrared radiation in
Na ($2\omega_1+\omega_2$ to yield ω_3) at 3321 Å by BLOOM et al[20], and in third
harmonic generation of ruby laser light in Cs by LEUNG et al[21].
Resonant enhancement also made possible the generation of tunable
VUV radiation by four-wave sum mixing ($2\omega_1+\omega_2 \rightarrow \omega_3$) in Sr vapor,
reported by HODGSON et al in 1974[22]. They employed two tunable
dye lasers, one tuned to a two-photon allowed transition and the
other to a frequency such that $2\omega_1+\omega_2$ corresponded to a transition
from the ground state to an autoionizing state, and succeeded in
generating continuously tunable radiation from 1778 to 1817 Å and
from 1833 to 1957 Å. Similar methods have been used at the University of Toronto to generate tunable radiation in Mg vapor,
from 1400 to 1600 Å[23], and recently this has been extended to
~1200 Å. It has also been possible to obtain four-wave mixing

in molecular vapors, notably in NO, with tunable emission in broad regions centred at 1510, 1430, 1360 and 1300 Å[24].

It is the purpose of this paper to briefly review the process of four-wave sum mixing, with application to the generation of tunable coherent radiation in the VUV region. Brief descriptions of apparatus will be given, followed by specific examples taken from the work of the groups at IBM, Stanford and Toronto.

II Theoretical Framework

It is well known[9] that the polarization of a medium in the presence of a monochromatic or quasimonochromatic field, namely, $\bar{E}(r,t) = \sum_i \bar{E}(\omega_i)$, can be written as

$$\bar{P}(\omega_i) = \chi^{(1)}(\omega_i) \cdot \bar{E}(\omega_i) + \sum_{jk} \chi^{(2)}(\omega_i = \omega_j + \omega_k) : \bar{E}(\omega_j) \bar{E}(\omega_k)$$
$$+ \sum_{jk\ell}^i \chi^{(3)}(\omega_i = \omega_j + \omega_k + \omega_\ell) : \bar{E}(\omega_j) \bar{E}(\omega_k) \bar{E}(\omega_\ell) + \dots \quad (1)$$

where $\chi^{(n)}$ are the susceptibility tensors of nth order. For third harmonic generation and four-wave sum mixing, we need only be concerned with $\chi^{(3)}$, whose principal term may be written as

$$\chi^{(3)}(\omega_0 = \omega_1 + \omega_2 + \omega_3) \ \alpha \cdot \frac{<g|r|a><a|r|b><b|r|c><c|r|g>}{(\Omega_{cg} - \omega_1 - \omega_2 - \omega_3)(\Omega_{bg} - \omega_1 - \omega_2)(\Omega_{ag} - \omega_1)} \quad (2)$$

Here, $<g|r|a>$ is the electric dipole matrix element between the ground state g and an excited state a, having a lifetime Γ_a, and $\Omega_{ag} = \omega_{ag} - i\Gamma_a/2$ is the energy difference between the states a and g (see Fig.1).

Equation (2) shows that resonance enhancement occurs whenever the applied frequencies ω_1, ω_2, ω_3 are such that the real part of the resonance denominator vanishes, namely when $(\Omega_{ag} - \omega_1) = 0$, $(\Omega_{bg} - \omega_1 - \omega_2) = 0$, or $(\Omega_{ag} - \omega_1 - \omega_2 - \omega_3) = 0$, corresponding to one, two, or three photon resonance, respectively. If any of ω_1, ω_2, or ω_3 is set equal to a resonance frequency (Ω_{ag}, etc.), $\chi^{(3)}$ will be resonantly enhanced but the incident light will be strongly absorbed. Also if $\omega_0 = \omega_1 + \omega_2 + \omega_3$ equals a resonance frequency, the generated radiation will be absorbed. If, however, $\omega_1 + \omega_2$ is equal to a two photon transition (Ω_{bg}), the incident light at $\omega_1 + \omega_2$ will be relatively weakly absorbed by the weak two photon transition, while the resonant enhancement of $\chi^{(3)}$ could be just as strong as for the single-photon resonances.

4

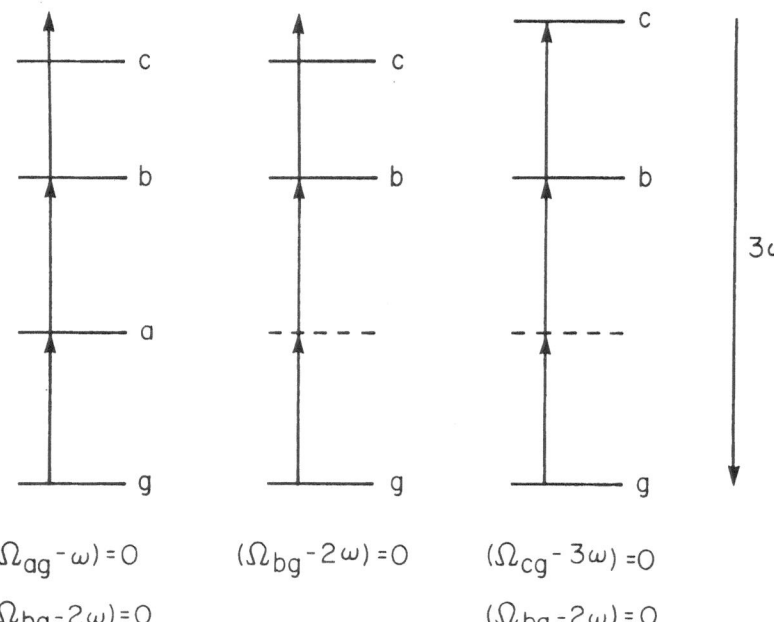

$(\Omega_{ag}-\omega)=0$ $(\Omega_{bg}-2\omega)=0$ $(\Omega_{cg}-3\omega)=0$

$(\Omega_{bg}-2\omega)=0$ $(\Omega_{bg}-2\omega)=0$

Fig.1. Energy level diagram representing some resonant processes leading to third harmonic generation.

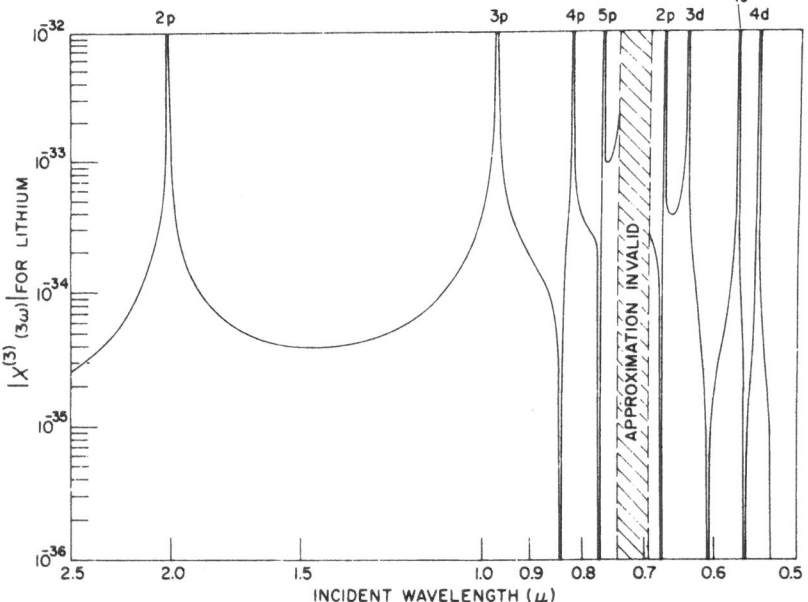

Fig.2. Variation of $\chi^{(3)}$ (3ω) as a function of wavelength for Li[19].

For third harmonic generation, $\chi^{(3)}$ simplifies to

$$\chi^{(3)}(\omega_0=3\omega) \quad \alpha \quad \frac{<g|r|a>etc.}{(\Omega_{cg}-3\omega)(\Omega_{bg}-2\omega)(\Omega_{ag}-\omega)} \tag{3}$$

When ω, 2ω, or 3ω approaches resonance, $\chi^{(3)}$ undergoes strong enhancement, as shown in Fig.2 for lithium. By tuning 2ω to such a resonance, the nonlinear polarization $\chi^{(3)}|E|^3$ in vapors can be made to approach that of solids.

For efficient third-harmonic generation, collinear phase-matching is a necessity, that is, the refractive index $n(3\omega)=n(\omega)$, or $\chi^{(1)}(3\omega)=\chi^{(1)}(\omega)$, in order to yield a longer effective interacting length. In a vapor, this is achieved by selecting an appropriate buffer gas and varying its density to satisfy the phase matching relation $\chi^{(1)}_{vapor}(\omega)+\chi^{(1)}_{buffer}(\omega)=\chi^{(1)}_{vapor}(3\omega)+\chi^{(1)}_{buffer}(3\omega)$. An example is shown in Fig.3 for a mixture of Rb vapor and Xe buffer gas. This was demonstrated to have an efficiency of 10 to 20%[25] in converting 1.06 µm radiation to 3547 $\overset{0}{A}$ radiation.

For generation of tunable VUV radiation, the process $\omega_0=2\omega_1+\omega_2$ is of interest: $\chi^{(3)}$ becomes

$$\chi^{(3)} \quad \alpha \quad \frac{<g|r|a>etc...}{(\Omega_{cg}-2\omega_1-\omega_2)(\Omega_{bg}-2\omega_1)(\Omega_{ag}-\omega_1)} \tag{4}$$

In this case, strong enhancement is achieved by tuning $2\omega_1$ to a parity-allowed two-photon resonance frequency, Ω_{bg}. Moreover, third harmonic generation can be eliminated by using circularly polarized ω_1 and ω_2 radiation, since angular momentum will not be conserved for frequency tripling in an isotropic medium under these circumstances.

Tunability can be obtained by selecting ω_1 and ω_2 such that $2\omega_1+\omega_2$ corresponds to an energy beyond the ionization limit, and in particular to a broad autoionized level, as demonstrated by HODGSON et al[22] for Sr (Fig.4). SOROKIN et al[26] have discussed the contribution of the autoionizing state and of the continuum with which it interacts, to the four-wave susceptibility, namely

$$\chi^{(3)} \quad \alpha \quad \frac{<g|r|a><a|r|b>}{(\Omega_{bg}-2\omega_1)(\Omega_{ag}-\omega_1)} \int \frac{<g|r|\Psi_\nu><\Psi_\nu|r|b>d\nu}{(\Omega_{\nu g}-2\omega_1-\omega_2)} \tag{5}$$

Here $<g|r|\Psi_\nu>$ is the matrix element of the dipole moment operator from the ground state to the unperturbed continuum state Ψ_ν.

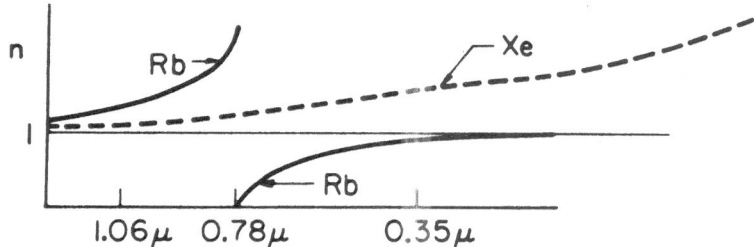

Fig.3. Variation of refractive index of a Rb:Xe mixture with wavelength near a resonance line[17].

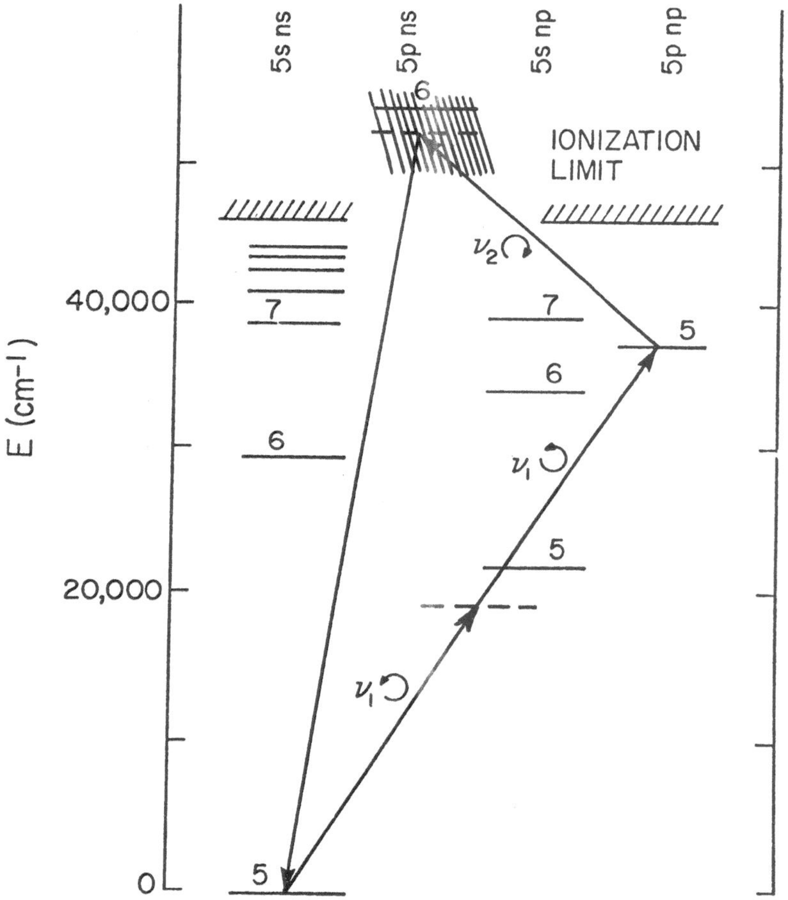

Fig.4. Energy level diagram of Sr showing resonances with $2\nu_1$, and with $2\nu_1 + \nu_2$ to a broad autoionizing level.

Resonance enhancement also plays a role here since the integral
has a large contribution from the broad region of the autoionizing
state ($\Omega_{\nu g}=2\omega_1+\omega_2$). Thus by keeping $2\omega_1$ fixed to the resonance
Ω_{bg}, and varying ω_2, so that $2\omega_1+\omega_2$ scans through the autoionizing
states or the continuum, reasonably intense, tunable VUV radiation
can be generated.

III Experimental Considerations

The pioneering work of Harris and his co-workers on harmonic
generation, phase-matching, and resonance enhancement, led to the
first announcement of efficient harmonic generation into the VUV
in 1972[16]. Their primary laser source for mixing experiments is
shown in Fig.5. A mode-locked Nd:YAG laser and amplifier produce
1.06μm radiation in the form of several pulses each of 50 psec duration, with total energy ~10 mJ and peak power of 20 MW. This radiation is frequency doubled by a KDP crystal to 5320 Å (at up to
80% efficiency). The 5320 Å radiation is then used directly, or
again doubled or mixed with 1.06 μm radiation in a second KDP crystal, to produce radiation at 2660 Å or 3547 Å (with up to 10% efficiency). In this way, high power radiation at many fixed frequencies is readily available for further mixing experiments in metal
vapors and inert gases, for example, the tripling of 3547 Å radiation in a Cd:Ar mixture to yield 1182 Å radiation (KUNG et al[16]),
and tripling of 2660 Å in Xe:Ar to yield 887 Å radiation(HARRIS et
al[17,27]).

Tunability was added to this source[18] by using the 2660 Å radiation to pump a parametric generator (consisting of an ADP crystal). The resulting broad-band fluorescence is spatially and frequency filtered, and amplified in a second ADP crystal. Variation
of the crystal temperatures from 50 to 105°C then provides a tuning
range of 4200 to 7200 Å, with >100μJ output and bandwidth of 5 to
20 Å. This tunable source was used by KUNG[18] to generate tunable
picosecond VUV radiation in Xe, as discussed below.

HODGSON, SOROKIN, and WYNNE[22] obtained tunable VUV radiation
with the use of N_2-laser-pumped dye lasers as the primary source
(Fig.6). In this scheme, the narrow bandwidth radiation is tuned
to approach the doppler profile of the atomic energy levels of the
nonlinear medium, thus making use of specific resonance enhancement

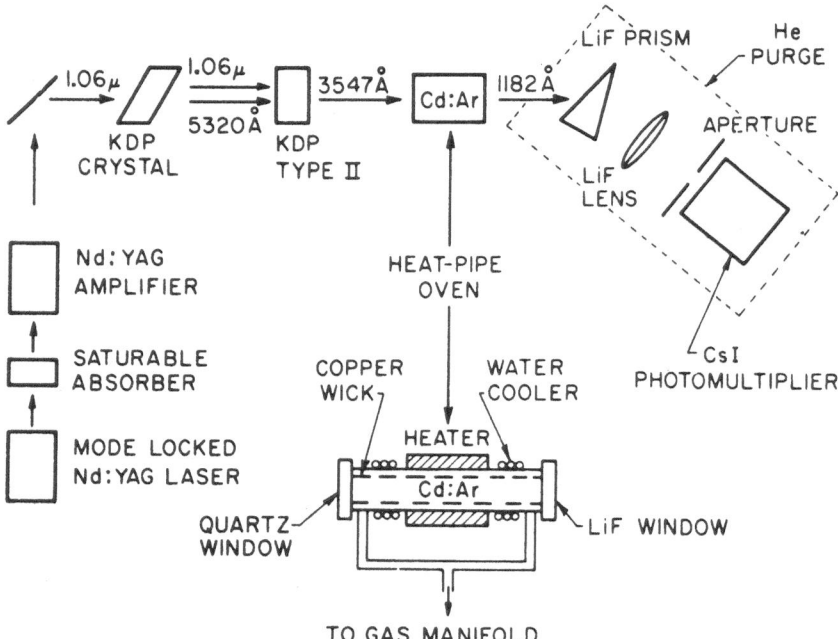

Fig.5. Method of generating coherent VUV radiation with
a high power laser[16].

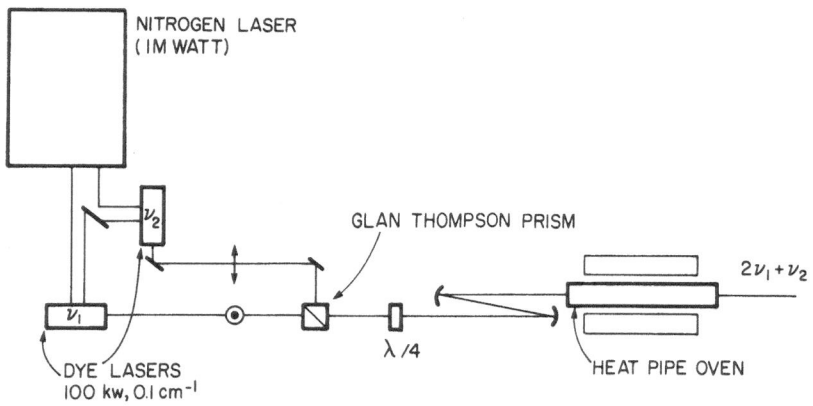

Fig.6. Method of combining beams from two dye lasers to
produce tunable, coherent VUV radiation[22,23].

and thereby requiring significantly lower laser powers to increase
$\chi^{(3)}$. We have built a similar apparatus at Toronto, following the
general design of HODGSON et al[22]. A N$_2$ laser emitting ~10 nsec
pulses of 1 MW peak power is used to pump, simultaneously, two dye
lasers of the HÄNSCH design[28]. These yield up to 50 KW output,
typically ~0.1 cm^{-1} in linewidth, and tunable from ~4000 to 7000 Å,
depending on the dye or dye mixture used. The two laser beams are
orthogonally polarized, then spatially overlapped in a Glan-Thomp-
son prism, and circularly polarized in opposed senses by a quarter-
wave plate. The collinear beams, of frequencies ν_1 and ν_2 are
then focused into a heat-pipe oven containing the atomic metal
vapor. Typically, one dye laser is fixed at half the frequency
of an appropriate double-photon transition, and the other is of
variable frequency, so as to generate tunable radiation at the sum
frequency $2\nu_1+\nu_2$. In this way, HODGSON et al[22] used Sr vapor to
generate the first tunable VUV radiation. This achievement, and
the recent results of WALLACE and ZDASIUK with Mg[29], and of INNES
et al with NO[30], at Toronto, will be reviewed below.

IV Results and Discussion

Xenon. The pump, signal, and idler radiation generated by the
high power parametric source was used by KUNG[18] with Xe to produce pulsed VUV radiation tunable in the range 1180 to 1946 Å.
The incident radiation was tightly focused in Xe, and the many
third-order nonlinear processes provided by sum, difference, and
harmonic generation of the pump, signal, and idler radiation yielded the range of tunability shown in Fig.7 (along with the predicted range). It was possible to tune continuously from 1631 to
1946 Å, and to sample portions of the range from 1180 to 1470 Å.
Measured output was typically 1 W, corresponding to ~10^7 photons
in a 20 psec pulse. For this specific example, there was no possibility for resonance enhancement nor for phase matching: hence
only such a powerful laser source could be used for VUV generation, and of necessity the resulting bandwidth was 5 to 20 Å.
Nevertheless, this would be a useful source for studies in picosecond spectroscopy and chemical reactions.

Strontium. The initial experiments in Sr vapor[22] were the first
experimental verification that resonantly enhanced four-wave mixing

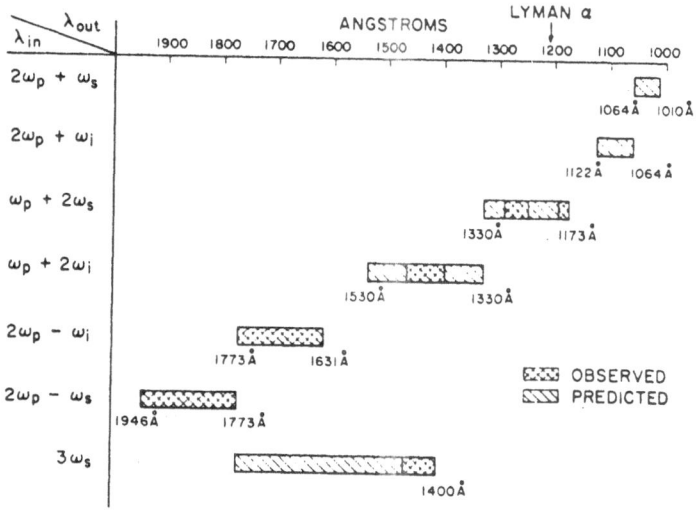

Fig.7. Tuning ranges available by nonlinear sum, difference, and third harmonic generation using a parametric generator[18]. (ω_s, ω_p, ω_i are the signal, pump and idler frequencies, respectively.)

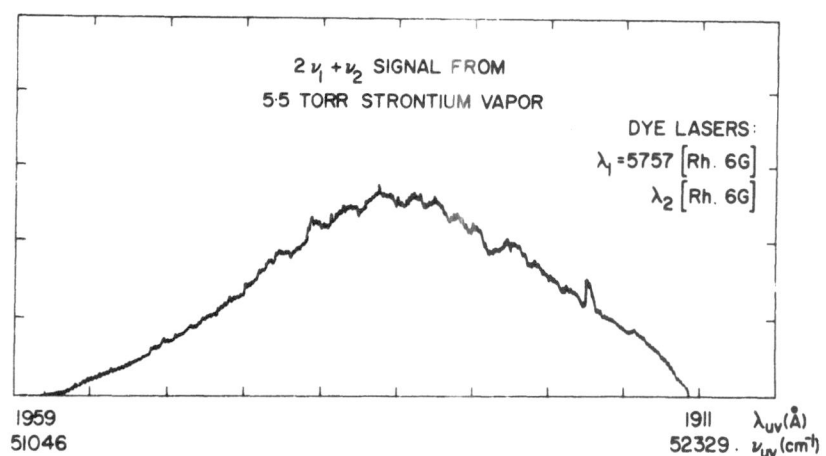

Fig.8. An example of the broad range of VUV generated in Sr[26].

could produce a tunable, single-frequency source of coherent VUV radiation, using quite modest powers of ~10 KW at the fundamental frequencies. Here, radiation of frequency ν_1 was tuned so that $2\nu_1$ was in resonance with various excited even-parity states of $^1D_2$ symmetry, and ν_2 was swept to enable $2\nu_1 + \nu_2$ to reach into the ionization region. As already noted in II, extra resonant enhancement in $\chi^{(3)}$ is produced by autoionizing states embedded in the continuum, which correspond in energy to the sum-frequency generated (recall the term $\Omega_{vg} = 2\omega_1 + \omega_2$ in Eq.5). The technique of using opposing senses of circular polarization for ν_1 and ν_2 makes the process $\nu = 2\nu_1 + \nu_2$ the exclusive nonlinear effect to be observed. A summary of the tuning ranges that can be covered using various combinations of four commonly used dyes are shown in Table II. The two-photon resonance states (Ω_{bg} in Eqs.(4) and (5) are shown, as well as the dye laser wavelengths. The ranges underlined in Table II show the actual tunable radiation observed[22].

An example of the intensity variation of VUV generated in Sr is shown in Fig.8. Such "spectra" yield new information about autoionizing states and bound-continuum matrix elements in general [26]. The autoionizing levels appear as very intense peaks because they gain intensity at the expense of the continuum oscillator strength. However, the consequence of this effect is that, in the presence of autoionizing levels, the range of continuous wavelength tunability is somewhat limited, and there are often wide regions of low conversion efficiency between each state. Fortunately, the autoionizing levels of lowest principle quantum number, are usually very broad so that regions as broad as 500 cm^{-1} are easily covered at wavelengths immediately above the ionization limit. Other metal vapors showing similar behaviour to Sr, are Ca, Eu, and Hg. Table III summarizes the available data of VUV radiation generated by these systems.

As the strong effects of the autoionizing transitions on $\chi^{(3)}$ demonstrate, one-photon resonant-enhancement terms ($\Omega_{cg} - \omega_0$) and ($\Omega_{ag} - \omega_1$) can significantly contribute to harmonic intensities if the linear absorption is not too large. Examples of additional intensity due to the term ($\Omega_{ag} - \omega_1$) have been observed in Eu[26] and Sr vapor[29] where intercombination lines provide states of

the appropriate oscillator strength. An experimental difficulty
with this approach lies in the fact that three dye lasers are
required for tunable harmonic generation. Thus in Sr one laser
is tuned to the $5s5p(^3P_1)$ state, the second to the $5p^2(^3P_2)$ two
photon state, and only the third frequency can be freely varied.
An increase of ~10 in harmonic intensity can be obtained in this
manner.

Magnesium. Significantly wider regions of wavelength tunability
have been obtained in Mg vapor[23]. In this example the continuum
is unperturbed by autoionizing levels over a 10,000 cm^{-1} region
from the ionization limit, resulting in the broad range of tunability shown in Fig.9. The solid curve in Fig.9 was obtained by
using the $3s3d(^1D_2)$ state (see Fig.10) for resonant enhancement
$(2\nu_1)$, and tuning many different dye lasers (ν_2) to cover the
region of the ionization continuum from 1400 to 1600 Å. The conversion efficiency in Mg vapor is the highest observed for these
systems, namely 0.2% (or $2x10^{11}$ photons/pulse) at 30 KW incident
dye laser power. The exact form of the "spectrum" in Fig.9 is
determined primarily by the bound-continuum matrix elements from
the $3s^2$ and 3s3d states to the p-wave continuum. For Mg, it is
realistic to evaluate these matrix elements using quantum defect
theory[30], and the theoretical curve given by the dashed line
in Fig.9 agrees quite well with experiment. The physical basis
for the position of maximum harmonic intensity is simply the relative phasing of the bound and continuum wavefunctions. Since
the susceptibility depends on the integral over the entire continuum, it should be recognized that nodes in the bound electronic wavefunctions will result in interference effects and
in significant reduction of the third-order nonlinear susceptibility because of cancellation in the integral. In Mg, the
3s electron is quite hydrogenic (quantum defect ~2) and the 3d
radial wavefunction asymptotically approaches zero, so that the
only important factor is the quantum defect of the continuum
wavefunction. More recent studies at Toronto in the region of
the autoionizing transitions at 1300 Å, also give strong harmonic
radiation, but the intensity in the 1200 to 1300 Å region is reduced by a factor of 10 over that at the longer wavelengths.

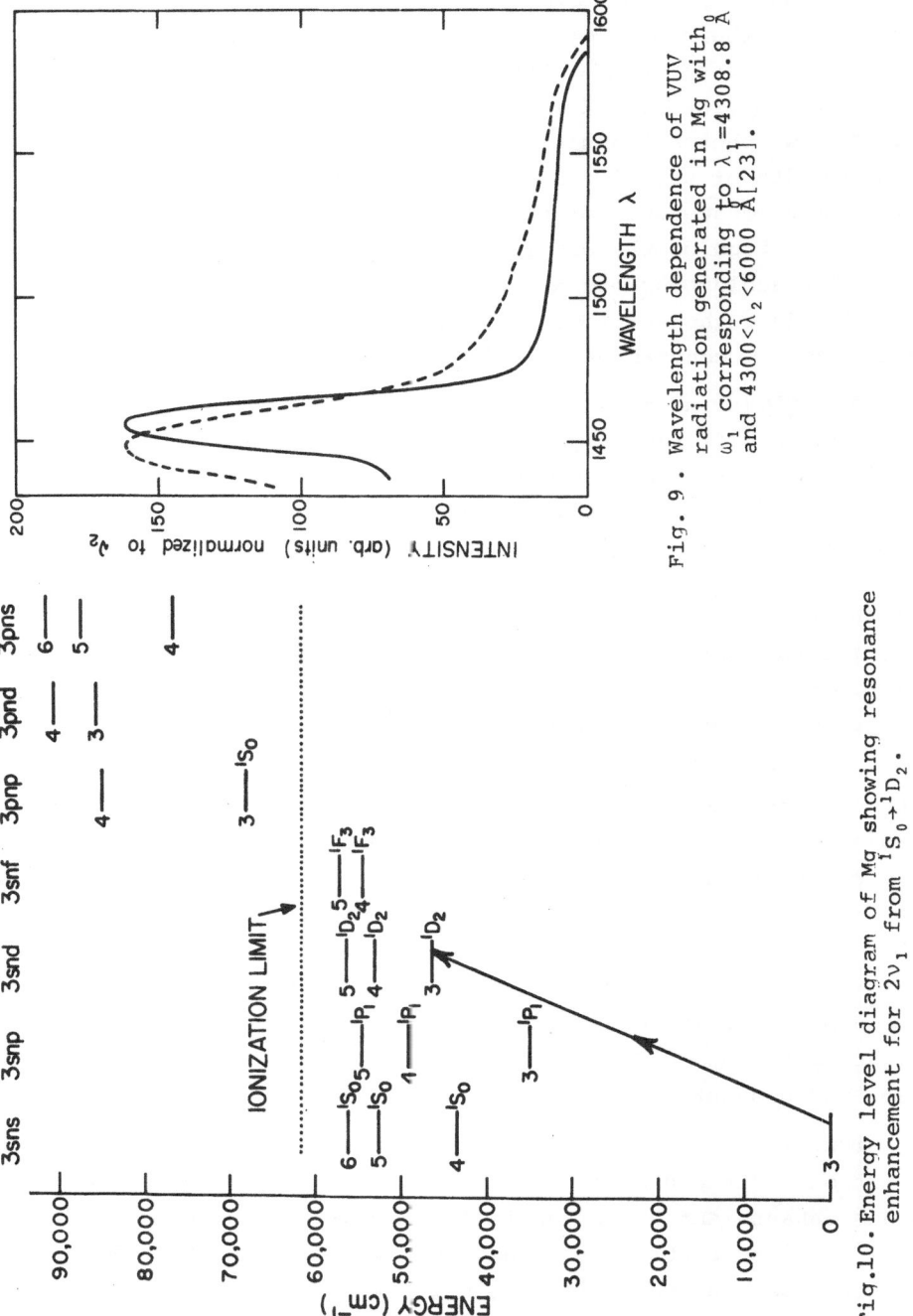

Fig. 9. Wavelength dependence of VUV radiation generated in Mg with ω_1 corresponding to $\lambda_1=4308.8$ Å and $4300<\lambda_2<6000$ Å[23].

Fig.10. Energy level diagram of Mg showing resonance enhancement for $2\nu_1$ from $^1S_0 \rightarrow {}^1D_2$.

The onset of saturation was observed in Mg vapor with the departure of the cubic dependence of output on the incident intensity, at powers >30 KW. This appears to be in agreement with an approximate criterion for such effects that $W^{(2)}\tau=1/2$: here, $W^{(2)}$ is the two-photon transition rate and τ is the lesser of the laser pulse width or the lifetime of the two-photon state[34]. The important consequence of this relation is the restriction on pulse duration of the driving radiation, with the result that the use of flashlamp-pumped dye lasers (with τ typically ~400 μsec) would reduce the saturation limited power by two orders of magnitude and the conversion efficiency by 10^6.

Nitric Oxide. As an alternative to metal vapors, INNES and the authors[24] recently explored the use of molecular systems as nonlinear media, and found that NO was admirably suited to VUV generation. Nitric oxide offered several advantages for these experiments. Detailed information was available on its electronic states and on oscillator strengths, and earlier work[35] had identified a strong two-photon transition which proved to be suitable for resonant enhancement of $\chi^{(3)}$. Finally, a simple gas cell replaced the heat-pipe oven used with the metal vapors, and high gas pressures could be readily obtained.

As shown in the energy level diagram (Fig. 11) one laser was tuned so that $2\nu_1$ was in resonance with the $X^2\Pi \rightarrow A^2\Sigma^+$ transition. Tuning of the second dye laser to the $C^2\Pi$ manifold, produced a rich rotational structure in the VUV output, part of which is shown in Fig. 12. VUV radiation was generated in bands of breadth ~600 cm^{-1} in the regions 1510, 1430, 1360, and 1300 Å, corresponding to the (0,0), (1,0), (2,0) and (3,0) γ-bands[36]. Although the susceptibility of NO is much lower than in atomic systems, the higher gas density (~90 torr) permitted photon yields of ~10^7 photons/pulse at 20 KW incident laser powers.

Significant pressure broadening occurred at 10 atm, and the rotational structure of the two-photon transition used for resonance enhancement was essentially eliminated, as shown in Fig. 10. This provided continuously tunable VUV radiation by simple third harmonic generation, using a single dye laser. Thus with NO, we have in the VUV, the equivalent of the frequency-doubling crystal

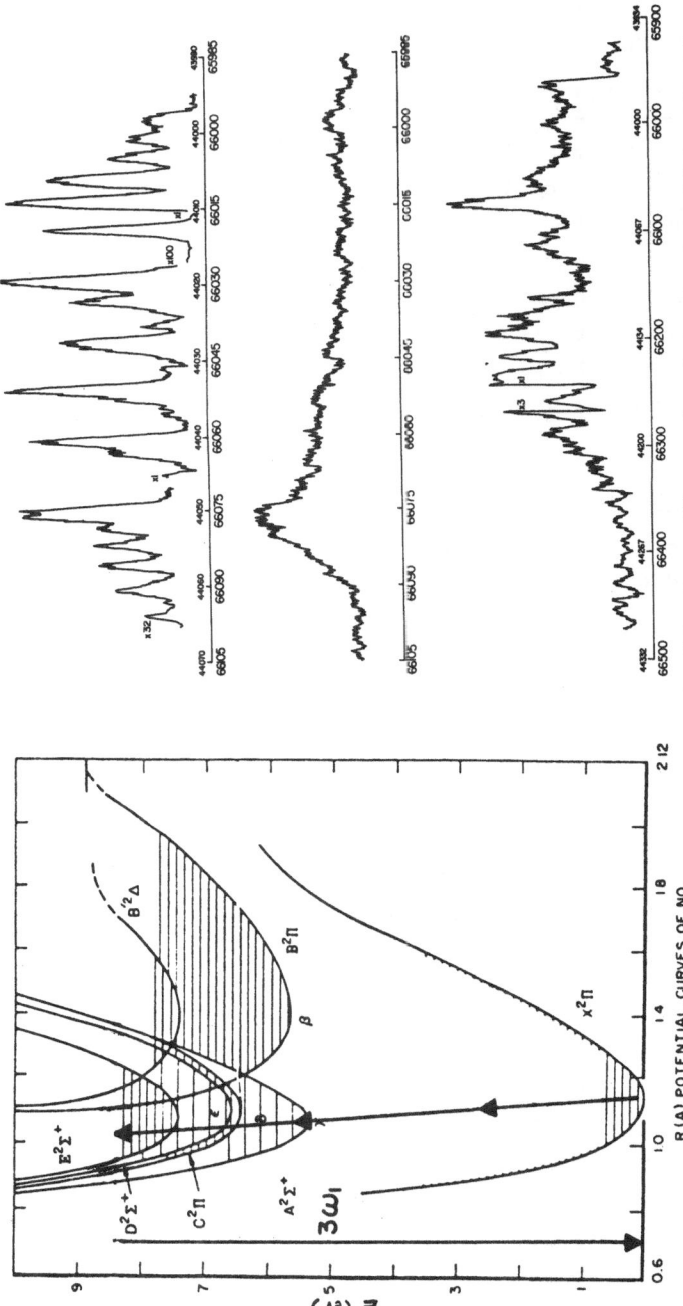

Fig.12. Recordings of coherent, tunable radiation generated in NO[24]. The top and middle show the radiation near 1500 Å, at pressures of 50 torr and at 10 atm, respectively. At the bottom is shown the full tunable range of 600 cm⁻¹ obtained when the γ(0,0) band of the $A^2\Sigma^+$-$X^2\Pi_{3/2}$ system is used for resonance enhancement.

Fig.11. Energy level diagram of NO showing the resonances used in third harmonic generation[24].

commonly used for UV generation from visible radiation.
Moreover, it appears that CW third harmonic generation may
be possible with NO.

V Summary and Conclusions

Tunable, coherent VUV radiation has been generated in several
atomic vapours and in a diatomic gas. Tunability is available
over wide portions of the VUV from ~2000 $\overset{\circ}{A}$ to ~1000 $\overset{\circ}{A}$.
The intensity is sufficient for use in spectroscopy, and the
effective resolution in some regions is equal to or better
than that available with the best grating spectrographs.
CW. operation appears to be feasible, and extension to even
shorter wavelengths may soon be possible with the new excimer
lasers.

Acknowledgements

We are pleased to acknowledge the very important contributions
of K.K. Innes and G. Zdasiuk to the results on NO and Mg
discussed here.

Table I. Summary of VUV Laser Radiation

Classification	Wave-length ($\mathring{A}$)	Gas	Laser Power (KW)	Energy (mJ)	Line-width (cm^{-1})
Molecular Dissociation	1726	Xe	2×10^3	0.5	600
	1695-1745	Xe	20	0.005	150
	1724	Xe	10^2	1	600
	1724	Xe	6×10^4	100	600
	1457	Kr	$\leqslant 10$	-	600
	1261	Ar	-	~200	-
	1933	ArF	2×10^3	10^5	200
Molecular Hydrogen	1098-1240	p-H_2	3.2 at 90 K	~.01	<1
	1150-1240	D_2	4.8 at 90 K	~.01	<1
	1240-1613	H_2	3×10^2	~1	<1
Carbon Monoxide	1811-1970	CO	15	~.03	<1
Carbon Ion	1548,1551	CIV	~1	~10^{-3}	<1

Table II. Tuning Ranges in the VUV Obtainable by Sum Frequency
Mixing in Sr[22,26]. (The underlined entries show
the tuning ranges achieved.)

Dye for Laser at ν_1	7-diethylamino-4-Methylcoumarin	Mixture Coumarin 6 and Coumarin 102	Sodium Fluorescein	Rhodamine 6G
Resonant State	$5s7d\ ^1D_2$	$5s6d\ ^1D_2$	$5p^2\ ^1D_2$	$5s5d\ ^1D_2$
λ_1	4779 A	5032 A	5409 A	5757 A
Dye for Laser at ν_2 and its Tuning Range				
7-Diethylamino-4-methyl-coumarin 4648-4925 A	1578-1609 A	1632-1666 A	1710-1746 A	<u>1778-1817 A</u>
Mixture coumarin 6 and coumarin 102 5000-5350 A	1617-1651 A	1674-1712 A	1756-1797 A	1827-1872 A
Sodium fluorescein 5337-5710 A	1657-1685 A	1710-1747 A	1795-1836 A	<u>1870-1914 A</u>
Rhodamine 6G 5680-6111 A	1682-1718 A	1744-1783 A	<u>1833-1875 A</u>	<u>1907-1957 A</u>

Table III. Summary of Tunable Coherent VUV Radiation

System	Wavelength Range (Å)	Photons/pulse (at 10KW)≡F	Linewidth (cm^{-1})	Remarks	Ref.
Sr	1600-1950	$10^5 \leqslant F \leqslant 10^7$	0.1	autoionizing transitions	22,26,31
Ca	1760-1770	4×10^8	0.1		31
Ca$^+$	1278	–	0.1	——	32
Eu	1855	–	0.1	——	26
Mg	1400-1600	$10^8 \leqslant F \leqslant 10^{10}$	0.1	continuum	23
	1200-1330	$10^7 \leqslant F \leqslant 10^9$	0.1	autoionizing transitions	30
Hg	1203	5×10^7	1100	2×10^9 photons/pulse at normal power	33
Xe	1180-1940	10	44	5×10^7 photons/pulse at normal power	18
NO	1520,1430, 1360,1300	$10^6 \leqslant F \leqslant 10^7$	0.1	each range tunable over 600 cm^{-1} with one dye laser	28

References

* Research supported in part by the Connaught Fund and the University of Toronto

1. R.T. Hodgson: Phys. Rev. Lett. $\underline{25}$, 494 (1970); R.W. Waynant, J.D. Shipman, R.C. Elton, and A.W. Ali: Appl. Phys. Lett. $\underline{17}$, 383 (1970)

2. R.W. Waynant: Phys. Rev. Lett. $\underline{28}$, 533 (1972); R.T. Hodgson and R.W. Dreyfus: Phys. Rev. Lett. $\underline{28}$, 536 (1972)

3. R.T. Hodgson: J. Chem. Phys. $\underline{55}$, 5378 (1971)

4. R.W. Waynant: Appl. Phys. Lett. $\underline{22}$, 419 (1973)

5. J.B. Gerardo and A.W. Johnson: IEEE, J. Quantum Electron. $\underline{QE-9}$ 746 (1973); P. Hoff, J.C. Swingle, and C.K.Rhodes: Appl. Phys. Lett. $\underline{23}$, 246 (1973); S.C. Wallace, R.T. Hodgson, and R.W. Dreyfus: Appl. Phys. Lett. $\underline{23}$, 672 (1973); W.M. Hughes, J. Shannon, and R. Hunter: Appl. Phys. Lett. $\underline{24}$, 488 (1974)

6. S.C. Wallace and R.W. Dreyfus: Appl. Phys. Lett. $\underline{25}$, 498 (1974); D.J. Bradley, D.R. Hull, M.H.R. Hutchinson, and M.W. McGoech: Opt. Comm. $\underline{14}$, 1 (1975)

7. J.M. Hoffman, A.K. Hays, and G.C. Tisone: Appl. Phys. Lett. $\underline{28}$, 538 (1976)

8. P.A. Franken, A.E. Hill, C.W. Peters, and G. Weinreich: Phys. Rev. Lett. $\underline{7}$, 118 (1961)

9. J.A. Armstrong, N. Bloembergen, J. Ducuing, and P.S. Pershan: Phys. Rev. $\underline{127}$, 1918 (1962)

10. J.A. Giordmaine: Phys. Rev. Lett. $\underline{8}$, 19 (1962); P.D. Maker, R.W. Terhune, M. Nisenoff, and C.M. Savage: Phys. Rev. Lett. $\underline{8}$, 21 (1962)

11. P.D. Maker, R.W. Terhune, and C.M. Savage: Proc. III International Quantum Electronics Conference, edited by P. Grivet and N. Bloembergen (Columbia Univ. New York, 1963)p. 1559

12. P.D. Maker and R.W. Terhune: Phys. Rev. $\underline{137}$, A801 (1965)

13. G.H.C. New and J.F. Ward: Phys. Rev. Lett. $\underline{19}$, 556 (1967); Phys. Rev. $\underline{185}$, 57 (1969)

14. S.E. Harris and R.B. Miles: Appl. Phys. Lett. $\underline{19}$, 385 (1971)

15. J.F. Young, G.C. Bjorklund, A.H. Kung, R.B. Miles, and S.E. Harris: Phys. Rev. Lett. $\underline{27}$, 1551 (1971)

16. A.H. Kung, J.F. Young, G.C. Bjorklund, and S.E. Harris: Phys. Rev. Lett. $\underline{29}$, 985 (1972)

17. S.E. Harris, J.F. Young, A.H. Kung, D.M. Bloom, and G.C. Bjorklund: Laser Spectroscopy, edited by R.G. Brewer and A. Mooradian (Plenum, New York, 1973) p. 59

18. A.H. Kung: Appl. Phys. Lett. 25, 653 (1974)

19. R.B. Miles and S.E. Harris: IEEE J. Quantum Electron. QE-9, 470 (1973)

20. D.M. Bloom, J.T. Yardley, J.F. Young, and S.E. Harris: Appl. Phys. Lett. 24, 427 (1974)

21. K.M. Leung, J.F. Ward, and B.J. Orr: Phys. Rev. A9, 2440 (1974)

22. R.T. Hodgson, P.P. Sorokin, and J.J. Wynne: Phys. Rev. Lett. 32, 343 (1974)

23. S.C. Wallace and G. Zdasiuk: Appl. Phys. Lett. 28, 449 (1976)

24. K.K. Innes, B.P. Stoicheff, and S.C. Wallace: (1976) unpublished

25. D.M. Bloom, G.W. Bekkers, J.F. Young, and S.E. Harris: Appl. Phys. Lett. 26, 687 (1975)

26. P.P. Sorokin, J.J. Wynne, J.A. Srmstrong and R.T. Hodgson: Annals N.Y. Acad. Sci. 267, 30 (1976)

27. A.H. Kung, J.F. Young, and S.E. Harris: Appl. Phys. Lett. 22, 301 (1973)

28. T. Hänsch: Appl. Opt. 11, 895 (1972)

29. S.C. Wallace: (1976) unpublished

30. M.J. Seaton: Proc. Phys. Soc. 88, 801 (1966)

31. G. Zdasiuk: M.Sc. Thesis, Generation of Tunable Coherent VUV Radiation, Univ. Toronto (1975), unpublished

32. P.P. Sorokin, J.A. Armstrong, R.W. Dreyfus, R.T. Hodgson, and J.J. Wynne: 2nd Laser Spectroscopy Conference (1975) unpublished

33. K.S. Hsu, A.H. Kung, L.J. Zych, J.F. Young, and S.E. Harris IEEE J. Quantum Electron. QE-12, 60 (1976)

34. S.E. Harris and D.M. Bloom: Appl. Phys. Lett. 24, 229 (1974)

35. R.G. Bray, R.M. Hochstrasser, and J.E. Wessel: Chem. Phys. Lett. 33, 1 (1975): P.M. Johnson, M.R. Berman and D. Zakheim: J. Chem. Phys. 62, 2500 (1975)

36. K. Dressler and E. Miescher: Astrophys. J. 141, 1266 (1965)

HIGH EFFICIENCY UV LASERS

J.J. Ewing and C.A. Brau
Avco Everett Research Laboratory, Inc.
2385 Revere Beach Parkway
Everett, Massachusetts 02149, USA

Abstract

The potential of the rare gas halide lasers for high efficiency is briefly reviewed. The spectroscopy and kinetics of these lasers is emphasized.

Introduction

Within the past year it has become clear that UV lasers having high efficiency are practically realizable. This paper will briefly review the research that is leading to this revolution in UV lasers. The paper will also emphasize the underlying physics that determines the efficiency of electronic transition lasers. We will draw examples from recent work on rare gas halide lasers.

The biggest gains towards the goal of high efficiency at short wavelengths have been made in a new class of molecules and molecular lasers, the rare gas monohalides. The spectra of some of these species were first taken within the past two years[1, 2, 3, 4, 5] and a solid understanding of the spectroscopy and the formation mechanisms is developing.[3, 4, 5, 6, 7, 8] Specifically, lasing on XeBr at 282 nm,[9] XeF at 354 nm,[10] XeCℓ at 308 nm,[11] KrF at 248 nm,[11] ArF at 193 nm[12] and KrCℓ at 222 nm[13, 14] has been observed. Electric discharge pumping of some of these species has also been observed, on KrF[15] and XeF[16] up to now, although one assumes that some of the other members of the class can be pumped by discharges as well.

A complementary set of lasers operating on halogen molecules formed in certain rare gas/halogen mixtures has also been demonstrated. I_2 lases at 342 nm,[17] Br_2 at 292 nm,[18] and $C\ell_2$ is projected to lase near 260 nm,[19] although it has not lased in our laboratory or others[20] despite substantial effort. Discharge production of Br_2^* and I_2^* lasers has also been achieved.[21, 22]

Although the halogen systems are very efficient as flourescence sources,[23, 24] they have not yet performed as well as the best rare gas halide lasers. The reasons for this decreased performance are not yet quantitatively established.

The most efficient lasers to date in this class are the KrF and XeF lasers. High intrinsic efficiencies[25,26] as well as high wall socket efficiencies[26,27] have been obtained. The KrF laser has also produced the highest total energy outputs to date although the ArF laser is also capable of comparably high energy outputs.[12]

Most of these exciting results in novel high efficiency UV lasers have developed out of lasers utilizing ground state dissociation, so called "excimer" lasers.[28] As we will discuss later, excimer lasers have the advantage that the kinetic extraction efficiency can be made large because the intra cavity flux can be large without causing bottlenecking. Although this is the current state of the art, we caution that two other methods of lower level removal, predissociation and collisional quenching, can in principle provide sufficiently rapid lower level quenching to allow a high cavity flux to efficiently extract upper laser levels.

What is High Efficiency

The words "high efficiency" mean different things for various laser applications. However, a definition of the term within the context of a particular application can serve as an important guideline in pursuing the goal. This is so because various applications require different pulse energies and average powers, and very often the application itself dictates the techniques needed to achieve the desired overall laser system efficiency. Two examples illustrate this point. Some laser isotope separation schemes require higher powers and efficiencies than currently available. However, for the near term and the long term, a high efficiency laser for UV uranium isotope separation probably only needs to be of order 1% efficient, operate at about the 1J per pulse level, with ultimate power levels of order 5 kW.[29] This efficiency and power level is clearly higher than what is currently available in the UV or visible. However, it does not place severe constraints on either the excitation methods or the nature of the laser medium and a variety of solutions appear feasible. As a contrast, for the projected laser needs for an ultimate high average power laser for a laser fusion application, energy outputs per laser amplifier of order 2000 J or more are of interest.[30] Economics presumably will dictate ultimate efficiencies as high as possible, (>10%) as well as some modest pulse repetition frequency. Clearly a combination of laser medium and excitation scheme applicable to the isotope separation effort need not be directly applicable to the higher power fusion application.

An important, if obvious point is that the overall efficiency of a laser is a product of efficiencies. To maximize the overall laser efficiency it is important that each element of the efficiency be high simultaneously. Roughly speaking the overall efficiency is a product of a quantum efficiency, an upper laser level production efficiency, an extraction efficiency, an efficiency for producing the initial excited states by the pump, and an energy coupling efficiency which describes the efficiency with which energy gets from a wall socket into the gas medium. The extraction efficiency is comprised of both spectroscopic and kinetic extraction efficiencies depending on the ratio of net gain to loss and the ratio of rates of stimulated emission of upper laser levels to the quenching and spontaneous loss of upper levels. Certain rare gas halide lasers have potential for high power and high efficiency because they have high efficiency in each of the above mentioned elements. Depending on the application requirements, one may accept a lower efficiency of one of these elements if the compromise allows for some simplification of the laser. Typically this trade off usually involves coupling and excited state production efficiencies.

Let us briefly summarize some of the means available for exciting UV or visible laser transitions and the ultimate efficiencies they can yield. For the very high powers and high efficiencies, as in the fusion application, one clearly requires a volumetrically scalable excitation technique. Three broad approaches are currently possible for making large volume UV or visible devices: E-beam (or possibly UV) ionized electric discharges, pure E-beam pumping, and optical excitation.[31] Of the above, experience with IR lasers shows that the discharge approach offers the potential for the highest efficiencies (> 10%), with high energy outputs and average powers. However, the discharge pumping approach imposes certain constraints on the laser medium. One key constraint of larger volume electric discharge lasers is that the coupling efficiency improves as the pumping duration increases. An excited medium having constant impedance is also a desirable feature for a high coupling efficiency. The pure E-beam pumping technique is versatile, but is limited in terms of overall systems efficiency and repetition rate. Overall efficiencies up to about 10% are probably possible in suitable media. Optical excitation schemes are inherently limited by the efficiency and repetition rate of the optical pump being used. However, as the overall efficiencies of UV and visible lasers and incoherent pumps improve, the potential and scope of optically pumped lasers will broaden. One can project optically excited high energy visible lasers with efficiencies of order 1-3% based on a hypothetical 10% efficient ArF or KrF UV laser.

At lower average power levels, all of the above methods or combinations thereof are suitable and can yield efficiencies of order 1% or greater. More convenient for efficiencies near 1%, however, is the use of self sustained discharges such as the fast pulse Blumlein devices which are performing so well with the XeF laser.[27] These devices typically run with an unstable discharge, one which becomes an arc after a short time, and the principal source of inefficiency lies in the "coupling" efficiency. These lasers do not lend themselves to scaling to large volumes and as such are limited in total energy outputs. Currently, these devices are excellent for producing high peak powers and short pulses. Since similar pumping techniques yield higher efficiencies in CO_2 IR lasers, improving the performance of short pulse UV lasers is currently an active area of research.

In general, it is true that the longer pulse, larger volume, highest efficiency excitation schemes place the most stringent constraints on the intrinsic aspects of the laser medium. For example, there is one very important difference in the required laser kinetics of an allowed UV or visible laser transition excited by an E-beam stabilized discharge for a large volume device, contrasted to a short pulse, small scale discharge. As noted above the coupling efficiency of a large scale, very efficent laser improves with long excitation pulses, $t \gtrsim 300$ ns. The fast pulse discharges can operate well with very short excitation times, $t \lesssim 50$ ns. Because the radiative lifetimes of typical allowed electronic transitions are roughly equal to the transient turn on time of a large scale discharge, the long pulse devices must have rapid removal of the lower laser level for a laser operating on an allowed transition. Obviously one cannot expect to have an efficient large scale device if the laser medium shuts off before the electrical circuitry is efficiently coupled to the laser plasma. The long pulse high efficiency lasers require one to look for means of removing the lower laser level, a natural venue for the excimer concept. The short pulse devices, can, however, work quite well on transitions that bottleneck such as the 510.6 nm Cu laser line, or the

337.1 nm N_2 (C→B) transition. A second kinetic constraint typical of the long pulse electric discharges derives from the preference for constant laser plasma impedance, implying constant electron number density. In the rare gas halide lasers this constraint is neatly managed by using electron attachment by the halogen species to balance the avalanching of electron number density caused by the high steady state number density of low ionization potential rare gas excited states.[32]

Since the highest efficiency pumping schemes by and large place the most stringent constraints on the laser medium, our discussion of new high efficiency lasers will be oriented towards E-beam and E-beam controlled discharge excitation of laser media. For each of these pumping techniques a set of guidelines can be drawn up that allows one to focus on key elements required to obtain overall high efficiency. Listed below are a set of such guidelines convenient for discussing potential as high efficiency electric discharge lasers.

Guidelines for an Efficient UV Electric Discharge Laser

- Quantum Efficiency

    Maximize Quantum Efficiency
    Usually a Fixed Parameter within a Pumping Scheme

- Upper Laser Level Production

    Minimize Branching into the Useless Channels

- Extraction

    Net Gain Much Greater than Loss (Minimize Excited State
    Absorption)
    Cavity Flux Large Enough to Beat Quenching and Spontaneous
    Emission

- Initial Excited State Production

    Minimize Production of Ions and Useless Excitations

- Coupling

    Stable, Constant Impedance Laser Plasma, Long Excitation Pulses

For a pure E-beam laser or an optically pumped laser the guidelines within the first three areas will be the same. For E-beams there is little control over the excited state production efficiency, whereas with optical excitation one can hope to choose the best wavelength for exciting the medium. The coupling efficiency for optical or E-beam pumping will be limited by desired homogeneity of excitation and the fraction of the primary electron range or optical depth utilized. The coupling efficiency will also include the efficiency with which the E-beam or laser beam itself is produced from electrical energy drawn from the wall socket.

In the following sections we will consider the spectroscopy and kinetics of the rare gas halide lasers and show how various components of the overall efficiency of each laser compares to the "ideals" stated in the above guidelines. Similar comparisons for other classes of UV or visible lasers can be drawn.

Rare Gas Halide Spectroscopy

The potential for high efficiency of the rare gas monohalides derives from both kinetic and spectroscopic aspects. Since the spectroscopy is somewhat better understood than the laser kinetics, we will describe it first. The spectroscopy impacts on the above mentioned efficiency areas in several ways. Obviously with the choice of an excitation scheme, the laser wavelength determines the quantum efficiency. The upper laser level production efficiency is primarily a kinetic phenomena, but certain aspects of high branching ratios can be traced directly to the almost unique spectroscopic features of these molecules. Spectroscopy impacts the extraction efficiency in two ways. The position of the molecular potential energy curves determines the amount of excited state absorption competing with stimulated emission. Obviously, the less absorption one has the better the extraction. Secondly, the shape of the lower level potential energy curve determines to a first approximation the emission spectrum bandwidth and as such impacts the stimulated emission coefficient and, correspondingly, stimulated emission rates. It is important to note that among UV excimer lasers the stimulated emission cross sections for the sharp bands in the rare gas monohalides are very large, $\sigma \approx 10^{-16}$ cm^2. This large cross section makes it possible to extract upper laser levels on time scales of order 1 ns with flux levels of order 5 MW/cm^2. The rapid stimulation of the upper laser levels effectively competes with kinetic quenching and is a key component of attaining high efficiency with these species.

Figure 1 shows a schematic molecular potential energy level diagram for a rare gas monohalide. One can see that these lasers utilize lower level dissociation. Unlike the rare gas excimer lasers there are two lower branches. For the simplified case shown here emission is centered near two wavelengths, λ_1 and λ_2. The two dissociative levels are designated as a $^2\Sigma$ state and a $^2\Pi$ state. Spin orbit splitting of the lower Π states is neglected but is important in rare gas bromides and iodides. The $^2\Sigma$ lower state corresponds to having one halogen "hole" in a p orbital on axis with the rare gas atom. The $^2\Pi$ state places the partially occupied atomic p orbital perpendicular to the molecular axis. The $^2\Pi$ state is strongly repulsive at the equilibrium internuclear configuration of MX*, R_o, because more electrons are between the halogen and rare gas nuclei. Emission terminating on the $^2\Pi$ state, near wavelength λ_2, is characterized by a relatively broad continuum band width. The $^2\Sigma$ lower level has a fairly flat potential energy curve at R_o, and the emission bandwidth for this transition is narrow. Assuming comparable radiative transition rates to both states, the gain on the transition near λ_1 is higher because the bandwidth is lower. To date rare gas monohalide lasers have operated on the sharper, higher gain bands. As noted above, a "high gain" excimer laser which utilizes a fairly flat lower laser level has many significant advantages because the higher stimulated emission cross section increases the rate at which stimulated emission can remove the upper laser level. This allows laser emission on the sharp band to effectively compete with quenching of the excited state. This benefit would not accrue as readily on the broad, lower cross section bands that terminate on very repulsive lower laser levels. The dissociative nature of the relatively flat lower level still allows for the rapid removal of the lower level, thus maintaining the population inversion.

The upper laser level, denoted as MX^*, is bound with respect to dissociation into an inert gas atom M and an excited halogen atom, X^*. It is also bound with respect to $M^* + X$. The ionic nature of these excited states has been discussed.[1,2,3,4,5] The excited state is nothing more than a positively charged inert gas ion, M^+, and a negative halogen ion, X^-, held together by coulombic rather than covalent forces. Predications of various properties of these excited species are based on the similarity of these excited states to the ionic ground states of the nearly isoelectronic alkali halide ground states.[3] These predictions are accurate to within a few percent.[4,5] The similarity of the excited ionic states of the rare gas monohalides to ground state alkali halides derives from the fact that a rare gas halide ionic excited state differs by only one electron from an alkali halide. KrF^* is simply the ion pair $Kr^+ F^-$. Kr^+ is only different by one electron from Rb^+. Thus, the properties of KrF^* are very close to those of the ionic RbF molecules. A number of other useful analogies exist as well. For instance, the rare gas excited states, such as Kr^* have low ionization potentials and as a result the chemistry of Kr^*, both kinetically and generically, is very similar to that of Rb.

The simplified potential curves shown in Fig. 1 do not illustrate the other ionic excited states, which lie very close to the ionic $^2\Sigma_{1/2}$ state. Two other curves having symmetry $^2\Pi_{3/2}$, $^2\Pi_{1/2}$ lie within 1 eV of the upper laser level. The splitting depends on the amount of spin orbit coupling, being largest for the Xe halides. A priori calculations show that a substantial amount of mixing occurs between the $\Omega = 1/2$ states, and as such the $^2\Sigma$ notation given in Fig. 1 is an over simplification.[33] Emission from the higher ionic states has been observed.[5,8,13] In fact, the broad bands are probably super positions of emission from the $^2\Sigma_{1/2}$ and the $^2\Pi_{3/2}$ states to the repulsive $^2\Pi$.

Note the high quantum efficiency that is intrinsic to the rare gas monohalide lasers. Assuming the process begins with production of an excited metastable, such as Kr^*, quantum efficiencies of order 50% can be achieved. The lost energy goes into chemical potential by dissociation of a weakly bound halogen source, such as F_2, and excess vibrational energy initially invested in producing the rare gas monohalide excited state. Discharge excitation of Kr^* is capable of fairly high excited state production efficiencies. Boltzmann code calculations[32,34] suggest that the efficiency of producing excited rare gas atoms could be as high as 70% for an excitation level sufficiently high to produce useful laser gain on rare gas halides in a reasonable length device. Thus, before accounting for losses due to extraction and energy coupling, a discharge pumped rare gas halide laser could have an efficiency of order 35%.

For E-beam pumping the quantum efficiencies are not as high as with discharge pumping. Moreover, the production efficiencies are much lower than with discharge excitation because the primary electron beam looses roughly twice the energy of ionization of the dominant species to produce an ion with some atomic excitation.[35] Thus the product of production efficiency with quantum efficiency is about 25% for typical rare gas halides. Power coupling efficiency lowers the overall potential system efficiency for an E-beam device relative to a discharge.

Naturally, the development of new lasers and new molecules is leading to increased research on these species. The exact dependence on internuclear separation as well as the locations and assignments of the potential curves for many of the states of the rare gas halides remains to be done. The lowest $^2\Sigma$ covalent state, the lower laser level, is not truly dissociative for XeF and XeCl. The lasing transitions are, in fact, bound to bound,[5,6,36] for these species since these molecules have shallow wells in the lowest states. Because the wells are fairly shallow, depopulation of the lower level can take place via rapid collisional dissociation. The binding of the XeF ground state was not anticipated theoretically.[37] Refined calculations on the XeF ground state show that inclusion of the long range Vander Waals attraction is important in calculating the properties of the bound ground state of XeF.[38]

As seen in Fig. 1, there are higher lying potential curves that derive from $M^* + X$ and $X^* + M$. Not drawn are the potential curves for the ion MX^+, which are typically at higher energies. The exact energy and shape of the curves coming from $M^* + X$ has an important impact on the possibility of self absorption. Xenon bromide is probably a fine example of a rare gas haide that fluoresces very efficiently but lases very poorly. This is possibly due to self absorption as shown in Fig. 2. Note that the self absorption is a photodissociation process rather than photoionization which plagues the rare gas excimer lasers such as Xe_2^*. The positions of the upper and lower laser levels can be determined by the wavelength and shape of the $XeBr^*$ emission bands. The exact position of the potential curves and the wavelength dependence of the absorption bands relative to the stimulated emission cross section is not known. However, it is reasonably clear that absorption is possible in the $XeBr^*$ case since a 282 nm photon has ample energy to produce an $Xe^* + Br$ from the ionic upper laser level. Rough considerations of the atomic orbitals that the electrons occupy in the various states suggests that absorption plays a major role in decreasing gain and efficiency in this system. Recall that the upper laser level is an ion $Xe^+ Br^-$. In making a transition to the lower laser level an electron hops out of a p orbital of Br^- into the lowest vacant orbital of Xe^+, a 5 p orbital. This produces ground state Xe and Br atoms. However, in making a transition upward to $Xe^* + Br$ the electron hops from Br^- into a

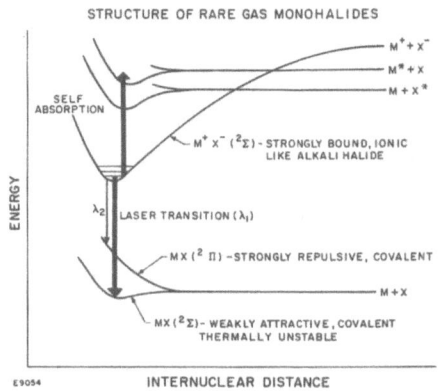

Fig. 1   Schematic of rare gas monohalide potential energy curves

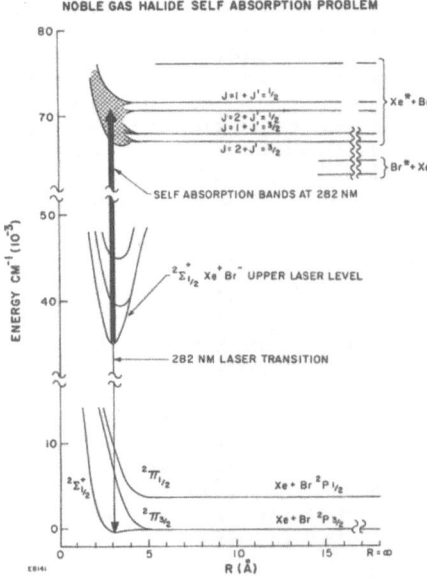

Fig. 2   Estimated potential curves for XeBr showing the posibility of strong absorption band overlapping the lasing transition

large 6 s orbital of Xe. By virtue of the fact that the final electron orbitals centered on Xenon for states of Xe^* are larger than those in which a ground state Xe atom is produced, one expects that the total transition dipole moment for absorption is larger than that corresponding to the lasing transition. The absorption is probably spread out over a broader band than the sharp laser transition, thus allowing for net gain to occur. Such absorption, as in the case of $XeBr^*$, obviously limits the spectroscopic component of the extraction efficiency as well as lowering the gain. A similar absorption can occur in other rare gas halide lasers, even in the high efficiency KrF laser. The KrF^* 248 nm photon barely allows the molecule to reach the energetic limit for forming $Kr^*+ F$. Such absorption would clearly limit the scalability and efficiency of a larger KrF laser. Small signal gain and loss measurements through the KrF laser band would shed light on this problem. That absorption is a problem in the KrF laser can be shown simply by measuring laser power out as a function of output coupling.[25] For $Ar/Kr/F_2$ mixtures the maximum efficiency is derived with large output coupling, for example when using the 15% feedback of a quartz flat as one of the laser reflectors. Similar results obtained with $Ar/Kr/NF_3$ mixtures suggest that the F_2 by itself was not the only absorbing species (NF_3 does not absorb at 248 nm). As mentioned above KrF^* self absorption could be the cause of this effect, or possibly photoionization of the excited atoms Ar^* or Kr^*, absorption by ArF, Ar_2^*, or Kr_2^* molecules.

The XeCl laser is another case where absorption strongly limits laser performance. With the use of Cl_2 as an oxidizer, absorption is clearly a problem because of its strong photodissociation band at 308 nm. However, use of alternate sources such as HCl, while improving performance relative to Cl_2, does not make the XeCl laser as efficient as either XeF or KrF.[6,39]

Spectroscopically speaking, even the best of the rare gas halide lasers is far from ideal. Quantum efficiencies, while excellent, are lower than that of the CO infrared laser. Spectroscopically efficient extraction requires high output coupling in the best of cases to date. The use of high output coupling lowers the intra cavity flux for a fixed total energy input. This can lower kinetic extraction efficiency of a KrF laser if the flux drops below that required for reasonable saturation. The simultaneous observation of efficient discharge pumping of excited states, high flux levels for efficient kinetic extraction, and high output couplings for good extraction relative to losses has yet to be shown for KrF.

Rare Gas Halide Kinetics

The mechanism for producing the excited states of the rare gas halide lasers are not yet fully understood or quantitatively documented. However, the general ways in which excited states can be produced can be given and estimates of pertinent rates made. Producing excited states from the primary excitation products, ions and excited atoms, is only a part of the overall laser kinetics. This part is becoming well understood. From an efficiency point of view, the kinetics are ideal, since branching ratios are near unity for many cases. Electron attachment rates are very important in electric discharge lasers because attachment is usually the dominant mechanism for removing electrons. Moreover, attachment provides a mechanism for maintaining a constant impedance discharge in strongly excited rare gas mixtures.[32,34] The kinetics of this process is critical to the attainment of high efficiency at high average powers. The quenching kinetics and radiative lifetimes of these species are poorly understood but are clearly important because they determine the flux level needed for a kinetically efficient extraction.

Table I gives a simplified mechanism for the KrF laser. This laser operates most efficiently in $Ar/Kr/F_2$ mixtures. Left out of this mechanism is any specification of the excited states. Since a manifold of states are excited and can react with different reaction rates for forming excited states, this is possibly an oversimplification. The three body loss of excited states into excimers is not tabulated, and this is an important reaction when the halogen number densities are below optimal. Three body loss will decrease the upper laser level production efficiency. Ionization of Ar and Kr excited states has been left out of the Table, and this is a very important rate for efficient discharge pumped rare gas halide lasers.[32, 34]

With the above mention of the simplifications in the mechanism, we discuss the reaction kinetics as they are understood at this stage. We begin with a discussion of the "classical" route to rare gas halide excited states, reactions of rare gas excited states with halogenated species.

The first public mention of the potential of these new molecules as "excimer" lasers was a result of fundamental rate constant measurements for the quenching of inert gas metastables.[1, 2] Groups in Kansas and Cambridge, England, were attempting to look at the energy partioning in these reactions by monitoring product fluorescence. Both observed new UV and VUV luminescence due to quenching reactions of rare gas excited states with halogens. These new, bright emission spectra were attributed to bound-free continua of rare gas monohalides. The prototypical reactions are

$$Ar^* + Cl_2 \rightarrow ArCl^* + Cl$$
$$\longrightarrow Ar + Cl + h\nu \quad (172 \text{ nm}) \qquad (1)$$

and

$$Xe^* + Cl_2 \rightarrow XeCl^* + Cl$$
$$\longrightarrow Xe + Cl + h\nu \quad (308 \text{ nm}) \qquad (2)$$

These reactions all tend to proceed with large cross sections. This is not surprising since the corresponding reactions of alkali metal atoms with halogens and halogenated species also have large cross sections. As mentioned earlier, a noble gas in its excited state is very similar in properties to an alkali ground state. A low ionization potential is the key to the similarity of reaction chemistry and chemical compounds.

Species such as KrF^*, are nothing more than short lived "pseudo alkali halides", i.e., diatomic molecules in which one electrical charge has been transferred from the rare gas atom to the halogen. The formation of alkali halide molecules and rare gas halide excited states can all proceed along chemically similar pathways. This similarity derives from the fact that metastable rare gas atoms, Kr^* for example, or any highly excited state of an atom have low ionization potentials and chemically will behave like ground state atoms that have low ionization potentials, viz the alkali atoms. The chemical kinetics of alkalis reacting with halogen molecules has a long history and this chemistry is understood in much detail. The mechanism for such alkali/halogen reactions is discussed in detail elsewhere.[40] The reaction of a Kr^* with a halogen containing molecule should produce an ionic species. A graphic explanation of these reactions, thought to proceed by way of an ionic-covalent curve crossing, is called the "harpooning mechanism.[1, 2, 40] In such reactions the low ionization potential of the

Table I

Simplified mechanism for rare gas halide lasers
($Ar/Kr/F_2$ mixture)

EXCITATION REACTIONS

Discharge Pumping

- Metastable Formation

$$e + Ar \rightarrow Ar^* + e$$

$$e + Kr \rightarrow Kr^* + e$$

- Reaction with Halogen

$$Ar^* + F_2 \rightarrow ArF^* + F$$

$$Kr^* + F_2 \rightarrow KrF^* + F$$

- Displacement

$$Kr + ArF^* \rightarrow KrF^* + Ar$$

E-Beam Pumping

- All of the Above

- Ion Recombination

$$e + F_2 \rightarrow F + F^-$$

$$F^- + Ar^+ + M \rightarrow ArF^* + M$$

$$F^- + Kr^+ + M \rightarrow KrF^* + M$$

LOSS REACTIONS

- Quenching

$$KrF^* + \begin{Bmatrix} Ar \\ Kr \\ F_2 \\ e \end{Bmatrix} \rightarrow Kr + F + \begin{Bmatrix} Ar \\ Kr \\ 2F \\ e \end{Bmatrix}$$

$$KrF^* + e \rightarrow Kr, Kr^* + F^-$$

- Radiation

$$KrF^* \rightarrow Kr + F + h\nu$$

$$h\nu + KrF^* \rightarrow Kr + F + 2h\nu$$

---

alkali or excited species and the high electron affinity of the halogen molecules allows an ionic potential curve of the form $M^+ + X_2^-$ to cross that of $M*$ + X_2 at fairly large internuclear separations, distances of order 5 Å. The X_2^- ion formed by an electron hop, when the ionic and covalent potentials become degenerate, is unstable with respect to dissociation into X^- + X in the presence of the large electric field of the M^+ ion. As a result, the $M^+X_2^-$ temporary triatomic species fall apart into an ion pair M^+X^- leaving behind an X atom. The term "harpooning" mechanism derives from considering the electron that hops from the $M*$ over to X_2 as the harpoon that creates an extremely large coulombic force that pulls the old X_2 molecule apart and forms the new ionic, M^+X^- species. The prime difference between reactions of halogen molecules with alkali metal atoms and the corresponding chemistry with excited states such as $Kr*$ is that in the alkali reactions only one electronic state can be formed while several potential product channels can exist in certain excited state/halogen molecule reactions.

The cross sections and exit channel branching ratios for a number of excited metastable inert gas atoms reacting with halogens are now being measured.[7,8] The bulk of this work is being done at low pressures. The spectra observed are similar but not identical to those at high pressure as in E-beam or discharge excited lasers. A brief compendium of some of the relevant reaction rates for reaction of metastable inert gases with halogen containing compounds is given in Table II. The key phenomenon, from an efficiency viewpoint, is that the kinetic branching ratios for forming rare gas halide excited states are unity for several of the most important reactions: viz $Xe*$ + F_2, $Xe*$ + NF_3, $Xe*$ + Cl_2, $Kr*$ + F_2 and $Kr*$ + Cl_2. Thus, if one makes rare gas excited states and does not lose excitation into excimer states such as Xe_2*, Kr_2*, or Ar_2*, one has a very efficient means of producing the upper laser level from a species which can be produced directly, and hopefully efficiently in a discharge. At typical F_2 mole fractions, 0.3%, and total pressures of 2 atm, the characteristic time for $Ar*$ or $Kr*$ reaction with F_2 is about 10 ns. The Kr_2* excimers formation time is longer than this because of the use of Ar/Kr mixtures. Some Ar_2* excimers are made from $Ar*$ initially produced, but because of the large cross section for producing $Kr*$ upon collision of Kr with Ar_2* excimers, the formation of Ar_2* does not constitute a kinetic branching loss mechanism. Thus for suitable mixtures of rare gases and halogens the production channel involving rare gas metastables can provide upper laser levels with branching ratios that appear to be unity.

The reactions tabulated above pertain, to some degree, to discharge pumping of a rare gas halide laser. In a discharge, however, one can have several other excited states produced, and the reaction kinetics of other states may differ, possibly even having larger cross sections. The low pressure E-beam excited $XeBr*$ experiments of Searles and Hart suggest this since their modeling of the production and decay of $XeBr*$ from E-beam excited Xe/Br_2 mixtures gives a rate constant for an undefined ensemble of $Xe*$ excited states reacting with Br_2 that is a factor of 4 times larger than the rate measured out of Xe metastables.[9]

Not all reactions of rare gas metastables give high yields of rare gas halide excited states. In fact, a number of reaction pairs, such as $Ar*$ + Br_2[7] give halogen atom emission.

Table II

Some kinetic rate constants for
rare gas metastable/halogen reactions

		k_Q [a.]	σ_Q [b.]	Ref.	Comments
$Xe^*(^3P_2)$ +	Cl_2	6.5	193	(2)	
	Br_2	6.0	202	(2)	
	CF_3I		184	(2)	
	F_2	7.3	156	(8)	Produces XeF^* with unit quantum yield.
	NF_3	.86	23	(8)	
$Kr^*(^3P_2)$ +	F_2	8.1	163	(8)	Produces KrF^* with unit yield.
	Cl_2	6.0	147	(8)	
	NF_3	1.6	39	(8)	
$Ar^*(^3P_2)$ +	Br_2	6.5	147	(7)	Produces Br^*
	Cl_2	7.1	142	(7)	Produces $ArCl^*$ and Cl^*
	F_2	8.5	148	(8)	Should produce ArF^* with near unit yield.
	Kr	.06	1.3	(51)	
	Xe	1.8	40	(51)	

a. Rate constants, k_Q, given in units of $10^{-10} cm^3/$ molecule sec.

b. Cross sections, σ_Q, given in units of $10^{-16} cm^2$.

A simple understanding of why some reactions produce halogen atom excited states and others produce rare gas halide emission can be gained by inspecting approximate potential curves for a rare gas halide. In the measured case of $Ar^* + Br_2$ chemistry, the $ArBr^*$ is presumably formed with energy up to that available from the combining reactants, in this case $E(Ar^*) - DE(Br_2)$ where DE is the dissociation energy of the halogen bearing fuel.[1] Sketching up the $ArBr^*$ potential curves shows that the energy available by this reaction is sufficient to populate a number of exit channels that can yield Br^* by predissociation. Making a statistical argument, which has been found excellent in the analogous alkali dimer/halogen atom reactions,[41,42,43] one would say that every state accessible via some curve crossing will be produced. For the case of $Ar^* + Br_2$ or $Ar^* +$ an iodine containing compound such as I_2 or HI there are many more accessible states that put the energy into electronically excited halogen atoms rather than rare gas halide excited states. For the case of ArI^*, whose potential energy curves are schematically shown in Fig. 3, no net binding of ArI^* relative to $Ar + I^*$ is expected[3] and one predicts that $Ar^* + RI$ reactions should produce primarily high lying iodine atom excited states. In marked contrast to this is the reaction of Ar^*, Kr^*, and Xe^* with F_2 and Xe^* with Cl_2. The rare gas halides formed in these cases <u>cannot</u> <u>predissociate</u> into any F^* or Cl^* excited states since the rare gas halides produced by reactions all have lower energy than any halogen atom excited state.[3] Reactions of these species should then lead primarily to rare gas halide emission. This, of course, is one reason for the high efficiency of the KrF laser.

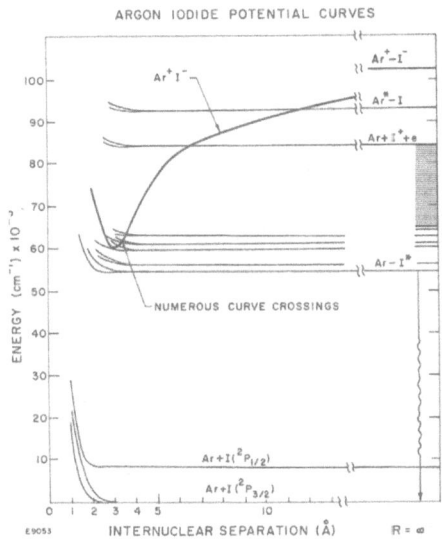

ARGON IODIDE POTENTIAL CURVES

Fig. 3   Gross estimate of the potential curves for ArI. Note that the ionic state is not bound relative to Ar + I*. As a result quenching reactions of Ar* with Iodine containing molecules such as CF_3I, or ion recombination of Ar$^+$ with I$^-$ should yield I* excited states

Since the upper levels of rare gas halide lasers are ionic in nature, positive-negative ion recombination can also feed the upper laser level. In E-beam excited Ar/F_2 mixtures the Ar^+ ions are formed by the relativistic electrons. F^- ions can be formed by attachment of the secondary electrons. Recombination then yields ArF^*. For those rare gas halides that have very large radii curve crossings with $M^* + X$ or $M + X^*$,[3] theory and experiments with alkali halide dissociation show that these species should recombine entirely into ions.[44,45] For the recombination of $Ar^+ + I^-$ or $Ar^+ + Br^-$, however, ion recombination should simply yield electronic excitation of the various atoms. Because of the long range of the coulomb potential these ion recombination reactions have huge three body rates, $\sim 10^{-25}$ cm^3 sec.[46] These three body rates become diffusion limited two body rates at high pressures. The degree to which ion-ion recombination occurs rather than ion-electron recombination (typically by way of Ar_2^+ + e at high pressure) depends on the rate constants for dissociative attachment

$$e + RX \rightarrow R + X^- \tag{4}$$

where R is anything bonded to a halogen. For mixtures containing F_2, HI, and CF_3I attachment is fast and negative ion-positive ion recombination dominates for mixtures with halogen pressures of order 3 torr.[47] Br_2 and I_2, however, attach slowly[48] and rare gas mixtures with these halogen sources will be dominated by neutral chemistry.

Displacement reactions should be very important in rare gas halide lasers, although little is yet known quantitatively. These come about when a light rare gas halide excited state collides with a heavier rare gas atom:

$$ArF^* + Kr(Xe) \rightarrow KrF^* (XeF^*) + Ar \tag{5}$$

ArF^* is simply an ion pair, $Ar^+ F^-$. Kr and Xe have lower ionization potentials than Ar. ArF^* has a higher energy than KrF^* or XeF^*. These facts suggests that these exothermic "knockout" reactions could have large reaction rate constants, at least as large as 3×10^{-10} cm^3 sec^{-1}. These reactions must be accounted for to properly model the rare gas halide lasers. They will be important in both E-beam pumped lasers where ArF^* will form by ion recombination, and also in discharge pumped KrF lasers where ArF^* forms by way of $Ar^* + F_2$ reactions. Electron-beam pumped KrF lasers, using F_2 as the flourine source, are excited primarily by ion recombination. Thus the branching ratio for the reaction $ArF^* + Kr \rightarrow KrF^* + Ar$ must be unity since Ar^+ is the dominant ion formed, and since the intrinsic efficiency of the KrF laser is so high.[25]

The excited states of these species can also be pumped by neutral recombination, for example, Br^* recombining with Ar, Kr, or Xe. Such reactions are important in E-beam pumped mixtures where the attachment rate is low leading to Ar^* or Ar_2^* chemistry and a primary reaction leads to an excited halogen rather than a rare gas halide.

The radiative lifetimes of these species are not well known. Searles modelled the kinetics of his low pressure, E-beam excited Xe/Br_2 system to give a radiative lifetime of 17 ns.[8] This lifetime is quite reasonable for an allowed charge transfer transition corresponding to an oscillator strength of about 0.1. Preliminary measurements at AERL have suggested an upper limit of 40 ns for the radiative lifetime of XeF.[49] This corresponds to an oscillator strength of about 0.05, which also is consistent with experience with alkali halides or the recently measured lifetime of an analogous charge transfer band in I_2.[50] Since the radiative lifetime is needed to determine the gain and saturation parameters of these lasers it is anticipated that calculations and measurements of radiative lifetimes will appear in the next year. Dunning and Hay have given a theoretical calculation of the KrF^* lifetimes for the various ionic excited states.[33] For the intense laser band a lifetime of about 6 ns is calculated. This corresponds to a stimulated emission cross section, $\sigma \sim 4 \times 10^{-16}$ cm^2.

The quenching kinetics of the upper laser levels of rare gas halides and halogens is not yet completely understood. However, it is known that the laser efficiency of KrF is much higher than its fluorescence efficiency. High cavity fluxes should in principal be capable of competing with any quenching, but from an academic point of view it is important to identify the quenching mechanisms. The obvious quenchers are the rare gas atoms present in the excited mixtures. Quenching in the KrF laser occurs about four times faster than radiation.[25] This implies a quenching frequency of about 5×10^8 sec^{-1}. If the quenching is ascribed wholly to Ar or Kr, rate constants of 5×10^{-12} and 8×10^{-11} cm^3 sec^{-1} would be required. This seems rather large for what must be an E → T process with almost 4 eV of energy appearing in the relative translational energy of 2 Kr atoms and an F atom. F_2 could also quench KrF* and at the typical mole fractions used this would correspond to a rate constant of 8×10^{-10} cm^3 sec^{-1}. This seems at first to be too large also, but is within the range of rate constants one can have for a long range dipole interaction.[51] This is not unreasonable since F_2 does absorb in the 248 nm band and a continuum of final states is accessible. The cross section for such a process is about 200 $Å^2$. Larger cross sections are known for the transfer of energy from Kr_2^* and Ar_2^* excimers to Xe by a similar mechanism.[51] Finally electrons can quench these excited states. Recall that the excited state in both halogen and rare gas halide lasers is an ion pair. The permanent electric dipole moment of rare gas halides is hugh, ~10D. The charge-dipole interaction then at large distances, ~5 $Å$, can exert a significant force and initial and final states can be mixed together by passing electrons. Also, a dissociative attachment process such as

$$KrF^* + e \rightarrow Kr^* + F^- \tag{6}$$

can occur with electrons of energy ~1 eV. If the entire quenching of excited states was due to electrons, a rate constant of order 2×10^{-7} cm^3 sec^{-1} corresponding to a cross section of ~200 $Å^2$ is inferred. This is a little bit larger than most dissociative attachment rates, but not outside the realm of possibility. The total quenching is probably due to contributions from each of the above, and detailed modeling will require definitive measurements of these quenching rates individually.

In terms of the guidelines given earlier, the kinetics of the rare gas halides is nearly ideal. Coupling efficiencies can be large for several reasons. Attachment of electrons to halogens balances the avalanching of electron number density. The net rate of ionization, and consequent change in plasma impedance, and the state population of excited rare gas atoms can also be minimized because the halogens react rapidly with excited states. Because the stimulated emission cross-sections are sufficiently high, excited state number densities can be kept lower than those required for broader band excimer lasers, such as Xe_2^* or Hg_2^*. Naturally, since the lower levels, excepting possibly XeF, are rapidly removed, the inversion can be maintained while the discharge circuitry is being switched. Upper laser level production efficiencies are large. Both ion recombination and direct reactions can yield certain rare gas halides with unit branching ratios. The displacement reaction apparently also has a high branching ratio. Loss into other excimer channels is minimized. At sufficiently low excitation levels, $< 10^{-2}\%$, the initial excited states can be produced at roughly 80% efficiency. [32, 34] The area of extraction efficiency is the only aspect of rare gas halide lasers that critically taxes the kinetics of the system. Quenching of excited states does occur, but reasonable flux levels can be used to extract efficiently. Since the lower level is dissociative, no bottlenecking occurs even when the stimulated rate is significant.

Conclusions

The growth of research in the rare gas/halogen systems has been pheno-
menal over the past year. This is not surprising since the potential of
these species as lasers and as fluorescence sources is quite large. The
rare gas monohalide molecules themselves are intrinsically interesting
because of their novelty. It is not hard to predict that practical lasers
based on the schemes discussed here will evolve over the next few years.
If indeed practical UV lasers with overall efficiencies well in excess of
1% become available, UV photochemistry, and other areas requiring
coherent UV light, should be dramatically enhanced.

Nothing with similar laser potential has been demonstrated yet as a
coherent visible source. However, the potential for higher efficiency
clearly exists. Aside from finding systems analogous to the rare gas
halides that might work in the visible, one can hope to find systems with
higher efficiencies by overcoming some of the shortcomings of the rare
gas halides. The rare gas halides do have fairly low quantum efficiencies
and have problems with spectroscopic extraction, and these areas would
have to be improved on if one hopes to find visible or UV lasers with even
higher efficiency.

There is no reason, however, to doubt that practical, high efficiency
lasers can also be found in the visible, as well as the UV and IR.

Acknowledgments

The authors are indebted for numerous stimulating conversations with his
co-workers at Avco Everett: J. D. Daugherty, H. Hyman, J. Jacob,
and J. Mangano. Also they thank D. W. Setser for providing preprints
of very important rate constant measurements prior to their publication.
A variety of other workers in the field have provided the authors with
preprints and comments prior to final publication and these are acknowledged: R. Gordon, Y. T. Lee, J. Murray, M. McCusker, M, Krauss,
A. K. Hays, S. Suchard, N. Djeu, and R. Hunter. Support of ARPA/
ONR through Contract No. N00014-75-C-0062 is acknowledged.

References

1. M. F. Golde and B. A. Thrush, Chem. Phys. Lett. 29, 486 (1974).

2. J. E. Velazco and D. W. Setser, J. Chem. Phys. 62, 1990 (1975).

3. J. J. Ewing and C. A. Brau, Phys. Rev. A12, 129 (1975).

4. M. F. Golde, J. Mol. Spectry 58, 261 (1975).

5. C. A. Brau and J. J. Ewing, J. Chem. Phys. 63, 4640 (1975).

6. J. Tellinghuisen, J. W. Hoffman, G. C. Tisone, and A. K. Hays, J. Chem. Phys. 64, 2484 (1976).

7. L. A. Gundel, D. W. Setser, M. A. A. Clyne, J. A. Coxon, and W. Nip, J. Chem. Phys. 64, 4390 (1976).

8. J. E. Velazco, J. H. Kolts and D. W. Setser, "Quenching Rate Constants for Metastable Argon, Krypton and Xenon Atoms by Fluorine Containing Molecules and Branching Ratios for XeF and KrF Formation," to be published.

9. S. K. Searles and G. A. Hart, Appl. Phys. Lett. 27, 243 (1975).

10. C. A. Brau and J. J. Ewing, Appl. Phys. Lett. 27, 435 (1975).

11. J. J. Ewing and C. A. Brau, Appl. Phys. Lett. 27, 350 (1975).

12. J. M. Hoffman, A. K. Hays, and G. C. Tisone, Appl. Phys. Lett. 28, 538 (1976).

13. J. R. Murray and H. T. Powell, "KrCl Laser Oscillation at 222 nm," to be published.

14. S. Searles, private communication.

15. J. A. Mangano and J. H. Jacob, Appl. Phys. Lett. 27, 495 (1975).

16. R. Burnham, N. W. Harris and N. Djeu, Appl. Phys. Lett. 28, 86 (1976).

17. J. J. Ewing and C. A. Brau, Appl. Phys. Lett. 27, 557 (1975).

18. J. R. Murray, J. C. Swingle, and C. E. Turner, Appl. Phys. Lett. 28, 530 (1976).

19. C. H. Chen and M. G. Payne, Appl. Phys. Lett. 28, 219 (1976).

20. J. R. Murray, private communication.

21. J. J. Ewing, J. H. Jacob, J. A. Mangano and H. Brown, Appl. Phys. Lett. 28, 656 (1976).

22. J. Jacob, private communication.

23. M. V. McCusker, R. M. Hill, D. L. Huestis, D. C. Lorents, R. A. Gutchek and H. H. Nakano, Appl. Phys. Lett. 27, 363 (1975).

24. M. V. McCusker, private communication.

25. C. A. Brau and J. J. Ewing, "Rare Gas Monohalide Lasers: Performance and Spectroscopy," Second Summer Colloquium on Electronic Transition Lasers, Woods Hole, Massachusetts, Sept., 1975.

26. M. L. Bhaumik, R. S. Bradford, Jr., and E. R. Ault, Appl. Phys. Lett. 28, 23 (1976).

27. R. Burnham, F. X. Powell and N. Djeu, Appl. Phys. Lett. 29, 30 (1976).

28. The term "excimer" is standard chemical nomenclature for dimers which are bound in the excited state and free in the lower state. In standard nomenclature dimer means two of the same thing. Thus, Xe_2^*, Ar_2^*, etc. are excimers. Species that are heteronuclear such as $LiXe^*$ or KrF^* can also have bound excited states and dissociative lower states. The accepted chemical nomenclature for such species is "exiplex" short for excited complex. Unfortunately, the laser community has not uniformly utilized good scientific English and typically calls all species with dissociative lower states, and even some like XeF with bound lower states, "excimers."

29. J. H. Birely, D. C. Cartwright and J. Marinuzzi, "Application of High Power Lasers to Problems in the Nuclear Fuel Cycle," to be published.

30. K. Brueckner and S. Jorna, Rev. Mod. Phys. 46, 325 (1974).

31. Chemical production of excited states and a corresponding visible chemical laser is a volumetrically scalable excitation technique. However, chemical pumping of a UV or visible laser has not yet been demonstrated.

32. J. D. Daugherty, J. A. Mangano and J. H. Jacob, Appl. Phys. Lett. 28, 581 (1976).

33. T. H. Dunning, Jr., and P. J. Hay, Appl. Phys. Lett. 28, 649 (1976).

34. J. H. Jacob and J. A. Mangano, Appl. Phys. Lett. 28, 724 (1976).

35. D. C. Lorents and R. E. Olson, Stanford Research Institute Project PYU-2018 Semi annual Report, 1972 (unpublished).

36. J. Tellinghuisen, G. C. Tisone, J. M. Hoffman and A. K. Hays, J. Chem. Phys. $\underline{64}$, 4796 (1976).

37. D. H. Liskow, H. F. Schaefer III, P. S. Bagus and B. Liu, J. Am. Chem. Soc. $\underline{95}$, 4056 (1973).

38. M. Krauss, private communication.

39. J. J. Ewing and C. A. Brau, unpublished.

40. Keith J. Laidler, "Theories of Chemical Reaction Rates," McGraw Hill Book Co., New York (1969).

41. J. R. Krenos and J. C. Tully, J. Chem. Phys. $\underline{62}$, 420 (1975).

42. W. S. Struve, J. R. Krenos, D. L. McFaddlen and D. R. Herschbach, J. Chem. Phys. $\underline{62}$, 404 (1975).

43. D. O. Ham. J. Chem. Phys. $\underline{60}$, 1802 (1974).

44. J. J. Ewing, R. Milstein, and R. S. Berry, J. Chem. Phys. $\underline{54}$, 1752 (1971).

45. A. Mandl, J. Chem. Phys. $\underline{55}$, 2918, 2922 (1971).

46. D. R. Bates and M. R. Flannery, Proc. Roy. Soc. (London) $\underline{A302}$, 367 (1968).

47. L. G. Christophorou and J. A. D. Stockdale, J. Chem. Phys. $\underline{48}$, 1956 (1968).

48. D. W. Trainor, private communication.

49. C. A. Brau and J. J. Ewing, "Radiative Lifetime of XeF," unpublished.

50. D. L. Rousseau, J. Mol. Spectry. $\underline{58}$, 481 (1975).

51. A. Gedanken, J. Jortner, B. Raz and A. Szoke, J. Chem. Phys. $\underline{57}$, 3456 (1972).

52. L. G. Piper, J. E. Velazco and D. W. Setser, Chem. Phys. Lett. $\underline{25}$, 197 (1974).

TUNABLE VUV EXCIMER LASER SYSTEMS

D.J. Bradley, M.H.R. Hutchinson and C.C. Ling
Optics Section, Physics Department, Blackett Laboratory
Imperial College, London SW7 2BZ, U.K.

Abstract

The present state of development of coaxial-diode electron-beam pumped, narrow-band, frequency-tunable, high-power VUV Xe_2 lasers is summarized and their performance characteristics are discussed. The production of coherent radiation in the XUV at 57nm (the shortest wavelength to date) by third-harmonic generation in Argon is described as an example of the capabilities of these lasers for extending the range of nonlinear mixing, selective excitation and photoionization investigations into the VUV and XUV spectral regions.

Introduction

Since the laser pumping power requirement scales as the fourth power of the frequency, for a Doppler broadened lasing transition, the development of VUV lasers had to await the arrival of sufficiently intense pumping sources. The availability of high-current, high-energy electron-beam sources (1) has had a major effect on the technology of high-pressure gas lasers particularly for systems operating in the UV and VUV spectral regions. There is now no difficulty in obtaining the very fast rate of excitation needed for short wavelength laser action so as to overcome the effects of the short fluorescence lifetimes of the upper-energy lasing levels. As a result, it is now as easy to produce narrow-band, frequency-tunable, low beam-divergence lasers with mega-watt powers in the VUV (2), as with dye lasers operating in the visible. These new VUV lasers employ noble gases as the active media. In 1960 Houtermans (3) pointed out the potential advantages for high-efficiency laser action of excited diatomic molecules with unstable ground states (excimers). The most

intensive fluorescence, centred at $\sim$ 172nm, is produced by Xenon.
Stimulated emission of this transition was first demonstrated in
liquid Xenon pumped by an electron-beam (4). Since then laser
action has been produced in gaseous Xenon (5), Krypton (6), Argon
(7) or mixtures of these gases (6,8) by transverse excitation with
high-energy (0.5 - 2MeV) electron-beams having current densities
of hundreds of amperes per square centimetre. The detailed atomic
and molecular processes involved in excitation, and the laser
kinetics are now much better understood (9,10) although some areas
of ambiguity still exist in connection with the lifetimes of energy
storage in the excited states.

To achieve the high efficiencies potentially obtainable with
this "topside-down" excitation by relativistic electron-beams,
it is necessary to obtain good coupling of the electron-beam to
the high-pressure gaseous medium. The introduction (2,11) of
the coaxial electron-beam diode arrangement has permitted
excitation of the laser medium with a high degree of uniformity
and with $\sim$ 50% of the electron energy deposited in the working
volume of the laser cavity. As a result the laser pumping threshold
is   considerably reduced and megawatt output powers are obtainable with pumping energies of a few joules. Also high-repetition
rate operation is now feasible. With a uniformly excited
cylindrical volume of gas, the coaxial diode-laser operates with
a low beam-divergence. This permits substantial frequency-
narrowing and spectral tuning, over the broad fluorescence bandwidth,by the insertion of a single intra-cavity dispersive
element (2). An incidental advantage is that investigation of
the laser parameters is easily carried out.

Since the first coaxial-diode laser was reported (12) the
arrangement has been adopted for $Ar-N_2$ (13) and noble gas halide
(14,15) lasers to produce high-energy, high-efficiency operation
in the far UV (16). This pumping arrangement makes electron-beam
excitation highly competitive in convenience, size and efficiency
with the long established travelling wave and electrical discharge
methods. Much larger coaxial diode systems, with modular units
to overcome the pinch effects of self-magnetic fields, are in
the design stage (17).

The Xe_2 laser will be described as the prototype VUV system,
since most work has been carried out to date on this laser, for
which the materials problem is less severe than for the shorter
wavelength Kr_2 and Ar_2 lasers. It is likely that with sufficient
optical engineering and improvements in gas purification and
handling techniques, all three noble-gas VUV lasers could be
brought to a level of performance at least comparable to that
already achieved with the Xe_2 laser. Finally the generation of
coherent radiation in the XUV at 57nm, by frequency tripling
of the Xe_2 laser beam in Argon gas, employed as an isotropic
nonlinear medium (18), is described. Since the starting frequency
is already in the VUV a single-stage of harmonic generation is
adequate to reach the XUV spectral region. It is clear that by
this technique coherent radiation could be generated at even
shorter wavelengths by employing the noble gas excimer lasers
and appropriate nonlinear media. The tunability of these VUV
lasers permits resonant enhancement of the harmonic generation

by tuning the fundamental frequency to a two-photon allowed transition (19).

The Coaxial-Diode Laser

The efficient and convenient pumping system of the coaxial electron-beam diode is shown schematically in Fig. 1. The field emission diode consists of a thin-walled ($\sim 25 \mu$m) stainless steel tubular anode, maintained at earth potential. The diameter of the diode is determined both by the electron range in the gas and the impedance of the high voltage power supply (17). The electron range in turn depends upon the gas density and the energy of the transmitted electrons. For 600 keV electrons a tube of 4mm diameter gives uniform pumping of Xenon at a pressure of 10 kTorr. These were the parameters chosen for the first coaxial diode system (11,12) which employed a Febetron (Hewlett Packard) 706 generator, considerably modified to produce a 20J, 5ns pulse of 500 keV electrons. Calorimetric measurements showed that a total energy of 10J was transmitted into the gas contained in the 10cm long anode cylinder (5.5J cm^{-3}) and the distribution of pumping energy along the tube was measured to have a variation of $<$ 10%. With a 25cm long laser resonator terminated by 93% and 20% reflectivity mirrors, respectively, up to 15mJ was produced in a 3ns pulse (5MW peak power) (20). The addition of an intra-cavity fused-quartz prism reduced the lasing bandwidth to 0.1nm and rotation of the prism provided continuous tuning of the narrowed bandwidth over a range of 2500 cm^{-1}. The beam divergence was 1 milliradian when a wire spiral was inserted inside the anode tube walls to prevent grazing incidence reflection. Otherwise the divergence was $\sim$5 mR.

Since a narrower bandwidth is achieved with longer pumping pulses and more resonator transits, a second 50cm long coaxial diode was constructed. Preliminary results have been reported (20) and since then this laser has been developed into a narrow-band, high-power system for four-wave mixing experiments. The diode was designed to operate with an available Febetron 705

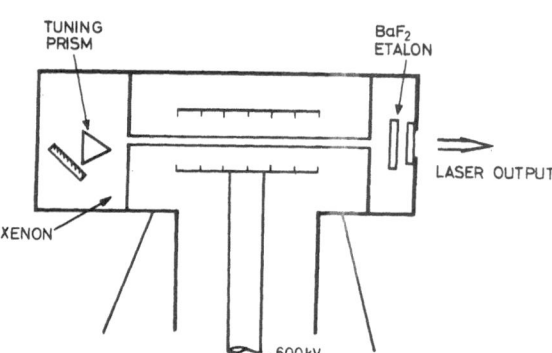

Figure 1. Small coaxial-diode tunable VUV laser arrangement

power supply (2.3 MeV, 400J, 50ns) with the laser diode connected
in series with a voltage-dropping copper sulphate resistor, to
produce a working voltage of 600 kV. The energy deposited in
the laser gas is ∼35J, corresponding to an energy density of
2.5Jcm⁻³. The laser resonator length is 75cm and the cathode is
50cm long and 10cm in diameter (Fig. 2). The anode diameter is
6 mm and the general design, apart from size, is similar to that of
the smaller diode laser shown in Fig. 1. Employing a 20% reflectivity mirror, produced a total output energy of 55mJ, with a peak
power of ∼5MW. The Xenon operating pressure can be varied from
4 kTorr to 20 kTorr with optimum performance at ∼6 kTorr. The
insertion of a fused-quartz prism reduced the peak power to
∼1MW and the laser pulse duration to 10ns. The laser bandwidth
was then 0.15nm. Operating at lower powers (∼0.3MW) the laser
linewidth narrowed to 0.015nm and was tunable over a range of
∼1000 cm⁻¹ around λ172nm. The spectral bandwidth obtained
after deconvolution of the instrumental function is close to the
resolution limit (∼0.014nm) of the 1 metre spectrograph (Fig. 3).
Fine tuning of the lasing frequency is conveniently obtained by
varying the pressure of the Xenon in the laser. In this manner
the dispersion properties of the prism are changed by altering
the refractive index of the surrounding gas. At a working pressure
of 5 kTorr, the tuning rate with change of pressure is 1nm per
kTorr. Coarse tuning is produced by rotation of the prism or the
high reflectivity mirror. This ease of tuning close to a desired
resonance is another attractive feature of high-pressure broadband laser media.

Figure 2.   Photograph of cathode structure of 50cm long coaxial-
            diode

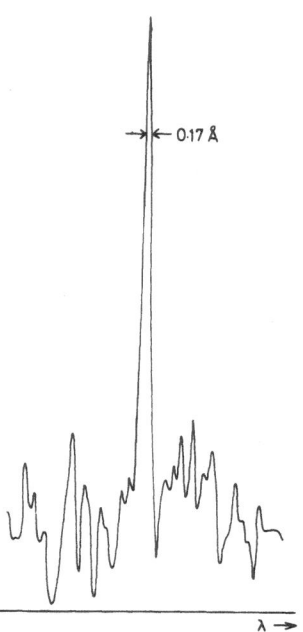

— 0.17 Å

λ →

Figure 3.   Microdensitometer trace of spectrum of output beam
            of Xe_2 laser pumped by 50cm diode. Peak power = 1MW

Generation of Coherent XUV Radiation

Third harmonic generation in noble gas was first demonstrated
by WARD and NEW (18). Since then third-order nonlinear processes
in isotropic media have been employed for the production of
tunable (19) and fixed frequency (21) coherent VUV radiation.
The shortest wavelengths thus generated previously were at 118.2
nm, obtained by tripling 354.7nm radiation in a phase-matched
mixture of Xenon and Argon (21), and at 887Å by tripling 226nm
radiation in Argon (22). In both of these experiments the
starting frequency was a higher-harmonic of the 1.06μm fundamental
frequency of picosecond pulses from a mode-locked Nd-YAG laser.
With the high-power, tunable frequency Xe_2 laser we have been
able to extend the range of generation of coherent radiation into
the XUV at 57nm. Since the starting frequency is already in the
VUV, a single stage of tripling is sufficient. Also with a
tunable laser the efficiency of harmonic-generation can be
considerably enhanced by exploiting an intermediate two-photon
resonance (19). Since this experiment is an excellent
demonstration of the capability of the coaxial-diode laser for
extending nonlinear mixing, selective excitation (23) and photoionization (24) into the VUV spectral region, and of the general
performance of the laser system, it will be described is some
detail.

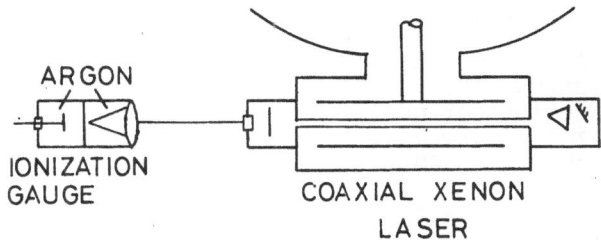

Figure 4.   Experimental arrangement for generation and detection
of third-harmonic radiation at 57nm in Argon

The experimental arrangement is shown in Fig. 4.   A 1MW
beam, tuned to λ =   170.94nm in resonance with the $3p^6$ - $3p^5 5s$
two-photon transition, (Fig. 5) was focussed by a BaF_2 lens,
of focal length 3 cm, into a cell containing Argon at a pressure
of 4 Torr. The focal point was located just in front of an
aluminium foil (4.5 μm thick) covering the window of an Argon
photoionization chamber. The gas pressure of the ionization
gauge was also kept at 4 Torr to minimize the differential

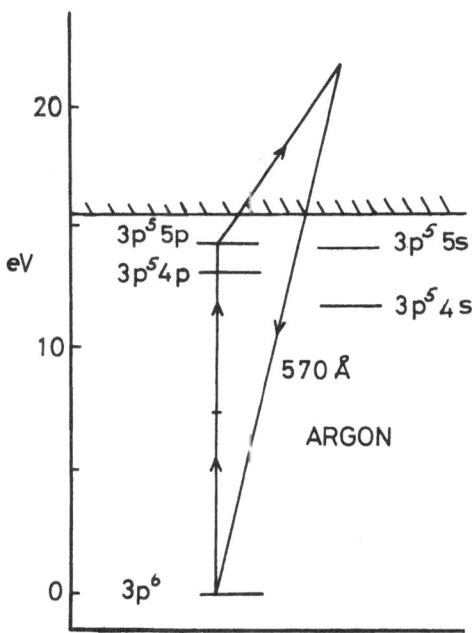

Figure 5.   Energy levels involved in third-harmonic generation
in Argon with the Xe_2 laser

pressure. The aluminium foil was needed to filter out the
fundamental radiation, so as to avoid photoemission from the
gauge electrode. The energy of the third-harmonic radiation
transmitted through the foil was obtained from the integrated
signal from the ionization chamber displayed on an oscilloscope.
The photoionization cross-section of Argon at 570nm has the
value $\sim$ 3 x 10^{-17} cm^2 and it was thus necessary to have the
focal point of the laser focussing lens close the aluminium
foil to prevent excessive absorption of the third-harmonic
frequency, before it entered the ionization chamber. The
experimental arrangement was checked by measurements taken with
and without Argon in the third-harmonic generation section,
while the laser was tuned through the two-photon resonance
wavelength. The size of the stray electrical background-signal
was determined by evacuating both the third-harmonic generation
and the photoionization chambers with the laser tuned to
resonance. Finally the laser was quenched by reducing the
Xenon gas pressure and records were again taken with and
without Argon in the two chambers. Fig. 6 shows the harmonic-
generation efficiency resonance curve centred at λ 170.94nm.
The halfwidth of $\sim$ 0.1nm was equal to the operating bandwidth
of the laser.

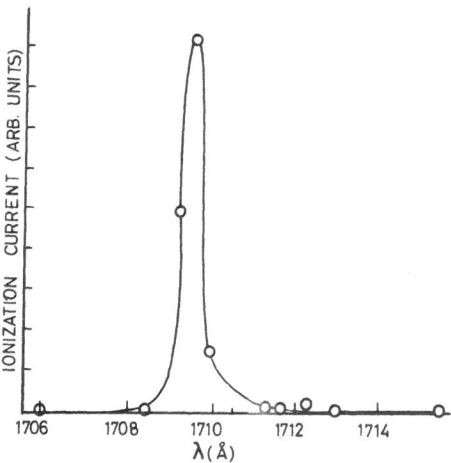

Figure 6. Third-harmonic generated energy as a function of the
Xe$_2$ laser wavelength showing the two-photon resonance
at λ 170.9nm

    To confirm that the coherent radiation was being generated at
57nm the spectra was recorded with the arrangement of Fig. 7.
The focal point was then located just in front of a 250μm
diameter aperture in the focal plane of a 1 metre, normal
incidence spectrograph. The Argon pressure was maintained at
4 Torr and the spectrograph pressure at $\sim$ 10^{-4} Torr by
differential pumping through the aperture. The third-harmonic

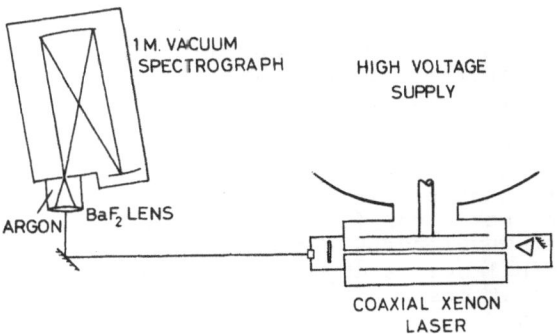

Figure 7.   Arrangement for recording spectra of coherent XUV
             radiation

spectrum was recorded on Kodak SC7 VUV film. The spectrograph
wavelength scale was calibrated by recording the λ 54.8nm line
of Helium, which also gave the instrumental resolution limit.
Comparison with the calibration spectrum shows (Fig. 8) that the
recorded bandwidth of the 57 nm radiation is instrument limited
and is narrower than that of the fundamental laser linewidth.

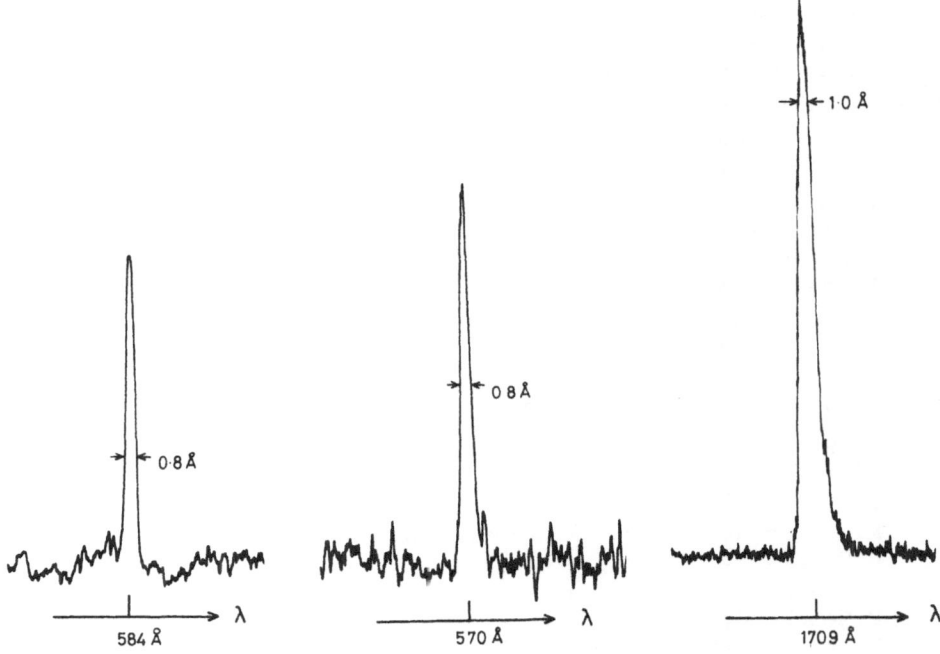

Figure 8.   Spectra of the fundamental laser light, the third-
             harmonic radiation and the   λ 54.8nm Helium line

Conclusion

It is clear that the Xe_2 laser and other excimer lasers are likely to play as important a role in VUV spectroscopy as dye lasers are currently playing at longer wavelengths. With the now demonstrated capability of generating harmonic frequencies in isotropic media this could well prove to be the easiest route to generating coherent light, at useful energies, at wavelengths of a few tens of nanometres. Electron-beam pumped noble gas lasers, and noble gas fluorescence could be used to efficiently pump longer wavelength lasers (15,17) in the same manner that the Nitrogen laser is employed to pump dye systems. For all these applications a high-repetition rate laser would have many advantages and a 10 pps e-beam pumped system is being constructed in our laboratory.

Acknowledgements

We wish to acknowledge the skilled technical assistance of Mr. A. JOHNSON and the assistance of, and useful discussions with, Professor H. MAHR. Financial support was provided by the Science Research Council and the UKAEA Culham Laboratory.

References

1   H. H. Fleischmann, Physics Today, 28, 35 (1975)

2   D. J. Bradley, D. R. Hull, M. H. R. Hutchinson and M. W. McGeoch, Opt. Commun., 7, 187 (1975)

3   E. G. Houtermans, Helv. Phys. Acta, 33, 933 (1960)

4   N. G. Basov, V. A. Danilychev, Yu. M. Popov and D. D. Khodkevich, Soviet Physics JETP Letters, 12, 329 (1970)

5   H. A. Koehler, L. J. Ferderber, R. L. Redhead and P. J. Ebert, Appl. Phys. Letters, 21, 198 (1972)

6   P. W. Hoff, J. C. Swingle and C. K. Rhodes, Appl. Phys. Letters, 23, 245 (1973)

7   W. M. Hughes, J. Shannon and R. Hunter, Appl. Phys. Letters, 24, 488 (1974)

8   A. Wayne Johnson and J. B. Gerardo, J. Appl.Phys., 45, 867 (1974)

9   A. Wayne Johnson and J. B. Gerardo, J. Chem. Phys., 59, 1738 (1973)

10   R. A. Gutcheck, R. M. Hill, D. L. Huestis, D. C. Lorents and M. V. McCusker, Stanford Research Institute Report, SRI, No. MP75-43, 1975

11   D. J. Bradley, D. R. Hull, M. H. R. Hutchinson and M. W. McGeoch, Opt. Commun., 11, 335, (1974); D. J. Bradley and M. H. R. Hutchinson, UK Provisional Patent No. 14102/74 (1974)

12   D. J. Bradley, D. R. Hull, M. H. R. Hutchinson and M. W. McGeoch. Post-deadline paper at VIII International Conference on Quantum Electronics, San Francisco June 1974.

13   E. R. Ault, Appl. Phys. Letters, 26, 619 (1975)

14   E. R. Ault, R. S. Bradford and M. L. Bhaumik, Appl. Phys. Letters, 27, 413 (1975)

15   M. L. Bhaumik, "High power gas lasers" Edited by E. R. Pike, Institute of Physics (London and Bristol) Conference Series No. 29, 122 (1975)

16   M. L. Bhaumik, R. S. Bradford and E. R. Ault, Appl. Phys. Letters, 28, 23 (1976)

17   L. P. Bradley, "High power gas lasers" Edited by E. R. Pike, Institute of Physics (London and Bristol) Conference Series No. 29, 58 (1975)

18   J. F. Ward, and G. H. C. New, Phys. Rev., 185, 57 (1969)

19   R. T. Hodgson, P. P. Sorokin and J. J. Wynne, Phys. Rev. Letters, 32, 343 (1974)

20   E. G. Arthurs, D. J. Bradley, C. B. Edwards, D. Domanski, D. R. Hull, C. C. Ling and M. H. R. Hutchinson, Proceedings of Intern. Conf. on Electron Beam Research and Technology, Albuquerque, 1975, Sandia Laboratories Report, SAND 76-5122, Vol. II, 193.

21   A. H. Kung, Appl. Phys. Letters, 25, 653 (1974)

22   S. E. Harris, J. F. Young, A. H. Kung, D. M. Bloom and G. C. Bjorklund, "Laser Spectroscopy" (Plenum Press, New York and London) 59, (1973)

DYE LASER TECHNOLOGY

F.P. Schäfer

Max-Planck-Institut für Biophysikalische Chemie,
D-3400 Göttingen, Fed. Rep. Germany

The practicability of such interesting projects as industrial enrichment of uranium with lasers are hinged on the availability, on the efficiency and not least the reliability of tunable lasers in the near ultraviolet and visible with very high average output powers for which at present dye lasers seem to be the only candidates. For these reasons this review will cover the various approaches leading to the development of dye lasers with average output power capabilities in the kW region.

Disregarding, for the moment, the active medium, i. e. the dye solution itself, the three main points that determine the properties of such a laser are:

1. pumping geometry and dye cell construction,
2. pumplight sources,
3. electrical circuitry.

There are three different pump geometries that have been applied until today to high power dye lasers:

a) the coaxial design, that has been used by several Russian groups,
b) the elliptical cylindrical cavity, used by the United Technology group, and
c) a novel illumination geometry, that is being used by our group.

One of the Russian workers' design is shown in Fig. 1. This laser [1] operated with an alcohol solution of rhodamine 6G, and 17 kJ electrical pump energy delivered an output pulse energy of 32 J with only 30 mℓ of solution, thus giving more than 1 J/cm^3 of solution. This is much higher than the maximum stored energy and indicates that each dye molecule is pumped many times during the relatively long pump pulse of 30 μsec. Recently BALTAKOV [2] and coworkers, scaling up this design, obtained an output of 400 J in a 10 μsec pulse with 50 kJ of pump energy.

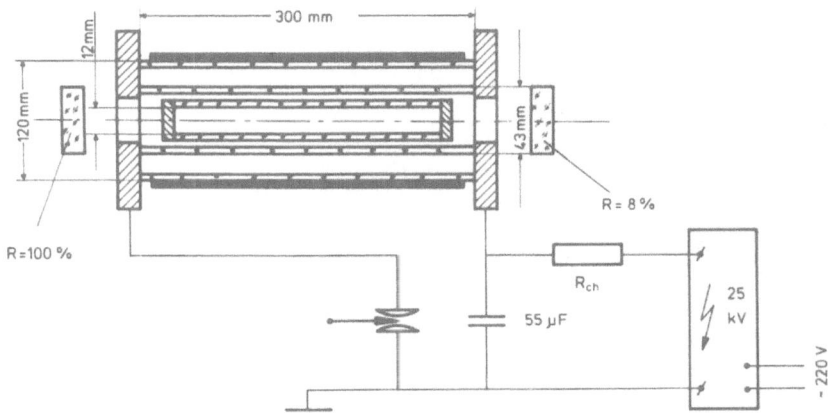

V.A. Alekseev et. al., Sov. J. Quant. Electr. 1, 643 (1972)

Fig. 1. Coaxial flashlamp dye laser (from [1])

That is an efficiency of 0.8 %. The peak output power was 40 MW.
In this case the authors used 1.7 ℓ of dye solution. They report
that the beam, which had a nearly Gaussian intensity distribution in the beginning, increased its beam divergence during the
pulse, up to 2 degrees full width at the end of the pulse. This
is mainly due to the reverberating shock waves from the discharge, which were observed by several groups and which are one
of the main drawbacks of coaxial lamps. One cannot hope to obtain a small bandwidth with this large beam divergence.

The authors only report on single shot measurements, but it
should be easy to reach at least a few Hertz repetition rate
and thus more than 1 kW average output power with a flow cell
and a power supply of at least 120 kW. The emission, however,
will only be broadband.

The elliptical cylinder is much better in that respect. In a
laser of 100 W average power designed by MOREY, GLENN, and
FERRAR of the United Technology (formerly United Aircraft) Research Center [3], the distance between flashlamp and dye cell
is 66 mm, so that the shock wave from the flashlamp travelling
at about 1 mm/μsec reaches the dye cell only after the 2 μsec
laser pulse is over. The only distortion that remains is the
thermally induced schlieren from the nonradiatively dissipated
pump energy. This was not serious since they obtained a beam
divergence of only 5 mrad using a 50 % water/50 % ethanol solvent and 70 cm resonator length.

To obtain 100 W average power with only 0.4 J pulse energy
they had to use 250 Hz pulse repetition frequency. The necessity
to fill the dye cell with fresh solution for successive pulses

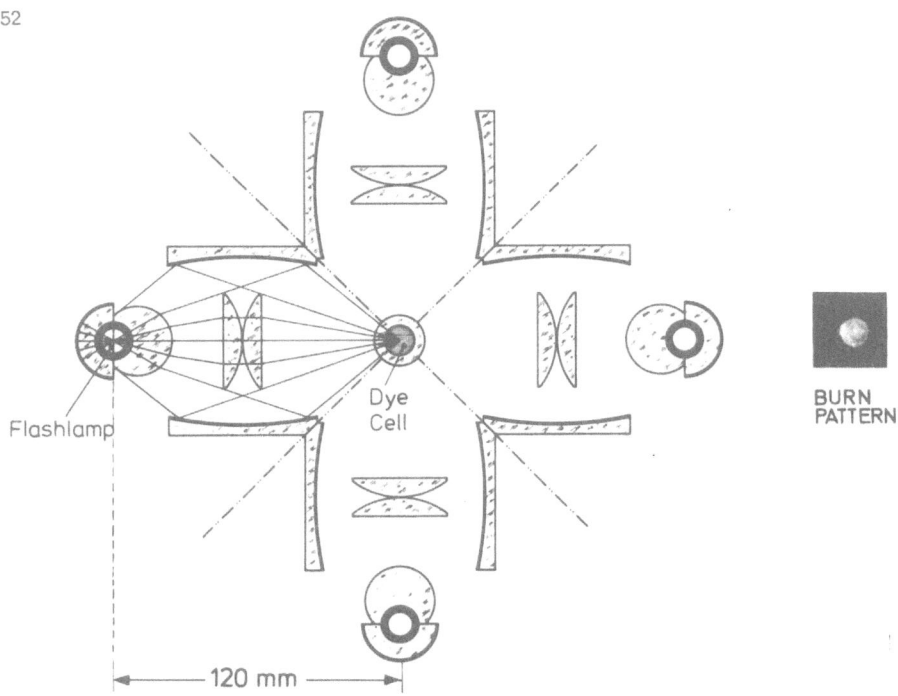

Fig. 2. Novel pumping system for high power dye lasers

to remove the heat created some problems with the flow rate. A
longitudinal flow at the necessary rate invariably created difficulties with cavitation and consequently air bubbles in the
cell. To avoid this, the authors then used a transverse flow
cell. A disadvantage of this design is the low coupling efficiency because of the missing part of the elliptical cylinder
and the highly non-uniform illumination and the rectangular
cross-section. The resulting burn patterns are highly irregular
and show that maximum output can only be obtained when the reflectors of the resonator are misaligned by 10 mrad.

Our approach to the design of a 100 W laser uses four flashlamps in the optical arrangement shown in Fig. 2. The rays parallel to the plane of the screen radiating from the flashlamp
into 360° are collected into an angle of only 85° or even less
by a back reflector, an aplanatic lens and an additional focusing lens-mirror system. The aberrations are very small and the
magnification is only 1.5. A great advantage of this pumping
system is the great uniformity of pumplight distribution in the
dye cell. This arrangement with its long distance of 120 mm between flashlamps and dye cell is also free from premature pulse
termination or distortion by acoustic shock waves. In addition,
the cooling liquid, which is a 2 mm thick layer of copper sulfate solution at the same time acts as a UV-absorbing filter
and prolongs the useful life of the dye solution by a factor of
ten. We use four commercial xenon flashlamps of 6 inch arc
length and 7 mm bore (type ILC 7F6) which together can take an
average load of 27 kW. At present we could only operate this

laser single-shot, because of long delivery delays of the large power-supply. We obtained 1.5 J per shot from a rhodamine 6G solution with a beam divergence of only ___ mrad.

The bandwidth was 2 to 3 nm and the introduction of an interference filter into the cavity reduced this to .1 Å with still 75 % of the braodband output at the peak of the tuning range at 595 nm. Of course, at high repetition rates one cannot use interference filters or mirrors which would be damaged by the high average power and we intend to use again a ring laser cavity with four Abbé prisms and uncoated etalons and a frustrated total reflection outcoupler, that we have used successfully in other pulsed lasers to reduce the bandwidth to less than a hundredth of an Ångström. Still another possibility that has been used recently by several groups is the injection of small bandwidth emission from a well-designed small oscillator into the high power laser then acting as a regenerative amplifier.

We use a longitudinal flow cell at present for low repetition rates up to 50 Hz. But if we want to extend the operation towards higher repetition rates we could easily use a transverse flow cell, shown in Fig. 3, with two input and two output ports

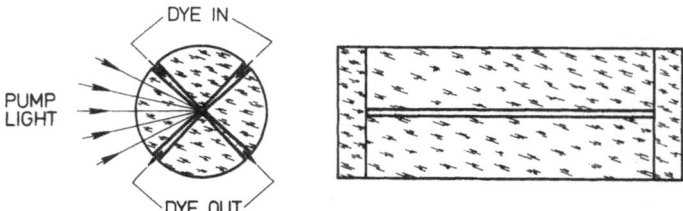

Fig. 3. Transverse flow cell (cross-sections) for use with pumping system shown in Fig. 2

in this optical arrangement without any obstruction of the pumplight. Here the cell proper is the free middle section in a polished cylindrical quartz block that is constructed from four slices and top and bottom plates so that the dye solution can flow in and out through the gaps, while for the pump light from the flashlamps the quartz block acts as a zero power lens illuminating the region around the axis of the cylinder.

This design is easily scalable, and one can calculate that e. g. 24 inch lamps with 19 mm bore should suffice to bring the average output power beyond 3 kW.

We chose commercial linear xenon flashlamps as pumplight sources because of several advantages compared to coaxial or vortex flashlamps. I mentioned already some disadvantages of coaxial flashlamps. The vortex-stabilized flashlamp developed by MACK [4] and used by the United Technology group in their laser is much better in this respect. An argon stream is blown

tangentially into a quartz glass tube, creates a vortex and
flows out through bores in the electrodes. The pressure differential always keeps the arc centered along the axis of the lamp.
One disadvantage is the high gas consumption rate of 5 to
10 ℓ/sec, another disadvantage is the bad noise molestation by
the shock waves from the exhaust ports. With forced nitrogen
cooling of the quartz tube and water cooling of the electrodes
the lamp is expected to handle 25 to 50 kW for long periods. At
present the authors have applied these power levels for only a
few seconds.

A better solution might be a water vortex lamp which we are
presently testing in our laboratory. Here the gas is replaced
by a thin streaming film of water running along the inside of
the quartz glass envelope which is spinning at a few hundred
revolutions per minute. The plasma then is created in a water
vapor atmosphere. Similar devices have been investigated in the
mid-fifties by MAECKER and coworkers [5] and were found capable
of power densities in the plasma of 10 MW/cm^3 even in continuous
operation. With these high power densities one should even be
able to pump a cw-dye laser incoherently.

Coming back to pulsed lasers, the pulse forms, as determined
by the electrical circuits, are of importance for the efficiency
and reliability of the laser. Figure 4 shows the circuit and
pulse forms of current and pumplight for the vortex lamp.

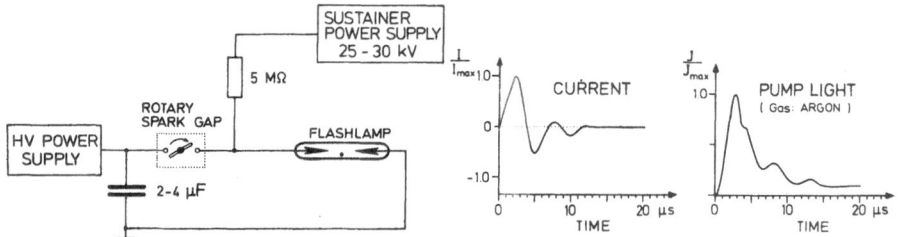

Fig. 4. Electrical circuit and pulse forms of a vortex lamp
        (adapted from [3])

A sustainer current maintains a constant ionization level between pulses and a rotary spark gap is used to discharge a 2 to
4 μF capacitor charged up to 10 kV through the lamp. One sees
that this gives a ringing discharge and the light output shows
several wiggles. These voltage reversals shorten the lifetime
of flashlamp and capacitor. An especially weak point is the
rotary spark gap, which introduces pulse and amplitude jitter,
is quickly worn out and at high powers does only quench the arc
when a strong gas jet directed at it.

In contrast we use a critically damped discharge as shown in
Fig. 5. You see that the light intensity smoothly follows the
current, without any long tail. Instead of a spark gap we use
thyristors, as we had done before in a 10 W laser [6]. Since
here we use 6 inch long flashlamps, we had to increase the capacitor voltage to 5 kV. In the beginning we had to connect

several thyristors in series, but recently developmental types with neutron-irradiated silicon tablets have become available that have 5 kV blocking voltage and 13 kA peak current, so that one thyristor per flashlamp is sufficient. An important point, when using thyristors, is to limit the current in the first μsec to about 100 A, to give the charge carriers time to diffuse evenly from the gate over the whole silicon tablet, which is easily done with properly dimensioned saturable reactors. After this first μsec the current rise may be much faster than 1 kA/μsec.

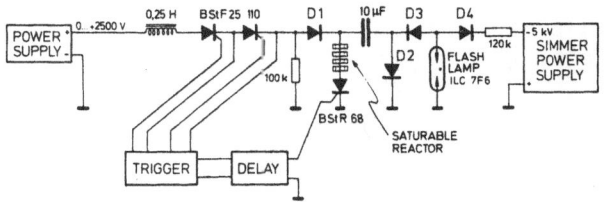

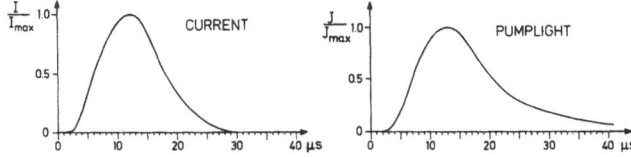

Fig. 5. Electrical circuit and pulse forms of our high power dye laser

One bad problem of high power dye lasers is the photochemical degradation of the dye solution. At present this problem can only be circumvented to some extent, if one uses a sufficiently high ratio of reservoir volume to dye cell volume. In a proto-type 10 W laser [6] we found a decrease in laser output power due to photodecomposition as shown in Fig. 6. With 1 mℓ dye cell

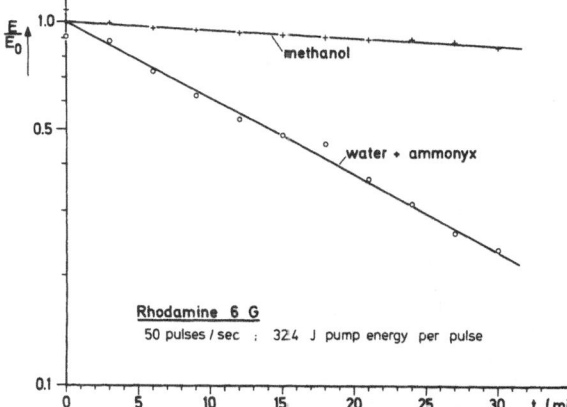

Fig. 6. Photodegradation of a rhodamine 6G solution. E is the laser output at time t, E_0 the laser output at zero time

volume and a 10 ℓ reservoir the time to fall to half the initial value was a quarter hour for a water-Ammonyx solution of rhodamine 6G and two hours for a methanol solution when operated with 32.4 J per pulse at 50 Hz repetition frequency. Recently, however, we have found a new class of dyes, the phenoxazones [7], which seem to be much more stable than xanthene dyes like the rhodamine 6G. One member of this new class of dyes, the nile-blue-A-phenoxazone, shown in Fig. 7, has nearly the same efficiency as rhodamine 6G but is six times more stable as shown by the diagram of normalized laser energy per pulse versus

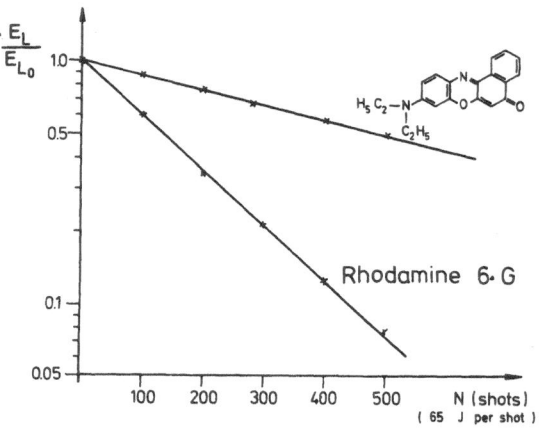

Fig. 7. Photodegradation of a phenoxazone compared to that of rhodamine 6G under identical conditions

the integrated pump energy. This class of dyes has also the nice property of strong solvatochromy, giving the large shifts in laser wavelength with change of solvent shown in Fig. 8.

As a last point I would like to make some comments on several possible ways to increase the efficiency of dye lasers. The first possibility is an enlargement of the width of the absorption spectrum of dyes in order to utilize the white light from the spectrum to a greater extent. An ideal case of high efficiency would be a black dye that absorbs all of the visible spectrum and emits in the near infrared with a quantum efficiency of the fluorescence of approximately unity. One way to approach this goal would be the linking of several chromophors by short saturated carbon chains as shown in Fig. 9 for one hypothetical example. In this way there is no direct overlap of the π-electron clouds, so that the chromophors can absorb independently, but an efficient radiationless energy transfer of the Förster type is now possible because of the short distance between the chromophors, and the fluorescence is emitted by the chromophor with the longest wavelength absorption. A second possibility would be the reduction of triplet losses by triplet quenchers which are chemically bonded to the chromophors in much the same way as the linking of several chromophors.

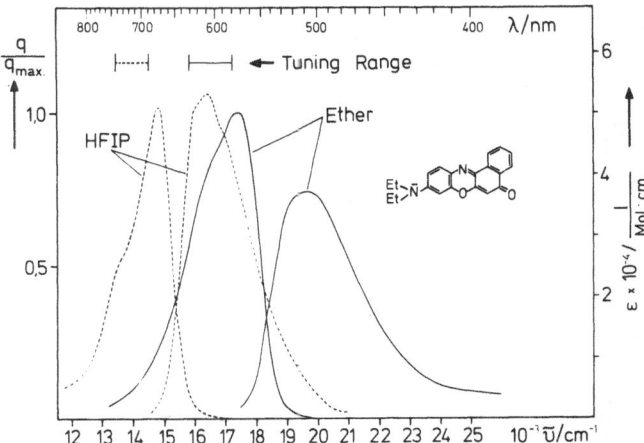

Fig. 8. Spectral shifts of absorption, fluorescence and laser
emission with change in solvent

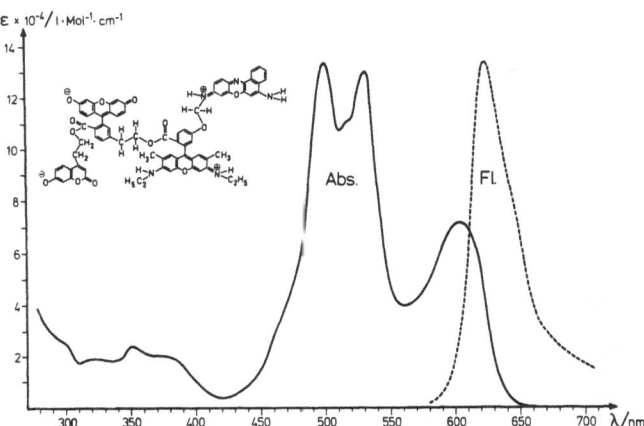

Fig. 9. Calculated absorption and fluorescence spectrum of the
hypothetical molecule

There are a lot of variations and ramifications possible a-
long these lines. As a consequence in the long run one would
expect dye laser efficiencies of about 5 %, which is to be compared to neodymium-glass lasers with at most 2.5 %.

Still another possibility that holds great promise for increasing, at least in the long run, the efficiency of the dye
laser especially in the blue and ultraviolet region of the spectrum is the vapor phase dye laser. Since we and other groups
found that vapours of suitable dyes can be pumped optically by
a nitrogen laser and lases at the same efficiency as in solu-

58

tion, we have tried to apply electron beam pumping as a next
step towards the goal of direct discharge pumping that should
finally result in the highest efficiency. We did not yet set up
a laser cavity around the vapour cell but rather looked at the
spontaneous fluorescence with various buffer gases added to the
dye vapour [8]. Some of the results are shown in Fig. 10 which

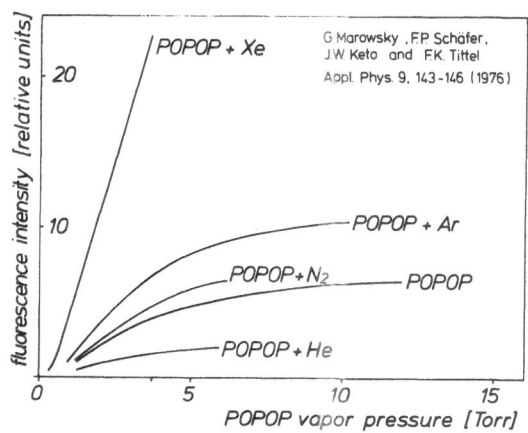

Fig. 10. Fluorescence intensity of POPOP-vapour with e-beam excitation as a function of vapour pressure with various
buffer gases added at 100 °C at a constant pressure of
0.5 atm. (from [10])

gives the spontaneous fluorescence intensity vs. the vapour
pressure of the dye POPOP for several buffer gases at about
1 atmosphere fill pressure. The important observation here is
the superiority of xenon as a buffer gas, which is only partially due to its high electron stopping power because of its
high atomic weight. In addition there seems to be a resonant
energy transfer from the lowest excited singlet state of the
xenon atom to the second excited singlet state of the POPOP-
molecule. This observation of energy transfer seems to be cor-
roborated by a recent investigation of WEBB and coworkers in
Oxford [9], who used a stream of helium metastables to excite
POPOP and perylene molecules and found a proportionality of
spontaneous fluorescence with the concentration of the metastables. It would be interesting to repeat their experiments
with xenon instead of helium, which according to our results
should give a much more efficient energy transfer.

In conclusion one can say that even though at present dye
lasers have found their most interesting applications as low
power devices, they have a great potential of development into
high power devices at least with regard to the average power
of pulsed dye lasers which could be especially useful for photochemical purposes.

References

1. V. A. Alekseev, I. V. Antonov, V. E. Korobov, S. A. Mikhnov, V. S. Prokudin, B. V. Skortsov: Sov. J. Quant. Electron. $\underline{1}$, 643 (1972)
2. F. N. Baltakov, B. A. Barikhin, L. V. Sukhanov: JETP Letters $\underline{19}$, 174 (1974)
3. W. W. Morey, W. H. Glenn, C. M. Ferrar: Technical Report, R 75-921617-13, ARPA Order No.: 1806 AMEND No. 16
4. M. E. Mack: Appl. Opt. $\underline{13}$, 46 (1974)
5. F. Burhorn, H. Maecker: Z. Physik $\underline{129}$, 369 (1951)
6. J. Jethwa, F. P. Schäfer: Appl. Phys. $\underline{4}$, 299 (1974)
7. D. Basting, D. Ouw, F. P. Schäfer: Opt. Commun., in print
8. G. Marowsky, F. P. Schäfer, J. W. Keto, F. K. Tittel: Appl. Phys. $\underline{9}$, 143 (1976)
9. T. Sakurai, I. M. Littlewocd, C. E. Webb: Appl. Phys. Letters $\underline{28}$, 533 (1976)

II. Tunable IR Laser Systems

SCALABLE TUNABLE IR LASERS*

A. Mooradian
Lincoln Laboratory, Massachusetts Institute of Technology
Lexington, Massachusetts 02173, USA

Introduction

This paper discusses some of the practical operating characteristics and limitations of tunable infrared sources including semiconductor lasers, spin-flip Raman lasers, and nonlinear mixers. In particular, those characteristics which are important for applications in high resolution spectroscopy and laser induced photochemistry and diagnostics are discussed. These fields are still in an early stage of their development with a substantial amount of work yet to be done in the laboratory using small to moderately powered laser systems. Industrial photochemical processes will require efficient and high average power laser sources at a specific frequency once a specific technique has been demonstrated in the laboratory. A significant impact in the above areas will occur when tunable sources that are simple to operate, inexpensive, and cover a broad wavelength range with pulsed and moderately high average (several watts) power output and high resolution capability are made available to the research community. Substantial progress in this direction is expected to occur in the next few years.

Semiconductor Lasers

Semiconductor diode lasers [1] have been used more extensively than any other tunable infrared laser source for high resolution spectroscopy. Power levels of ten to a hundred microwatts cw have been sufficient for high resolution linear absorption spectroscopy of molecules. Diode lasers are one of the few class of laser devices which can operate efficiently at very low output power levels and are particularly useful for applications where primary power drain is important such as in airborne systems. Increasing laboratory use of closed-cycle, variable temperature helium refrigerators for broadband tuning of diode lasers is being made and has greatly increased their utility. Overall power efficiencies of such systems are substantially reduced because of the power consumption (> 1 kW) of these cooling units. For most laboratory applications, this is relatively unimportant.

The recent development of lead-salt diode lasers that operate above 77 K [2] has greatly extended the gross tuning range [3] for a single device. Figure 1 shows

*This work was sponsored by the Department of the Air Force, NSF/RANN, and the U. S. Energy Research and Development Administration under a subcontract from the Los Alamos Scientific Laboratory.

the wavelength tuning for a PbSnTe double-heterostructure laser that operates continuously up to 114 K. When cooled by a variable-temperature, closed-cycle helium refrigerator, the useful range for spectroscopic applications is from 8.5 to 16 μm. Recently developed molecular beam epitaxial fabrication techniques [3] for these lead-salt devices have substantially increased the yield of cw lasers that operate somewhat above 77 K. Because of the stripe-geometry structure, the spectral output of these lasers occurs in a limited number of modes at both ends of the spectral operating range. Figure 2 shows the characteristic spectral output of a device similar to that of Fig. 1. The tuning range for a single mode is about one cm^{-1} before a mode jump occurs. Because the double-heterostructure confines the gain region to only a few micrometers, the divergence of the laser output is large; usually in an angle greater than f/1.

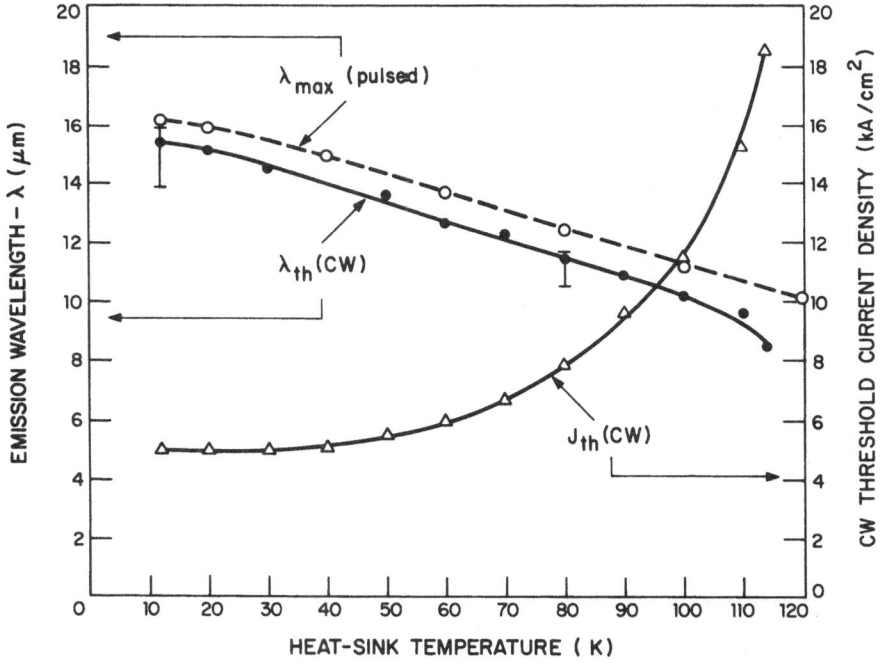

Fig. 1. Laser emission wavelength for a double-heterostructure $Pb_xSn_{1-x}Te$ diode laser (pulsed and cw operation) as a function of heat sink temperature. Also shown is the threshold current density. From ref. [3].

At drive current levels substantially above threshold, the output of most lead-salt lasers occurs in a number of modes. Use of an external cavity would greatly increase the utility of these devices by funneling most of the power into a single, well controlled, tunable mode. Recently [4], improvement of lead-salt laser spectral characteristics has been demonstrated using a distributed feedback grating structure incorporated along the junction region. Figure 3 shows the detailed structure of this device. Most of the output of this laser was centered at a single frequency with the resolution limited by the analyzing spectrometer. No spectral purity measurements have been reported for distributed feedback devices to date. The output frequency of this laser was temperature tuned several cm^{-1} with no frequency jumps. For practical spectroscopic applications, however, an external cavity device is probably the most useful approach since the full gain-bandwidth is usable at a fixed temperature and the cavity tuning elements can track the entire temperature tuning range.

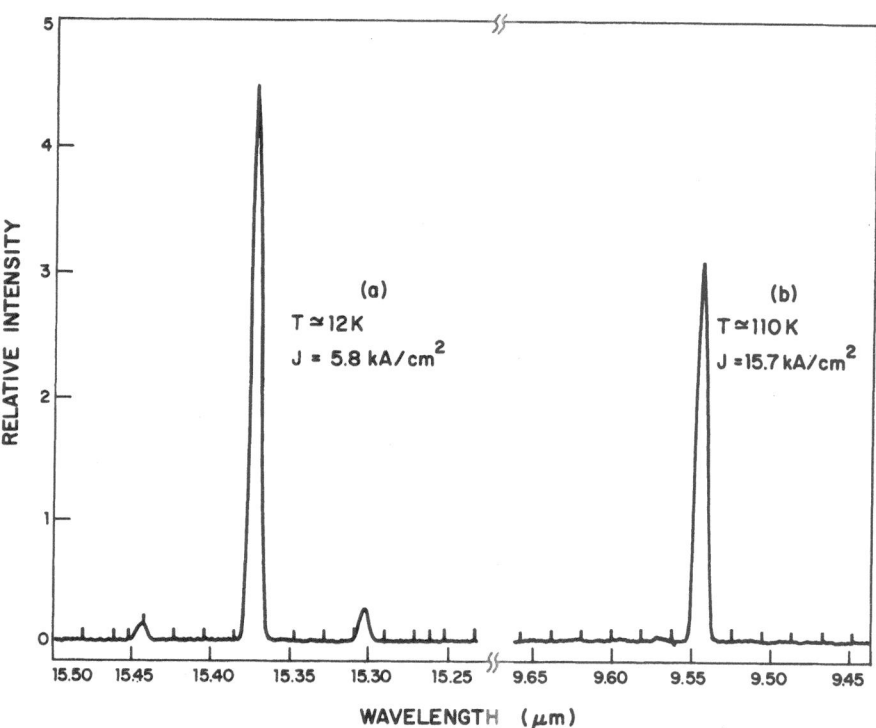

Fig. 2. Characteristic spectral output of a double-heterostructure $Pb_xSn_{1-x}Te$ diode laser at 12 K and 110 K taken using a conventional grating spectrometer. From ref. [3].

The peak output power from diode lasers has been limited in part by catastrophic degradation at the Fabry-Perot faces at the junction. In the case of GaAs diode lasers, degradation intensities in excess of 10^7 W/cm^2 have been observed at room temperature, while for lead-salt lasers, degradation intensities are only about 10^6 W/cm^2 at low temperatures. Peak output powers from simple junction lead-salt lasers are limited to about a watt, while double-heterostructure devices are limited to even less power. In fact, the cw and pulsed output power capabilities can be comparable for these devices. The maximum thickness of the laser region is usually limited by the diffusion length of injected carriers which is about 20-100 μm for lead-salt devices. The width of the laser region for a stripe geometry device is typically limited to 25-50 μm in order to prevent power being lost to bounce modes. Cavity lengths are about 500 μm and are limited by internal

absorption losses which are remarkably similar for most semiconductor lasers
($\sim$ 10-40 cm^{-1}).

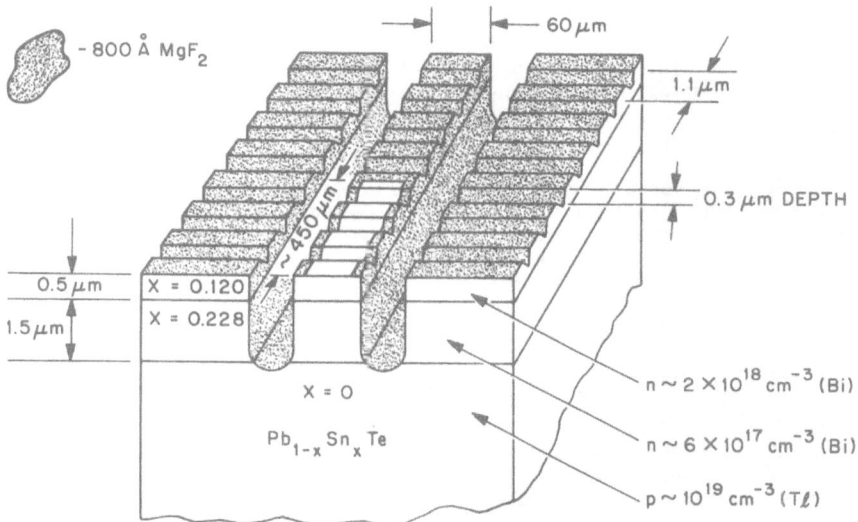

Fig. 3. Detailed structure of distributed feedback $Pb_x Sn_{1-x} Te$ double-heterostructure
diode laser .   From ref. [4].

Increasing the peak or average output power from a single lead-salt diode
laser substantially beyond the one watt level is probably unrealistic. By coupling
the electromagnetic radiation from adjacent diode lasers in an array, it would be
possible to generate a uniphase wave front in the far field with an output power
nearly equal to the sum of the individual output of each device. This has been demonstrated using a small number of GaAs diode lasers [5] in a linear array. Extending this technique to a two-dimensional array of diode lasers would be quite difficult.
A two-dimensional array fabricated from GaAs has been pumped by an electron
beam and operated with an external cavity that produced a uniform beam intensity
profile in the far-field [6]. This technique could, in principal, be applied to the
lead-salts. Volume excitation of semiconductor lasers could be achieved by one or
two-photon excitation with another laser. This has been done successfully in GaAs
[7] with the Stokes shifted wave in liquid nitrogen using a ruby laser and in InSb [8]
using a CO_2 laser. One of the main limitations in the lead-salts would be free carrier absorption of the pump radiation by the photo-excited carriers as well as the
reduction in external quantum efficiency from these internal losses. In addition,
scaling would be difficult because of the excessively large gains.

Spin-Flip Lasers

Spin-flip Raman lasers are one of the more relatively complicated tunable infrared
devices to operate and have not been used as extensively as say diode lasers for

spectroscopy. There are, however, some advantages as well as fundamental limitations to the spin-flip Raman laser.

The InSb spin-flip laser covers a substantial wavelength range in the middle infrared. Recently, both cw [9] and pulsed [10] spin-flip lasers have been operated with an external cavity, demonstrating the improved utility of these devices. Like the dye laser, the spin-flip laser must be operated with an external cavity in order to fully exploit its high resolution capability. Because the spin-flip spontaneous linewidth is smaller than the cavity mode separation, it would be relatively simple to continuously tune a single frequency over several cm^{-1} without the mode jumping that would normally occur.

One of the reasons for the complexity of spin-flip lasers is the wide range of parameters over which the device operates, including electron concentration, magnetic field, and pump frequency. The choice of one set of parameters usually excludes the others and limits the operating characteristics of the device. Probably the most optimum spin-flip laser system for spectroscopic applications would consist of a high repetition rate pulsed CO_2 laser with a frequency doubling crystal of $CdGeAs_2$ used as the pump source. Besides the use of a single pump laser, the frequency doubled CO_2 laser would be much easier to operate on a single line than a CO laser which has many closely spaced lines that are difficult to separate in a normal grating controlled laser. Since low temperature operation of the crystals is required, the use of a moderately sized superconducting magnet (~ 70 kG) would provide a broad frequency average and allow the use of high concentration ($\sim 10^{16}$ electrons/cm^3) samples for maximum power. Pulsed operation would also allow for higher efficiency in sum and difference frequency generation for covering other frequencies in the infrared.

The maximum reported cw output power from an InSb spin-flip laser is one watt in the first Stokes component using a CO laser as the excitation source [11]. By proper heat sinking of the crystal it should be possible to extract at least 10 watts cw from such a spin-flip laser in the 5-6 μm region. The limitation in this case is thermal heating of the crystal lattice.

The maximum output energy from an InSb spin-flip laser using a pulsed CO_2 laser as the excitation source is limited by free carrier absorption [12]. The CO_2 laser pump radiation excites electrons from the bottom of the band to high momentum states where they do not participate in the spin-flip process. The spin-flip Raman wave builds up much faster than the rate at which electrons are excited up in the band at the intensities normally used in pumping InSb (10^5-10^6 W/cm^2). The spin-flip output occurs with one or more relaxation oscillations [12] before quenching occurs. Figure 4 shows the CO_2 laser excitation pulse (top) obtained by chopping out a segment of an unmodelocked pulse with a CdTe electro-optic modulator, and the transient relaxation oscillations (bottom) of the spin-flip output. At high enough pump intensities, the spin-flip output occurred in only one single pulse with a duration of 3-5 nsec. Maximum internal energy conversion efficiencies in the single pulse case was about 60%. In the case where half of the available spins are flipped in the mode volume, the maximum output energy would be about 10 μJ/cm^3. By pumping with pulses only long enough to drive the spin-flip output, the electron gas heating can be minimized and pulse repetition rates on the order of 10^5 Hz would be possible with average output powers of several watts.

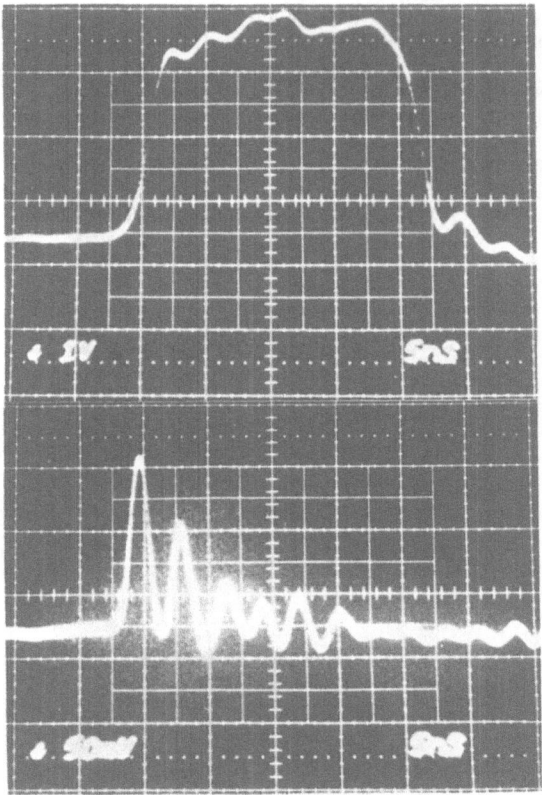

Fig. 4. Top: CO_2 laser pump pulse sliced out of unmodelocked CO_2 laser output using a CdTe electrooptic modulator. Time base is 5 nsec per small division. Bottom: Relaxation oscillation on the output of a spin-flip Raman laser on same time scale obtained using above pump pulse. Electron concentration $\sim 10^{16}$ cm^{-3}, T $\sim$ 20 K, H = 60 kG. λ_{pump} = 10.6 μm. From ref. [12].

Nonlinear Mixing

This section describes some recent results and scaling properties of nonlinear frequency mixing in the chalcopyrite $CdGeAs_2$ using infrared lasers. Because of its high nonlinear figure of merit and useful transmission range of 3-17 μm, it can be one of the most useful infrared nonlinear mixing crystals for high and low power device applications. The capability for downward scalability using cw lasers is particularly important for applications in heterodyne radiometry, very high resolution spectroscopy, and where compact and efficient systems are required.

66

A significant improvement in the output of second harmonic radiation from a CdGeAs$_2$ crystal operating near 77 K has recently been demonstrated using a cw discharge, high repetition rate, pulsed CO$_2$ laser [13], an average second harmonic output power at 5.3 μm of one watt has been obtained with output being limited by the pump laser. Typical second harmonic output power plotted as a function of the square of the input power is shown in Fig. 5. No sign of saturation of harmonic output was observed. Maximum average and peak power conversion efficiencies of 21 and 30%, respectively, were obtained for pulse repetition rates of 1.5 kHz where the maximum peak pump laser power (~ 1.8 kW) occurred. With appropriate heat sinking, several watts of average second harmonic radiation can be extracted from present crystals.

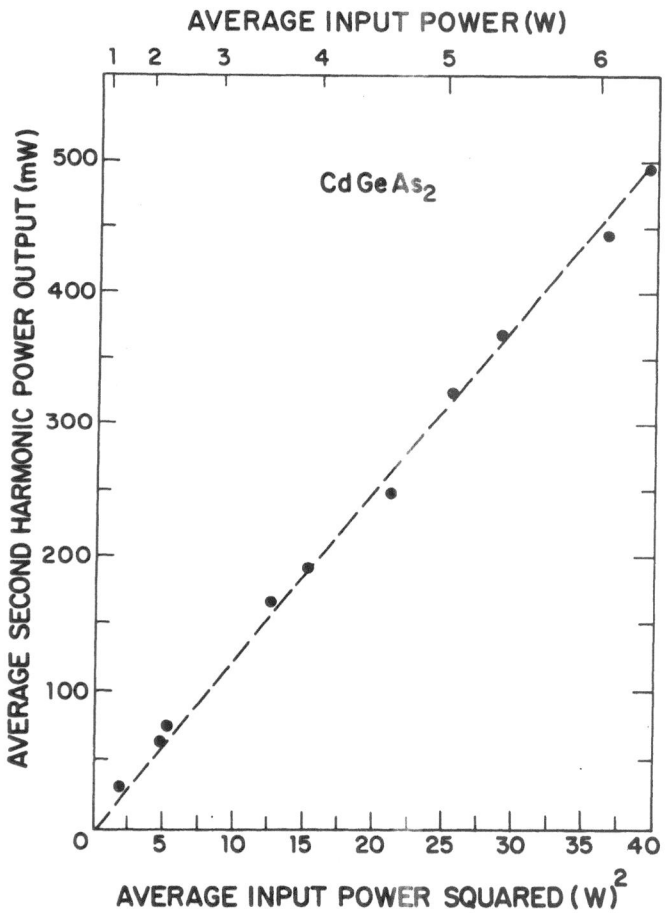

Fig. 5. Average second-harmonic power generated by CdGeAs$_2$ crystal as a function of the average Q-switched CO$_2$ pump power. Pulse repetition rate = 27 kHz. From ref. [13].

Figure 6 shows the maximum cw second harmonic output of about 75 mW obtained for a cw input power of 16 watts (intensity $\sim 10^5$ W/cm^2) from a CO_2 laser with no sign of saturation due to heating. Typical optical absorption coefficients (see Fig. 7) for the best crystals of $CdGeAs_2$ at 77 K are 0.1 and 0.4 cm^{-1} at 10 and 5 μm, respectively. Some additional improvement in optical absorption coefficient would allow placing the crystals inside the CO_2 laser cavity.

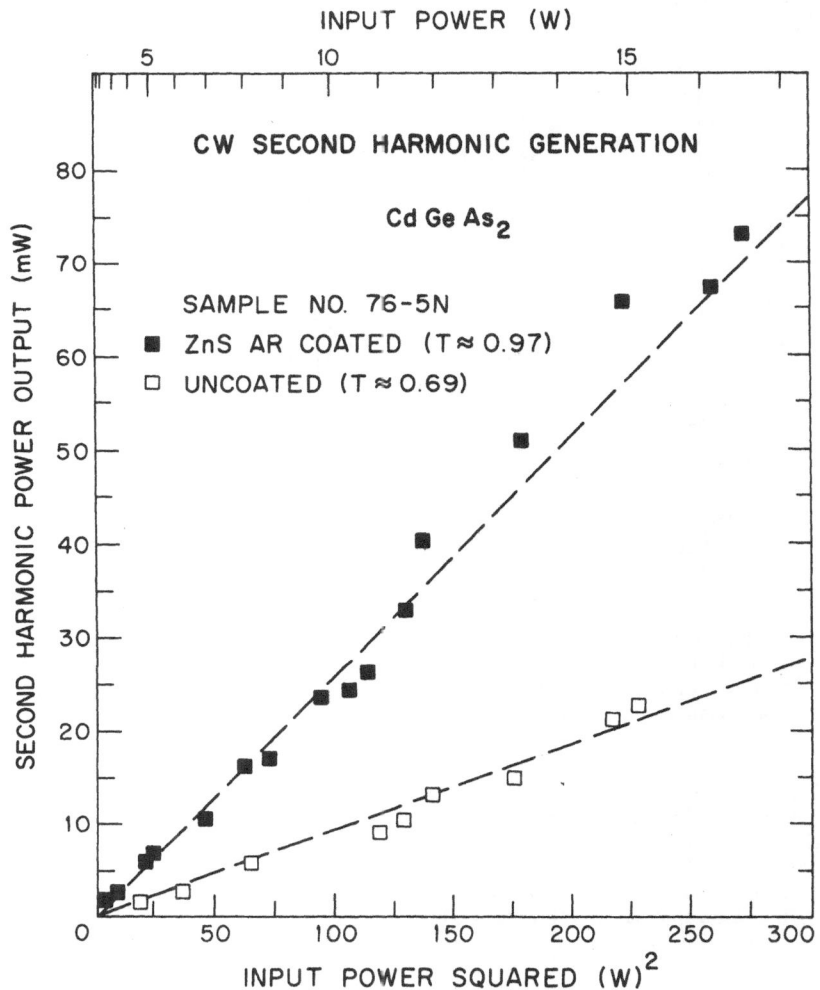

Fig. 6. CW second-harmonic power generated by $CdGeAs_2$ crystal as a function of CO_2 pump laser power. Lower set of points obtained with uncoated sample; upper set after anti-reflection coating. Dashed lines are related by ratio $(0.97/0.69)^3$. From ref. [13].

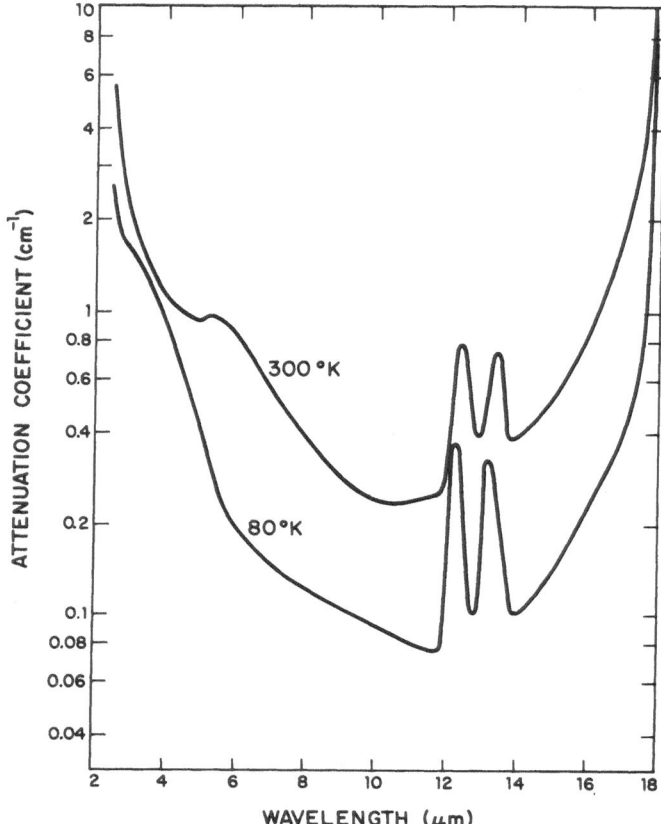

Fig. 7. Absorption coefficient for CdGeAs$_2$ at 300 and 80 K. Long wavelength
limit is due to multiphonon absorption while short wavelength absorption
is due to intervalence band transitions in p-type material. Band gap is
approximately 2.5 μm. Data taken by H. Kildal and G. W. Iseler.

The maximum pulsed second harmonic energy obtained to date has been 200 mJ
using nearly 4 J of multimode energy from a CO$_2$ TEA laser [14]. Energy conversion efficiencies of 27% have been measured in experiments with somewhat lower
input energies. Upward scaling by using one or more cooled matrices of CdGeAs$_2$
crystals is limited only by the size of the pump laser. The extension of the harmonic generation results to sum and difference frequency generation can be carried
out with comparable performance.

Wide spread utilization of CdGeAs$_2$ crystals is presently limited by the low
yield of high quality crystals. Material grown by the vertical Bridgman technique
has a tendency to crack throughout the boule while those crystals which are un-
cracked yield only a small number of crystals with optical absorption low enough
for high power applications. This is not a fundamental limitation.

REFERENCES

1. For a detailed review on tunable semiconductor diode lasers, see I. Melngailis and A. Mooradian in Laser Applications in Optics and Spectroscopy, edited by S. Jacobs, M. Sargent, M. Scully and J. Scott, (Addison-Wesley Company, 1975).

2. S. H. Groves, K. W. Nill and A. J. Strauss, Appl. Phys. Lett. 25, 331 (1974).

3. J. N. Walpole, A. R. Calawa, T. C. Harman and S. H. Groves, Appl. Phys. Lett. 28, 552 (1976).

4. J. N. Walpole, A. R. Calawa, S. R. Chinn, S. H. Groves and T. C. Harman, Appl. Phys. Lett. 29, 307 (1976).

5. J. E. Ripper and T. L. Paoli, Appl. Phys. Lett. 17, 371 (1970).

6. O. V. Bogdankevich, S. A. Darznek, A. N. Pechenov, B. I. Vasiliev and M. M. Zverev, IEEE J. Quantum Electron. 9, 342 (1973).

7. N. G. Basov, A. Z. Graziuk, V. F. Efimkov, I. G. Zabarev, V. A. Katalin and Ju'. M. Popov, Proceedings of International Conference on Physics of Semiconductors, Kyoto, 1966, J. Phys. Soc. Japan, vol. 21, supplement, 277 (1966).

8. S. R. Manlief and E. D. Palik, Appl. Phys. Lett. 22, 443 (1973).

9. S. D. Smith and R. B. Dennis in Proceedings of International Conference on Laser Spectroscopy, Megeve, France, edited by S. Haroche, J. C. Pebay-Peyroula, T. W. Hansch and S. E. Harris, (Springer-Verlag, 1975).

10. S. R. J. Brueck and A. Mooradian, IEEE J. Quantum Electron. 12, 201 (1976).

11. S. R. J. Brueck and A. Mooradian, Appl. Phys. Lett. 18, 229 (1971).

12. S. R. J. Brueck and A. Mooradian, Optics Commun. (to be published).

13. N. Menyuk, G. W. Iseler and A. Mooradian, Appl. Phys. Lett. to be published, October, 1976.

14. H. Kildal and G. W. Iseler, private communication.

PARAMETRIC OSCILLATORS

R.L. Byer

Edward L. Ginzton Laboratories, Applied Physics Dept,
Stanford University, California, USA

Introduction

Parametric oscillators have recently gained wider acceptance
as sources of visible and infrared tunable coherent radiation.
Their acceptance has come about due to improvements in the
laser pump sources and in the nonlinear crystals which have
led to wider tuning ranges at substantially higher peak and
average powers than previously available.

Since the operation of the first parametric oscillator in
$LiNbO_3$ by GIORDMAINE and MILLER [1], parametric oscillators
have been demonstrated in KDP, ADP and related isomorphs,
HIO_3 and $LiIO_3$, and Proustite and CdSe. The operation
characteristics of the devices is well understood and has
been reviewed by HARRIS [2], SMITH [3] and BYER [4], and
recently infrared parametric oscillator progress has been
reviewed and summarized. [5]

This paper describes a Nd:YAG pumped, computer controlled,
$LiNbO_3$ parametric oscillator. The oscillator is now well
engineered and offers operating characteristics expected of
a tunable coherent spectrometer source.

The 1.4 - 4.0 µm $LiNbO_3$ Parametric Oscillator Source

The Nd:YAG pumped $LiNbO_3$ parametric oscillator source was
described one year ago at the Megeve Tunable Laser Spectroscopy Conference. [6] The parametric oscillator tuner was
pumped by a TEM_{oo} mode Nd:YAG oscillator/amplifier system
and was angle tunable over a 1.4 - 4 µm range. Operating
energy conversion efficiencies up to 40% at six times above
threshold were measured. Of the three line narrowing methods
tried, the Littrow grating method with an additional tilted
etalon proved optimum. The measured linewidths were near
1 cm^{-1} with the grating only, and less than 0.1 cm^{-1} with
the grating plus tilted etalon for linewidth control. During
the past year progress has been made in $LiNbO_3$ crystal
quality, the Nd:YAG pumping source, and the automatic control
of the tunable source. At present, work is in progress on
frequency extension into the infrared using H_2 Raman down
conversion and in single axial mode tuning of the parametric
tuner.

A. LiNbO$_3$ Crystal Quality

The operation of the LiNbO$_3$ parametric tuner depends upon the availability of high optical quality LiNbO$_3$ crystals, approximately 1 cm in diameter and 5 cm in length, cut in the negative yz quadrant at 46^0 to the optic axis. Crystals meeting these specifications were first available following the growth of [01·$\overline{4}$] oriented LiNbO$_3$ boules. [7] The boule axis is 38^0 to the optic axis and the optical material proved to be of high optical quality. Recently we have extended the growth of off axis LiNbO$_3$ to the [01·$\overline{3}$] direction 47^0 to the optic axis and thus virtually parallel to the parametric tuner crystal cut. We have, therefore, been able to fabricate parametric oscillator crystals up to 20 mm diameter and over six centimeters in length. Much larger crystals could be cut, but 20 mm diameter and 5 cm length seems to be adequate for our present Nd:YAG laser system.

LiNbO$_3$ has in the past exhibited an OH$^-$ infrared absorption band at 2.85 μm which, in scme cases, completely prevented oscillation at that wavelength. During the past year we have found a method [8] of reducing the OH$^-$ absorption by a factor of more than five to less than 0.02 cm^{-1}. Improved optical quality LiNbO$_3$ crystals are now being fabricated for use in our present system.

B. Nd:YAG Source

The operation of a parametric oscillator requires a diffraction limited pump laser source. We have previously used a TEM$_{oo}$ mode Nd:YAG oscillator operating with a 780 μm spot size and 5-8 mJ of output energy to pump a Nd:YAG amplifier chain. The amplifier chain,which consisted of two 1/4" amplifiers and 3/8" amplifier and two Faraday rotator isolators for stability, provided over 380 mJ output energy at up to 10 pps. The output energy was limited by the small spatial filling factor of the Gaussian beam. To prevent mode clipping and resulting diffraction problems the beam spot size was chosen at the 1% intensity loss or the 3w_o = d condition, where w_o is the Gaussian beam spot size and d is the Nd:YAG rod diameter.

To more efficiently utilize the Nd:YAG rods, we have investigated a Nd:YAG unstable resonator oscillator. [9] Our first design, which took into account the focusing properties of the Nd:YAG rod, used a 1/4" diameter rod and an output magnification of 3. The resulting output coupling was slightly over 80%. The oscillator operated with up to 200 mJ output energy in a diffraction limited beam. Using this source operating at 150 mJ at 10 pps we were able to demonstrate 30 mJ of .532 μm output by second harmonic generation in Type II KD*P, 10 mJ of .3547 μm and 10 mJ of .266 μm . More importantly, the output pumped a LiNbO$_3$ parametric oscillator which showed a reduced threshold compared to the Gaussian intensity source at the same input energy.

Figure 1 shows the Nd:YAG unstable resonator/amplifier system now employed to pump the LiNbO$_3$ tuner. The available output energy after the 3/8" amplifier is 780 mJ at 10 pps. Thus the use of the unstable resonator oscillator allows considerable improvement in output energy and a simplification of the Nd:YAG pump system. The gain in energy is largely due to the better geometric filling factor provided by the unstable resonator mode. It should be noted that the mode has a plane wave front and in the near field still has the output mirror zero intensity hole in the center. In the far field the mode transforms to a modified Airy disk intensity function which can couple very well to Gaussian profile beams. The above results were obtained in the near field.

TUNABLE SYSTEM WITH UNSTABLE RESONATOR
AND DOUBLE PASS OPO

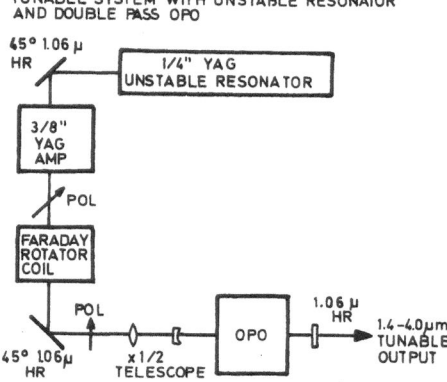

Fig.1. Schematic of the unstable resonator pumped LiNbO$_3$ tunable source

C. LiNbO$_3$ Parametric Tuner

The LiNbO$_3$ parametric oscillator operates over the 1.4 - 4 μm tuning range shown in Fig. 2 as a singly resonant oscillator (SRO) with the 1.4 - 2.1 μm signal wave resonated within a 13 cm long optical cavity formed by a grating and 50% reflecting output coupler. The grating provides linewidth control to less than 2 cm^{-1} even to degeneracy. In addition the grating provides an absolute frequency reference to within ±1 cm^{-1} over the tuning range of the oscillator.

The parametric tuner is controlled by an interactive software program using a PDP11E10 minicomputer acting through a CAMAC controller. Figure 3 shows a schematic of the control system. The software allows the initial wavelength scanning rate, and scan range to be set and also allows for tilted etalon control. Figure 4 illustrates a coherent anti-Stokes Raman spectra of H$_2$ taken with the LiNbO$_3$ tuner by mixing the tunable output

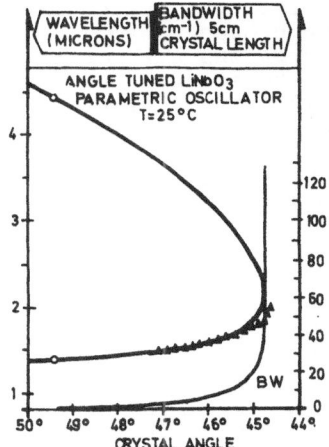

Fig.2. Tuning range and gain bandwidth for the angle tuned 1.06 μm pumped LiNbO$_3$ parametric oscillator tuner

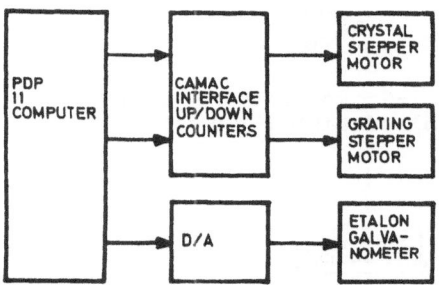

Fig.3. Schematic of the computer control system for the LiNbO$_3$ tuner

between 1.912 and 1.8939 μm to generate the red anti-Stokes output. The spectra clearly shows the 0.25 cm^{-1} grating steps over the 50 cm^{-1} tuning range. This spectra was taken single ended and serves to illustrate the stability of the source.

Comparison of the $Q_0(1)$ peak frequency relative to 1.064 μm showed that the grating was calibrated to within ±1 cm^{-1}. Furthermore, rescanned spectra showed that the grating resettability was ±.25 cm^{-1} or one motor step. The conversion efficiency in this CAR spectra was high enough to warrant four sheets of computer paper prior to the photo-multiplier to prevent saturation.

As an example of control of the system for spectroscopy, Fig. 5 shows the pressure broadened CO overtone absorption spectra. Again, the spectra was not normalized so that the 100% line increases over the scanned interval. To improve the resolution, the tilted etalon was inserted into the tuner and scanned with a square root function under computer control so as to give a linear output spectra. The spectra of the R(7) and R(8) lines at 1 atm pressure are shown in Fig. 6. In this case the resolution was limited to the 0.3 cm^{-1} unnarrowed Nd:YAG laser source since the non-resonated idler was used to generate the spectrum. Nevertheless, the spectrum illustrates the control designed into this tunable system.

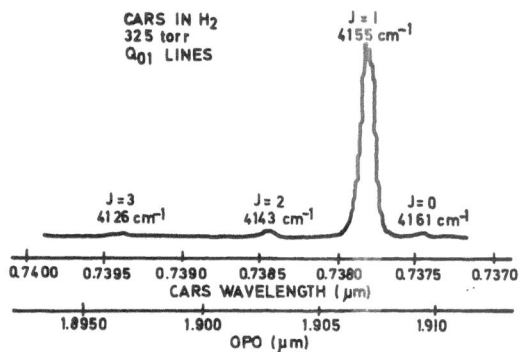

Fig.4. Coherent anti-Stokes Raman spectra of H_2 generated using the computer controlled $LiNbO_3$ source

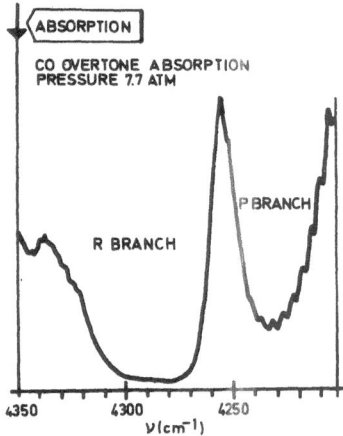

Fig.5. Absorption spectra of the CO overtone band at a CO pressure of 1 atm using the grating only

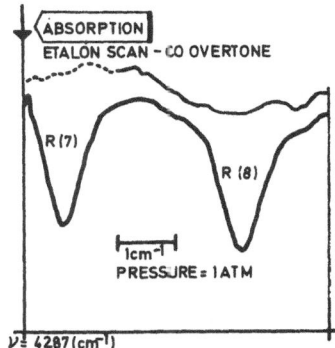

Fig.6. Absorption spectra of the CO overtone R(7) and R(8) lines using the tilted etalon for line narrowing

One of the difficulties in using the source in its present single pass configuration is the elimination of the transmitted 1.06 μm radiation prior to absorbing infrared optical components. The consideration of this problem led to the double pass SRO first analyzed by BJORKHOLM.[10] Upon reflecting 1.06 μm back through the LiNbO$_3$ crystal we observed the expected threshold reduction by a factor of two and an improvement in the conversion efficiency at the same input intensity. More importantly, the oscillator's stability improved markedly. The double pass SRO configuration was only possible due to the use of the Faraday rotator isolator following the 3/8" amplifier. Experiments are now in progress to determine the optimum output coupling and maximum conversion efficiency for the double pass SRO configuration. However, initial results suggest that its benefits far outweigh the disadvantages of adding one more mirror to the tuner system.

D. Frequency Extension by Raman Mixing

An alternative method to generating tunable radiation by a stimulated Raman process is to coherently scatter an input beam from a stimulated Raman cell. In this process, which is responsible for anti-Stokes generation for example, a pump beam generates stimulated Raman scattering in the Raman medium. The input beam at frequency ω_i scatters from the generated Raman polarization to generate output frequencies at $\omega_i \pm \omega_R$. Thus the Raman medium acts like a mixer with a local oscillator frequency ω_R .

Scattering from coherently driven Raman oscillation was first discussed by GIORDMAINE and KAISER [11] who demonstrated the process in calcite. Earlier GARMIRE et al., [12] correctly interpreted higher order Stokes and anti-Stokes generation by the mixing mechanism. Later DUARDO et al., [13] carried out experiments in H$_2$ and CH$_4$ and described the generation of new frequencies in terms of mixing from coherently driven Raman oscillation.

HARRIS [14] first proposed using mixing from coherently driven Raman oscillation as a source of tunable radiation. He pointed to the advantage of fixed Raman threshold for the pump while still obtaining tunable output with good conversion efficiency due to the mixing process. Later FLEMING [15] showed that under proper excitation conditions phasematching is not required. Similar results were arrived at by VENKIN et al., [16] who demonstrated frequency conversion by mixing.

To evaluate the conversion efficiency for coherent Raman mixing we assume four fields are present: the pump, the generated Stokes, the tunable input field at ω_i and the generated output field at ω_o. Substituting the fields and polarizations into the wave equation yields four simultaneous equations for the fields given by following equations

$$\frac{\partial E_p}{\partial z} = - \frac{\omega_p}{2cn_p} \chi_R^{\sim} \{ |E_2|^2 E_p + E_i E_o^* E_s e^{i\Delta kz} \} \tag{1a}$$

$$\frac{\partial E_s}{\partial z} = + \frac{\omega_s}{2cn_s} \chi_R^{\sim} \{ |E_p|^2 E_s + E_i^* E_o E_p e^{-i\Delta kz} \} \tag{1b}$$

$$\frac{\partial E_i}{\partial z} = - \frac{\omega_i}{2cn_i} \chi_R^{\sim} \{ |E_o|^2 E_i + E_p E_s^* E_o e^{-i\Delta kz} \tag{1c}$$

$$\frac{\partial E_o}{\partial z} = \frac{\omega_o}{2cn_o} \chi_R^{\sim} \{ |E_i|^2 E_o + E_p^* E_s E_i e^{i\Delta kz} \tag{1d}$$

where $\Delta k = - (k_p - k_s) + (k_i - k_o)$

In general the solution of these coupled equations requires numerical techniques. However, assuming minimal depletion of the pump and input waves the problem reduces to the solution of the two coupled equations, (1b) and (1d). If we assume solutions of the form

$$E_s = E_s e^{(\Gamma - \frac{i\Delta k}{2})z} \quad \text{and} \quad E_o = E_o e^{(\Gamma + \frac{i\Delta k}{2})z}$$

we then find

$$\Gamma_{\pm} = \frac{\Gamma_p + \Gamma_i}{2} \pm \frac{1}{2} \left\{ (\Gamma_p + \Gamma_i) - 4(\frac{\Delta k}{2})^2 + 4(\frac{i\Delta k}{2})(\Gamma_p - \Gamma_i) \right\}^{\frac{1}{2}} \tag{2}$$

where

$$\Gamma_p = \frac{\omega_s}{2n_s c} \chi_R^{\sim} |E_p|^2 \qquad (3a)$$

$$\Gamma_i = \frac{\omega_o}{2n_o c} \chi_R^{\sim} |E_i|^2 \qquad (3b)$$

are the field gain coefficients.

At exact phasematching the gain coefficient becomes $\Gamma = \Gamma_p + \Gamma_i$ Substituting the expression for the gain coefficient into the wave equation for E_o gives

$$\frac{E_o}{E_i} = \frac{\omega_o}{\omega_s} \frac{n_s}{n_o} \frac{E_s}{E_p} \qquad (4)$$

which results in a conversion efficiency for the coherent Raman mixing process given by

$$\frac{I_o}{I_i} = \left(\frac{\omega_o}{\omega_s}\right)^2 \frac{n_s n_p}{n_o n_i} \frac{I_s}{I_p} \qquad (5)$$

a result first derived by GIORDMAINE and KAISER.[11] Thus the coherent Raman mixing   efficiency equals the conversion efficiency from the pump to the Stokes field times a frequency factor $(\omega_o/\omega_s)^2$. This is the same conversion efficiency for mixing in nonlinear crystals if we identify I_s/I_p with the small signal gain or conversion efficiency for mixing.

In stimulated Raman scattering processes the photon conversion efficiency to the Stokes field is a maximum of 40%. Thus under optimum conditions the coherent Raman mixing process should allow generation of tunable down shifted output with a corresponding 40% photon efficiency.

The advantage of coherent Raman mixing is two fold. First a strong fixed frequency pump beam can be used to efficiently generate the stimulated Raman scattering and, secondly the mixing process efficiency is independent of the intensity of the tunable input beam to a first approximation.

The coherent mixing process does require phasematching to achieve optimum conversion efficiency. However, if the gain constant is large compared to the phase mismatch factor Δk, then the conversion efficiency is reduced to approximately

$$\frac{I_o}{I_i} = \left(\frac{\omega_o}{\omega_s}\right)^2 \frac{n_s n}{n_o n_i} \frac{I_s}{I_p} \left[\frac{1}{1 + \frac{\Delta k^2}{\Gamma_p}}\right] \qquad (6)$$

78

where $\Gamma_p \gg \Gamma_i$ is also assumed. Thus in H_2 gas pumped at 1.06 μm at 20 atmospheres pressure, $\Gamma_p/I_p \sim 4 \times 10^{-3}$ cm/MW. At 200 MW/cm² input intensity $\Gamma_p \sim .8$ cm⁻¹ . If ω_i varies from 4155 - 7000 cm⁻¹ (the tuning range of a 1.06 μm pumped LiNbO₃ parametric oscillator) and the generated output frequency varies from 0 - 2845 cm⁻¹ , then $\Delta k = .4 - .6$ cm⁻¹ . At 50 atm Δk increases to 1.2 - 1.5 cm⁻¹ . Thus for the above input intensity $\Delta k/\Gamma_p \sim .5 - .75$ and lack of phasematching does result in some conversion efficiency reduction. Since Δk is relatively constant over a wide range of generated frequencies, deviation from exact phasematching can be kept relatively small without adjusting the cell length or gas pressure. Thus phasematching is not a serious problem in coherent Raman mixing.

Figure 7 shows the tuning range available by coherent mixing in H_2 pumped at 1.064 μm with a 1.4 μm → 2.40 μm tunable LiNbO₃ parametric oscillator as the source at ω_i . Preliminary experiments have demonstrated a stimulated Raman efficiency to the first Stokes of 9% energy conversion and 16% photon conversion at two times above threshold. The pumping energy required at 1.06 μm to achieve the peak conversion efficiency was 30 mJ in an 80 cm long cell. The corresponding input intensity was 200 MW/cm² to give a total gain 2Γℓ of 128.

Coherent Raman mixing offers the possibility of generating tunable infrared radiation without the limitations inherent in nonlinear crystal mixing of phasematching, low damage threshold and difficulty in obtaining adequate size and quality crystals. Experiments are in progress to demonstrate the full capability of this method.

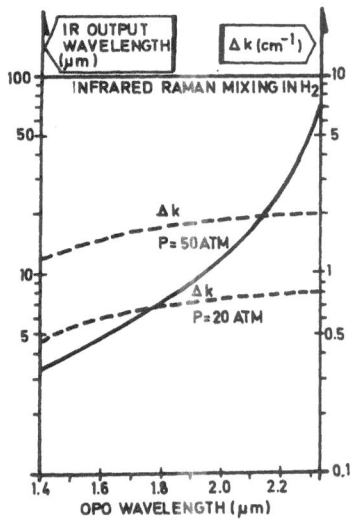

Fig.7. Extended tuning range in H_2 by coherent Raman mixing with the 1.06μm pumped LiNbO₃ parametric oscillator source

Conclusion

We have operated the LiNbO$_3$ parametric tuner under computer
control and have demonstrated its capability by generating a
CAR spectrum in H$_2$ gas and linear absorption spectra of CO.
With improvements in signal processing the system should
operate as a coherent spectrometer over a primary tuning range
of 1.4 - 4 μm with the possibility of an extended tuning range
to the visible and ultraviolet by SHG and into the infrared
by Raman mixing in H$_2$.

The operation of the LiNbO$_3$ tuner as a double pass SRO
has led to a reduced threshold by a factor of two and to an
improved conversion efficiency and stability. For example,
for 100 mJ input energy in a 2 mm spot size, the single pass
SRO operated at two times above threshold with an 8% conversion
efficiency and a 20% peak-to-peak stability. The double pass
SRO operated at four times above threshold with a 19% external
energy conversion efficiency with a 5% peak-to-peak stability.
Improvements in both stability and conversion efficiency are
expected with optimization of the output coupler reflectivity
and operation at higher intensities.

The future research effort on this tunable source will
concentrate both in frequency extension studies and in
frequency narrowing and control. Work is presently underway
to complete a pressure tuned LiNbO$_3$ tuner with single axial mode
tuning over a 3 cm^{-1} frequency range. [17]

The present high energy capabilities of the LiNbO$_3$ tuner
pumped with the unstable resonator Nd:YAG source make it an
attractive alternative as a tunable coherent source in the
infrared for applications to laser chemistry, nonlinear spectroscopy and remote monitoring.

ACKNOWLEDGEMENTS

The work reported here represents contributions by R.L. Herbst,
R.N. Fleming, S. Brosnan and H. Ito. The work was supported in
part by ERDA, ARO, NSF and EPRI.

REFERENCES

1.   J.A. Giordmaine and R.C. Miller, Phys. Rev. Letts,
     14, p.973, (1965).

2.   S.E. Harris, Proc. IEEE 57, 2096, (1969).

3. R.G. Smith, "Optical Parametric Oscillators", in Laser Handbook, ed. by F.T. Arecchi and E.O. Schulz Du Bois, p.837, (1972).

4. R.L. Byer, "Optical Parametric Oscillators", in Treatise in Quantum Electronics, vol. I, part B, ed. H. Rabin and C.L. Tang, New York, Academic Press, p.587-702, (1975).

5. R.L. Byer and R.L. Herbst, "Infrared Generation by Parametric Oscillation and Mixing", to be published in Topics in Applied Physics, ed. Y.R. She, Springer-Verlag, Berlin.

6. R.L. Byer, R.L. Herbst and R.N. Fleming, "A Broadly Tunable IR Source", published in Laser Spectroscopy, ed. by S. Haroche, J.C. Pebay-Peyroula, T.W. Hansch and S.E. Harris, Springer-Verlag, Berlin, p.207-225, (1975).

7. R.L. Byer, R.L. Herbst, R.S. Feigelson and W.L. Kway, Optics Commun. 12, p.427, (1974).

8. R.S. Feigelson, W.L. Kway, R.L. Byer and R.L. Herbst, (private communication).

9. R.L. Herbst, H. Komine and R.L. Byer, "A 200 mJ Unstable Resonator Nd:YAG Oscillator", (to be published).

10. J.E. Bjorkholm, A. Ashkin, R.G. Smith, IEEE J. Quant. Electron, QE-6, p.797, (1970).

11. J.A. Giordmaine and W. Kaiser, Phys. Rev. 144, p.676, (1966).

12. E. Garmire, F. Pandarese and C.H. Townes, Phys. Rev. Letts, 11, p.160, (1963).

13. J.A. Duardo, F.M. Johnson, L.J. Nugent, IEEE Journ. Quahnt. Elect. QE-4, p.396, (1968).

14. S.E. Harris and R.L. Byer, Progress Report No. 5 for the U.S. Army Research Office, (Durham), Contract No. DAHC-04-68-C-0048, December, 1970. Also available as Stanford University Microwave Laboratory Report No. 1918.

15. R.N. Fleming, "The High Energy Broadly Tunable Coherent Source", Ph.D. Thesis, Stanford University, 1976, available as Stanford University Microwave Laboratory Report No. 2521.

16. G.V. Venkin, G.M. Krochik, L.L. Kulyak, D.I. Malaev, Yu G. Khronopulo, JETP Letts. 21, p.105, (1975).

17. H. Ito, Tohoku University, Sendai, Japan, (presently at Stanford University), (private communication).

TUNABLE INFRARED GENERATION IN MOLECULAR GASES

J. Ducuing, R. Frey and F. Pradère
Laboratoire d'Optique Quantique du C.N.R.S.
Ecole Polytechnique
Route de Saclay, Palaiseau, France

I. Introduction

In recent years the use of nonlinear optical processes in gases to generate tunable infrared radiation has received renewed interest. Although gases, in opposition to solids, show no second order dipolar nonlinearity their third order susceptibility can be considerably enhanced through resonances and lead to strong interactions. Furthermore gases offer many advantages over solids : higher damage threshold, broader transmission range, better optical quality and larger volume. Until now, most of the work on tunable generation has concentrated on metal vapours [1-3], which near resonance exhibit very large susceptibilities. However room temperature molecular gases which are considerably simpler to deal with, can also support strong interactions, as evidenced long ago by the easy appearance of stimulated Raman scattering in H_2, CH_4 and a number of other gases [4]. H_2 is of special interest, as it shows no electric dipole absorption in the whole infrared spectrum. In the following we discuss the generation of tunable medium infrared through 4-wave interactions in this gas and present results obtained in the range 5-16.5 μm. We also consider a scheme for the efficient generation of tunable far infrared through resonantly enhanced stimulated Raman scattering in HF.

II. Four-Wave Interactions in H_2

The interaction is schematically shown in Fig.1. A strong wave at ω_1 beats with two waves at ω_2, ω_3 to generate the infrared frequency $\omega_4 = \omega_1 - \omega_2 - \omega_3$. The fixed frequencies ω_1 and ω_2 are chosen such that $\omega_1 - \omega_2 \simeq \omega_{ba}$ where ω_{ba} corresponds to the Q(1) line (v = 0, J = 1 → v = 1, J = 1) of H_2. This condition ensures the resonant enhancement of the corresponding susceptibility. The tunability is provided by the adjustable ω_3 frequency. (Note that ω_3 is not restricted to be smaller than ω_{ba}).

The process can be described as a resonantly enhanced 4-wave interaction, but can also be viewed as resulting from the driving of the polarisability [5] at frequency $\omega_1 - \omega_2$ by the ω_1, ω_2 fields. In this picture the wave at ω_4 results from the coherent scattering of ω_3 by the polarisability modulation [6]. Generation of fixed frequency infrared radiation by this process has been demonstrated by AKANAEV and PETSELT [7].

Assuming fields linearly polarized along the same direction the Fourier component at ω_4 of the source polarisation can be expressed in terms of the Fourier components E_j of the field as :

$$P(\underline{r}, \omega_4) = 6\chi^{(3)}(\omega_1, -\omega_2, -\omega_3) E_1 E_2^* E_3^* + 6\chi^{(3)}(\omega_3, -\omega_3, \omega_4)|E_3|^2 E_4$$

where the first term corresponds to the 4-wave interaction and the second to two photon transitions between a and b involving ω_3 and ω_4. According to the sign of $\omega_4 \simeq \omega_{ba} - \omega_3$ this last process will correspond to attenuation of

82

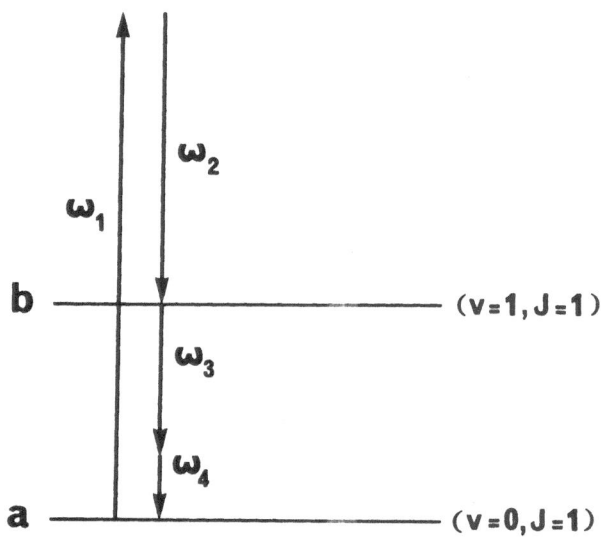

Fig.1 Four-Wave Interaction in H_2

ω_4 via two photon absorption ($\omega_4 > 0$) or amplification via stimulated Raman scattering ($\omega_4 < 0$). With the usual notation, P being the dipole moment operator, $\Delta\omega$ the half width of the Raman transition, Ω the resonance detuning : $\Omega = \omega_1 - \omega_2 - \omega_{ba}$ and N the number of molecules per unit volume, the resonant part of the 4-wave susceptibility is :

$$\chi^{(3)} (\omega_1,-\omega_2,-\omega_3) = \frac{-N}{6\hbar^3(\Omega+i\Delta\omega)} \sum_{i,j} <a|P|i><i|P|b><b|P|j><j|P|a>$$

$$\times \left(\frac{1}{\omega_4 - \omega_{ia}} + \frac{1}{\omega_3 - \omega_{ia}}\right) \left(\frac{1}{\omega_1 - \omega_{ja}} - \frac{1}{\omega_2 + \omega_{ja}}\right)$$

In H_2 the contributing intermediate states are electronic and lie above the photon energies which can be neglected in the denominator. Accordingly :

$$\chi^{(3)} (\omega_1,-\omega_2,-\omega_3) = - \frac{\Delta\omega}{6(\Omega+i\Delta\omega)} \chi_0 \tag{1}$$

where

$$\chi_0 = \frac{4}{\Delta\omega} \frac{N}{\hbar^3} \left| \sum_i \frac{<a|P|i><i|P|b>}{\omega_{ia}} \right|^2 \tag{2}$$

A similar evaluation of $\chi^{(3)}(\omega_3,-\omega_3,\omega_4)$ leads to :

$$\chi^{(3)}(\omega_3,-\omega_3,\omega_4) = \sigma(\omega_4) \times \chi^{(3)}(\omega_1,-\omega_2,-\omega_3) \tag{3}$$

where $\sigma(\omega_4) = 1 \ \omega_4 > 0$
$= -1 \ \omega_4 < 0$

For plane waves interaction, the growth of the wave at ω_4 is described by :

$$\frac{\partial A_4}{\partial z} + \beta A_4 = \alpha \ e^{i\Delta kz} \tag{4}$$

where :

$$\alpha = + \frac{i4\pi^2}{\lambda_4} \ 6\chi^{(3)}(\omega_1,-\omega_2,-\omega_3) \ A_1 A_2^* A_3^* \tag{5}$$

$$\beta = - \frac{i4\pi^2}{\lambda_4} \ 6\chi^{(3)}(\omega_3,-\omega_3,\omega_4) \ |A_3|^2 \tag{6}$$

We have introduced the field complex amplitudes defined by $E_j = A_j e^{ik_j z}$ and the wave vector mismatch $\Delta k = k_1-k_2-k_3-k_4$. Assuming negligible depletion of the waves at ω_1, ω_2 and ω_3, equation (4) gives :

$$|A_4|^2 = \frac{|\alpha|^2}{\beta'^2+(\beta''+\Delta k)^2} \left\{ (1-e^{-\beta' z})^2 + 4e^{-\beta' z}\sin^2 \frac{(\beta''+\Delta k)z}{2} \right\} \tag{7}$$

where β' has the sign of ω_4. This incorporates the effect of either stimulated Raman gain or two photon absorption. At low conversion efficiency these effects are negligible in our experiments. In this case, the intensity at ω_4 is an oscillatory function of z with period $\frac{\pi}{\Delta k} = 2 \ l_c$ and maximum value :

$$I_{4,m} = \frac{2}{\pi^3} \ c|\alpha|^2 \ l_c^2 \tag{8}$$

In the following we consider more specifically the case $\omega_2 = \omega_3$ and assume exact resonance. Then equation (8) yields a simple expression for the intensity conversion ratio :

$$\frac{I_{4,m}}{I_1} = \frac{l_c^2}{\pi^2} \ |2\beta|^2 \tag{9}$$

where $|2\beta|$ the ω_2-induced gain or absorption coefficient at ω_4 is readily evaluated from the measurements of BLOEMBERGEN et al [8]. Hydrogen shows a very low dispersion and consequently offers very long coherence lengths which vary little with the generated frequency (Fig.2). The conversion ratio varies with density. For a given frequency it is proportional to $(l_c \chi^{(3)})^2 \propto \frac{1}{\Delta\omega^2}$. In distinction with stimulated Raman scattering, advantage can be taken here of the strong Dicke-narrowing of the Q(1) line [9] which occurs around 4 Amagat.

We have conducted a preliminary experiment using a dye laser system [10] to produce ω_1 and ω_2 through stimulated Raman scattering in H_2. This system consists of a transversally ruby pumped dye during 30 ns and two high power amplifiers pumped for only 2 ns. The main characteristic of the dye laser and its two Stokes components are listed in Table 1. As ω_2 results from Raman scattering in H_2 the process is automatically at resonance and tuning is provided by the variation of ω_1. Using a cell of variable length at a given pressure we have observed directly the influence of phase matching and measured the coherence length. The measured intensities were in reasonable agreement with the predictions of eq. 9 . Thus for the situation $\omega_4>0$ at a pressure of 30 atm ($l_c = 5$cm), with an intensity $I_2 = 130$ MW/cm^2 and a generated wavelength $\lambda_4 = 7.5$ µm we measure a conversion ratio of 3.10^{-4} whereas the predicted value is $1.5 \ 10^{-3}$. The discrepancy is probably due in part to the divergence of the ω_1,ω_2 beams which are a few times the diffraction limit. For multimode beams a reduction of the efficiency below the value given by Eq. 9 is expected [11]. The imperfect time overlap of the ω_1,ω_2 pulses further decreases the ω_4 energy. We have been able to tune the generated wavelength down to 16.5 µm. This present limit is not imposed by the efficiency of the interaction process but by the sensitivity of our HgCdTe detector. No

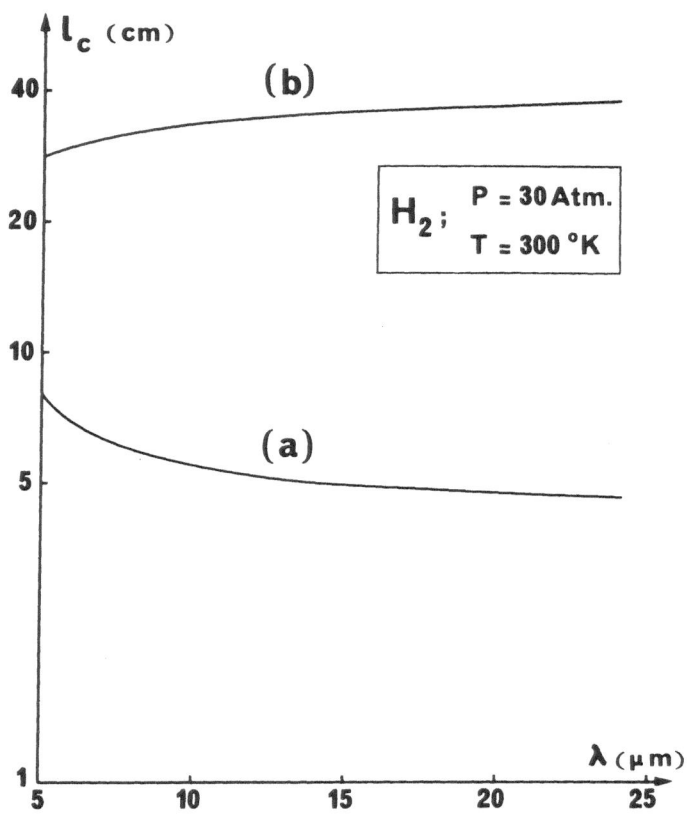

Fig. 2 Coherence Length for the 4-Wave Process in H₂ (a) $\omega_4>0$ (b) $\omega_4<0$

Table 1

	Output Power	Tuning Range	Linewidth	Pulse Duration
Dye Laser	500 - 800 MW	.72 - 1.07 μm	.07 cm^{-1}	2 ns
First Stokes ω_1	100 - 200 MW	1. - 1.9 μm		2 ns
Second Stokes ω_2	15 - 45 MW	1.7 - 5.3 μm		

major difficulty is expected in extending the tunability range well beyond this value.

The conversion ratio can also be substancially increased. Much larger coherence lengths can be obtained (see Fig.2) by chosing $\omega_2 > \omega_{ba}$. Note also that in that case a focused geometry can be used [12]. Furthermore, as we remarked earlier for the interaction of plane monochromatic waves the conversion ratio is inversely proportional to the square of the Raman linewidth $\Delta\omega$. However in practice the minimum value of $\Delta\omega$ is limited by the spectral width of the sources. Better monochromaticity will allow lower working pressures and increase the conversion ratio.

III. Low Frequency Stimulated Raman Scattering

Other things being equal, the stimulated Raman gain is inversely proportional to the stokes wavelength. This makes difficult to use SRS in H_2 to generate radiation beyond 10 μ. In a similar (although less critical) way, the efficiency of a 4-wave parametric process is proportional to λ_4^{-2} (see eq.8 et 5) and will be severely reduced in the far infrared range. A low frequency resonant enhancement of the susceptibility may help to overcome this difficulty. This can be provided by pure rotational transitions. Consider for instance (Fig.3) in an heteronuclear molecule the Raman transition $Q(J)$ $(v=0,J \rightarrow v=1,J)$ and assume that the pump frequency ω_1 lies close to the $R(J-1)$ and/or $R(J)$ lines. Under these conditions the rotational states belonging to the levels $c_1 \equiv (v=1,J+1)$ and $c_2 \equiv (v=0,J)$ will dominate the intermediate state contribution to the susceptibility expansion. The Raman gain will be given by :

$$\gamma_R = \frac{8\pi^2}{\lambda_2 \, \hbar^3 \Delta\omega} \times \frac{N_J}{2J+1} \sum_{M=-J}^{M=+J} \left| \frac{<a,M|P|c_1,M><c_1,M|P|b,M>}{\omega - \omega_{c_1 b}} \right.$$

$$\left. + \frac{<a,M|P|c_2,M><c_2,M|P|b,M>}{\omega_{ac_2} - \omega} \right|^2 |A_1|^2 \tag{10}$$

where $|a,M>$ designates a state with azimuthal quantum number M and with total angular momentum and vibrational quantum number of level a as indicated on Fig.3. N_J is the total equilibrum population of level $(v=0,J)$ and $\Delta\omega$ the half line width of the Raman transition. Assuming for the sake of simplicity that one of the denominators, say $\omega - \omega_{c_1 b}$, is appreciably smaller that the other, eq. 10 reduces to :

$$\gamma_R = N_J \frac{8\pi^2}{\lambda^2 \hbar^3 \Delta\omega} \frac{\mu_{01}^2 \, \mu_{11}^2}{(\omega - \omega_{c_1 b})^2} \, u(J) \, |A_1|^2 \tag{11}$$

where μ_{01} and μ_{11} are respectively the vibrational transition dipole moment and the permanent dipole moment of the molecule, and :

$$u(J) = \frac{J+1}{15} \frac{4J^2+8J+5}{(2J+1)^2(2J+3)}$$

Consider the case of the Raman transition $Q(3)$ in HF with a pump around 2.4 μm which corresponds to a stokes wavelength in the range of 60 μm. Assuming the pressure broadening to be the same as that of the $R(3)$ line [13] $\Delta\bar{\nu} = \frac{\Delta\omega}{2\pi c} = 5.3 \ 10^{-4}$ cm^{-1}/torr and taking [13] $\mu_{01} = 10^{-19}$ e.s.u with

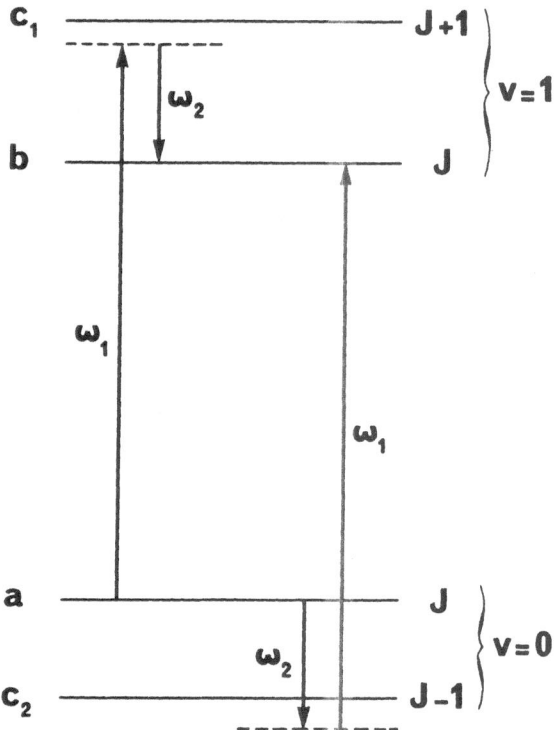

Fig. 3 Resonantly Enhanced low Frequency Raman Scattering in HF

$\mu_{11} = 2.10^{-18}$ e.s.u, we obtain a gain coefficient at room temperature

$$g_R = 4.7 \ 10^{-2} \{\omega_2 - \omega_{c_1 b})_{cm^{-1}}\}^{-2} \ (MW)^{-1} cm$$

For a detuning $\omega_2 - \omega_{c_1 b} \simeq 10cm^{-1}$ this gives a gain coefficient of the same magnitude as that for Raman amplification of 3µ radiation. In HF consecutive rotational lines are separated by $2B \simeq 40$ cm^{-1}. Considering that for a given Q(J) transition, there are two possible resonant terms (eq. 10), corresponding to the R(J-1) and R(J) lines, one sees that broad parts of the far infrared spectrum could be covered by such a source.

The authors are very much indebted to Dr VERIE and Dr KOZACKI (CNRS-Bellevue) and Dr CHRISTOFF (S.A.T.) for the loan of the infrared detectors.

References

1   P.P. Sorokin, J.J. Wynne and J.R. L ankard Appl. Phys.Lett $\underline{33}$, 1183 (1974)

2   J.J. Wynne and P.P. Sorokin, Laser Focus $\underline{11}$, 62 (1975)

3   D. Cotter, C. Hanna and R. Wyatt, Opt.Comm. $\underline{16}$, 256 (1976)

4   R.W. Minck, R.W. Terhune and W.G. Rado, Appl.Phys.Lett., $\underline{3}$, 181 (1963)

5   E. Garmire, E. Pandarese and C.H. Townes, Phys.Rev.Lett., $\underline{11}$, 160 (1963)

6   J.A. Giordmaine and W. Kaiser, Phys.Rev, $\underline{144}$, 676 (1966)
    J.A. Giordmaine Bul.Am.Phys.Soc., $\underline{10}$, 34 (1965)

7   B.A. Akanaev and Y. Petselt, Zh ETF Pis'ma $\underline{3}$, 327 (1966)

8   N. Bloembergen, G. Bret, P. Lallemand, A. Pine and P. Simova, IEEE J.
    Quantum Electron. QE-3, 197 (1967)

9   P. Lallemand, P. Simova and G. Bret, Phys.Rev.Letters $\underline{17}$, 1239 (1966)
    J. Murray and A. Javan, J.Mol.Spectrocopy, $\underline{29}$, 502 (1969)
    J.L. Buijs and H.P. Gush, Can.J.Physics, $\underline{49}$ (1971) 2366

10 R. Frey and F. Pradère, Opt.Comm. $\underline{12}$, 98 (1974)
    R. Frey and F. Pradère, Infrared Physics, $\underline{16}$, 117 (1976)

11 J. Ducuing and N. Bloembergen, Phys. Rev, $\underline{133A}$, 1473 (1964)

12 G.C. Bjorklund, IEEE J.Quantum Electron. $\underline{QE-11}$, 287 (1975)

13 G.A. Kuipers, J.Mol.Spectroscopy $\underline{2}$, 75 (1958)

TUNABLE HIGH POWER RAMAN LASERS
AND THEIR APPLICATIONS

A.Z. Grasiuk and I.G. Zubarev
P.N. Lebedev Physical Institute, Ac. Sci.,
Leninsky prospect 53, Moscow, USSR

1. Introduction

Laser photochemistry (1,2), isotope separation (3-7), laser spectroscopy (8) and other types of selective interaction of radiation with matter call for the application of new types of lasers. Such lasers must have high power (≥ 1 MW), short pulses (10 - 100 ns), narrow linewidth ($\sim 10^{-2}$ cm^{-1}), tuning and high efficiency. It is also preferable to have a low beam divergence to achieve high intensity by focusing. Lasers of this type are necessary throughout the spectral range and particularly in the far IR because today only CO_2 and CO lasers are available in the far IR range, tuning in the regions 9.2 - 10.9 μm and 5 - 7 μm, respectively.

Raman lasers are a promising type of high-power tunable lasers operating in new ranges (9). They are high efficiency convertors which often give improved values of intensity and divergence. As a result, tunable laser emission of high brightness can be achieved in spectral ranges where no lasers were previously available.

This paper describes some results in recent research, development and application of tunable IR liquid N_2- and compressed H_2-Raman lasers. These lasers have been developed for the selective interaction with molecules, particular for the isotope separation and photochemistry. The results cover the work performed in the laboratory of Quantum Radiophysics of P.N. Lebedev Physical Institute, Moscow.

2. Pumping Source

A Nd-glass laser is used as the pump source. The laser consists of a master oscillator and power amplifiers. The master oscillator (Fig. 1) includes a free-running narrowband tunable oscillator (10) which locks the frequency of a Q-switched laser (11). The pulse of the master oscillator (with energy about 0.5 J) then passes through several stages of amplification. As a result narrow band tunable radiation is achieved. The central frequency is 9460 cm^{-1}, tuning range $\pm$ 50 cm^{-1}, linewidth $5 \cdot 10^{-2}$ cm^{-1}, pulse energy up to 100 J, and pulse duration 50 ns.

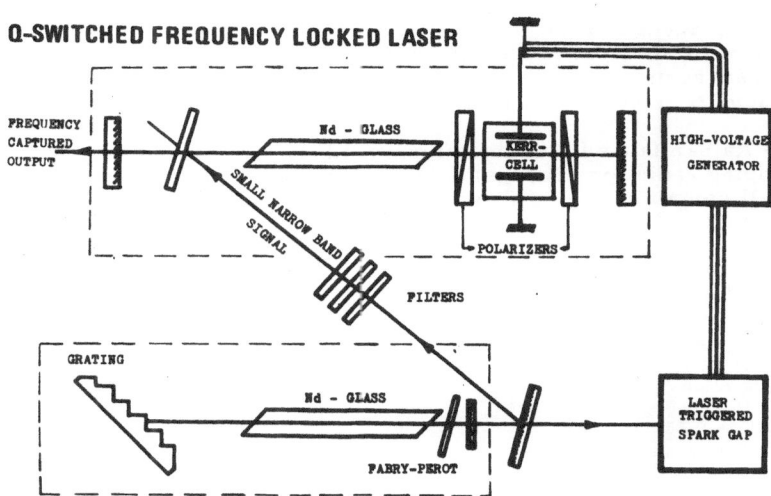

Q-SWITCHED FREQUENCY LOCKED LASER

FREE RUNNING TUNABLE NARROW BAND LASER

Figure 1   Block diagram of tunable master oscillator

3. Design of Raman Lasers

It should be emphasized that spatial homogeneity of the exciting radiation is extremely important in order to obtain proper operation of the Raman laser.   Therefore, the standard use of spherical lenses in the focusing system is inefficient.

To achieve a spatial homogeneity of pump intensity, we have put into practice a special raster focusing system (12) in connection with a lightguide (Fig. 2).   The raster consists of a number of square prisms.   The combination of prisms splits the pumping beam into many beams.   Each of these

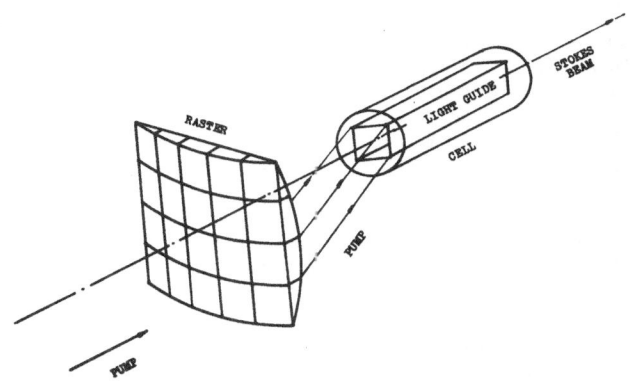

Figure 2   Optical system which provides spatial homogeneity of pumping radiation

beams has a square cross-section, and passes through a focal plane which coincides with the input of the lightguide. The lightguide is 70 cm long and has a cross-section of 1 x 1 cm. As a result, high spatial homogeneity of pump intensity is achieved throughout the length of the lightguide.

The Raman laser based on compressed H_2 is shown in Fig. 3 (13). The pumping radiation propagates through the input window (d = 70 mm) into a cell containing H_2 at pressures 50 - 60 atm.

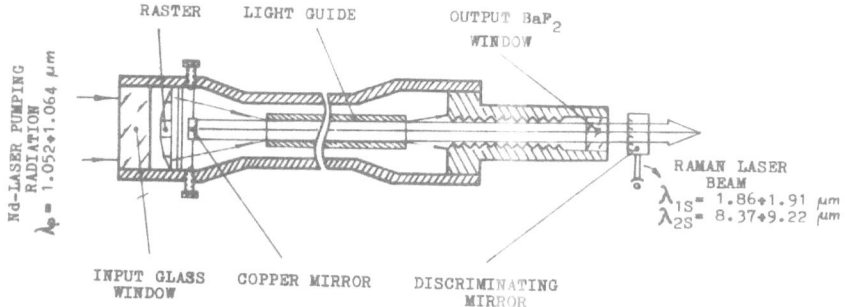

Figure 3   Compressed H_2-Raman laser

Behind the raster a 1 x 1 cm square copper plane mirror is installed. The BaF_2 output window of the cell is located 40 cm from the lightguide output, in order to avoid damage by the pumping radiation. Behind the output window a dichroic mirror is placed. The mirror has reflection coefficients of 100% and 10% on the 1st and 2nd Stokes components, respectively. Such a selective resonator allows the buildup of the 1st Stokes radiation and, as a result, the threshold pump intensity of the 2nd Stokes decreases.

A typical construction of the liquid N_2-Raman lasers uses the same principles. The cryogenic cell has adjustable windows which are used to define an optical cavity (9, 14, 15).

4. Principle Parameters

The principle parameters of the liquid N_2-Raman lasers have been partly investigated earlier (9). It should be borne in mind that the high pumping intensity results in Stokes and antistokes components which propagate together with the exciting beams rather than colinearly with the resonator axis. Such an effect decreases the conversion efficiency of the Raman generation.

Typical time histories of Raman pulse generation at various pumping intensities are shown in Fig. 4. The characteristic dip on the 1st Stokes pulses is caused by generation of the 2nd Stokes component. This phenomenon limits the quantum efficiency of the 1st Stokes component to about 50%.[*]

[*] To achieve higher conversion efficiencies in the 1st Stokes it is necessary to exclude buildup of the 2nd Stokes component. For this purpose a Raman amplifier operating in the saturation regime can be used (9, 15).

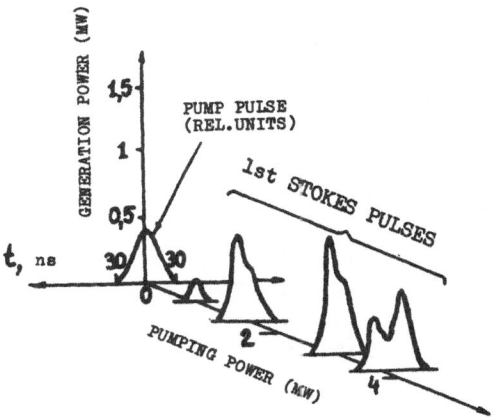

Figure 4   Dynamics of Raman laser

Figure 5 and Table 1 show the principle parameters of the liquid N_2-and compressed H_2-Raman lasers. One can see that the relative tuning range increases for down-conversion. For the 2nd Stokes component in H_2 the relative tuning range is about 10% (Fig. 5).

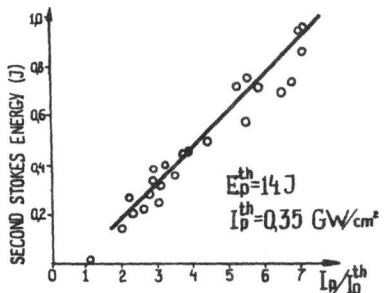

$E_p^{th}=14 J$

$I_p^{th}=0.35 \ GW/cm^2$

$\nu_p^{Nd} = 9460 \pm 50 \ cm^{-1}, \ \tau_p = 50 \ ns$

ACTIVE MEDIA	STOKES	TUNING RANGE μm	PULSE ENERGY J	PEAK POWER MW	EFFICI-ENCY %	QUANT YIELD %
COMPRESSED HYDROGEN P = 50 atm T = 300 K	1st	1.87–1.91	5	100	10	25
	2nd	8.33–9.1	0.8	30	1	20
LIQUID NITROGEN T = 77 K	1st	1.39–1.41	2	50	30	70
	2nd	2.06–2.1	1.2	30	20	50

Figure 5   Pulse energy of the 2nd Stokes component (ν_{ss} = 1100 - 1200 cm^{-1}) as a function of pump intensity I_p normalized to the threshold intensity I_p^{th} of the H_2-Raman laser

Table 1   Parameters of compressed H_2 and liquid N_2-Raman lasers pumped by tunable Nd-glass laser

5. Applications

a) Two-photon absorption spectrum in GaAs (16)

Figure 6 shows the dependence of the two-photon absorption coefficient β (cm/MW) on photon energies in a semiconductor crystal of GaAs.

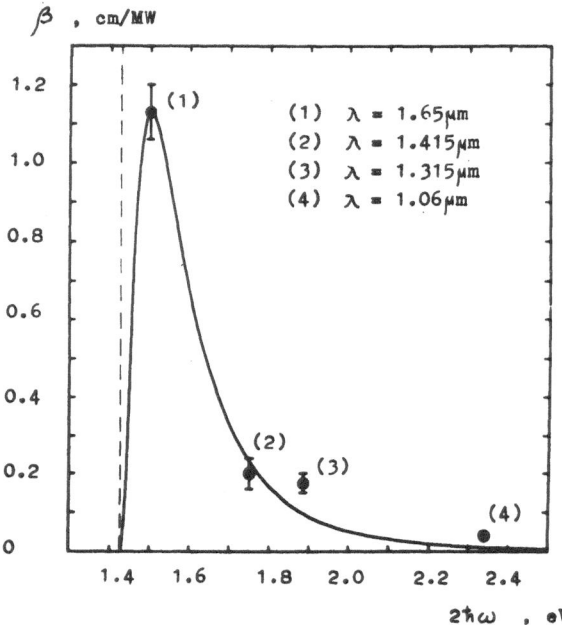

Figure 6   Spectrum of two-photon absorption coefficient β (cm/MW) in GaAs

Along with the liquid N_2 Raman laser ($\lambda_s = 1.4$ μm), excited by a Nd-glass laser (see Table 1) we used a liquid O_2-Raman laser ($\lambda_s = 1.65$ μm). The latter was pumped by an iodine flash photolysis laser with wavelength $\lambda_s = 1.315$ μm. In this case the frequency "tuning" was achieved by changing the active medium. It is interesting to note that in contrast to the earlier theory a sharp increase of the two-photon absorption coefficient occurs near the two-photon resonance. This increase is due to deep levels in the forbidden gap of GaAs (16).

b) Figure 7 a, b shows an application of H_2-Raman laser for the study of atmospheric absorption spectra (9). Fig. 7a shows a scan of the atmosphere obtained by a conventional spectrometer which has a resolution of 0.5 cm^{-1}. Fig. 7b shows fine structure in the absorption coefficient of the indicated single line near 5300 cm^{-1} at high resolution $5 \cdot 10^{-2}$ cm^{-1}. The results of Fig. 7b were obtained on a 4 m long path by using the tunable H_2-Raman laser, operating on its 1st Stokes component with line-width $5 \cdot 10^{-2}$ cm^{-1}. Large difference between these spectra is obvious because the spectral resolution of the Raman laser spectrometer is 10 times better than that of the conventional one.

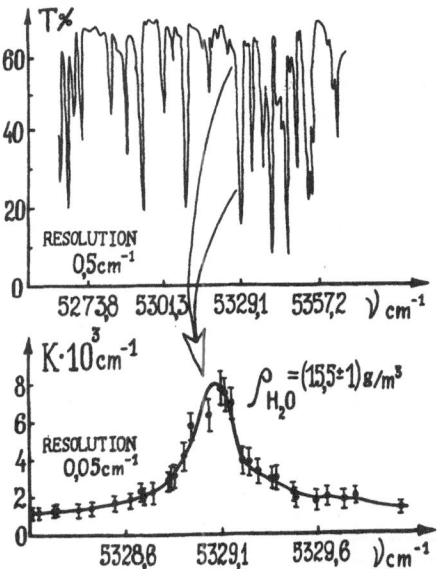

Figure 7   a) Scan of the atmosphere at resolution $0.5 \, \text{cm}^{-1}$.
           b) Fine structure of the line near $5300 \, \text{cm}^{-1}$ obtained at resolution of $5 \cdot 10^{-2} \, \text{cm}$ using H_2-Raman laser.

c) Sulphur isotope enrichment (17).

The 2nd Stokes of the tunable H_2-Raman laser ($_{ss} = 1100 - 1200 \, \text{cm}^{-1}$) has been used for sulphur isotope enrichment using selective dissociation in a strong IR field. The Raman laser was tuned to the frequency $\nu_{ss} = 1138 \, \text{cm}^{-1}$, which coincides with the combination band $\nu_4 + \nu_5 = 1138 \, \text{cm}^{-1}$ of the molecule $^{32}SF_6$. The isotope shift $^{32}SF_6 - {}^{34}SF_6$ for ν_4 is $4 \, \text{cm}^{-1}$. The Raman laser radiation passed through two successive cells (Fig. 8) filled with SF_6 at a pressure of 0.13 torr.

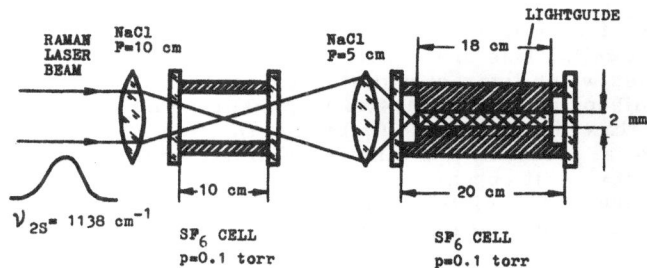

Figure 8   Block diagram of the apparatus for sulphur isotope enrichment using tunable H_2-Raman laser

The laser pulse (with 20 ns duration and divergence of 3 mrad) was focused in the first cell (10 cm long), and then again focused in the second cell (20 cm long). A cylindrical lightguide with L = 18 cm and i.d. 2 mm was placed in the second cell. It allowed us to increase the interaction volume of the SF_6 gas with the strong IR field, and thereby increase the ratio between excited volume and total volume of the cell.

The enrichment coefficient in residual SF_6 gas (K ($^{34}S/^{32}S$)) as well as in reaction products K^* ($^{32}S/^{34}S$) were measured by a mass spectrometric analysis. The concentration ratios of the appropriate molecules before and after irradiation were recorded. The dissociation rate was calculated from the equation $W = (1/n)\ln(N_o/N)$, where N_o and N are the selected isotopic mass numbers before and after irradiation, and n is the number of laser pulses.

Table 2 shows the results obtained after 60 pulses.

PRINCIPLE DATA	1st CELL	2nd CELL (WITH THE LIGHTGUIDE)
NUMBER OF RAMAN LASER PULSES	60	60
TOTAL ENERGY (J)	25	22
FOCAL INTENSITY (GW/cm^2)	10	10
ENRICHMENT COEFFICIENT OF RESIDUAL $^{34}SF_6$	1.06 ± 0.02	1.09 ± 0.02
ENRICHMENT COEFFICIENT OF PRODUCED $^{32}SOF_2$	1.25 ± 0.05	
RATIO BETWEEN EXCITED VOLUME AND TOTAL VOLUME OF THE CELL WITH SF_6	1 : 10	1 : 4

Table 2   Sulphur Isotope Enrichment (Experimental Results)

One can see that the enrichment coefficient in the lightguide cell is larger than in the first cell. This difference is accounted for by the larger volume of SF_6 molecules which interact with the strong laser field. It should be noted that the dissociation rate (1.4×10^{-3}) observed here for the $\nu_4 + \nu_5$ band is higher than that of the $\nu_2 + \nu_6$ band (18). Note that the intensity of the line $\nu_4 + \nu_5$ at 1138 cm^{-1} is almost 100 times weaker than that of $\nu_2 + \nu_6$ at 983 cm^{-1}. It appears that such an increase in the dissociation rate is due to higher IR radiation intensity because of lower beam divergence and shorter pulse duration than was used in (18).

References

1. N. G. Basov, E. P. Markin, A. N. Oraevsky, A. V. Pankratov, Dokl. Akad. Nauk SSSR, 198, 5, 1043 (1971)
2. N. G. Basov, A. N. Oraevsky, A. V. Pankratov, "Chemical and Biochemical Applications of Lasers" (Ed. B. C. Moore), ch. 7, p. 207, Acad. Press, New York (1974)
3. R. V. Ambartzumian, V. S. Letokhov, J. Quant. Electr. QE-7, p. 305 (1971)
4. R. V. Ambartzumian, N. V. Chekalin, V. S. Dolzhikov, V. S. Letokhov, E. A. Ryabov, Chem. Phys. Lett. 25, 515 (1974)
5. J. Lyman, R. J. Jensen, J. Rink, C. P. Robinson, S. D. Rockwood, Appl. Phys. Lett. 27, 87 (1975)
6. N. G. Basov, E. M. Belenov, V. A. Isakov, E. P. Markin, A. N. Oraevsky, V. I. Romanenko, N. B. Ferapontov, Kvantovaya Elektronika 2, 5, 938 (1975)
7. R. V. Ambartzumian, Yu A. Gorokhov, V. S. Letokhov, G. N. Makarov, JETP 69, 1956 (1975)
8. V. S. Letokhov, V. P. Chebotaev, "Nonlinear Laser Spectroscopy" (in Russian), Moscow, 1975; Springer Verlag Heidelberg, 1976 (in English)
9. A. Z. Grasiuk, Kvantovaya Elektronika 1, 3, 485 (1974)
10. I. G. Zubarev, V. F. Mulikov, Kvantovaya Elektronika No. 3 (9), 13 (1972)
11. I. G. Zubarev, S. I. Mikhailov, Kvantovaya Elektronika 1, 3, 625 (1974)
12. A. Z. Grasiuk, V. F. Efimkov, V. G. Smirnov, Pribory i Tekhnika Experimenta (PTE), 1 (1976)
13. A. Z. Grasiuk, I. G. Zubarev, A. V. Kotov, S. I. Mikhailov, V. G. Smirnov, Kvantovaya Elektronika 3, 4, 1062 (1976); II Symp on Gas Lasers Novosibirsk, 1975
14. V. V. Bocharov, A. Z. Grasiuk, I. G. Zubarev, A. V. Kotov, V. G. Smirnov Kvantovaya Elektronika 1, 10, 2185 (1974)
15. A. Z. Grasiuk, I. G. Zubarev, V. I. Mishin, V. G. Smirnov, Kvantovaya Elektronika No. 5 (17), 27 (1973)
16. A. Z. Grasiuk, I. G. Zubarev, A. B. Mironov, I. A. Poluektov, Fizika Tverdogo Tela 10, 2, 262 (1976)
17. R. V. Ambartzumian, Yu A. Gorokhov, A. Z. Grasiuk, I. G. Zubarev, A. V. Kotov, A. A. Puretsky, VIII USSR Conf. on Coherent and Nonlinear Optics, Tbilisi, 1976 (Digest, vol. II, p. 102)
18. R. V. Ambartzumian, Yu A. Gorokhov, V. S. Letokhov, G. B. Makarov, A. A. Puretsky, JETP Lett. 22, 374 (1975)

EFFICIENT HIGH-POWER 8.62 μm INFRARED RADIATION SOURCE FOR URANIUM ISOTOPE SEPARATION IN UF₆*

R.L. Aggarwal [†], N. Lee and B. Lax [†] [‡]
Francis Bitter National Magnet Laboratory
Massachusetts Institute of Technology
Cambridge, Massachusetts 02139, USA

Introduction

This paper examines various aspects of the generation of infrared radiation at 8.62 μm by noncollinear phase-matched four photon mixing of CO_2 laser beams in germanium at room temperature, as a potential infrared source for the uranium laser isotope separation in UF_6 .

The success of the two-step laser isotope separation in UF_6 will depend, ultimately, on the development of efficient, high-power narrow-band tunable infrared (IR) radiation sources. In this laser isotope separation scheme, the $U^{235}F_6$ molecule is first selectively excited to a higher vibrational energy level by the absorption of an IR photon followed by the absorption of a higher-energy photon in the ultraviolet (UV) region. The important selective IR absorption bands in UF_6 occur at 15.9 μm, 12.1 μm, 8.62 μm, and 7.74 μm, which correspond, respectively to the ν_3 , $\nu_3 + \nu_5$, $\nu_3 + \nu_2$, and $\nu_3 + \nu_1$ vibrational transitions [1, 2]. Here ν_3 is an IR-active mode and ν_1 , ν_2 and ν_5 are the Raman-active modes of UF_6 [3]. According to JENSEN et al.[1], the pulse energy requirements for the IR laser source range from several millijoules at 15.9 μm to tens of millijoules at 8.62 μm with a linewidth of less than 0.05 cm^{-1} . For economic feasibility of the enrichment program, the efficiency of the IR source should be in excess of 0.1% at a pulse repetition exceeding 500 pps.

So far, emphasis has been placed on the generation of the 15.9 μm radiation since the cross-section for absorption at the 15.9 μm fundamental band is about a hundred times larger than that at the 12.1 μm, 8.62 μm, and 7.74 μm combination bands. Unfortunately, various methods employed to-date have provided 15.9 μm radiation pulses with energy in the 100-μJ level which is about an order of magnitude below the energy requirement of several mJ [1]. On the other hand, the present 4-photon mixing scheme appears to satisfy not only the pulse energy requirements but also the efficiency, repetition rate, frequency and linewidth requirements.

*Supported in part by the Massachusetts Institute of Technology Funds and in part by the Advanced Research Projects Agency through the Office of Naval Research
†Also Department of Physics, Massachusetts Institute of Technology
‡Supported by the National Science Foundation

Noncollinear 4-Photon Mixing

Recently we have reported [4] the achievement of phase-matching in 4-photon mixing of CO_2 laser beams in germanium using the noncollinear mixing geometry. The wave vector diagram for noncollinear phase-matching is shown in Fig. 1. Here $\vec{k}_1$ and $\vec{k}_2$ are the wave vectors for the input laser beams of frequencies ω_1 and ω_2 respectively; $\vec{k}_3$ is wave vector for the radiation generated at the frequency $\omega_3 = 2\omega_1 - \omega_2$ for $\omega_1 > \omega_2$. Using the refractive index data for germanium [5], the angles θ and φ are determined to be less than 1^0. Using this method approximately 1 mJ of energy per pulse was obtained at 8.7 μm from an 8.3 cm long crystal of germanium with 1 cm^2 cross-section and 3 MW/cm^2 peak input intensity from each of two CO_2 TEA lasers operating at the frequencies ω_1 and ω_2. The output energy can be scaled up through the use of higher input intensities and longer crystals of germanium.

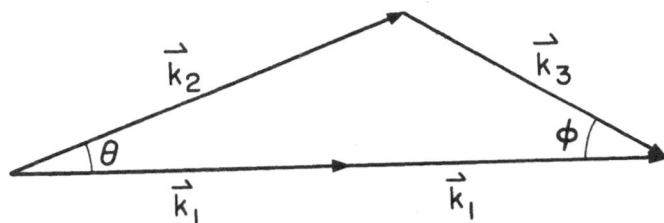

Fig. 1.   Wave vector diagram for noncollinear phase-matched 4-photon mixing

Since the damage threshold of germanium is approximately 100 MW/cm^2, one can conservatively increase the input intensity to 10 MW/cm^2 for each of the two input beams. The output energy which is proportional to the third-power of the input intensity will scale up to 37 mJ per pulse. Another factor of 5 increase in the output energy to 185 mJ can be readily obtained by anti-reflection coating of the input and output surfaces of the germanium crystal.

Under the phase-matched condition, the maximum effective length of the crystal is limited by the absorption of the crystal. The absorption coefficient α of optical grade germanium, as shown in Fig.2, is less than 0.03 cm^{-1} in the 8 μm to 11 μm region. So one should be able to obtain useful mixing lengths of about 30 cm. Therefore the output energy can be further increased to ~ 400 mJ by increasing the length of the mixing crystal from 8.3 cm to 20 cm. This would correspond to ~ 1% efficiency for converstion from the electrical power into the 8.62 μm infrared radiation, assuming 5% efficiency for the CO_2 lasers. Such a source is well above the minimum energy requirement of tens of mJ and minimum efficiency of 0.1% for the uranium enrichment program. Furthermore, it is obvious to see that the output power can be linearly scaled up with input beams of the same intensity but larger cross-section.

A typical atmospheric TEA laser has a gainwidth of 3 GHz or 0.1 cm^{-1} and its output consists of several longitudinal modes. Thus the resulting output of the 4-photon mixing with these typical CO_2 TEA lasers may exceed the 0.05 cm^{-1} linewidth requirement for the selective isotope excitation of the UF_6 molecule. However, single longitudinal mode operation in CO_2 lasers can be obtained by using the hybrid laser scheme [6] in which the laser cavity contains both a low pressure discharge tube and a high pressure TEA section. Due to the narrow linewidth of the low pressure medium, the hybrid system is forced to oscillate in a single longitudinal

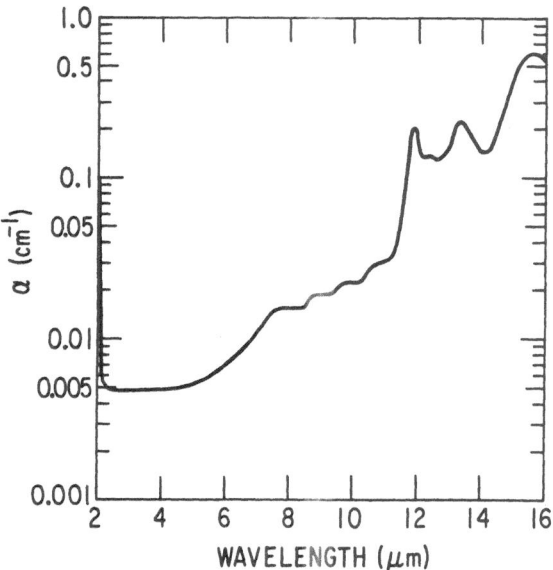

Fig. 2.   Absorption coefficient α of optical grade germanium at room
temperature, as supplied by Eagle-Picher

mode. There is an alternative scheme to obtain single longitudinal mode action in
a CO_2 TEA laser. Recently, we have demonstrated [7] single longitudinal mode
operation with fine tuning capability over the pressure broadened gainwidth of the
CO_2 laser by using a tilted intracavity GaAs etalon. The advantage of this scheme
is that it does not only provide single longitudinal mode to satisfy the linewidth
requirement but can also provide fine frequency tuning as large as 0.3 cm^{-1} which
would be useful to optimize the frequency of the 8.62 μm radiation for the uranium
separation. However, the losses due to the insertion of an intracavity etalon and
laser operation at a frequency away from the center of the gain curve would reduce
the efficiency of the CO_2 laser system.

In Fig. 3, we show a number of lines which can be obtained in the 8.62 μm region by mixing two CO_2 laser lines. If one uses three input CO_2 lasers for mixing
instead, on the order of a thousand different combinations can be obtained in that
same small frequency range of Fig. 3. Therefore it should be possible to select
the desired wavelength in the 8.62 μm region by using two or three CO_2 lasers.

For the high repetition rate of 500 pulses per second, one has to consider heat
dissipation capability of the crystal. For a projected total of 2 to 3 joules per pulse
CO_2 laser input energy, and a maximum absorption coefficient of 0.03 cm^{-1}, the
power absorbed by a 20 cm long crystal would be 600 W at the repetition rate of
500 pps. If the crystal is held between two copper blocks maintained at room temperature, the rise in temperature, ΔT, at the center of the crystal with respect to the
temperature of the copper blocks is given by

$$\Delta T = \frac{tQ}{8\,K\,A} \tag{1}$$

where t is the thickness of the germanium crystal, Q is the power dissipated by the two surfaces, K is the thermal conductivity of the germanium crystal at room temperature and A is the area of contact between the crystal and the cold finger. Using the values $t = 1\,cm$, $Q = 600\,W$ and $K = 0.5\,W\,cm^{-1}\,K^{-1}$, (1) predicts $\Delta T = 6.4^{\circ}K$, assuming perfect thermal contact between germanium and copper. Thus it appears that germanium should be able to handle high repetition rates of several hundred pulses per second.

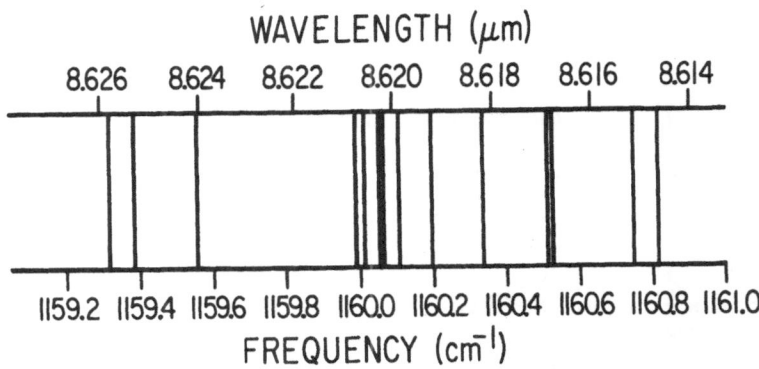

Fig. 3.   Infrared lines obtainable in the 8.62 μm region by 4-photon mixing of two CO_2 laser lines

In summary, the noncollinear 4-photon mixing of CO_2 lasers in germanium can provide an infrared source consistent with the requirements for the uranium isotope separation program in UF_6.

Acknowledgments

We wish to thank Mr. J. Reising of Eagle-Picher for supplying the absorption spectrum of germanium.

References:

1. R.J. Jensen, J.G. Marinuzzi, C.P. Robinson and S.D. Rockwood, Laser Focus, May 1976, p. 51.
2. J.W. Eerkens, Appl. Phys. 10, 15 (1976).
3. For the normal modes of an octahedral XY_6 molecule, see for example, G. Herzberg, Molecular Spectra and Molecular Structure, Vol. II (Van Nostrand Reinhold Company, New York, 1945), p. 51.
4. N. Lee, R.L. Aggarwal and B. Lax (Submitted to Appl. Phys. Lett.)
5. C.D. Salzberg and J.J. Villa, J. Opt. Soc. Am. 47, 244 (1957).
6. A. Gondhalekar, E. Holzhauer, and N.R. Heckenberg, Phys. Lett. 46A, 229 (1973).
7. N. Lee, R.L. Aggarwal, and B. Lax (unpublished).

III. Isotope Separation and Laser Driven Chemical Reactions

LASER CHEMISTRY AT SURFACES

M.S. Djidjoev, R.V. Khokhlov, A.V. Kiselev, V.I. Lygin,
V.A. Namiot, A.I. Osipov, V.I. Panchenko and B.I. Provotorov
University of Moscow, USSR

So far all the studies of laser stimulation of chemical reactions have been undertaken with a homogeneous medium which is in most cases a gas [1, 2]. At the same time, surface reactions, where at least one of the reacting media is absorbed at a two-medium interface, play a great role in chemistry and, particularly, in chemical technology. For example, the majority of catalysts used in industry are solids of high area and small particle size. The possibility of laser stimulation of such reactions looks very attractive.

By laser radiation it seems to be possible to selectively affect the desorption process and to change the molecular concentration at a surface; it is probably possible to separate molecules by masses (laser chromatography), to affect the surface diffusion, to control the surface chemical reactions, to change the catalysis properties of adsorbents, etc.

The laser stimulation of surface reactions can be realized in two ways. The first is the excitation of molecules adsorbed at a surface and the second way is to excite the molecules in the gaseous phase situated in the volume above the surface which then react with the surface. It has to be taken into account that the adsorbing ability of the excited molecules is quite different from that of the molecules in the ground state.

Recently, investigations of both ways of heterogeneous chemical reactions under laser excitation were started at Moscow University and at the Lebedev Institute [3]; this paper is devoted to the first results obtained in the first way.

The mechanisms of the laser action on adsorbed groups of atoms can differ depending on the range of radiation frequency. If the radiation frequency is in IR range, its action leads to the molecular vibration in the adsorbtion potential, if the light frequency is in UV range, the radiation excites the electronic transitions thereby changing the potential curve; the interaction between molecules and surface is also changed. In such a way, it is possible directly to transform, for example, a weak physical adsorbtion into strong chemical adsorbtion, and vice versa.

In the absence of laser irradiation, the adsorbtion kinetics can be described in the first approximation by the LANGMUIR theory [4]. This theory is based on equation

$$\frac{dN}{dt} = \alpha (N^x - N) - \beta N \quad ,$$ (1)

where N and N^x are the numbers of adsorbed particles and adsorbing centers at the surface unit respectively; α is determined by the probability of the molecular adhesion to surface and is proportional to pressure and depends on molecular parameters; β is the desorption coefficient. The first term in the right-hand side describes the number of the molecules that come to the surface from the gas in unit time interval and the second describes the desorption process.

It is essential to note that under certain conditions the temperature dependence of parameters α, N^x, and β is determined by relations [4]:

$$\alpha = \alpha_o \exp \left[- \frac{E_1}{kT} \right] , \quad N^x = N_o^x \exp \left[- \frac{E_2}{kT} \right] , \quad \beta = \beta_o \exp \left[- \frac{q}{kT} \right] .$$ (2)

This dependence suggests that the adhesion and the desorption processes occur by overcoming of a potential barrier and it is possible to make it easier to overcome this barrier by laser radiation of an appropriate frequency. This was a basic idea in starting our study. Highly dispersive silica powder compressed into thin plates was chosen as an object of experimental study. It was chosen because it has a transmittance band in the 10 μm range [5] and because it is extensively used as an adsorbent in processes of the gas purification and separation and as a carrier of catalysts.

The silica structure is schematically shown in Fig. 1. Ordinarily, the silica surface is highly covered by the hydroxyl groups formed due to adsorbtion of water [6].

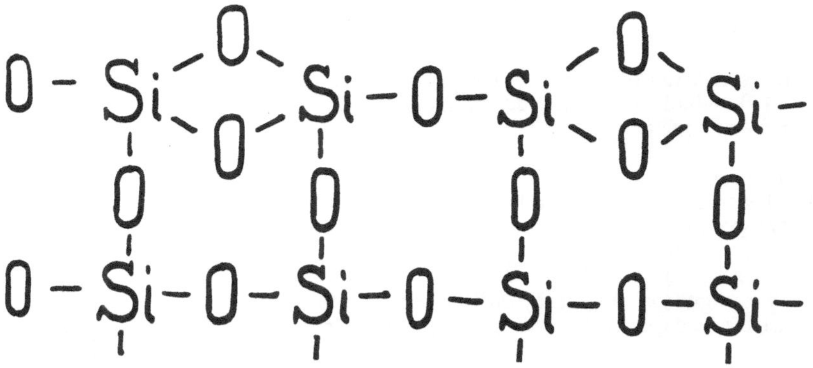

Fig. 1. Silica structure.

The silica transmittance band and the generation range of the CO_2 laser overlap a 950-970 cm^{-1} absorption band which is due to Si-OH stretching oscillations coupled between themselves [7]. This band is removed during evacuation of 400°C. At higher temperatures, a spectral line at 987 cm^{-1} related to uncoupled oscillations of hydroxyl groups [5] can be observed.

The experiment consisted of laser stimulation of a surface reaction between hydroxyl groups with formation of water

$$-\underset{|}{\overset{|}{S}i} - \underset{|}{\overset{\overset{\displaystyle H}{\overset{|}{O}}}{}} + -\underset{|}{\overset{|}{S}i} - \underset{}{\overset{\overset{\displaystyle H}{\overset{|}{O}}}{}} \longrightarrow \underset{/|\backslash}{S}i\overset{O}{\diagup \diagdown} \underset{/|\backslash}{S}i + H_2O \qquad (3)$$

and also laser stimulation of the reactions between hydroxyl groups and amino groups NH_2 and NH. These groups can be formed at surface by heating the silica in an ammonia atmosphere under certain conditions accordance with the reactions [5], [8]:

$$-\underset{|}{\overset{|}{S}i} - \underset{}{\overset{\overset{\displaystyle H}{\overset{|}{O}}}{}} + NH_3 \longrightarrow -\underset{|}{\overset{|}{S}i} - \underset{}{\overset{\overset{\displaystyle H_{\diagdown}N\diagup H}{\overset{|}{}}}{}} + H_2O \ , \qquad (4)$$

$$\underset{/|\backslash}{S}i\overset{O}{\diagup \diagdown} \underset{/|\backslash}{S}i + NH_3 \longrightarrow -\underset{|}{\overset{|}{S}i} - \underset{}{\overset{\overset{\displaystyle H_{\diagdown}N\diagup H}{\overset{|}{}}}{}} + -\underset{|}{\overset{|}{S}i} - \underset{}{\overset{\overset{\displaystyle H}{\overset{|}{O}}}{}} \ , \qquad (5)$$

$$-\underset{|}{\overset{|}{S}i} - \underset{}{\overset{\overset{\displaystyle H_{\diagdown}N\diagup H}{\overset{|}{}}}{}} + -\underset{|}{\overset{|}{S}i} - \underset{}{\overset{\overset{\displaystyle H}{\overset{|}{O}}}{}} \longrightarrow \underset{/|\backslash}{S}i\overset{\overset{\displaystyle H}{N}}{\diagup \diagdown} \underset{/|\backslash}{S}i + H_2O \ . \qquad (6)$$

Silica covered by amino groups is extensively used for air purification, for ferment grafting, etc.

The oscillation frequency of NH_2 groups relative to the surface atom Si is 932 cm^{-1} [5], which is also inside the transmission band of silica and inside the CO_2 laser generation range.

Any surface chemical reaction, including dehydroxylation, stimulated by conventional heating or by laser radiation goes through highly excited vibrational states of the atom groups situated at the tail of the Boltzmann distribution. The surface mobility of these excited vibrational states is much higher than the mobility of ground states because the height of the potential barrier for them is relatively small. When the atom groups move they collide between themselves, or, under other conditions, with other (possibly stationary) groups, in the same manner as in gases in volume in the three-dimensional case. When the groups collide, the probability of reaction arises, and in some cases they react between themselves.

It is possible to use the surface mobility stimulated by the resonance laser radiation for selective control of a surface chemical reaction. Let the mobility be intrinsic at low temperature and the desired atom group be selectively excited to states where the surface mobility is high enough. In the case of such excitation, only these atom groups can participate in surface chemical reactions.

The main result of our experimental studies is that (1) different reactions occur if different atom groups are preferentially excited by laser radiation and (2) the laser-induced reactions differ essentially from the similar reactions stimulated by heating.

The experimental set-up consisted of a CO_2 laser, a vacuum cell with the studied sample and an IR spectrometer. Frequency tuning of CO_2 laser was done by a grating. The cw laser power was quite low and equalled $10 \, W/cm^2$. The cell design is shown in Fig. 2. The cell body is a quartz tube flattened in its lower part. This part has a hole covered by an IR transparent window. The sample holder is mounted inside the cell. The sample is positioned on a platinum frame fixed on the holder. The cell is inserted into the oven.

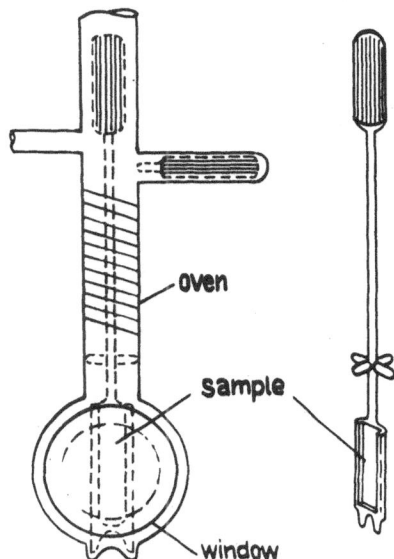

Fig. 2. Schematic view of the cell.

The silica powder was compressed to form a sample plate of 8 × 30 mm size, 10 mg/cm^2 weight, and thickness of about 50 μm. The surface area of this highly dispersive sample was 180 m^2/g. The samples were evacuated at 650^{0}C in order to remove the physically adsorbed water and for partial dehydroxylation with formation of siloxane bridges.

The hydroxyls and amino groups were detected by observing the OH, NH, and NH$_2$ group spectra in the 3 μm range where silica is completely transparent and these groups exhibit the well known spectral lines due to OH, NH_2 and NH stretching modes.

The spectrum of OH groups in this range is shown in Fig. 3. The absorbtion peak magnitude is proportional to the OH group concentration at surface.

Dehydroxylation is extensively used for the silica surface activation. The removal rate of the surface hydroxyl groups depends on the sample porous structure and on the presence of impurities [9]. The OH groups can be almost completely removed by evacuating the sample for several hours at temperatures higher than $1100^{\circ}C$ [8].

Dehydroxylation under laser irradiation at a frequency of 950 cm^{-1} is considerably faster. Figure 3 shows the result of a sample irradiation for several minutes. It can be seen from the figure that more than half of the hydroxyl groups are removed. Thus, the laser stimulation of the dehydroxylation process differs essentially from the pure heating stimulation.

The next type of surface reactions studied was the reactions with amino groups. Figure 4 shows the spectra of the sample covered by NH, NH_2 and OH groups.

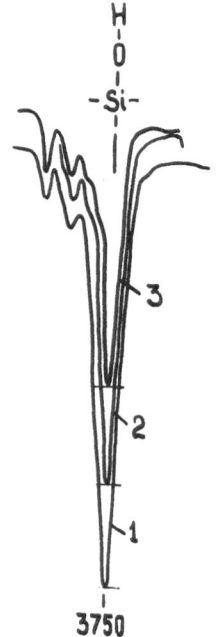
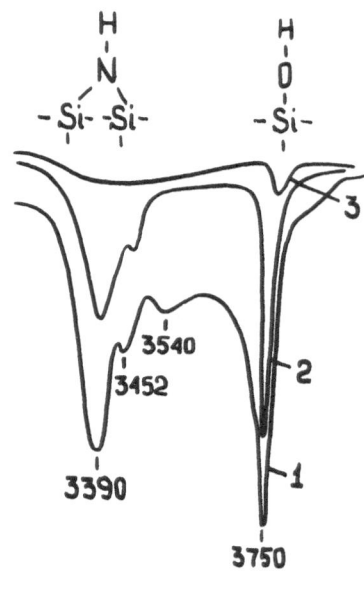

Fig. 3. Laser irradiation at 950 cm^{-1}. 1 - Silica treated in vacuum at 650°C. 2 - Sample irradiated for 2,5 min. 3 - Sample irradiated for 15 min.

Fig. 4. Laser irradiation at 950 cm^{-1}. 1 - Silica irradiation in NH_3. 2 - Sample irradiated in vacuum for 10 sec. 3 - Sample irradiated for 80 sec.

Curve 1 presents the spectrum of sample irradiated for 40 sec in ammonia atmosphere. It can be seen that the silica surface amination proceeds at a much higher rate than the process under conventional heating. The spectrum consists of the intense line of NH groups (ν = 3390 cm^{-1}) and weak lines of NH$_2$ groups (ν = 3452 cm^{-1} and ν = 3540 cm^{-1}) [10]. Irradiation for 10 sec leads to the partial removal of the amino groups hydroxyl groups. After irradiation for 80 sec, the OH and NH groups are almost completely removed, which shows a very high rate of dehydroxylation of the samples covered by amino groups.

Another result can be obtained if the sample covered by amino groups is irradiated at the resonance frequency of the NH$_2$ oscillating mode (ν = 932 cm^{-1}). Under the irradiation for 5 sec the concentration of NH$_2$ groups decreases while the concentration of OH groups increases. The decrease rate of NH$_2$ groups is several times higher than the rate with irradiation at ν = 950 cm^{-1}, and at least two orders higher than with pure heating. It is interesting to note that at this stage the concentration of NH groups is maintained at the same level. If the irradiation interval increases, all amino groups are removed together with the hydroxyl groups.

Thus, the irradiation of the amined sample by laser radiation with different frequencies leads to the essentially different concentrations surface species.

There are two possible explanations of the observed anomalous features of the dehydroxylation and deamination processes of the silica surfaces. The first explanation is to assume that a spatial temperature gradient exists. This gradient, which is equal to several tens of degrees in the order of magnitude, results in thermodiffusion of the adsorbed atom groups to the sample edge. The gradient usually arises at laser heating, and this is the basis of differences between the laser and conventional heating.

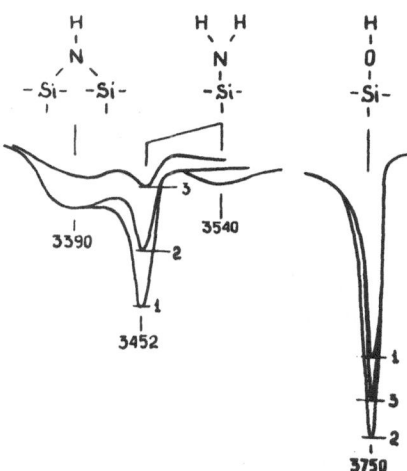

Fig. 5. Laser irradiation at 932 cm^{-1}. 1 - Silica treated in NH$_3$ at 650°C. 2,3 - Sample irradiated in vacuum.

The observable peculiarities of the dehydroxylation and the deamination processes can be explained by applying the ideas and images developed in the laser chemistry of gases [11], [12], [13] the key point of these ideas is the difference between the temperatures of optical phonons and temperatures of acoustic phonons under the resonance laser irradiation.

It is necessary to note that an exchange of vibrational energy between the Si-NH_2 bond and the N-H bond is quite possible. In this case, the migration of H ion (proton) can lead to the observable intensity increase of OH groups, whereas the intensity of NH groups fails to decrease (in contrast to Fig. 4).

Let us forget for a moment the surface reactions and consider the resonance laser stimulation of gaseous chemical reactions. Under certain conditions, the amount of vibrational energy in the laser pumped mode can be much higher than the equilibrium level. The latter is determined by the temperature of translational motion connected directly with heating. The fact of such difference forms the basis of the laser stimulation of gaseous chemical reactions.

Returning to the surface reactions, we can assume that the amount of vibrational energy in the Si-NH_2 modes corresponds to a temperature higher by several tens of degrees than the temperature of acoustic phonons connected directly with the oven heating.

The difficulty with this and similar explanations is that for such a difference between the temperature of acoustical phonons and the vibrational temperature to exist it is necessary that the lifetime of excited states be not less than 10^{-8} sec. In fact, however, if one makes an estimate from the linewidth, it is equal to 10^{-11} sec. The large lifetime is, nevertheless, quite possible. It is known, for example, that nitrogen molecules in excited state can live at a surface for a very long time of up to 10^{-3} sec [14]. There are also purely theoretical reasons in favour of the possibility of lifetimes longer than 10^{-11} sec under certain conditions; these conditions are satisfied in the considered cases.

So, it is quite possible that the temperature difference between the oscillational modes and acoustic phonons is realized in the cases considered.

Future studies will show which of these (or other) explanations is correct. But independently of explanations, we are quite confident that the action of the laser on adsorbed molecules cannot be reduced to pure oven heating.

REFERENCES

1. N. V. Karlov, I. A. Karpov, Yu. I. Petrov and O. M. Stelmakh, ZhETF 64, 196 (1973).

2. I. G. Basov, E. M. Behetov, E. N. Markin and A. I. Oraevsky, ZhETF 64, 485 (1973).

3. K. S. Gogelashvili, I. V. Karlov, A. I. Orlov, R. P. Petrov, Yu. I. Petrov and A. M. Prokhorov, JETP Lett. 21, 640 (1975).

4. F. F. Volkenshtein, Physical Chemistry of Semiconductor Surfaces, (Nauka, Moscow, 1973).

5. B. A. Morrow, J. A. Cody and Lydia S. M. Lee, J. Phys. Chem. 79, 2405 (1975).

A. V. Kiselev and V. I. Lygin, Infrared Spectra of Surface Compounds, (Nauka, Moscow, 1972).

A. V. Kiselev and V. I. Lygin, Kolloid Zh. 21, 581 (1959).

V. M. Kiryutenko, A. V. Kiselev and V. I. Lygin, Zh. Fiz. Khim. (in press).

R. S. McDonald, J. Phys. Chem. 62, 1168 (1959).

J. B. Peri, J. Phys. Chem. 70, 2937 (1966).

N. D. Artamonova, V. T. Platonenko and P. V. Khokhlov, ZhETF 58, 21 (1970).

B. F. Gordiets, A. I. Osipov, E. V. Stuynochenko and L. A. Shelepin, Ispekhi Fiz. Nauk 108, 655 (1972).

B. F. Gordiets, A. I. Osipov and V. L. Panchenko, ZhETF 65, 894 (1973).

G. Black, H. Wise, S. Schecter and R. L. Sharpless, J. Chem. Phys. 60, 3226 (1974).

THE PHOTOPHYSICS AND PHOTOCHEMISTRY OF FORMALDEHYDE

A.P. Baronavski, A. Cabello, J.H. Clark, Y. Haas,
P.L. Houston, A.H. Kung, C.B. Moore, J. Reilly,
J.C. Weisshaar and M.B. Zughul
Chemistry Department, University of California, USA
and
Materials and Molecular Research Division
Lawrence Berkeley Laboratory, Berkeley, Ca 94720, USA

Formaldehyde is a particularly interesting molecule for fundamental photophysical and photochemical studies. Its structure
and spectroscopy have been thoroughly studied for the three
lowest electronic states. It is small enough so that spectra
are fully resolved and laser excitation sources can populate
single excited states. The goal of the research described below is a detailed knowledge of the non-radiative transition
rates, and of the intermediate and final states involved in
predissociation. This knowledge would provide considerable
qualitative insight into the nature of radiationless transition
processes in small molecules and allow a severe quantitative
test of theoretical treatments.

The laser techniques which have been brought to bear on formaldehyde typify many of the needs of chemical physicists for
tunable sources. Isotopic separations with formaldehyde illustrate the needs for lasers in this practical application.

The energy states of formaldehyde and its dissociation products are diagrammed in Fig. 1. The overall photochemistry resulting from excitation of the first excited singlet state, S_1,
is

$$H_2CO(S_0) + h\nu \longrightarrow H_2CO(S_1,v') \qquad (1)$$

$$H_2CO(S_1,v') \xrightarrow{k_M} H_2 + CO \qquad (2)$$

$$\xrightarrow{k_R} H + HCO \qquad (3)$$

Since the direct dissociation of S_1 leads to energetically
inaccessible product states, a radiationless transition must
occur. The nature of this transition, the mechanism of dissociation, the branching between molecular and radical product
channels, the distribution of the available energy among the

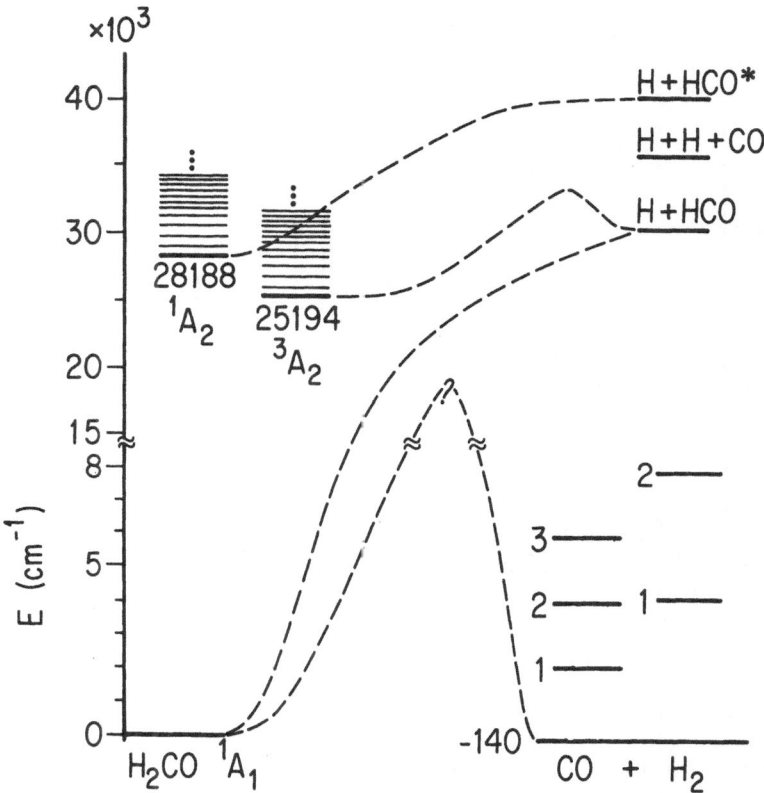

Fig. 1 Energy level diagram of formaldehyde and its dissocia-
        tion products. Broken lines show correlation of each
        state to dissociated fragment states. Maxima indicate
        barriers known to exist but of uncertain height.

product degrees of freedom, the role and behavior of the triplet
state and the subsequent chemical reactions of H and HCO are a
few of the questions which must be addressed. Many of the
answers are a strong function of the exciting wavelength. Some
depend strongly on the pressure as well. Thus, this simple
molecule displays a very rich and sometimes bewildering photochemistry.

Formaldehyde photochemistry is currently of practical interest in its application to isotope separation, for its activity
in polluted urban atmospheres and its role in the interstellar
medium. In this paper we outline recent progress and some prospects for future research.

I. Lifetimes of optically excited states

The spectroscopy of the first excited singlet state (S_1) of
formaldehyde is very well understood [1]. The vibronic bands

are nicely separated (Fig. 2) and the assignments are essentially complete.

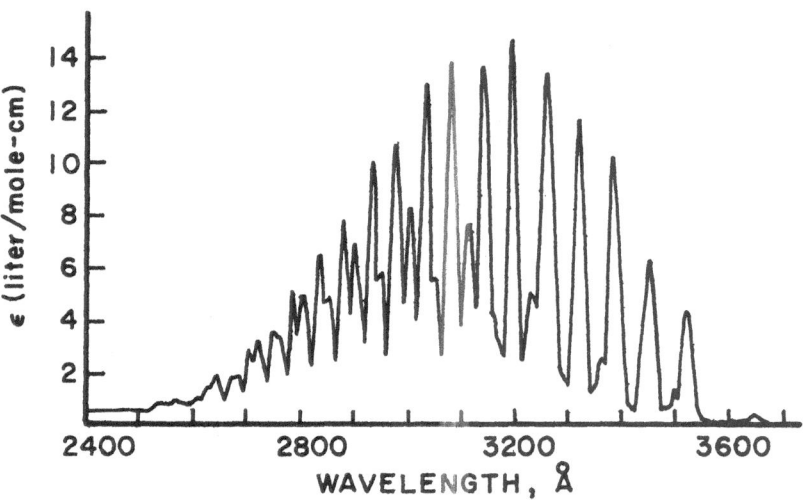

Fig. 2 $S_0 \rightarrow S_1$ absorption spectrum showing resolved vibrational
structure.

The transition to S_1 is only vibronically allowed; thus the
absorption is weak and the radiative lifetime is $\sim 10^{-5}$ sec[2].
Selective excitation of a single vibronic band can be accomplished using a dye laser of a few Å spectral width. The rotational structure of many bands is well-resolved and analyzed.
Single rovibronic levels could be selectively excited in
favorable cases with a laser of less than 0.1 cm^{-1} spectral
width. So the molecule provides an excellent opportunity for
selective photochemistry.
    The first triplet state (T_1) is also well studied [1]. The
triplet origin is only 3000 cm^{-1} below that of S_1, so that the
T_1 level density is very low near the S_1 origin. There are
some isolated cases of singlet-triplet perturbations [3].
    Several experimental techniques have been used to obtain
lifetime information [4-10]. The most direct technique monitors
fluorescence intensity as a function of time following excitation
of formaldehyde with a tunable laser pulse [4,5], a N_2-laser
[5-7] or a nanosecond lamp [8]. Experimental parameters include
the exciting wavelength, the gas pressure, and the isotopic
species of formaldehyde. The weak absorption and low fluorescence quantum yields make the experiments difficult; dispersion of the fluorescence has been possible only in the case of
direct excitation with a 1MW nitrogen laser. In these experiments lifetimes of a few nanoseconds and longer are measured.

Two-photon excitation of S_1 has been accomplished [4]; the selection rules allow different vibronic bands to be reached.

Below $\sim$ 2800 Å, line broadening due to predissociation
exceeds the Doppler broadening of 0.06 cm^{-1} and it becomes
possible to determine lifetimes by measuring the spectral widths
of individual rotational lines. Lifetimes in the diffuse region
were thus determined from high resolution (600,000) absorption
spectra taken on a 3.4 m spectrograph [9].

Another spectroscopic technique involves measurement of the
relative fluorescence quantum yield as a function of wavelength.
The experiment simultaneously monitors laser intensity, fluorescence intensity, and transmitted light intensity while the wavelength of the narrowband ($\sim$ 0.1 cm^{-1}) exciting laser is scanned
(Fig. 3). The relative quantum yield is given by the ratio of

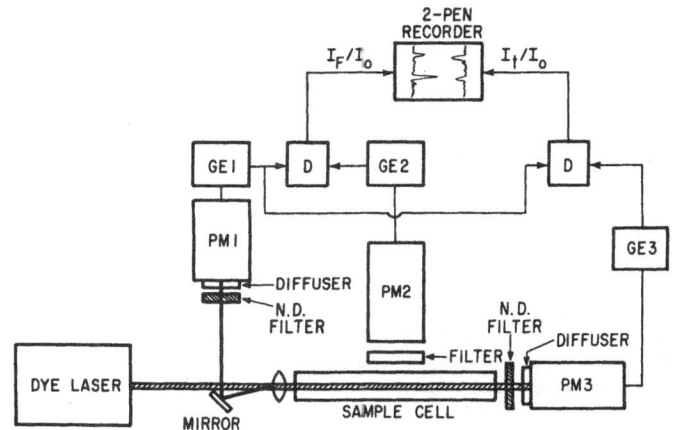

Fig. 3   Relative fluorescence quantum yield apparatus. The ratio
of signals I_t from PM3 to I_0 from PM1 gives the sample
absorbance and the ratio of I_F from PM2 to I_0 gives normalized fluorescence intensity. In strongly absorbing
samples PM2 is located close to the entrance window.
Dissimilarities in the two spectra indicate changes in
fluorescence quantum yield and hence in lifetime with
vibration-rotation level.

fluorescence intensity to absorption intensity, each normalized
to the laser power.

The lefthand part of Fig. 4 shows the results of fluorescence
lifetime measurements for single vibronic levels of H_2CO, HDCO
and D_2CO as a function of excess energy above the origin of S_1
[4,7]. The lifetimes shown are zero-pressure extrapolations.
Since the radiative lifetime of formaldehyde is estimated to be
$\sim$ 5 μsec and the measured fluorescence lifetimes with the
exception of the lowest levels of D_2CO are much shorter, the

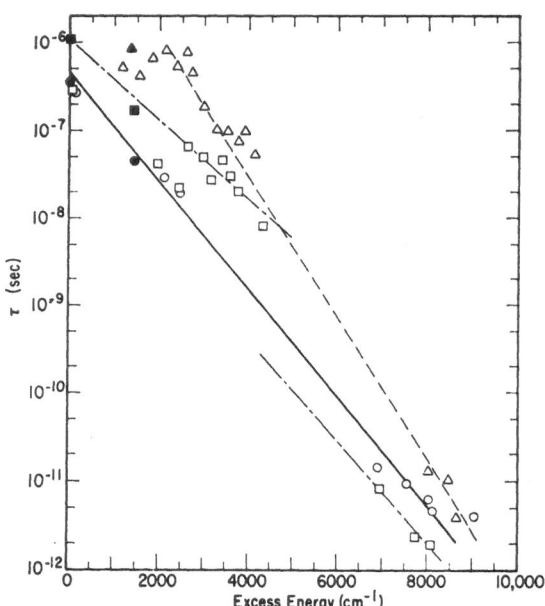

Fig. 4 Lifetimes of vibronic levels of S_1 vs vibrational energy.
○,● H_2CO; □,■ HDCO; and △,▲ D_2CO.

experiment essentially measures the rate of non-radiative decay
out of S_1. The general trend of roughly exponential decrease of
lifetimes with increasing excess energy is clear. A large
deuterium isotope effect is observed; for example, the vibra-
tionless lifetimes are 5 μsec, 1 μsec, and 366 nsec for D_2CO,
HDCO, and H_2CO respectively [7].

The results from linewidth measurements in the diffuse region
are shown on the righthand side of Fig. 4 [9]. Lifetimes in
the picosecond regime are obtained above 7000 cm^{-1} excess energy.
The HDCO lifetimes are shorter than the H_2CO lifetimes, and they
do not appear to extrapolate back to the long wavelength HDCO
results. By contrast, smooth extrapolations are possible for
H_2CO and D_2CO.

Luntz and Maxson [10] have measured the phosphorescence decay
of the triplet state of D_2CO after direct excitation with a
tunable laser. Their lifetime of 16.9 μsec extrapolated to
zero pressure is much shorter than the estimated radiative lifetime of ∿ 2 msec, so it appears that the triplet also decays nonradiatively. All of the measurements were made for P > 0.2 torr
or more than 30 collisions/lifetime. No emission from H_2CO
triplet molecules was observed.

Measurements of S_1 fluorescence lifetimes as a function of
pressure can indicate the relative importance of collision-induced processes as compared to intramolecular processes. Collision-induced predissociation rates are generally very fast, often
several times gas kinetic rates [4]. It is clear that early
photochemical studies at pressures of a few torr were dominated

by collisional processes. At low enough pressures, curvature
in the Stern-Volmer plots (Fig. 5) has been observed for many
of the bands [7].

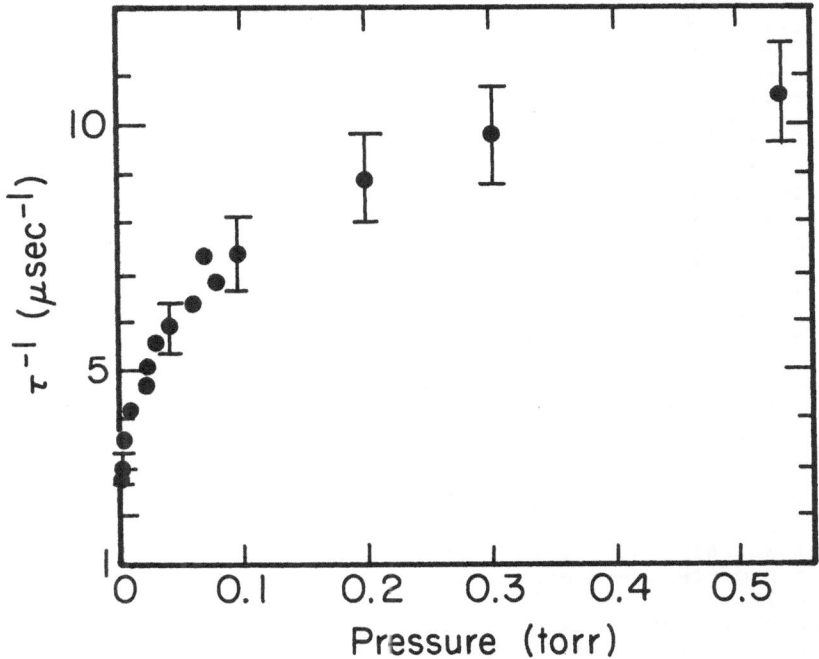

Fig. 5 Stern-Volmer plot of reciprocal lifetime vs pressure
for H_2CO with one quantum of out of plane bending excitation. In the ground state, at 0.04 torr, collisions
inducing rotational line broadening would occur once
per lifetime. (See [11]).

Stern-Volmer plots cannot separate energy transfer effects
from collision-induced predissociation. If the fluorescence
can be dispersed before observation, the time evolution of a
single vibronic level of S_1 can be followed. In addition,
changes in the intensities of fluorescence spectral features
with pressure can be used to detect vibrational relaxation
within S_1. For excitation of D_2CO with a nitrogen laser, fluorescence spectral resolution of ~ 6 Å was possible. The results
show that vibrational relaxation is important at pressures as
low as 0.25 torr. In addition, the time dependence of the
fluorescence from an apparently single level may not necessarily
be a simple exponential decay, even at 0.1 torr.

114

Preliminary results from rotationally resolved studies of relative fluorescence quantum yield as a function of wavelength seem to indicate that the non-radiative decay rate within a vibronic band depends on the rotational state. If the radiative lifetime is essentially constant over a vibronic band, then the fluorescence quantum yield should be proportional to the lifetime of an individual rotational level. The fluorescence excitation spectra do not seem to be proportional to the absorption intensity as the wavelength is scanned. Thus, it appears that the lifetimes depend on the rotational state. Assuming small variations of the radiative lifetime with vibronic level, it should be possible to use the above technique to estimate vibronic lifetimes in the gap between 4000 and 7000 cm^{-1} excess energy in S_1. (Fig. 2)

With a more powerful narrowband laser, it will be possible to selectively excite single rotational levels of formaldehyde and directly measure the fluorescence lifetimes as a function of rotational quantum number. The vibronic state lifetimes discussed above are presumably complicated averages over rotational states. Those states strongly mixed with the triplet may show anomalous behavior if the triplet plays a significant role in S_1 decay. It should be noted that radiationless transition theories for this type of molecule have ignored rotational effects.

Our theoretical understanding of the initial collisionless decay from S_1 is quite weak. Computed lifetimes for the various S_1 levels of D_2CO and H_2CO using the Franck-Condon factors for a model in which S_1 decays initially to high vibrational levels of S_0 reproduce the overall trend of lifetimes with excess energy and the large isotope effect [12]. Detailed variations of lifetimes with vibronic level are not well reproduced. A more serious difficulty is that the discovery of a long lived intermediate challenges the assumption of a quasicontinuous set of S_0 levels.

Freed [13] has recently proposed a theory of collision-induced non-radiative decay for intermediate case molecules. He predicts that decay rates should initially increase linearly with pressure, largely due to the increased number of allowed final rotational states. A major role of collisions thus involves taking up angular momentum. This effect saturates at higher pressures, and thus may provide an explanation for the observed curvature of Stern-Volmer plots at very low pressures.

II. Kinetic spectroscopy of the dissociation products

Absorption spectroscopy with laser sources may be used in order to follow the formation and decay of photolysis products. If the photolysis source is a laser, it must be considerably more powerful than needed for fluorescence lifetime studies. This is particularly true for studies of collision-free processes or processes induced by a small number of collisions. The source for following absorption as a function of time should ideally be narrower in frequency than the linewidth of the transition and of essentially constant intensity during the transient behavior following the photolysis laser pulse. A probe laser with a pulse duration short compared to the kinetics of interest but with a broad spectrum may be used to record an entire spectrum at a particular instant in time. In the formaldehyde system we are interested in observing the rate of formation of and distribution of energy among the products H_2, CO, HCO and H. The formation and decay of any intermediate states such as T_1 or vibrationally excited S_0 would be particularly important. Finally, the chemical reactions of H and HCO are important in determining the final reaction products and the extent of isotopic scrambling.

A. CO kinetic and vibrational distribution

The CO photochemical product of UV laser excitation of formaldehyde has been monitored by its infrared fluorescence and by its absorption of a cw CO laser [14]. In the former case, fluorescence from CO($v = 1$) was observed using an Hg:Ge detector, while, in the latter case, the ultraviolet and CO lasers were made to overlap spatially in a 1-m cell of formaldehyde. Pulsed excitation of the formaldehyde then produced CO which absorbed the CO laser and caused a change in signal intensity at an Au:Ge detector. By tuning the CO laser to various vibrational transitions and observing the relative signal intensity at each line, a measure of the vibrational distribution of the nascent CO product could be obtained. This method is particularly good for CO since the CO laser may be tuned to many CO transitions and is easily stabilized on line center. It is possible to make some use of accidental laser-molecule coincidences as well e.g. SF_6 and BCl_3 with the CO_2 laser [15]. Now the increased availability of laser diodes makes the method very attractive and general. In principle a diode may be tuned over the Doppler profile of an absorption thus giving the translational as well as vibrational and rotational distributions. Three types of information come from these studies.

1. The vibrational distribution of the CO product was found to vary with the ultraviolet excitation wavelength (Table I). The ratio of the vibrational excitation of the CO product to the total energy available to the molecular products was always found to be rather low, on the order of 1 - 5%.

Table I.   **CO Vibrational Distributions**

λ(nm)	H_2CO	v=0	1	2	3	4	5	6	E_v(%)
347.2	4^3_0	0.900	0.100						0.7
337.1	$4^2_0 6^1_0$	0.870	0.110	0.020					1.1
317.0	$2^2_0 4^3_0$	0.731	0.213	0.049	0.005	0.002			2.3
314.5	$2^3_0 4^1_0$	0.663	0.195	0.071	0.032	0.021	0.012	0.005	4.1
303.6	$2^4_0 4^1_0$	0.571	0.262	0.103	0.037	0.015	0.008	0.004	4.5
294.0	$2^5_0 4^1_0$	0.609	0.264	0.099	0.024	0.004			3.5

2. The yield of CO was found to be linear in ultraviolet pulse energy and formaldehyde pressure. The latter of these facts indicates that the quantum yield for CO production does not change in the pressure region studied, 0.1-10 torr. Product yields of CO were also measured as a function of addition of foreign gases. For nitrogen and argon, the CO yield decreases slowly with increasing pressure, indicating the possibility of some quenching mechanism. However, for NO and O_2 the yield increases rapidly and then levels off. This sensitivity to NO and O_2 will be discussed in section III.

3. The CO product formation time was inversely proportional to pressure and much longer than the S_1 decay time. Thus S_1 must decay to some intermediate state which is subsequently collisionally induced to dissociate to H_2 and CO.

$$H_2CO(S_1,v') \xrightarrow{\quad} I \xrightarrow{\quad} H_2 + CO \qquad (4)$$
$$\longrightarrow H + HCO \qquad (5)$$

The probability that an H_2CO collision will induce dissociation of I to $H_2 + CO$ is 0.15. In D_2CO dissociation of I occurs with a probability per collision of 0.09 for D_2CO and 0.012 for Ar [14]. The identity of this intermediate is not known. Possibilities include vibrationally hot S_0, T_1 and geometrical isomers such as the radical HCOH [16]. Ideally it should be possible to observe and identify I spectroscopically. The calculated potential surfaces of Morokuma and coworkers [17] will be valuable in establishing possible dissociation mechanisms.

B. HCO product detection

The mechanism of dissociation to radical products and their subsequent chemical reactions may be studied by monitoring absorption by the HCO radical. Intracavity dye laser spectroscopy is extremely sensitive to small concentrations of absorbers [18]. It has been used before to detect HCO produced by a 10^3 Joule flashlamp [19]. We now obtain comparable spectra with a monochromatic 10^{-3} J pulse of UV laser light with the apparatus shown in Fig. 6.

A frequency-doubled,flashlamp-pumped dye laser (Chromatix CMX-4) is focussed into a photolysis cell contained inside the cavity of a second flashlamp-pumped dye laser. This probe laser is operated broadband; its output is focussed through the entrance slit of a spectrometer and exposes a photographic plate. A variable delay between the photolysis pulse and the probe pulse allows HCO concentration vs time to be measured. A series of spectra of the characteristic HCO band near 614.4 nm is shown in Fig. 7. The spectra from 10 torr H_2CO alone and with 175 torr Ar added are similar. The HCO disappears much more rapidly with 0.5 torr O_2 or NO added. Rough estimates of the HCO + NO and HCO + O_2 rate constants can be extracted from these data. More accurate measurements will require calibration of the intracavity enhancement as a function of absorber concentration. However, there is evidence that the relation between dye laser quenching and in-cavity absorber concentration can be roughly linear [20]. In any case, these preliminary results shed light on some of the post-dissociation photochemistry discussed in

INTRACAVITY DETECTION OF HCO

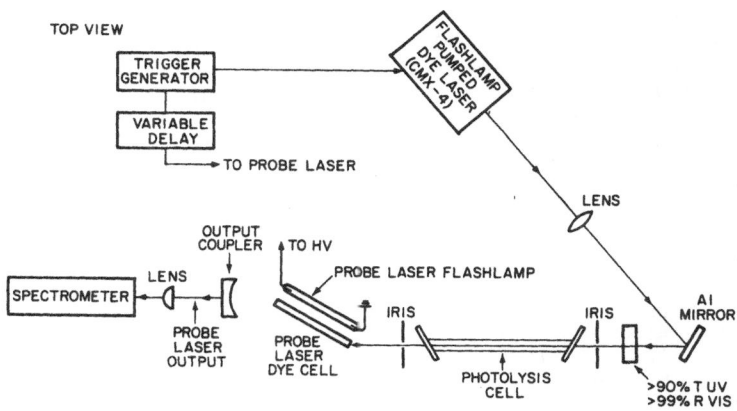

Fig. 6 Schematic diagram of the experimental apparatus for intracavity detection of HCO.

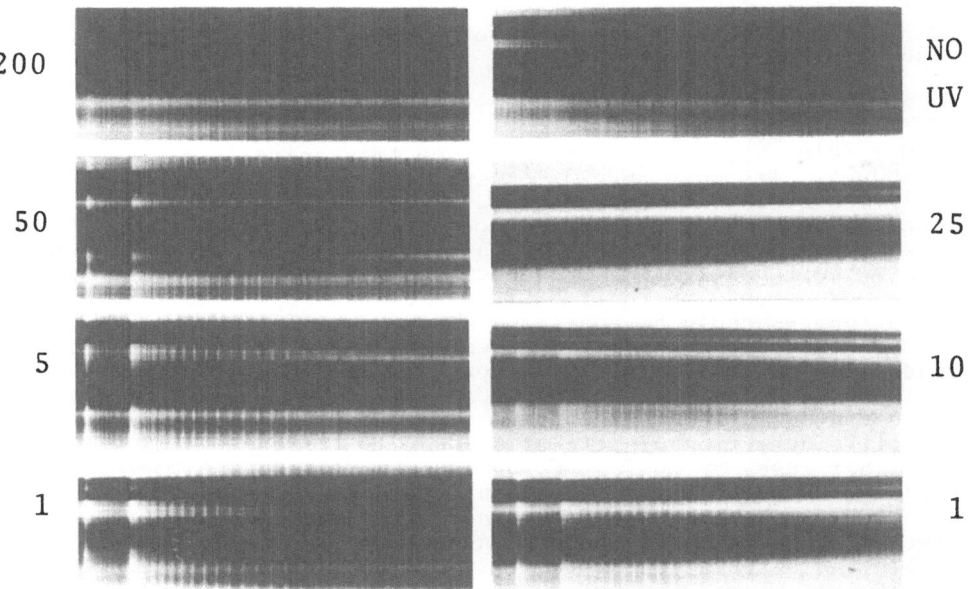

Fig. 7 Characteristic HCO absorption near 6144 Å following photolysis of 10 torr H_2CO (left), and 10 torr H_2CO with 0.5 torr added O_2 (right). The numbers beside each spectrum indicate the time delay in microseconds between the photolysis pulse and the probe laser pulse. The uppermost righthand trace is a blank spectrum, where the UV laser was not fired.

Section III. Kinetic spectroscopy of the HCO radical is now
being directed toward 1) measurement of the molecule to radical
branching ratio,k_M/k_R, by probing for both HCO and DCO; 2) measurement of the $H^{12}CO + H_2^{13}CO \rightarrow H^{13}CO + H_2^{12}CO$ exchange rate;
3) monitoring the appearance of secondary reaction products such
as HNO and DNO. Even these few examples illustrate the power
of coupling isotopically selective excitation with isotopically
selective product analysis in elucidating the detailed nature
of heretofore unstudied processes.

C.   H-atom detection

H-atoms may be detected by the techniques of resonance absorption and resonance fluorescence using continuously operating
Lyman-α resonance lamps [21, 22]. This technique is extremely
sensitive; H-atom densities on the order of 10^{10} cm^{-3} may be
observed on timescales of microseconds and longer. Resonance
fluorescence has been used widely in kinetics for species such
as H,D,Cl, Br, I, and O [23]. The development of pulsed laser
sources at resonance frequencies could improve time resolution
and sensitivity greatly over conventional sources.
For formaldehyde, observation of H atoms will allow direct
determination of the production time for radical products
without the ambiguities introduced by possible changes in energy
distribution with time which may occur with HCO. The kinetics
of H and D atom reactions with formaldehyde and added gases may
also be studied.

D.   H$_2$ detection

Molecular hydrogen is the most spectroscopically elusive pho-
toproduct. Its only absorption spectrum is in the vacuum ultraviolet where intense sources are scarce. Recent development in
coherent vacuum ultraviolet sources seems quite promising. Generation of radiation around 1200 Å can be achieved in metal vapor
and/or inert gases using frequency upconversion techniques [24].
With improvements on their efficiencies and amplitude stability,
these sources could be applied. Another strong possibility is
to use Raman spectroscopy. Conventional Raman spectroscopy is
generally low in sensitivity. However, using coherently driven
anti-Stokes Raman spectroscopy (CARS) [25] the signal-to-noise
ratio can be improved by several orders of magnitude, allowing
sensitive detection of molecules inert to infrared excitation.
In either case, by tuning the optical sources to the appropriate
frequencies the initial vibration-rotation energy distribution
of the H$_2$ product can be determined. Insertion of a variable
time delay between the photolyzing and probing radiation will
enable the observation of the time dependence of each level's
population.

III. Photochemical processes and products

The formation and removal of radical products in the photolysis of H$_2$CO presents several fundamental questions of practical importance for laser isotope separations. Past measurements
of the branching ratio between radical and molecular products

[26], (2) and (3), cannot be accurately interpreted because of kinetic isotope effects [27]. The relative quantum yield Φ_{H_2}/Φ_{CO} = 1 is accurately known [26]. However, the absolute quantum yields for near UV photolysis are not well known [28]. While it is known [29] that H atoms are removed in 10^3 collisions by

$$H + H_2CO \longrightarrow H_2 + HCO, \tag{6}$$

even the products of HCO removal

$$2HCO \overset{k_7}{\longrightarrow} H_2 + 2CO \tag{7}$$

$$\overset{k_8}{\longrightarrow} H_2CO + CO \tag{8}$$

are uncertain. The reaction may take place in the gas phase or on the walls of the reaction vessel. The rapid disappearance of HCO with added NO and O_2, II. B, shows that reactions like

$$HCO + O_2 \longrightarrow HO_2 + CO \tag{9}$$

remove HCO in $\sim$ 10 microseconds at 1 torr added gas. They like-wise give the increased CO absorption noted in II. A. These same additives may remove H atoms by

$$H + O_2 + M \longrightarrow HO_2 + M. \tag{10}$$

The HO_2 radicals may then give

$$HO_2 + HO_2 \longrightarrow HOOH + O_2. \tag{11}$$

Some mechanistic questions may be answered directly by end product quantum yield studies. For example molecular dissociation, (2), yields one CO and cne H_2 for each H_2CO dissociated. For radicals, (3) followed by (6), gives two HCO per H_2CO dissociation. If reaction (7) follows, then two molecules of CO and two of H_2 are produced. If reaction (8) follows, then one each of CO and H_2 are produced. The quantum yield in a wavelength region where radicals are produced distinguishes (7) from (8) immediately. If (7) dominates, then the radical molecule branching ratio can be determined at each photolysis wavelength. If (8) dominates then sufficient O_2 may be added to scavenge the radical products following (9) and (10). The result is only one molecule of CO and no H_2 per H + HCO pair produced, and branching ratios may be determined.

The use of lasers in the measurement of absolute quantum yields for photochemical processes offers some important advantages over the use of incoherent sources. The spatial properties of the laser beam make possible improved measurements of the absolute photon flux, since even with relatively small area detectors one can intercept 100% of the beam. Corrections for geometrical factors, common in experiments employing incoherent sources, are eliminated. Long absorption paths, and consequently low pressures and small absorption coefficients may be used. With cell length as an available variable, it is possible to work at transmissions near 50%, so that the accuracy of the absorption measurement is optimized. Commercially available

tunable dye lasers offer milliwatts of average power throughout
the UV. With photolysis times on the order of 30 minutes, micro-
moles of product can be produced. Finally, the tunability
offered by lasers allows the quantum yield to be measured as a
function of internal energy of the dissociating molecule.

The study of photophysical and photochemical processes for
individually selected rovibronic states of molecules promises
to be richly rewarding in terms of new phenomena and of a more
completely developed and tested theoretical framework. These
experiments are made possible by laser sources sufficiently
powerful and tunable to produce chemically and spectroscopically
observable quantities of molecules in single energy states. The
continued development of more versatile and reliable lasers for
excitation sources will lead to much broader applications.
Equally vital will be the development of spectroscopic probing
sources for the observation of small and rapidly changing concentrations of molecules in individual energy states. We can
expect the applications of lasers in chemistry to expand much
more rapidly now than in the past. Their uses in molecular beam
scattering experiments and in the more classical chemical pursuits
of organic and inorganic photochemistry promise to· open up many
new avenues of research. The practical application of lasers
to the separation of isotopes helps point the direction of possible future applications in chemical separations and synthesis.

Acknowledgments

Parts of this research were supported by the National Science
Foundation, the Energy Research and Development Administration,
the University of California Board of Patents and the Army Research Office. A.C. thanks the Program of Cultural Cooperation
between the U.S.A. and Spain. J.C.W. thanks the National
Science Foundation and M.B.Z. thanks the Agency for International
Development for predoctoral fellowships.

References

1. D.C. Moule and A.D. Walsh, Chem. Rev. 75, 67 (1975).
2. J.W. Sidman, J. Chem. Phys. 29, 644 (1958), and references
cited therein.
3. J.C.D. Brand and C.G. Stevens, J. Chem. Phys. 58, 3331 (1973)
4. E.S. Yeung and C.B. Moore, J. Chem. Phys. 58, 3988 (1973);
E.S. Yeung and C.B. Moore in Fundamental and Applied Physics, Proceedings of the Esfahan Symposium, M.S. Feld, A.
Javan and N.A. Kurnit editors, Wiley-Interscience (1973).
5. A.P. Baronavski, J.H. Clark, Y. Haas, P.L. Houston and C.
B. Moore in Laser Spectroscopy, S. Haroche et al. editors,
(Springer-Verlag, 1975) p. 259.
6. T. Aoki, T. Morikawa and K. Sakurai, J. Chem. Phys. 59,
1543 (1973).
7. A.P. Baronavski, Ph.D. Dissertation, University of California, Berkeley (1975).
8. R.G. Miller and E.K.C. Lee, Chem. Phys. Lett. 33, 104 (1975).

9.   A.P. Baronavski, A. Hartford, Jr., and C.B. Moore, J. Mol. Spectrosc. 60, 111 (1976).
10. A.C. Luntz and V.T. Maxson, Chem. Phys. Lett. 26, 553 (1974).
11. D.V. Rogers and J.A. Roberts, J. Mol. Spectrosc. 46, 200 (1973).
12. E.S. Yeung and C.B. Moore, J. Chem. Phys. 60, 2139 (1974).
13. K.F. Freed, J. Chem. Phys. 64, 1604 (1976).
14. P.L. Houston and C.B. Moore, J. Chem. Phys. (in press).
15. J.I. Steinfeld, MTP International Review of Science, Physical Chemistry, ed. J.C. Polanyi 9, 247 (Butterworth, London) (1973).
16. J.A. Altmann, I.G. Csizmadia, K. Yates and P. Yates, J. Am. Chem. Soc. (in press).
17. D.M. Hayes and K. Morokuma, Chem. Phys. Lett. 12, 539 (1972).
     R.L. Jaffe, D.M. Hayes and K. Morokuma, J. Chem. Phys. 60, 5108 (1974).
18. N.C. Peterson, M.J. Kurylo, W. Braun, A.M. Bass and R.A. Keller, J. Opt. Soc. Am. 61, 746 (1971).
     T.W. Hansch, A.L. Schallow and P.E. Toschek, I.E.E.E. J. Quant. Elect. QE-8, 802 (1972).
19. G.H. Atkinson, A.H. Laufer, and M.J. Kurylo, J. Chem. Phys. 59, 350 (1973).
20. R.B. Green and H.W. Latz, Spectrosc. Lett. 7, 419 (1974).
     R.A. Keller, J.D. Simmons and D.A. Jennings, J. Opt. Soc. Am. 63, 1552 (1973).
     M. Maeda, F. Ishitsuka and Y. Miyazoe, Opt. Commun. 13, 314 (1975).
21. W. Braun and M. Lenzi, Disc. Faraday Soc. 44, 252 (1967).
22. M.J. Kurylo, N.C. Peterson and W. Braun, J. Chem. Phys. 53, 2776 (1970).
23. D. Davis and W. Braun, Appl. Opt. 7, 2071 (1968).
     W. Braun, J. Chem. Phys. 53, 4244 (1970).
     W. Braun, A.M. Bass and D. Davis, J. Opt. Soc. Am. 60, 166 (1970).
24. K.S. Hsu, A.H. Kung, L.J. Zych, J.F. Young, S.E. Harris, "1202.8 Å Generation in Hg Using a Parametrically Amplified Dye Laser," IEEE J. Quant. Elect. QE-12, No. 1, 60 (Jan. 1976).
25. R.F. Begley, A.B. Harvey, R.L. Byer, "Coherent anti-Stokes Raman Spectroscopy," Appl. Phys. Lett. 25, No. 7, 387 (1 Oct. 1974).
26. R.D. McQuigg and J.G. Calvert, J. Am. Chem. Soc. 91, 1590 (1969).
27. J.R. McNesby, M.D. Scheer, and R. Klein, J. Chem. Phys. 32, 1814 (1966).
     D.D. Macdonald and E. Tschuikow-Roux, J. Chem. Phys. 49, 5345 (1968).
     P.M. Guyon and M. Tronc, J. Chim. Phys. 66, 35 (1969).
28. R.G.W. Norrish and F.W. Kirkbride, J. Chem. Soc. 1932, 1518.
29. A.A. Westenberg and N. de Haas, J. Phys. Chem. 76, 2213 (1972).
     B.A. Ridley, J.A. Davenport, L.J. Stief, and K.H. Welge, J. Chem. Phys. 57, 520 (1972).

FUTURE APPLICATIONS OF SELECTIVE LASER PHOTOPHYSICS AND PHOTOCHEMISTRY

V.S. Letokhov

Institute of Spectroscopy, Academy of Sciences USSR,
Moscow, Podol'skii rayon, Akademgorodok, 142092, USSR

1. Selective Laser Photophysics and Photochemistry

The basic processes of selective action on matter by laser radiation, which is often called "selective laser photophysics and photochemistry", are as follows:

I.   Selective separation of substances at atomic - molecular level

II. Selective chemical reactions of atoms or molecules of desired sorts (photochemical separation) or in a desired direction (photochemical synthesis)

III. Selective detection of atoms, molecules or molecular bonds

The difference in the absorption spectra of atoms or molecules which allows selective excitation of atoms, molecules or molecular bonds of a desired sort is fundamental for selective action of laser radiation on matter. The difference in the absorption spectra is conditioned by differences in any of the following characteristics of atoms or molecules:

1   chemical composition
2   spatial structure
3   isotopic composition
4   isomeric composition of nuclei
5   mutual orientation of nuclear spins

Using laser radiation it is possible to realize processes I to III (separation, chemical reaction, detection) in substances at atomic - molecular level selectively with respect to various chemical elements, molecular bonds, isotopes, nuclear isomers, molecular stereoisomers, ortho- and para-molecules, etc.

Though primary attention is being given now to isotopically-selective chemical reactions and isotope separation, the processes under development can be applied, in principle, to all the listed objects of selective laser photophysics and photochemistry. What is more, apart from selective

chemical reactions and selective separation of substances at atomic -
molecular level, selective detection of atoms, molecules and molecular
bonds is also a very important process. The variety of processes of selective laser photophysics and photochemistry is drawn very conditionally in
Fig. 1, where the types of selective processes are given down and their
objects across. One must keep in mind though that this classification is

I. SEPARATION	PURIFICATION OF SUBSTANCES		ISOTOPICALLY SELECTIVE PHOTOPHYSICS		LEFT-AND RIGHT-MOLECULES SEPARATION	
II. CHEMICAL REACTIONS	PHOTOBIO-CHEMISTRY		ISOTOPICALLY SELECTIVE PHOTO-CHEMISTRY			
III. DETECTION	DETECTION OF COMPLEX MOLECULES	LASER ION (PROTON) MICROSCOPY		DETECTION OF SINGLE EXCITED NUCLEA		
	CHEMICAL ELEMENTS AND COMPOUNDS	MOLECULAR BONDS	ISOTOPES	NUCLEAR ISOMERS	MOLECULAR STEREO-ISOMERS	ORTHO-AND PARA-MOLECULES

Fig. 1   Types of selective laser photophysics and photochemical processes

rather conventional. For instance, selectivity in photochemical reactions
may mean selective participation of chosen atoms or molecules in a reaction or selective participation of some molecular bond for specific chemical
synthesis. The classification of objects of selective processes under molecules and molecular bonds is also conventional but very important for macromolecules.

The isotopically-selective photophysical and photochemical processes
taking two shaded sections in Fig. 1 are developing rapidly at present.
The reports to this conference witness to a progressing advancement in
this field. I doubt whether there is anyone who does not rely upon great
prospects of this trend for science and technology (see review [1]).
Therefore in my report, the purpose of which is to outline the future and
less obvious applications of selective laser photophysics and photochemistry,
I would like to omit, with your permission, isotopic applications. I am
going to consider the prospects of development of some nonisotopic selective photoprocesses outlined in Fig. 1. So I restrict myself to selective
photoprocesses, the realization of which, I believe, offers quite new possibilities to some relative subjects, such as nuclear physics, spectroscopy,
molecular biology, chemical technology, biochemistry. Some of the processes considered below, such as laser identification of nucleotide sequence
in desoxiribonucleic acids (DNA) and laser selective biochemical processes,
are no less important than isotope separation by lasers.

2. Elementary Selective Processes

In view of the problem of isotope separation, a good deal of elementary processes have been recently suggested, discovered and successfully demonstrated for selective action of laser radiation on matter in various states of aggregation: 1) atomic and molecular gases, 2) condensed medium, 3) heterogeneous medium (gas-condensed medium boundary). Most of the experiments however have been conducted in a gas phase.

For atomic gases there are three quite different approaches (Fig. 2): 1) photochemical reaction, 2) ionization, 3) velocity change or trajectory deviation of selectively excited atoms. The photochemical approach has

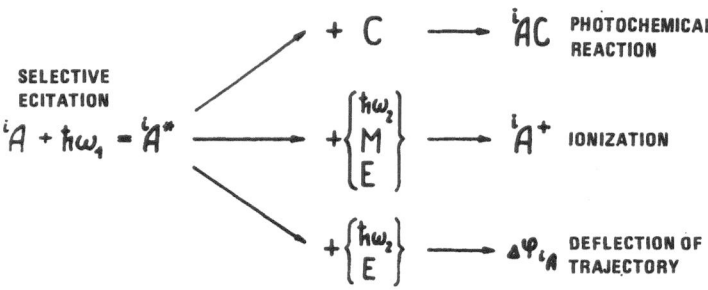

Fig. 2   Different selective atomic approaches

been known since the 1920-30's. The photoionization approach to atoms was proposed by the author as early as in 1969 and first realized in 1971 [2,3]. In the case of molecular gases, apart from the methods applied to atoms there are two new ones: the method of selective molecular photodissociation suggested by the author in 1969 and experimentally realized in 1971 [3,4], and molecular photoisomerization (Fig. 3). The method of selective

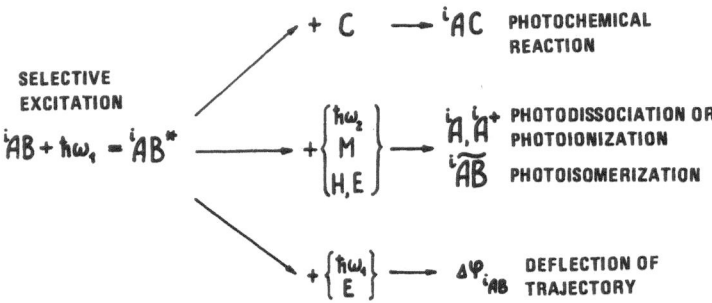

Fig. 3   Different selective molecular approaches

photodissociation has proved to be particularly efficient and has been accomplished in several different versions: two-step photodissociation by IR and UV radiation [3,4], one-step photopredissociation [5,6], collisionless and collisional dissociation by intense IR radiation [7], dissociation in a two-frequency IR field with separation of selective excitation and dissociation functions [8]. I wish to add that the technique of two-step selective molecular photoionization suggested quite a long time ago [9] has been realized just recently [10]. Selective molecular photoisomerization as well as photodissociation is possible not only under UV radiation [11] but also directly under an intense IR field [12].

Selective processes in <u>condensed media</u> are as yet imperfectly understood and seem to be less promising because of line broadening at normal temperatures and high-rate relaxation of vibrational excitation. This dictates certain stringent experimental conditions to selective action: 1) excitation of the electron states, the relaxation rate of which is lower than that of vibrational ones, 2) use of low temperatures at which the spectral line broadening and V-T relaxation rate drop drastically, 3) use of ultrashort pulses. For the time being we have only one example of an isotopically-selective photoprocess [13], but the case of condensed medium seems to be of prime importance in realizing selective photobiochemical processes [14].

In the case of a heterogeneous medium it is possible to act selectively both on the particles in the gas phase and on the atoms or molecules on the condensed medium surface. In the first case such processes as photochemical reactions, photoadsorption and condensation of excited atoms or molecules can be used. As the atoms or molecules on the condensed medium surface are excited, their selective breaking off, photodesorption and evaporation are possible. Selective breaking off of particular atoms or molecules from the surface [15] is a very important though the most hard-to-realize and undeveloped process. Selective breaking off of electrons or protons from certain parts of macromolecules may prove to be rather important for direct identification of the spatial-chemical structure of biological molecules [16].

3. Materials Technology at Atomic - Molecular Level

The methods of selective photophysics and photochemistry being developed to separate isotopes make it possible to elaborate a new approach to materials technology at atomic - molecular level, when by laser radiation one can directly manipulate atoms and molecules of a particular sort, that is, collect macroscopic amounts of a substance "by one atom, by one molecule". The most important process of laser atomic-molecular universal technology of materials is, no doubt, production of highly pure substances in atomic state, alloys and molecular compounds. The processes of selective atomic photoionization and selective molecular photodissociation may be used to produce highly pure substances or to purify a substance. Of course, the possibilities and fields of applications for both approaches differ greatly.

Selective atomic photoionization. This approach to materials technology is the most universal anf flexible. An optimal scheme for selective atomic photoionization under the action of two (or more, in principle) laser beams with properly tuned frequencies and chosen intensities enables every atom to be ionized within 10^{-5} to 10^{-8} sec. When 20% of the energy of the radiation with an average power of 10^3 W is used to photoionize atoms with an ionization potential $E_i \simeq 10$ eV, it is possible to ionize about one mole of substance per hour. Thus a comparatively moderate scale setup may ensure production of several tons of a pure substance per year. The method of selective atomic ionization in combination with tunable lasers, with average output from 100 to 1000 W, can therefore be considered as a sufficiently efficient method for fine substance separation at atomic level.

Laser purification of substances by selective ionization [17] must have a number of material advantages over the existing methods of purification on the basis of differences in any chemical or physical properties of a substance and its impurities:

1   High selectivity or high degree of purification in single-stop process. The degree of purification of a desired element from any admixtures may be higher than 10^3. This value depends on the process of exchange of charge during collisions between the ion of the given element and the neutral atom of the impurity. Basically, by decreasing the atomic density in the beam we may achieve a separation selectivity much higher than 10^3, with the efficiency reduced respectively. In particular, if we take a mass-production material with its purity of $10^{-7}\%$, it is possible to purify it up to $10^{-10}\%$ by the method of selective atomic ionization.

2   Universality. Selective ionization may be realized through proper selection of laser beam frequencies on any element, independent of their physical and chemical properties (melting and boiling temperatures, reactivity, etc.). When a substance is to be purified of one or more specified elements, it is possible to ionize selectively its impurities only and to remove them from the atomic beam of the substance. Under this regime a maximum efficiency of the method is achieved with a minimum of coherent light energy.

3   Flexibility which makes it possible to use ion beams directly to produce pure films or to implant ions into a homogeneous substance (ion implantation). Ion beams can be directed onto the substrate surface to produce a pure film of a specified element, as shown in Fig. 4. We think it is possible simultaneously to carry out independent selective ionization of two or three elements in different atomic beams and deposit these elements on the same surface. Thus, it will probably be possible to produce films of complex atomic compounds with their stoichiometric composition controlled by the intensity of photoion beams. The whole process of selective atomic ionization, extraction of ions from the atomic beam and their deposition on the substrate can be brought about in a high vacuum. It is not necessary for the process that the substance to be purified should make contact with any reagents or material apart from the substrate, for which we may always use a material without undesirable impurities.

We should concentrate our attention on the use of selectively formed photoions of boron, arsenic, phosphorus and other elements in setups for ionic implantation in semiconductors [18]. Electrodeless laser production of certain ions eliminates, first of all, the necessity of using an electromagnetic mass-separator and, secondly, makes it possible to insulate the high-temperature source of atoms from the ionizer. The latter is of no small importance since this enables the atoms to be photoionized near the

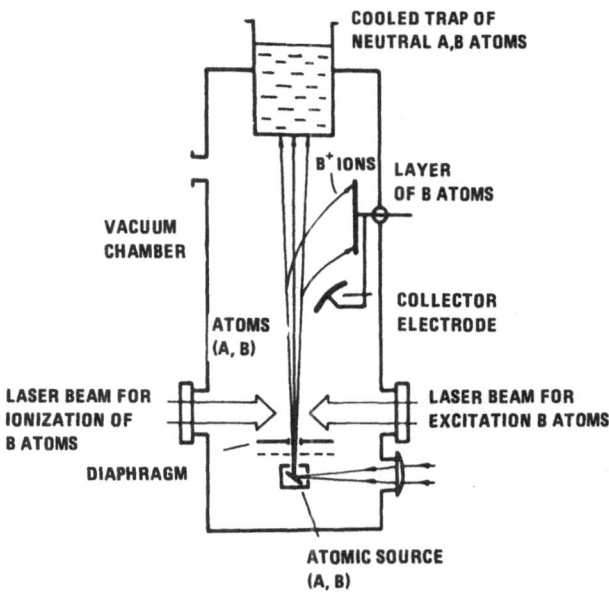

Fig. 4   Possible scheme of atomic purification by selective photoionization of atoms

high-voltage electrode and hence the construction of electrostatic ion
accelerators with an energy of about MeV and over to be substantially simplified.

The basis for successful development of the photoionization method is
elaboration of optimal schemes for multistep selective ionization of various
elements and also creation of rather efficient tunable lasers in the UV and visible range with a high average power and long lifetime. The multistep resonant excitation of the states near the ionization limit and subsequent auto-
ionization of highly excited atoms by a pulsed electric field suggested in
[19] is a universal and optimal ionization scheme providing a high excitation
cross-section and high ionization yield. The first experiments [20,21]
have proved the feasibility of this approach. As to lasers, the main difficulty we have when operating with UV lasers will be overcome probably by
the use of excimer lasers [22].

Selective molecular dissociation. The process may be used to purify
a substance in a gas phase of molecular impurities, the removal of which
by standard techniques is not efficient. Purification by the dissociation
method is based on differences in the physical - chemical properties of the
basic substance and the dissociation products. This enables us to use the
standard techniques of purification at the end of the process, after the mixture is irradiated.

The possibility of substance purification in a gas phase through dissociation of admixed molecules by intense IR radiation has been recently demonstrated experimentally [23] where arsenic trichloride ($AsCl_3$) was

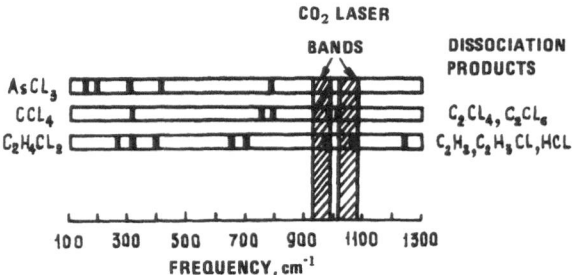

Fig. 5 Purification of $AsCl_3$ by selective multiple photon dissociation of impurities (CCl_4, $C_2H_4Cl_2$)

purified of 1,2-dichloroethane ($C_2H_4Cl_2$) and carbon tetrachloride (CCl_4). The minimum content of these impurities given by the standard techniques of purification is of the order $10^{-2} - 10^{-3}\%$. The absorption bands of the admixed molecules $C_2H_4Cl_2$ and CCl_4 fall within the oscillation region of a CO_2 laser where there are no absorption bands of the basic substance molecules $AsCl_3$ (Fig. 5). Therefore the effect of dissociation of polyatomic molecules in an intense field of CO_2 laser [7] can be used for selective dissociation. The final dissociation products were identified from IR absorption spectra (for $C_2H_4Cl_2$) and mass spectrum (for CCl_4). In experiments [23] selective dissociation of $C_2H_4Cl_2$ and CCl_4 admixed in $AsCl_3$ was clearly observed, the pressure of $AsCl_3$ being about 10 torr. The initial content of the admixed molecules was comparatively high, which was conditioned not by the limitations of the method but only by detection sensitivity. In case of 1,2-dichloroethane the final products differ greatly from $AsCl_3$ in their physical properties, and this enables them to be separated easily and $AsCl_3$ to be purified.

The method of selective molecular dissociation seems to be applied not only to technology of pure materials but also to removal of toxic and canceregeneous substances from gas mixtures, that is, to selective atmospheric photochemistry. If dissociation of such impurities converts them to inactive forms, the method becomes rather simple and independent.

4. Selective Laser Biochemistry

Selective action of laser radiation on complex molecules in a condensed medium is a promising possibility for molecular biology, which however has not been carefully studied and is therefore uncertain. In a condensed medium at normal temperatures the discrepancy between the requirements for excitation selectivity and conservation of selectivity becomes more aggravated. The electron transitions of biomolecules are grouped in the UV region, and a good excitation selectivity of particular molecules in a mixture can hardly be expected. On the other hand, electron excitation is conserved, as a rule, within a time ($\sim10^{-9}$ sec) that is sufficient for a photochemical reaction with an appreciable quantum yield [24]. Vibrational excitation of molecules is more selective but it relaxes to heat in a short time (at 300 $^{\circ}$K the relaxation time $T_1{}^{vib} \gtrsim 10^{-11}$ sec). Since the energy of one vibrational quantum is just a few times greater than the thermal energy kT, the contribution to the biochemical reaction rate by short-lived vibrational selective excitation cannot be as large as that by permanent thermal nonselective excitation.

Successful experiments [14] on isotopically-selective photoprocesses in gas media show that there are at least two ways of eliminating this discrepancy: 1) combination of selective vibrational excitation with subsequent electron excitation from vibration-excited states, that is, two-step IR - UV excitation, 2) multiple photon vibrational excitation in an intense IR field. In both cases the process, unlike the case of gas medium, should of course be realized under the action of picosecond laser pulses so that a molecule can absorb a considerable energy of several eV prior to thermal relaxation of vibrational excitation. The process of two-step IR - UV excitation, which offers additional advantages compared to vibrational or electron excitations, is believed to be especially universal. In each particular case, of course, we must first prove the potential feasibility of the two-step IR - UV selective photoprocess for a chosen molecule in the solution or for the molecular bond in a macromolecule. The scanty spectral information available on excited states, in particular for biomolecules, makes the answer to this question far from trivial. Below we shall consider several potentialities for different molecules and bonds. But besides this general question we have to deal with other "tricky" problems, i.e.: 1) heating of the medium during selective photoexcitation, 2) absorption of IR radiation by solvent molecules. These problems are particularly essential for experiments "in vivo".

The estimations in [14] show that it is possible to eliminate heating when the molar concentration of molecules in a solution is below 10^{-3} M and the absorption of IR radiation by solvent molecules is rather weak. To reduce the IR absorption by molecules in water, a typical solvent, in experiments "in vivo" one may try to excite the overtones and compound vibrations in the near IR region, even though the selective absorption cross-section is decreased. In the case of a homogeneous molecular solvent one can probably reduce the absorption when operating in the regime of "self-induced transparency" [25] for the solvent molecules. It is also possible to excite selectively the vibrational levels by the stimulated Raman process in the field of two-frequency visible laser radiation, which is quite transparent for the solvent, as has been done in experiments [26]. The estimations show that the required intensities of IR and UV pulses from 10^{-11} to 10^{-12} seconds in duration range between 10^{8} and 10^{9} W/cm^{2}. No doubt, cooling of molecules decreases the V-T relaxation rate and so makes all these difficulties less problematic.

Let us now consider some specific possibilities in laser selective biochemistry. They seem quite obvious "on paper" but may prove to be most difficult in experiments.

Selective excitation of DNA bases. Polynucleotide chain molecules DNA and RNA are important objects of investigation as far as both the possibility of selective laser mutations and their simple structure are concerned. Both molecules, despite their large dimensions, contain five repetitive nucleotides: guanine (G), cytosine (C), thymine (T), adenine (A) and uracil (U). All these five bases are purine and pyrimidine rings which have similar UV spectrum with two maxima. Table 1 presents the values of

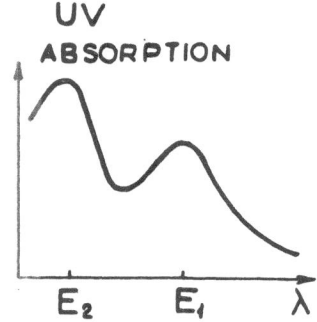

BASE	E_1(eV)	E_2(eV)
GUANINE	4.20	4.90
CYTOZINE	4.30	6.10
THYMINE	4.70	5.90
ADENINE	4.90	6.00
URACIL	5.10	6.00

Table 1   Energy excitation of the two first intense absorption UV bands
of nucleotide bases

energies for UV band maxima from [27]. We think that it is possible to
bring about selective electron excitation of the long-wave bands of the
bases of G and C which corresponds to excitation of the π-electron system
of the base pair "G-C" [28]. At the same time we cannot excite A- and
T-nucleotides without exciting the others.

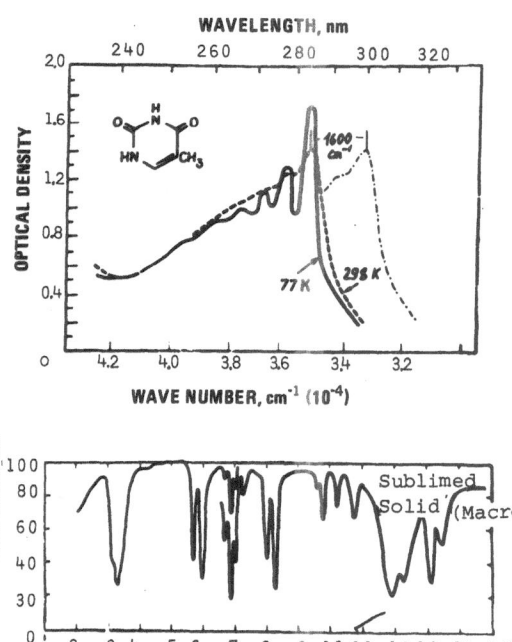

Fig. 6   UV and IR absorption spectra of thymine (from [24] and [29])

The difference in the molecular structure of these nucleotides gives us
hope of finding for each of them a specific vibrational band showing itself
in the UV absorption spectrum. Fig. 6 gives, for example, the UV and IR
spectra of thymine. It shows also a possible shift of the limit of UV absorption through excitation of the IR band in the region of 1600 cm^{-1} caused by
vibrations of the double bonds C=C and C=O in the ring. We wish to say
that the IR spectra of DNA are yet imperfectly understood [30], so we cannot specify now the frequencies for all the nucleotides. In our opinion, the
excitation of DNA nucleotide vibrations by picosecond IR tunable pulses
and simultaneous probing of changes in the UV spectrum may be a good
method to study and resolve the vibrational spectrum of DNA as well as a
necessary intermediate stage in studying the possibilities of selective action
on DNA bases.

Selective excitation of the double bond of lipid molecules is of basic
interest for studying the function and molecular structure of biological mem-
branes. In particular, selective excitation of the double bond - C = C- in
unsaturated fat acids would make it possible to investigate the energy migration along the lipid chain (photon tunneling along chain) and oxygenation by
an excited double bond (lipid photooxygenation). But natural lipids containing usually only isolated (unconjugated) double bonds have no absorption
bands in the near UV region, with the exception of a rather weak band in the
range between 200 and 210 nm [31]. On the other hand, isolated cis-double
bonds have a distinct band of IR absorption in the region of 2·14 mcm [31].
Therefore, if we provide selective absorption of two or three IR photons, we
may observe a band of UV absorption of a vibration-excited double bond in
the range from 250 to 270 nm. Similarly, the presence of IR absorption bands of methylene end groups within 1.62 and 2.10 mcm enables their
selective excitation by the scheme: "multistep selective IR excitation +
electron excitation".

Selective excitation and breaking of hydrogen bonds in DNA. The double
helix of DNA is formed by hydrogen bonds between the bases (guanine-cyto-
sine and adenine-thymine). The breaking of the hydrogen bonds must result
in splitting of the double helix into two single helixes and subsequent replication of DNA. Selective excitation of hydrogen bonds and their selective
breaking seem to be of interest for laser control over the process of DNA
replication. I think that this is essential not only as a potentiality of laser stimulation of a biological process rate but also as a basically new possibility
for an external controlled "start" of the DNA replication process, which has
not yet been studied in detail.

Two pairs of bases in DNA have slightly different hydrogen bonds. The
pair A-T is linked by two hydrogen bonds N-H...O, and an energy of about
7.0 kcal/mol is required to break this pair of bases. The pair G-C is linked
by two hydrogen bonds N-H...O and one bond N-H...N, and about 9.0 kcal/
mol must be consumed to break them [28].

An IR absorption band of about 1720 cm^{-1} corresponds to the bond G-C
in native DNA and a band of about 1700 cm^{-1} to the A-T bond. In DNA de-
naturation when the hydrogen bonds are broken, both bands disappear [30].
In [14] it was therefore proposed to act by picosecond powerful IR pulses
on these frequencies (5.814 mcm and 5.888 mcm) to stimulate hydrogen
bond breaking. There are, of course, many more possibilities for experiments "in vivo", especially with regard to the choice of more convenient
wavelengths not absorbed by the solvent.

132

The potential function of a hydrogen bond has two characteristic minima corresponding to two potential spatial and energy positions of a proton. (Fig. 7). Though not experimentally revealed, the energy levels of a proton in the hydrogen bonds N-H...O and N-H...N have been calculated. The transition of a proton (or rather of two protons in a pair of the next hydrogen bonds) to a higher energy tautomeric state is believed to result in mutations (Löwdin mechanism [32]). It is evident that using a laser at the wavelength of 1.8 mcm we may try to transfer a proton to an excited state and thereby stimulate its tunneling to a higher energy minimum. This possibility has been discussed in [28] and with the progress in picosecond tunable IR lasers must become a subject of experimental studies. Proton excitation must also show itself in the spectrum of electron UV absorption and thus can be used in two-step selective processes by the scheme: "selective IR excitation of a proton + UV excitation of an electron".

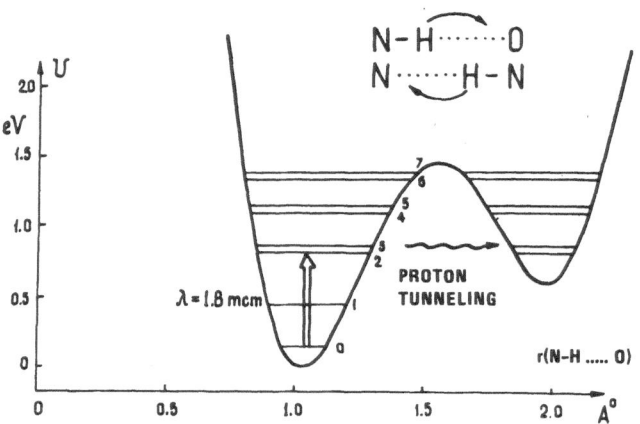

Fig. 7   Potential function of hydrogen bond in DNA

5. Selective Detection of Nuclei, Atoms, Molecules and Molecular Bonds

The methods of selective laser photophysics solve the problem of physical extraction of a particular atom or molecule from mixtures the chemical properties of which are very similar to those of other atoms and molecules. The primary and relatively simpler part of this problem is selective detection of single atoms and molecules. Selective two- (or multi-) step photoionization of atoms, molecules and molecular bonds is best suited to this purpose. Let us consider some of these potentialities.

Detection of excited nuclei. At present, excited (metastable) nuclei are being detected in the process of their radioactive decay. But the specific features of an excited nucleus affect not only nuclear transitions but also the hyperfine structure of optical transitions of the electron shell around the nucleus. As the isomeric structure usually considerably exceeds the Doppler broadening of spectral lines, we think, it is quite possible to selectively ionize not only nuclei of a particular isotopic composition but also excited nuclei with a specific nuclear spin and quadrupole moment. This possibility has been discussed as far as separation of isomeric nuclei and preparation of the active medium of the future γ-laser [34] are concerned. I would like to point here toward the possibility of selective detection of excited nuclei as a new approach to studying and searching for metastable nuclear levels. It is possible to define the quantum nuclear state from the electron shell of the nucleus without its deexcitation in the process of detection. After each selective extraction of an electron and its detection we may exchange the charge of the ion and thus repeat the whole process many times.

Detection of transuranium atoms. An active search for and studies of new heavy elements with atomic number over 100 and especially super-heavy ones in the region about 114 is being conducted at present [34]. The optical spectrum of an atom is its "passport" which ensures the uniqueness of detection and identification of a new element. Spectroscopy of electron transitions of single atoms by the method of selective photoionization must give new possibilities in this field (monoatomic laser spectroscopy [35].

Detection of complex molecules. As known, selective detection of trace amounts of polyatomic complex molecules is a very difficult problem as yet not solved by physical methods. Mass-spectral analysis is now a common method for detection and identification of complex molecules but its sensitivity is inadequate and there is practically no selectivity of detection for complex molecules which differ only in spatial structure. Therefore the development of new methods to solve this problem is today very urgent.

The method of selective molecular ionization by laser radiation may be used as the basis for the so-called laser mass-spectrometer [3, 36] illustrated schematically in Fig. 8. A laser with the tunable frequency ω_1 excites selectively the vibrational (and electronic for some molecules) state of molecules. Due to such an excitation the edge of the molecular

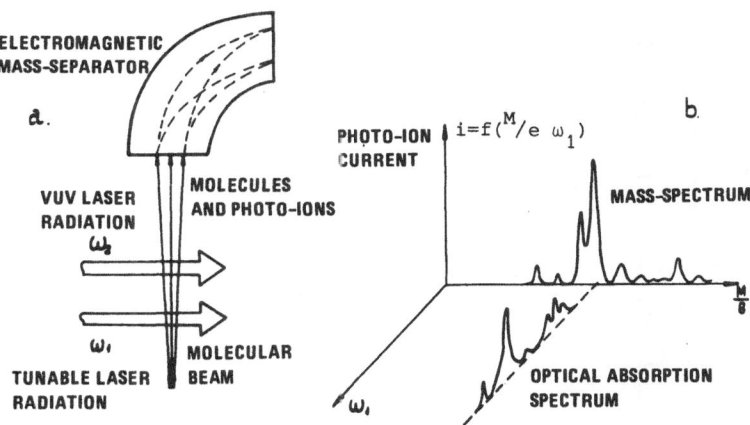

Fig. 8   Possible laser optical-mass spectrometer

photoionization band lying usually in the VUV region is shifted by a small magnitude. The second VUV laser brings a molecular photoionization, and its frequency is selected at the point of a maximal slope of the photoionization band edge. The preliminary selective excitation of the molecules by the tunable laser even by a comparatively small value $E_{exc} \simeq 0.1 - 0.5$ eV results in a marked change of the photoionization cross-section ($10^{-1} - 10^{-2}\%$ with regard to the molecular distribution over the rotational states), that is, a change in the photocurrent. Photoions are then sent into a common mass-spectrometer which measures the mass spectrum, i.e. $i = f(M/e)$. In addition to this, with this version of mass-spectrometer we can measure the photocurrent amplitude for a fixed value of M/e as a function of the tunable laser frequency ω_1. In this case the IR spectrum of trace amounts of complex molecules is measured since, with the ω_1 frequency and that of molecular absorption coincident, the molecules pass into the excited state and hence the ionic photocurrent amplitude varies. Thus the laser mass-spectrometer with selective molecular ionization, instead of the usual nonselective ionization by an electron beam (or by continuous VUV radiation), will produce at the same time the optical (IR and visible) absorption spectrum and the mass-spectrum. This method enables one to obtain information on the spatial structure of molecules with the same mass, etc.

We began systematic studies of this method of molecular detection after the simple VUV H_2-laser had been elaborated. First we carried out one-step photoionization of dimethylaniline and methylaniline molecules by H_2-laser radiation in the region of 1600 Å [38] as well as NO molecules within 1200 Å. Recently we have realized two-step photoionization of a H_2CO molecule by joint action of the radiation pulses of N_2-laser with $\lambda_1 = 3371$ Å exciting the states $^1A_2$ and of H_2-laser at $\lambda_2 = 1600$ Å photoionizing the excited molecules (Fig. 9a). The ionization potential of a H_2CO molecule is $E_i = 10.87$ eV, and the total energy of two laser quanta $\hbar\omega_1 + \hbar\omega_2 = 3.7 + 7.7$ eV $= 11.4$ eV, which is quite enough to photoionize this molecule. The time delay of a H_2-laser pulse (its duration is shorter than 1 ns) is varied with respect to the N_2-laser pulse (its duration about 2 ns), and this enabled us to measure the dependence of photoion yield on delay time (Fig. 9b). The

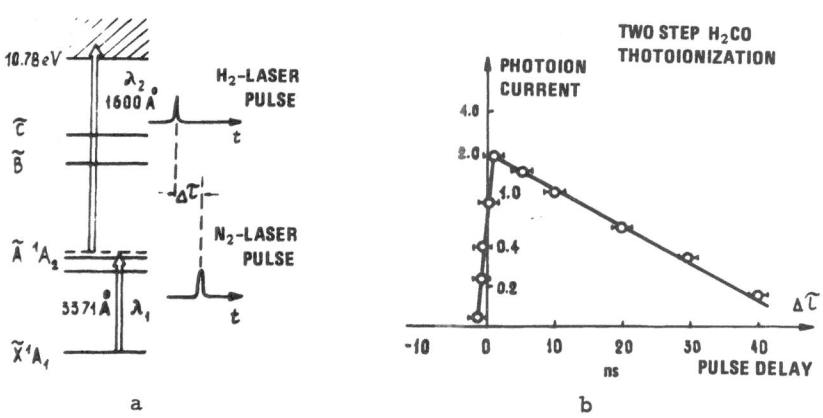

Fig. 9 Two-step photoionization of H_2CO molecule

experimental curve is similar to the exponential one with the decay constant 15 ± 2 ns, which equals the lifetime of a H_2 CO molecule in the excited $^1A_2$-state. During the reverse order of succession of N_2- and H_2-laser pulses there is no signal indicating H_2CO ionization.

At present we are experimenting on molecular beams with a tunable laser. In particular, we have lately managed [39] to realize one-step photo-ionization of dimethylaniline and other molecules in a molecular beam of the mass-spectrometer by radiation of H_2-laser at 1600 Å which increases the amplitude of photoionic current by several orders. Analogous experiments are being carried out with preliminary selective excitation of molecular electron states by a tunable dye laser in the visible and UV range. For selective detection of complex molecules we shall subsequently study two-step molecular ionization with excitation of the intermediate vibrational states by a tunable IR laser.

The infrared - mass - spectrometer may become a universal highly selective and sensitive detector of complex molecules in solving a large number of scientific and applied problems. Indeed, since detection of excited molecules through their photoionization is highly sensitive, we may hope to measure the IR spectrum of extremely small quantities of substances, much smaller than in the best classical and laser IR spectrometers existing now. It should be said here that it is possible at the same time to realize an extremely high spectral resolution which is determined, in principle, only by the residual Doppler broadening because of angular divergence of the molecular beam. We think that here we are approaching the solution of the principal problem in developing physical methods to detect and identify trace quantities of organic admixtures with the perfection that can be attained by the organs of smell of human beings and animals.

Spatial localization of molecular bonds. Selective action of laser radiation on molecular bonds opens up a principal possibility for spatial localization of particular molecular bonds, that is, for macromolecular "mapping". The idea of this approach can be understood from the so-called photoelectronic (photoionic) laser microscope [16] which is schematically shown in Fig. 10. Unlike the common field-electron of field-ion Müller projectors [40], an electron or an ion breaks away selectively here from the molecule under the selective action of laser radiation rather than because of non-selective field-induced ionization by a strong electric field. The only function left to the electric dc field is to transport electrons or ions along radial paths to the projector screen. Selective photoionization of certain molecular bonds in a macromolecule, being on the point of the projector, can be done by the multi-step scheme under the action of several picosecond laser pulses at special frequencies.

In case of selective breakaway of an electron it is possible to attain a resolution of about 25 Å which is limited by such fundamental causes as the presence of a tangential velocity component of emitted electron and the principle of uncertainties. In some cases after electron breakaway the resultant positive molecular ion becomes unstable and spontaneously gives off the proton [41]. If we change the polarity in the projector, it is possible to transport electrons instead of protons to the screen. The spatial localization of ionization point for protons must be $(m_p/m_e)^{\frac{1}{2}} \simeq 40$ times higher than for photoelectrons. Such a laser photoionic microscope may have a resolution quite sufficient to resolve atomic details in molecular structures. Further increase in the resolution of laser photoemission microscopy is

136

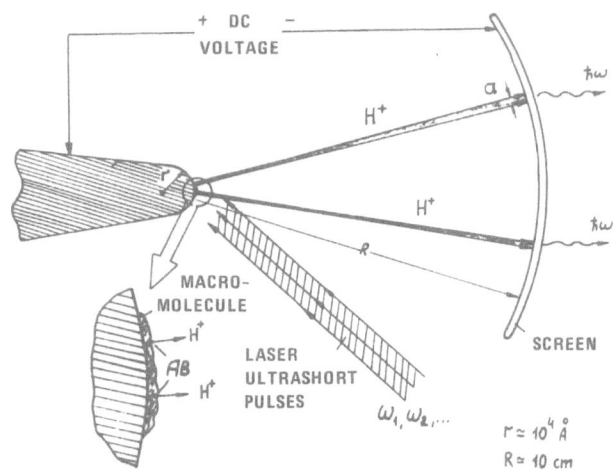

Fig. 10 Possible laser - ion microscope for spatial localization of molecular bonds

possible also due to selective photoionization of a molecule to heavier molecular ions.

The idea of this new approach to atomic-resolution microscopy is based on the combination of two important properties which usually belong to quite different methods. Corpuscular (electronic, for instance) microscopy ensures a high spatial resolution if the particle has a high energy. But in this case the "contrast" of image is lost automatically because a variation in the particle energy due to its interaction with an object being observed is difficult to distinguish against the background of the higher initial energy. Vice versa, as the particle energy drops, the image "contrast" may increase considerably and spectral selectivity becomes possible in observing the image details. But in this case the spatial resolution decreases. The method of selective photoionization makes it possible to combine the high selectivity (or contrast) of "optical channel" with the high spatial resolution of "corpuscular (ionic) channel".

Determination of the nucleotide sequence in a DNA molecule which carries hereditary information on every individual organism is one of the most important applications of the laser ion microscope. This problem consists in selective break-away of a proton (or a heavier molecular ion) either from the base pair A-T or G-C. Of course, we may try to realize the method based on two-step selective IR-UV excitation of an electron state through intermediate vibrational state of a particular nucleotide (see § 4) and subsequent ionization of the electron-excited state. For good distinction between the pair of bonds A-T and G-C it is essential that the selectivity of this process is adequate. Particular attention should be drawn to the principal possibility of preliminary chemical "marking" of either base pair to make selective excitation and ionization easier. For instance, it is known [42] that one of the bases of DNA (custosine) selectively reacts with hydroxylamine. The reaction is followed by changes in the UV spectrum of cytosine absorption the value of which is much larger than that of the typical difference between the spectra of the bases (see Table 1). In this case selective two-step photoionization through electron-excited state of the product of cytosine-hydroxylamine reaction may be enough.

When the magnification coefficient of the laser ionic projector $M = 3 \cdot 10^5$ and the screen dimension $R = 10$ cm, there will be imaged a section of linear DNA chain with about 10^3 nucleotides (adjacent nucleotides are spaced 3.3 Å apart), i.e. a length of about 3300 Å. To record the sequence of nucleotides in long DNA chains, we shall need, of course, successive projection and "sewing" the images of subsequent sections.

For successful realization of the laser ionic microscope in observing biological molecules we shall have to investigate a number of rather difficult problems: 1) selective excitation of macromolecules adsorbed on surface, 2) search for schemes of selective dissociative ionization resulting in breakaway of protons of heavier molecular ions, 3) spatial scanning of microscope needle point along the chain of a macromolecule, etc. The solution of the problem of direct observation of nucleotide sequence in genes, including those of man, would open up such large possibilities in heredity study and control that this field of selective laser photophysics seems to me to be worthy of our closest attention.

In conclusion I would like to express gratitude to my colleagues from the laboratories of laser spectroscopy and excited state spectroscopy for their joint experimental elaboration of the associated problems of selective laser photophysics and photochemistry.

References

1. V.S. Letokhov, C.B. Moore: Kvantovaya Elektronika (Russian) 3, 248, 485 (1976); Preprint LBL-4904 (English), Lawrence Berkeley Laboratory (1976)

2. R.V. Ambartzumian, V.P. Kalinin, V.S. Letokhov: Pis'ma Zh. Eksper. i Teor. Fiz. 13, 305 (1971)

3. R.V. Ambartzumian, V.S. Letokhov: IEEE Journ. Quant. Electr. QE-7, 305 (1971); Appl. Optics 11, 354 (1972)

4. R.V. Ambartzumian, V.S. Letokhov, G.N. Makarov, A.A. Puretzkii: Pis'ma Zh. Eksper. i Teor. Fiz. 15, 709 (1972); 17, 91 (1973)

5. V.S. Letokhov: Chem. Phys. Lett. 15, 221 (1972)

6. E.S. Yeung, C.B. Moore: Appl.Phys. Lett. 21, 109 (1972)

7. R.V. Ambartzumian, V.S. Letokhov, E.A. Ryabov, N.V. Chekalin: Pis'ma Zh. Eksper. i Teor. Fiz. 20, 597 (1974)

8. R.V. Ambartzumian, Yu.A. Gorokhov, V.S. Letokhov, G.N. Makarov, A.A. Puretzkii, N.P. Furzikov: Pis'ma Zh. Eksper. i Teor. Fiz. 23, 217 (1976)

9. J. Robieux, J.-M. Auclair: French Patent, N 1391738 (1965); US Patent, N 3 443 087 (1969)

10. S.V. Andreev, V.S. Antonov, I.N. Knyazev, V.S. Letokhov: Chem. Phys. Lett. (in press)

11. J. I. Brauman, T. J. O'Leary, A. S. Schawlow: Optics Comm. <u>12</u>, 223 (1974)

12. R. V. Ambartzumian, N. V. Chekalin, V. S. Doljikov, V. S. Letokhov, V. N. Lokhman: Optics Comm. (in press)

13. D. S. King, R. M. Hochstrasser: Report on Second Intern. Conf. on Laser Spectroscopy, 23 - 27 June, 1-95, Megeve, France

14. V. S. Letokhov: Journ. Photochem. <u>4</u>, 185 (1975)

15. R. V. Khokhlov: <u>Tunable Lasers and Applications</u> (Springer Verlag, Berlin Heidelberg - New York, 1976)

16. V. S. Letokhov: Kvantovaya Elektronika (Russian) <u>2</u>, 930 (1975); Phys. Lett. <u>51A</u>, 231 (1975)

17. V. S. Letokhov: Spectrocopy Letters <u>8</u>, 697 (1975)

18. V. S. Letokhov, V. I. Mishin, A. A. Puretzkii: In "Chemistry of Plasma" (Russian), Atomizdat, Moscow (in press)

19. L. N. Ivanov, V. S. Letokhov: Kvantovaya Elektronika <u>2</u>, 585 (1975)

20. R. V. Ambartzumian, G. I. Bekov, V. S. Letokhov, V. I. Mishin: Pis'ma Zh. Eksper. i Teor. Fiz. <u>21</u>, 595 (1975)

21. T. W. Ducas, M. G. Littman, R. R. Freeman, D. Kleppner: Phys. Rev. Lett. <u>35</u>, 368 (1975)

22. M. L. Bhanmik: Laser Focus <u>12</u>, 54 (1976)

23. R. V. Ambartzumian, Yu. A. Gorokhov, S. L. Grigorovich, V. S. Letokhov, G. N. Makarov, Yu. A. Malinin, A. A. Puretzkii, E. P. Filippov, M. P. Furzikov: Kvantovaya Elektronika (in press)

24. K. C. Smith, P. C. Hanawalt: <u>Molecular Photobiology</u> (Academic Press, New York and London 1969)

25. S. L. McCall, E. L. Hahn: Phys. Rev. Lett. <u>18</u>, 908 (1967); Phys. Rev. <u>183</u>, 457 (1969)

26. A. Laubereau, D. von der Linde, W. Kaiser: Phys. Rev. Lett. <u>28</u>, 1162 (1972)

27. L. B. Clark, I. Tinoco: J. Amer. Chem. Soc. <u>87</u>, 11 (1965)

28. Janos Ladik: <u>Quantenbiochemie für Chemiker und Biologen</u> (Akademiai Kiado, Budapest, 1972)

29. E. R. Blout, G. R. Bird, D. S. Grey: J. Opt. Soc. Am. <u>40</u>, 304 (1950)

30. H. Susi: In "<u>Structure and Stability of Biological Macromolecules</u>", Ed. S. N. Timasheff and G. D. Fasman (Marcel Dekker Inc., New York, 1969)

31. J. Hanna: In "<u>Instrumental Methods of Organic Functional Analysis</u>", Ed. S. Siggia (Wiley-Interscience, New York, 1972)

32. P. O. Löwdin: Adv. Quant. Chem. <u>2</u>, 213 (1968): In "<u>Electronics Aspects of Biochemistry</u>", Ed. B. Pullman (Academic Press, New York, 1964) p. 167

33. V. S. Letokhov: Optics Comm. <u>7</u>, 59 (1973)

34. G. N. Flerov, V. A. Druin, A. A. Pleve: Uspekhi Fiz. Nauk (Russian) <u>100</u>, 45 (1970)

35. V. S. Letokhov: Comments on Atomic and Molecular Physics (to be published)

36. V. S. Letokhov: Uspekhi Fiz. Nauk (Russian) 116, 199 (1976)

37. I. N. Knyazev, V. S. Letokhov, V. G. Movshev: IEEE Journ. Quant. Electr. 11, 805 (1975)

38. S. V. Andreev, V. S. Antonov, I. N. Knyazev, V. S. Letokhov, V. G. Movshev: Phys. Lett. 54A, 91 (1975)

39. I. N. Knyazev, V. S. Letokhov, V. G. Movshev, V. K. Potapov (in press)

40. E. W. Müller, Tien Tzon Tsong: Field Ion Microscopy (Amer. Elsveir Publ. Co., Inc., New York, 1969)

41. A. N. Terenin: Photonics of Dye Molecules and Allied Organic Compounds (Russian), (Nauka, Leningrad, 1967)

42. D. W. Werwoerd, H. Kohlhage, W. Zillig: Nature 192, 1038 (1961)

URANIUM ISOTOPE SEPARATION
AND ITS DEMAND ON LASER DEVELOPMENT

Prepared by Stephen Rockwood,
Work performed by Staff of Project JUMPer,
Los Alamos Scientific Laboratories, USA

The economic gains to be realized through a more efficient uranium isotope separation process have been described many times in the last few years and should be common knowledge to all members of this meeting. Similarly the demand for enriched uranium as a fuel source has been detailed and because of the magnitude of this demand a large through-put of natural material is required in any process. For this reason our research efforts are naturally directed towards implementation of processes employing uranium compounds for these highly bonded forms will have the highest volatility. Of all currently known uranium compounds the molecule UF_6 exhibits the highest vapor pressure and is an obvious candidate for study in regard to a uranium LIS plan.

For molecular forms the largest isotope shifts occur in the ir active absorption features involving motion of the atom of interest. For the case of UF_6 this focuses attention on the ν_3 and ν_4 modes which absorb at 16 and 53 μm, respectively. However, because of the large number of low frequency modes which may be thermally excited very little isotopic resolution may be obtained in the ir at temperatures where the vapor pressure is adequate to allow the required through-put of feed material. Thus nonequilibrium methods of cooling UF_6 must be exploited to achieve spectral simplification in the infrared and to uncover specific isotope effects.

The temperature required for significant spectral simplification can be found by examination of the statistical population distribution of vibrational energy in UF_6 displayed in Fig. 1 as a function of temperature. Notice that temperatures of less than 95°K must be reached before greater than 50% of the UF_6 resides in the vibrational ground state. The nonequilibrium cooling can be achieved by expansion of a UF_6/carrier gas mixture through a continuum flow supersonic nozzle as shown in Fig. 2. The carrier gas may be either a molecular form such as N_2 or an atomic form such as He depending on the $\gamma = C_p/C_v$ required for the mixture to achieve the desired final temperature,

$$T = T_0 \, (P/P_0)^{\frac{\gamma - 1}{\gamma}}.$$ The resulting cold gas moving at high velocity presents a unique spectroscopic medium where the vibrational, rotational, and translational degrees of freedom are in near equilibrium but the gaseous UF_6 is in a highly supersaturated state with respect to its solid/liquid/gas equilibrium.

Figure 3 displays the measured absorption spectrum of expansion cooled

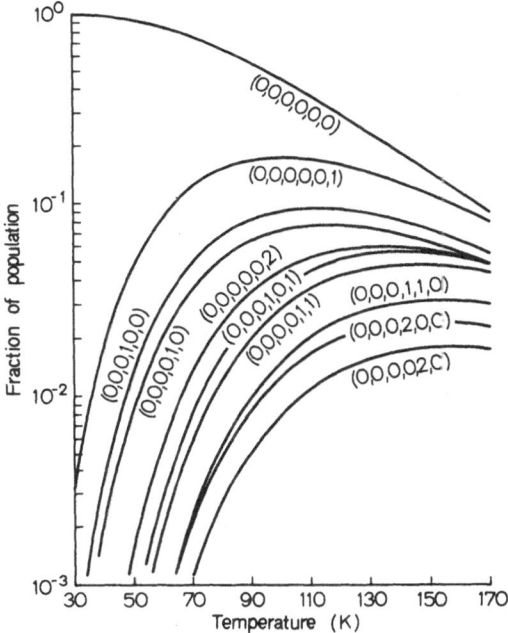

Fig. 1: Fraction of population in
the 10 lowest-lying vibrational states of UF$_6$ as a
function of absolute
temperature.

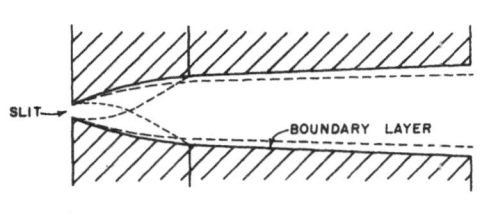

TWO DIMENSIONAL EXPANSION

NOZZLE

Fig. 2: Side view of supersonic gas
nozzle.

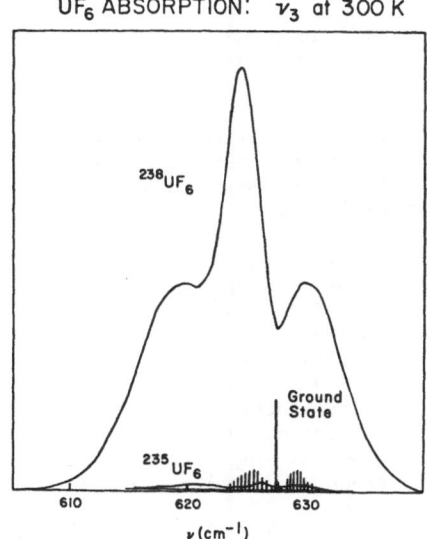

Fig. 3: Spectrum of ν_3 band of
^{238}UF$_6$ at room temperature compared with room
temperature ^{235}UF$_6$ ν_3 band
and flow cooled ^{238}UF$_6$.

UF$_6$ relative to the room temperature spectrum. Observe that the ν_3 Q-branch is significantly enhanced and shifted towards the blue as a consequence of the removal of vibrationally excited molecules. For similar reasons, the isotope shift between the ^{238}U$_6$ and ^{235}UF$_6$ is readily resolved; however, also note that the ^{235}UF$_6$ Q-branch resides among R-branch lines of the ^{238}UF$_6$, hence higher resolution spectroscopy must be accomplished before the actual isotopic resolution achievable with laser excitation can be known.

The development of tunable diode lasers at the MIT Lincoln Lab which operated in the vacinity of 16 μm permitted this question to be answered in early 1974. The results of this diode spectroscopy effort are shown in Fig. 4. In this scan of flow cooled UF$_6$ the ^{235}UF$_6$ Q-branch (peak 3) can be ob-

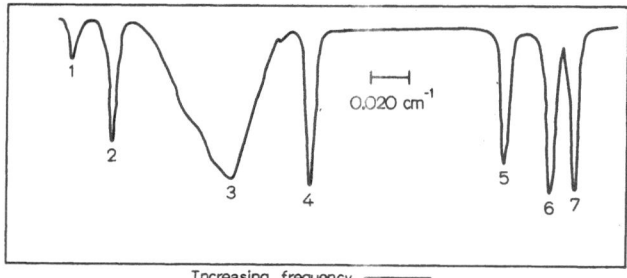

Fig. 4: Composite spectrum of flow cooled UF$_6$ and reference gases taken with a tunable diode laser. Peak 3 is ^{235}UF$_6$ Q-branch.

served to reside nicely between two R-branch lines of ^{238}UF$_6$. Thus very high discrimination is possible with the appropriate narrow band source. The bandwidth required of the ir source is observed to be on the order of .03 cm^{-1} with a frequency precision of $\sim 10^{-5}$ at 16 μm.

The application of tunable laser spectroscopy to polyatomic molecules such as SF$_6$ and UF$_6$ have revealed additional fine structure which previously only existed in theory as higher order terms in the vibration-rotation Hamiltonian. An example of this is shown in slide (5). Shown here are the P(17), P(18), and P(19) lines of SF$_6$ which under normal resolution would appear as a single feature but when observed with resolution comparable to laser line widths exhibits considerable structure due to the coriolis interaction. The correct analysis of this splitting has required a reformulation of the spin statistical weights of various symmetry species in the octahedral group as described in Ref. (1,2).

In fact, through analysis of high resolution spectra taken on flow cooled SF$_6$ it has been possible to establish the ground state molecular constants to sufficient precision to permit assignments of virtually all the major features observed by HINKLEY at room temperature.[3] This revolution in spectroscopy has proceeded as a natural partner to LIS because if the excitation is to be performed with the spectral resolution of a laser, then the details of the absorption mechanism must also be known with the same resolution. The present exception to this is light isotope enrichment through

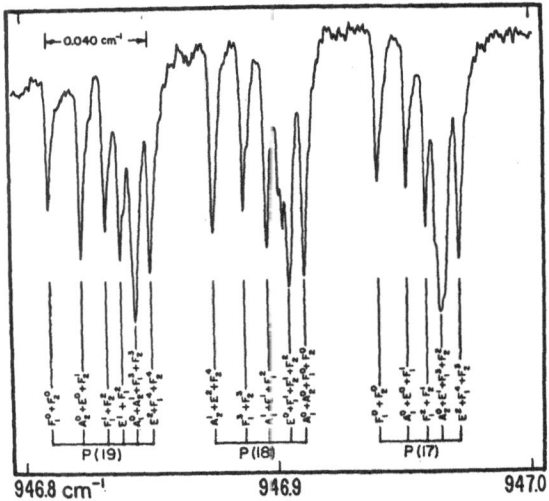

Fig. 5: High resolution spectrum of SF_6 P(17), P(18), and P(19) lines
taken with tunable diode laser.

multiple photon dissociation of polyatomic molecules.[4,5] In this case the
perturbation of the vibrational levels induced by the high intensity laser
fields (E $\gtrsim 10^6$ V/cm) will mask much of the fine structure in the absorption
features and is a contributing factor in compensating anharmonic effects.

The preceeding material has indicated a portion of the technology and
spectroscopic knowledge which now exists. These capabilities now make it
possible to perform LIS in molecular systems; however, the dominating factor
is still economics, hence most attention should be focused toward the development of reliable and efficient lasers systems.

At present it is not possible to tune through the spectral range (2.5 -
20 μm) occupied by molecular vibrations with the power and bandwidth offered
by dye lasers in the visible and near ir. Thus research on tunable ir laser
systems currently represents a very active field. The versatile system
being developed at Stanford University has been described by DR. BYER at the
Megeve conference last year.[6] For the duration of this talk I will describe the work on tunable ir systems being pursued both at LASL and through
contractual arrangements with other laser laboratories in the U.S. Because
of the committment to uranium enrichment virtually all of this research is
presently focused on the four wavelength regions of 7.74, 8.62, 12.1 and
15.9 μm. However, it is our hope that while meeting the present requirements we will also develop a set of laser systems which will give continuous
tunability throughout much of the middle and far ir. Our goal for the future is to explore isotope separation of all the elements as well as specific laser chemistry problems. Such laser tools are necessary prerequisites
for those studies.

The next figure, Fig. 6, presents in tabular form an overview of the
present research effort directed towards providing lasers for UF_6 excitation.

144

LASER RESEARCH FOR ISOTOPE SEPARATION

	EXPERIMENT	INVESTIGATOR	LOCATION
(I) 16 μM	(A) CO-CO_2 MIXING	BARNES/PILTCH	LASL
	(B) HF-OPO	ARNOLD/WENZEL	LASL
	(C) CO_2 BENDING MODES	OSGOOD	MIT LL
		MANUCCIA/STREGACK	NRL
		JONES/BUCHWALD	LASL
		NELSON/BYRON	MSNW
	(D) ELECTRONIC SRS	SZE	LASL
	(E) DOUBLED $^{14}CO_2$-CO_2	BARNES	LASL
	(F) MIXING IN CdTe	PILTCH	LASL
	(G) SRS OF CO_2	COLEMAN	U. OF ILL
(II) 12 μM	(A) $^{14}CO_2$	TELLE/BARNES	LASL
	(B) $^{13}CS_2$	NELSEN/BYRON	MSNW
		O'NEIL	NRL
		JONES	LASL
	(C) 4 WAVE MIXING	KURNIT	LASL
		ABRAMS	HUGHES
(III) 8.6 μM	(A) VIBRATIONAL SRS	LOREE	LASL
	(B) $O^{13}CS$	KILDAL	MIT LL
	(C) C_2HD	TENNANT	LASL
	(D) HCN	WITTIG	USC
(IV) 7.7 μM	(A) H_2O E-V TRANSFER	WITTIG	USC
	(B) HCN	WITTIG	USC
		FISHER/BYRON	MSNW

UV LASERS

(I)	KrF E-BEAM	MACE/WEINBRECHT/BLAIR	LASL
	KrF DISCHARGES	BLAIR/SZE/BIGIO/BEGLEY	LASL
	KrF DISCHARGES	JAVAN	MIT
	Hg_2 DISCHARGES	QUIGLEY	LASL
	CO	DEWEY	MIT
	DOUBLED DYE LASERS	BALOG/BUTCHER/SHERMAN	LASL

Fig. 6

The efforts have been grouped according to the particular wavelength region of current research, however several of the methods such as optical pumping and 4-wave mixing in gases are quite general in their application. There is insufficient time in this talk to review each of these projects individually; however, additional information is available in the proceedings of the joint LASL/ERDA conference held in April of this year and of course any of the individual investigators would be delighted to describe their own work. At this time I will discuss briefly the results of three individual systems under study at LASL.

The first is the generation of line tunable radiation at 15.9 µm through difference frequency generation of pulsed CO and CO_2 lasers. This particular approach has been pursued jointly at LASL and MIT Lincoln Labs for about two years.[7,8] Research on this approach has now been terminated and the results are summarized on the next figure, Fig. 7. The CO laser was constructed at MSNW and uses a plasma diode electron beam to sustain a discharge

---

SUMMARY OF MIXING RESULTS AND PROBLEMS

I.   HAVE ACHIEVED 100 µJ/PULSE AT THE REQUIRED WAVELENGTH.

II. A. THE CO LASER IS AN INHERENTLY LONG PULSE LENGTH DEVICE.

     B. THE TEMPORAL OVERLAP WITH THE CO_2 LASER IS QUITE SMALL.

III. A. THE MATERIALS NEEDED HAVE LARGE ABSORPTION COEFFICIENTS AND SCALING TO
         HIGH AVERAGE POWERS APPEARS DIFFICULT.

Fig. 7

in a 580 torr (a local Los Alamos atmosphere) mixture of 85% Ar, 10% CO and 5% N_2. The output is grating tuned to a line at 5.8 µm with an energy of 0.3 joules and a peak power of 10 kW. The CO_2 source is a Lumonics Model 103 modified for operation in a single longitudinal and transverse mode. It is grating tuned to a line near 9.2 µ and delivers 0.2 joules per pulse with a peak power of 2 MW. The difference frequency is generated in a crystal of $CdGeAs_2$ and 100 µJ per 100 ns pulse is obtained at 15.9 µm. The entire system operates at 1 pps.

This mixing system is now being replaced by a parametric oscillator pumped by an HF laser[9] as shown in the next figure, Fig. 8. The oscillator is operated on all lines and initiated prior to the amplifier in order to surpress parasitic oscillations in the amplifier. The OPO is pumped by the HF $P_2(7)$ line with external frequency selection achieved with the grating and spatial filters, as shown. Radiation is coupled into the CdSe crystal through a grating arrangement as shown in the next figure, Fig. 9. The pump radiation makes two passes through the medium while the 3.25 µm signal is in resonance between the grating and back mirror. Idler radiation at 15.9 µm is coupled specularly from the grating as shown. With this system an energy of 1 mJ per 100 ns pulse has been obtained with a bandwidth of ~ 0.3 cm^{-1} and a tuning curve as shown in the next figure, Fig. 10. This represents a very

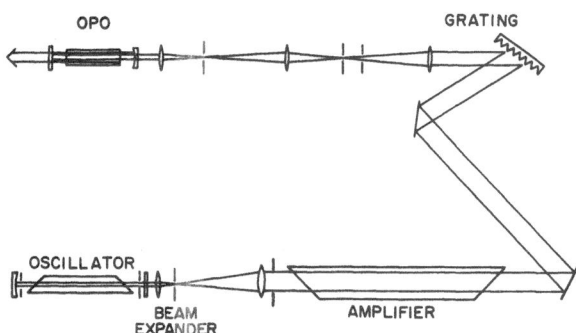

Fig. 8: Schematic diagram of HF oscillator-amplifier combination for pumping an optical parametric oscillator.

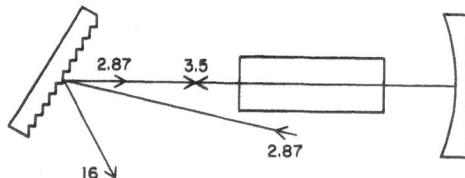

Fig. 9: Schematic diagram of optical parametric oscillator cavity using a CdSe crystal.

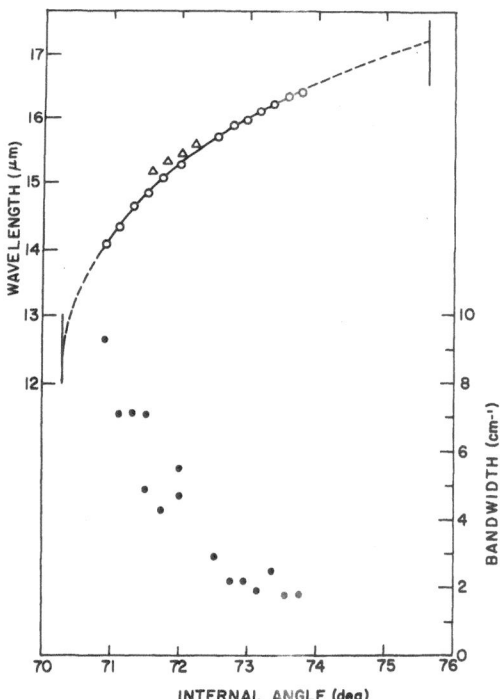

Fig. 10: Measured tuning curve and bandwidth of free running optical parametric oscillator pumped by HF $P_2(7)$ line.

HF OPTICAL PUMPING EXPERIMENT

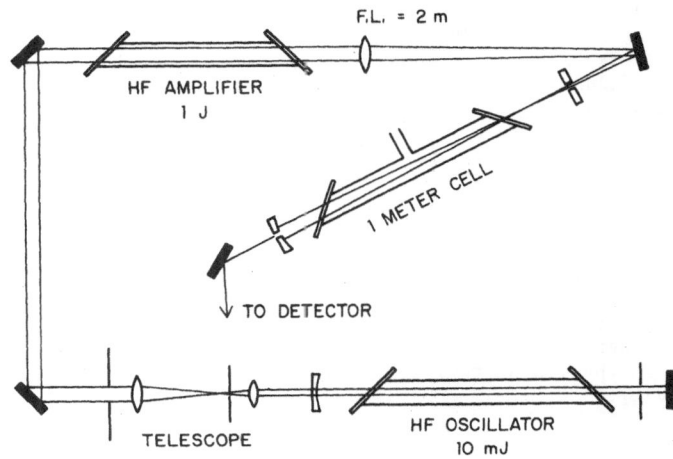

Fig. 11: Schematic diagram of HF oscillator-amplifier and optical pumped laser cavity.

LASING SYSTEMS

SAMPLE	PRESSURE RANGE TORR	LASING WAVELENGTHS μM
HF/$^{12}C^{16}O_2$	NEAR 1 TORR	10.6
$^{12}C^{16}O_2$	7 - 120	10.6
	0.5 - 10	4.3
$^{12}C^{16}O^{18}O$	2 - 50	10.6
AND $^{12}C^{18}O_2$	0.1 - 20	4.30-4.40, 12 LINES
MIXTURE	0.1 - 2	17.48, 16.75, 6 WEAKER LINES
$^{13}CO_2$	0.5 - 10	4.6
	10 - 100	11.3
$^{13}CS_2$	1.5 - 6	7.0 ONE LINE

Fig. 12

versatile far ir source and is now finding application in a number of experiments at Los Alamos.

While this system is very useful for nearterm experiments it does not have the potential for scaling to the high average powers which would be required in a commercial process. This requirement would best be met by a

148

gas laser system. In the remaining time I would like to describe experimental results from one gas phase laser resulting from the optical pumping of molecules by an HF laser. These experiments have been conducted jointly by Drs. R. JONES and M. BUCHWALD at LASL and Dr. H. FETTERMAN at MIT Lincoln Lab and Dr. H. SCHLOSSBERG at AFCRL. The schematic of the LASL experimental apparatus is shown in Fig. 11. The HF oscillator-amplifier combination is very similar to the one described previously for the parametric oscillator system. However, in this case all of the HF lines are transmitted into the pumped medium. Figure 12 summarizes the gas and lasing wavelengths observed to date. Our interest in the isotopic forms of CO_2 is to generate a very specific 16 µm wavelength transition. The next figure, Fig. 13, details some of the lasing transitions observed in $^{12}C^{18}O_2$. The mechanism of population inversion is direct absorption to the $02°1$ level from which lasing at 4.3 µm and 16 µm originate as indicated. Following the 4.3 µm emission optical cascading from the $02°0$ to the $01^1 0$ level results in further 16 µm emission. The details of these mechanisms have been supported by measurements of the tuning delay between various transitions as well as suppressing oscillation in the 4.3 µm region by placing a high pressure cell of HBr inside the cavity.

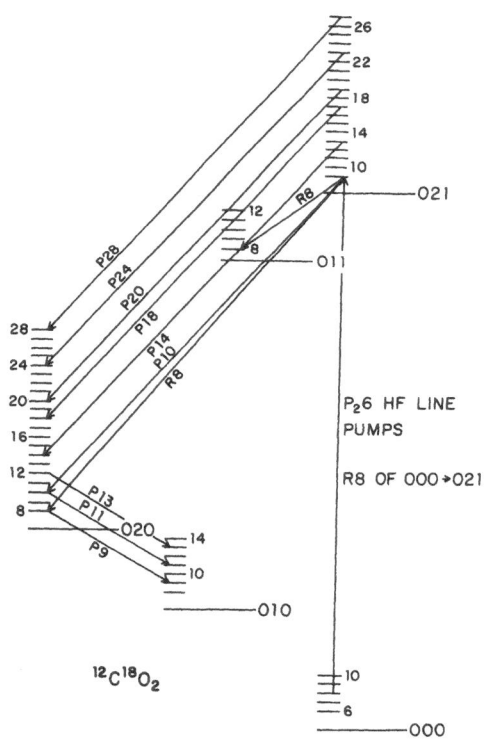

Fig. 13:   Selected vibration-rotation energy levels of $^{12}C^{18}O_2$ showing lasing transitions which have been identified following excitation by HF $P_2(6)$ line.

In the case of $^{13}C^{16}O_2$ we have observed an interesting intense emission at 4.3 μm. There is some experimental evidence that this 4 μm emission may be the Stokes frequency from a stimulated Raman scattering of the HF frequencies from $CO_2(100)$. The Raman cross section may be greatly enhanced by near resonances in the 10°1 and 02°1 levels. For example, the HF $P_1(7)$ and $P_2(4)$ lines fall respectively within 1 cm^{-1} of the R(14) and P(12) lines of $^{13}C^{16}O_2$. Calculations of the Raman gain using second order perturbation theory for these two lines indicate values on the order of several percent per centimeter per MW/cm^2 of HF laser power.[10] For the experimental conditions described powers of greater than 10 MW/cm^2 are reached on several HF transitions. Thus, the possibility of stimulated Raman scattering looks very real.

In conclusion, this talk has reviewed the spectroscopic prerequisites for performing isotopically selective excitation of UF_6. This data is now very well in hand and the emphasis is now turned towards developing the lasers to perform the work. This represents a new era for laser researchers in that they are now being asked to invent a laser to order with respect to wavelength, energy, and efficiency. It will be interesting to observe whether our knowledge of lasers has really progressed to the point that these requests can be met in a timely manner.

REFERENCES

1. J. P. Aldridge, H. Filip, H. Flicker, R. Holland, R. McDowell, N. Nereson, J. Molec. Spect. 50, 165 (1975).

2. C. D. Cantrell and H. W. Galbraith, J. Molec. Spect. 58, 158 (1975).

3. E. D. Hinkley and P. L. Kelley, Science 171, 635 (1975) theoretical analysis by R. S. McDowell et. al. to be published in Opt. Comm.

4. R. V. Ambartzumian, Y. A. Gorokhov, V. S. Letokhov, and G. N. Makarov, JETP Lett. 21, 171 (1975).

5. J. L. Lyman, R. J. Jensen, J. P. Rink, C. P. Robinson, and S. D. Rockwood, Appl. Phys. Lett. 27, 87 (1975).

6. R. L. Byer, R. L. Herbst, and R. N. Fleming, Laser Spectroscopy, S. Haroche et. al. eds., Springer-Verlag, Berlin 1975, p. 207.

7. H. Kildal and J. C. Mikkelson, Opt. Comm. 10, 306 (1974).

8. M. S. Piltch, J. Rink, and C. Tallman, Opt. Comm. 15, 112 (1975).

9. R. G. Wenzel and G. P. Arnold, Appl. Opt. 15, 1322 (1976).

10. R. Silver (private communication).

IV. Nonlinear Excitation of Molecules

DISSOCIATION OF POLYATOMIC MOLECULES
BY AN INTENSE INFRARED LASER FIELD

R.V. Ambartzumian
Institute of Spectroscopy
Academgorodok, Podolski r-n, Moscow, 142092 USSR

This talk will review most of the work performed at the Institute of Spectroscopy on dissociation of polyatomics by an intense infrared laser. The main contributors are: Yu. Gorokhov, G. Makarov, N. Furzikov and A. Puretzki.

The dissociation of polyatomics became very popular among scientists who are engaged with isotope separation after the publication of two of our works [1, 2] where we showed that such dissociation can be isotopically selective. This phenomenon is extremely interesting from the point of view of interaction of polyatomic molecules with the radiation field and especially from the point of view of the theory of monomolecular reactions. We note that the problems of dissociation by infrared radiation are very similar to the problems associated with monomolecular reactions.

At the Megève conference, we reported on extremely high enrichment factors in SF_6 [3]. While it is clear that the enrichment in the undissociated fraction of SF_6 can reach large values, the most interesting parameter is the selectivity of the dissociation process. To our regret we can not measure the dynamics of the process because of scrambling by nonselective chemical reactions which follow the dissociation. In our experiments, we measured the selectivity of dissociation by looking for enrichment in SOF_2 which formed in the reaction cell by the hydrolysis of SF_4 [4].

It is clear that the highest possible selectivity of dissociation is determined by the ratio of dissociation rates of each isotope and this ratio must depend on the laser frequency.

The rate of dissociation process we define as

$$W = n \, \ln (N_0/N) \tag{1}$$

where n is the number of irradiation pulses, and N, N_0 current and initial concentrations, respectively. Figure 1 shows the dependence of W on laser frequency together with the absorption spectrum of SF_6. The halfwidth of W (ω) is much broader than the linear absorption spectrum bandwidth and the maximum is shifted to the red. It is clear that if we draw the same dependence for $^{34}SF_6$, the point of interception of these two curves will give the frequency at which no enrichment will occur. The upper part of Fig. 1 shows the dependence of the enrichment factor in the products formed (SOF_2) as a result of dissociation. The frequency at which W_{34} intersects

W_{32} lies 1.5 cm^{-1} from the frequency where there is no enrichment in SOF$_2$. The highest enrichment measured in SOF$_2$ was 14, at a pressure of 0.05 Torr of SF$_6$. It should be emphasized that the real enrichment of sulfur is much higher than 14 since the measurements were made with a low resolution mass spectrometer which could not resolve the mass doublets ^{34}S^{16}OF$_2$ and ^{32}S^{18}OF$_2$. Taking this into account, we estimate the highest enrichment obtained in the dissociated products was 40-50. This is less than the figure 100 which we would derive from the data on dissociation rates.

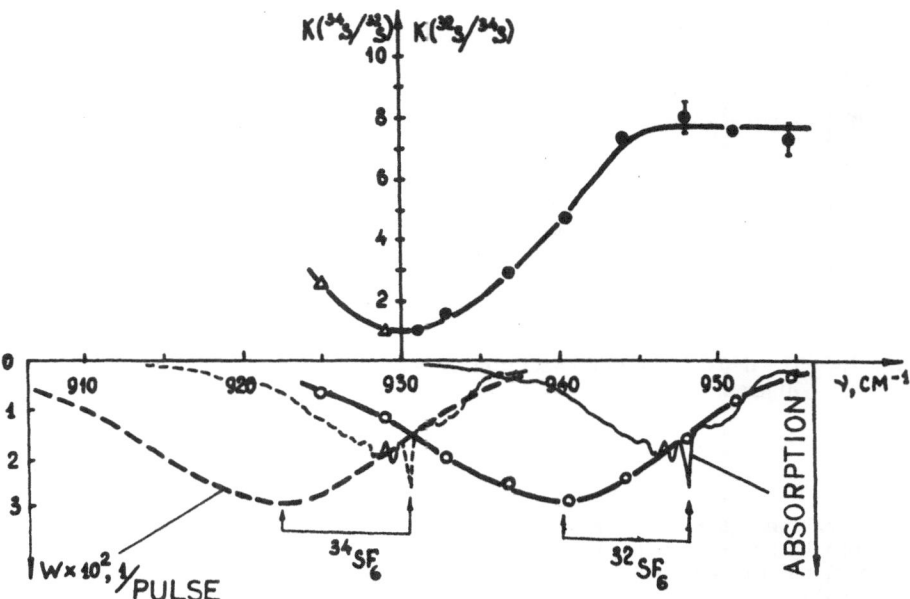

Fig. 1.   Dissociation rate of SF$_6$ versus laser frequency shown together with the absorption spectrum of SF$_6$. The upper part of the figure is the dependence of the selectivity of dissociation measured in SOF$_2$ (the enrichment factor) on laser frequency.

The highest possible selectivity of dissociation is defined by the width of the curve $W(\omega)$. The narrowing of the halfwidth of $W(\omega)$ is of extreme importance for separation where the isotope shift is smaller or comparable with the absorption bandwidth. To solve this problem one needs to understand better what happens to the molecule during the interaction with a strong monochromatic field.

In our studies of the dissociation of SF$_6$ by pumping the fundamental and combination bands it was found (Fig. 2) that the dissociation threshold (within the accuracy of our measurements) is the same for the ν_3 and $\nu_2 + \nu_6$ bands, even though the $\nu_2 + \nu_6$ band absorption is 100 times weaker than the ν_3 band; however, the dissociation rates for these two excitation processes are quite different, being proportional to μ_1, the matrix element of the transition [5].

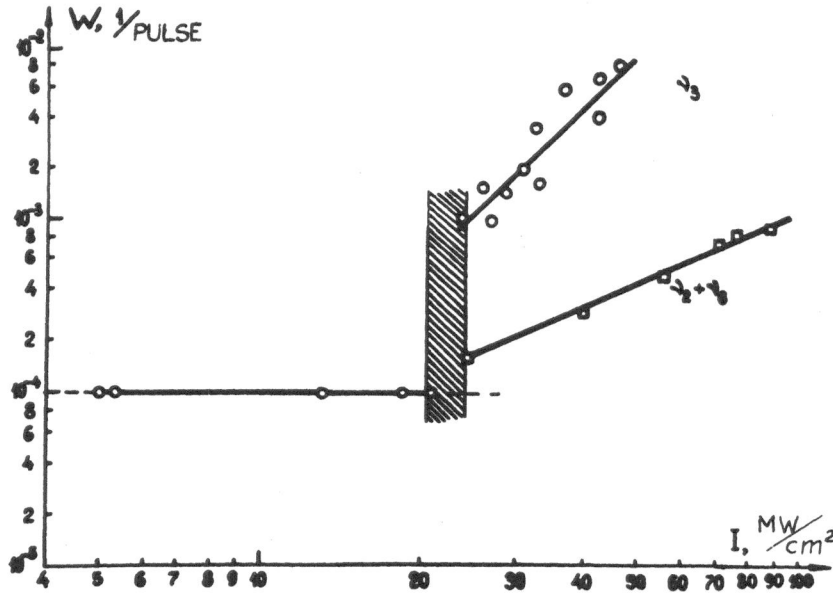

Fig. 2.   The dependence of dissociation rate on laser power. The pulsewidth is
90 nsec.

Combining this observation with the observation made in [3, 6] that the molecule absorbs large amounts of vibrational energy in a low intensity field (Fig. 3),
one concludes that the threshold is connected with saturation of some transitions
near the dissociation limit. Keeping in mind the idea of a "quasicontinuum" of vibrational levels, which in SF_6 starts at 0.3-0.4 eV above the ground state we can
divide the excitation process between two beams of different frequencies ω_1 and ω_2.
The low intensity beam, ω_1, selectively excites the molecules to high lying vibrational levels without any Stark broadening, while the second beam at ω_2, which is
off-resonance from the absorption of SF_6, helps the molecules to overcome the
threshold. The intensity of this beam can be very high without destroying the selectivity of excitation, since it does not produce any line broadening [4, 7, 8]. This
idea was checked in experiments where two synchronized TEA CO_2 lasers provided
two beams of different frequency, ω_1 and ω_2. The ω_2 was chosen so that no dissociation was detected without ω_1. Figure 4 shows the frequency dependence of dissociation rate W (ω_1) when ω_2 was fixed 130 cm^{-1} off resonance to the blue. Curve
1 was obtained at 300 K, and curve 2 at 190 K. Curve 3 is the single frequency case
and is shown here for comparison.

Comparison of curves 1 and 3 shows that considerable narrowing of the resonance occurs in the two frequency case and the major part of this narrowing is connected with almost total absence of dissociation in the R branch of the ν_3 mode.
The low frequency wing of the dissociation curve 3 did not change significantly in

the two-frequency dissociation case, bu: it narrowed strongly when the dissociation was performed at 190 K. The width of the curve W (ω_1) in this case was 5 cm^{-1} and this is four times narrower than the one-frequency case. The maximum of W (ω_1) is shifted from the Q branch of ν_3 by 4.5 ± 0.5 cm^{-1} compared with 7 ± 0.5 cm^{-1} in the one-frequency case [8]. From this comparison, it becomes clear that the dissociation in the single frequency region is connected with power broadening in the focal region of the beam. The comparison of curves 1 and 2 on Fig. 4 shows clearly that the low frequency wing of the dissociation curve W (ω_1) is explained by dissociation from the hot band absorption in SF$_6$.

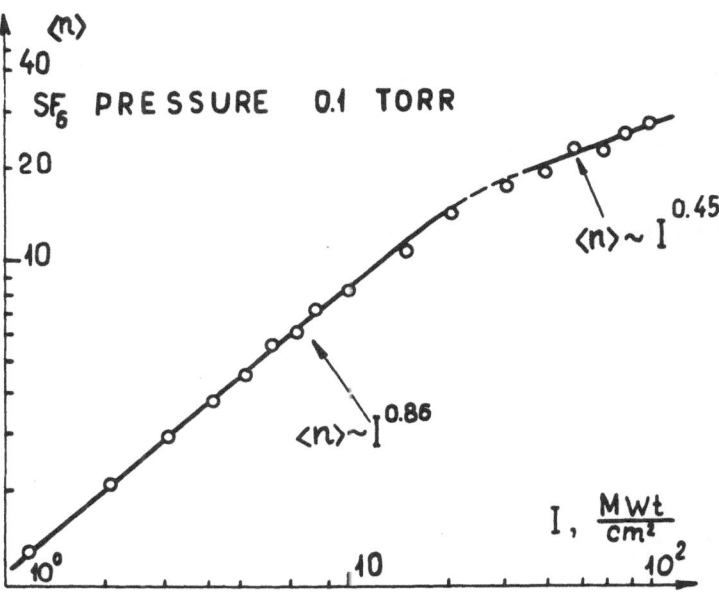

Fig. 3. The number of CO$_2$ quanta $<n>$ absorbed in irradiated volume versus laser intensity in SF$_6$, double log scale.

Figure 5 shows the dependence of dissociation rate, W, on the frequency of the dissociating beam at ω_2 while the frequency of the exciting beam was fixed (ω_1 = 942.4 cm^{-1}, P(22) line). The average intensity of the dissociating beam was held constant at all frequencies and was equal to 58 MW/cm^2, while the intensity of exciting beam was 4 MW/cm^2. In the absence of an exciting beam ω_2 could be set no closer than 20 cm^{-1} from the center of the ν_3 mode (towards the blue) if detectable dissociation was to be avoided (at 58 MW/cm^2). From Fig. 5 it is seen that when ω_2 approaches the ν_3 mode of SF$_6$, the dissociation rate increases greatly. Taking into account the fact that the nonresonant radiation at ω_2 is absorbed only in the vibrational quasicontinuum, it is clear that the dependence W (ω_2) respresents the dispersion characteristic of the vibrational quasicontinuum.

154

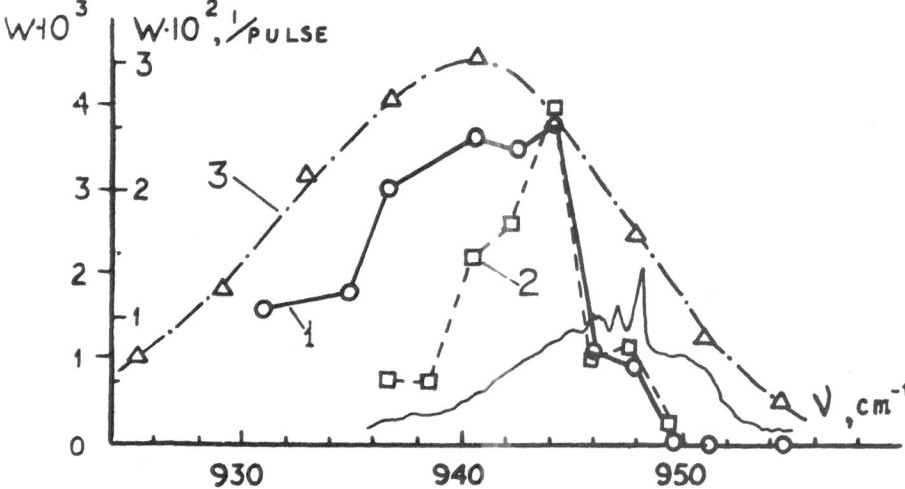

Fig. 4.   Dissociation rate of SF_6 versus exciting frequency; at room temperature $T = 300^{\circ}K$ (1) and (2) at $T = 190^{\circ}K$ (the left ordinate), two IR
pulses case.   Curve (3) shown for comparison represents the same
dependence in the case of dissociation of SF_6 in an intense single frequency field (the right ordinate).   The IR absorption spectrum of SF_6
is shown for comparison.

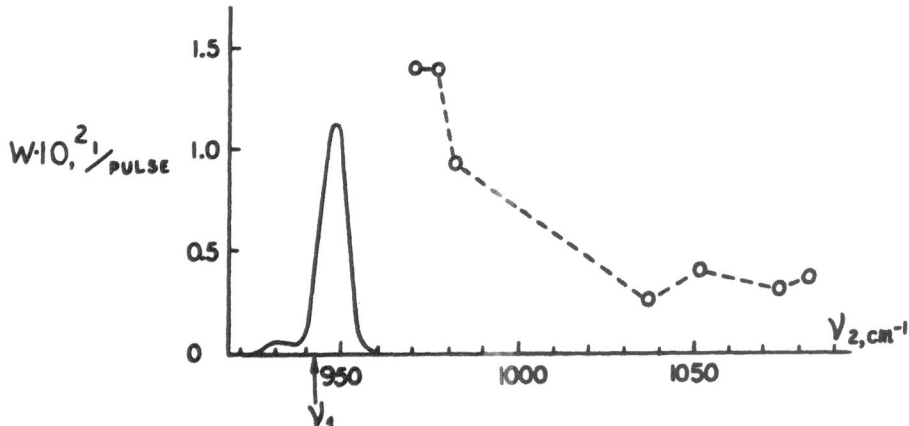

Fig. 5.   Dispersion characteristic of quasicontinuum - the dependence of dissociation rate of SF_6 on frequency of dissociating field ω_2.   The frequency
of exciting beam is fixed ($\omega_1 = 942.4$ cm^{-1}).

Figure 6 (solid line) shows the dependence of W on the intensity of the exciting beam I_1 (ω_1 = 942.4 cm^{-1}). The intensity of the dissociating beam was 60 MW/cm^2 and ω_2 = 1048 cm^{-1}. The dotted line in Fig. 6 represents the dependence of the number of absorbed quanta per molecule, <n>, at ω_1 on I_1. It is seen that both curves have the same slope which gives for <n> and W the same dependence: W, <n> ~ $I^{0.7}$. It is seen also that dissociation is significant at very low $I_1 \approx 60$ kW/cm^2. The correlation of <n> and W (I_1) indicates that the growth of W is connected mainly with a corresponding increase in the number of absorbed quanta, <n>, by SF_6. The obtained dependence, W (I), (Fig. 6) also proves that the dissociation threshold is not connected with overcoming the anharmonicity of vibrations but mainly with transitions within quasicontinuum. Therefore, the power dependence of W on I_2 must also have a sharp threshold, it must be the same as in the single frequency case, and W (I_2) must have the same dependence on power as the threshold is reached. The experiments gave threshold a value of 23 ± 2 MW/cm^2 and a cubic dependence on I_2 exactly as in the single frequency case.

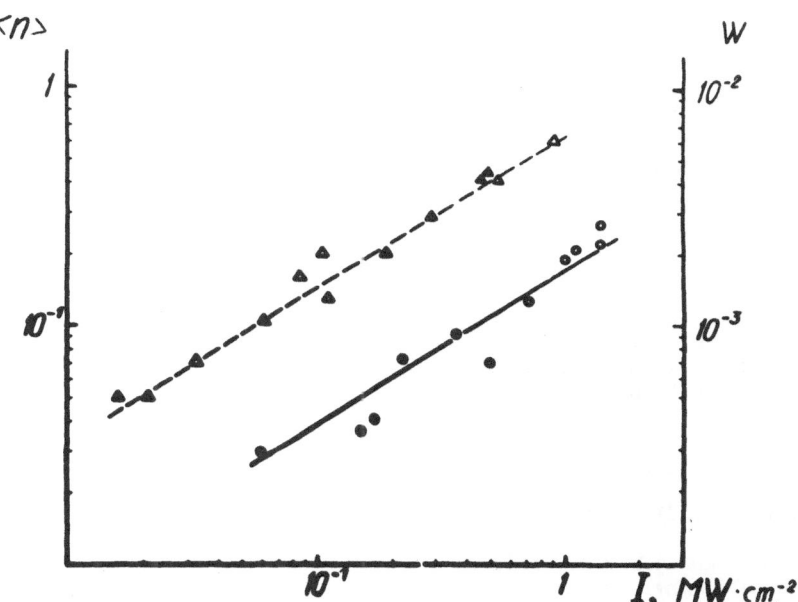

Fig. 6. Dissociation rate of SF_6 versus intensity I_1 of exciting beam (right scale). Average number of CO_2 quanta absorbed per molecule of SF_6 versus intensity of single frequency field ω_1 (left scale).

The two-infrared-frequency dissociation technique provides a unique opportunity for studying the V-V transfer and V-T relaxation rates for the molecules which are excited to the levels corresponding to the vibrational quasicontinuum. By a simple changing of delay between the exciting and the dissociating pulses, both rates can be determined easily. Figure 7 represents the dependence of enrichment factor measured in SOF_2 formed after dissociation when the delay between exciting pulse and dissociation pulse was varied. The frequencies and the intensities of those pulses were $\omega_1 = 942.4$ cm^{-1}, $I_1 = 1.5$ MW/cm^2, $\omega_2 = 1048.6$ cm^{-1}, and $I_2 = 60$ MW/cm^2. Initial pressure of SF_6 was 0.2 Torr. The enrichment falls exponentially with the delay giving $p\tau_{V-V} = 1$ μsec Torr.

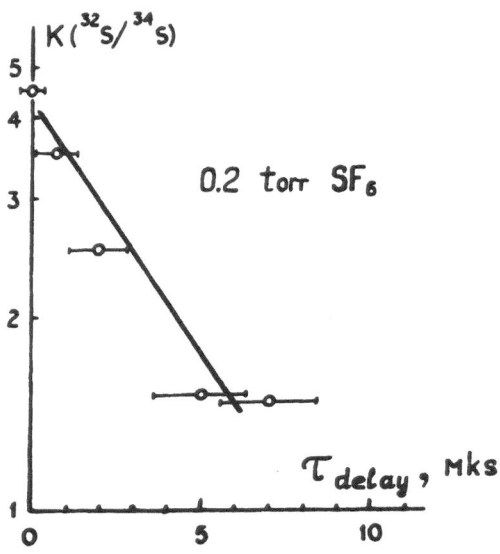

Fig. 7.   Enrichment factor measured in SOF_2 versus delay between exciting and dissociating pulses.

The Model of the Dissociation Process

The experimental data presented here are well explained by a simple model which includes: a) rotational compensation of anharmonicity [9] and b) broadening of high vibrational levels due to anharmonic coupling with the vibrational manifold (vibrational quasicontinuum). The rotational compensation means the possibility of sequential transitions for several first vibrational levels of the type:

$$[V = 0, J_{res.}] \xrightarrow{P(J_{res.})} [V = 1, (J_{res.} -1)] \xrightarrow{Q(J_{res.} -1)}$$

$$\longrightarrow [V = 2, (J_{res.} -1)] \xrightarrow{R(J_{res.} -1)} [V = 3, J_{res.}] \ ; \tag{2}$$

which might be fulfilled if

$$|2 B J_{res.} - \Delta\nu_{anh.}| < \Delta\nu_{broad} \ , \tag{3}$$

where B = rotational constant, $\Delta\nu_{anh.}$ = the shift of the level V = 1 due to anharmonicity, $J_{res.}$ = rotational quantum number [8]. Compensation of anharmonicity also occurs through transitions Q → R, P → R and therefore the dependence W(ω) might have several maxima with most intense maximum corresponding to transition (2). The anharmonic coupling of vibrational high lying levels broadens the modes. The broadening increases with vibrational number V and therefore the decay rate of that vibrational level into other vibrational modes increases with V also.

It is very attractive to explain the red shift of the maximum of W (ω) by stating that the maximum corresponds to J_{res} and $2B\,J_{res}$ equals the anharmonicity of the vibrational mode. But in fact the W (ω) reflects the spectrum of the same vibrational "quasicontinuum" because the W (ω) represents the product of two factors: distribution over the rotational levels and the spectrum of "quasicontinuum". If one normalizes W (ω) to a given number of molecules one obtains a W (ω) that looks as a straight line with the same slope as in Fig. 5. Reviewing the results on dissociation of SF_6 by infrared one can ask if the results are general and apply to other molecules. The answer is yes, as can be seen from the results on OsO_4 dissociation. OsO_4 is one of the most interesting molecules because of its electronic transition spectrum, which allows the direct exploration of most processes which occur in the dissociation process. At Megève we had reported about enrichment of Os isotopes and mentioned the observation of the collision free V-V energy transfer between ν_3 to ν_1 vibrational modes in the highly excited molecule. Here we shall present some detailed features of OsO_4 experiments.

The dissociation characteristics of OsO_4 were studied by examining the visible luminescence which accompanies the dissociation process [10]. Figure 8 shows the dependence of the luminescence intensity on the frequency of the dissociating beam. It is seen that in the R-branch there is no dissociation and the signal intensity has structure which explains our results on Os enrichment [9]. If the W (ω) would have the same appearance as in SF_6 (rather smooth wide curve) the results on enrichment could not be explained. The experiments on dissociation of OsO_4 by two infrared pulse technique showed even deeper structure of W (ω), (Fig. 9).

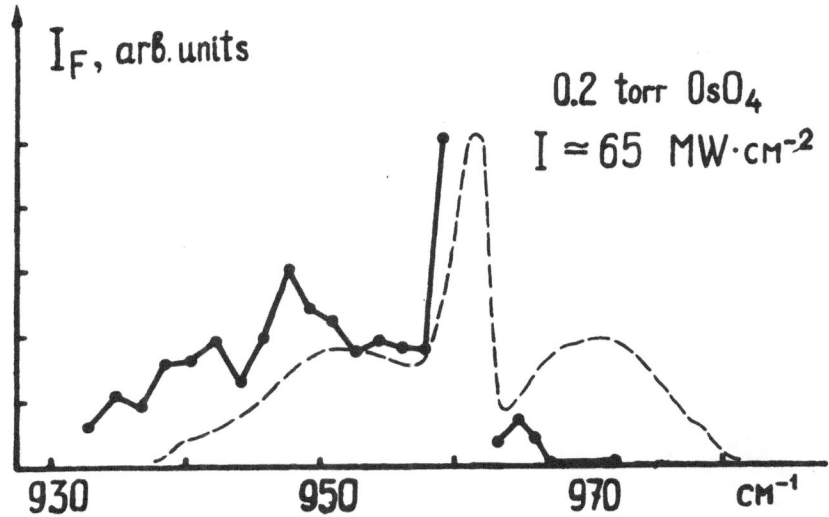

Fig. 8. The dissociation rate W (ω) in OsO_4 measured by fluorescence technique in single frequency case.

158

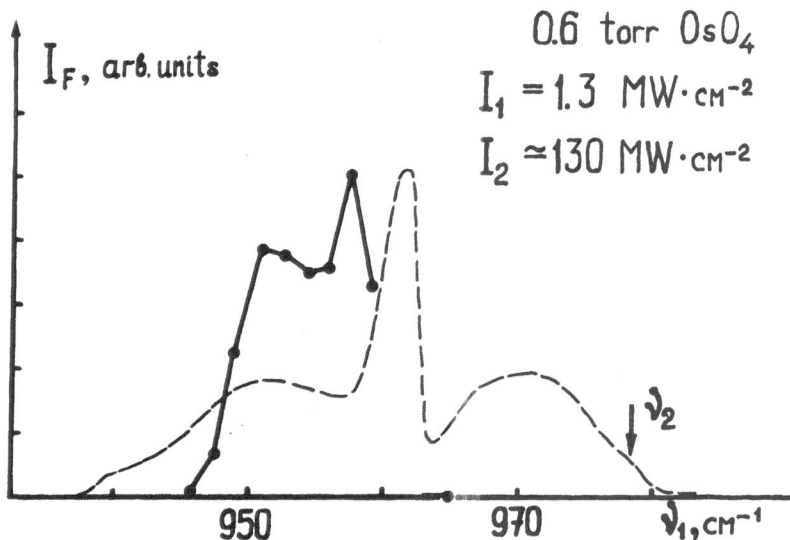

Fig. 9. The same dependence as at Fig. 8 but in the case of two IR pulses. The ω_1 was scanned, the ω_2 - fixed.

The absence of dissociation in the R-branch is caused by saturation of the absorption to a low value of $<n>$ (see Fig. 10, where the P branch absorption is also shown). This shows that the results obtained for SF_6 are rather general for all molecules.

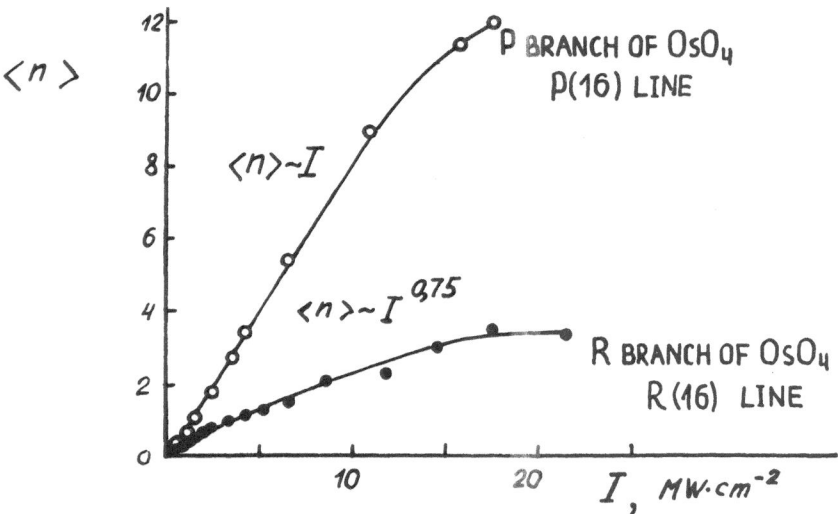

Fig. 10. The same dependence as in Fig. 3 measured in OsO_4.

The electronic absorption spectrum of OsO_4 represents an electronic-vibrational progression with the ν_1 mode present in it. By measuring the red shift of the absorbance of OsO_4 under infrared pumping one can examine the degree of vibrational excitation of the ν_1 mode population after the V-V process is over. Such V-V processes must occur because the laser pumps the ν_3 but not the ν_1 mode of the molecule. Examining the time behavior of the rise time of the signal corresponding to the appearance of excitation in the ν_1 mode one can evaluate the V-V rate between ν_3 and ν_1 modes. The apparatus is essentially the same as described in [9]. The results are shown on Fig. 11. An enormous red shift is seen; from this data one can conclude that the V = 10-20 levels of the mode ν_1 are populated after the V-V process is completed. In fact by increasing the sensitivity of detection system one can see larger red shifts. The rise time of the pulses gives information on energy flow between the modes. If we plot the inverse rise time versus OsO_4 pressure (Figure 11) we can see that even at zero pressure there is some energy exchange between the modes; this is attributed to collision free V-V energy transfer. The zero-pressure transfer time is 1.2 μsec. This was to our knowledge the first observation of such a process in small molecules.

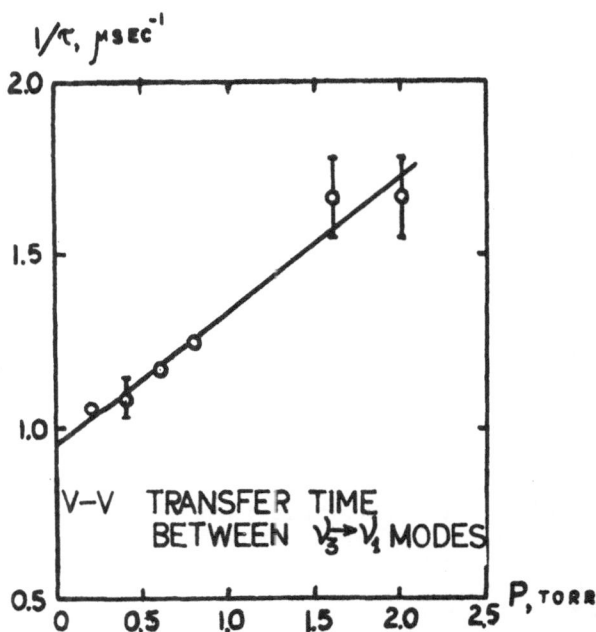

Fig. 11. The dependence of $1/\tau$, where τ is V-V transfer rate, versus OsO_4 pressure.

As is apparent from Fig. 2 the threshold in SF_6 is extremely sharp. An intensity 8 percent below the threshold causes no dissociation. The power dependence of dissociation rate in the shaded region of Fig. 2 must be higher than I^{29}. The results for most other molecules studied also show very sharp thresholds. In the case of OsO_4, however, we have a unique situation; Figure 12 shows dissociation at power levels as low as 3 MW/cm^2, and this figure is determined by the detectivity of the measurement system. The power dependence in the low intensity region is $W \sim I^7$ and it changes to I^4 above 20 MW/cm^2.

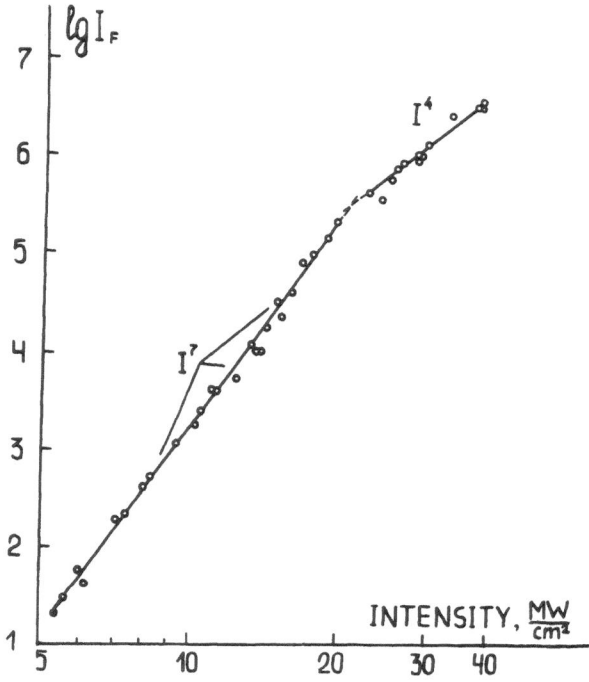

Fig. 12.   Threshold characteristics of dissociation in OsO_4.

The last item to be discussed is the question of thermal dissociation. As can be inferred from Fig. 3 and Fig. 10 the amount of absorbed energy is approximately equal from 2 to 3 eV per molecule. After all the relaxation processes are over the temperature in the irradiated volume should rise to extremely high values, and calculations for SF_6 show that very large part of molecules should dissociate thermally. But enrichment coefficients show that the thermal contribution is rather small. We succeeded in observing thermal dissociation in OsO_4 which occurred only at pressures of 0.8-1 Torr and higher. It followed the fast component of radiation induced dissociation approximately after 40-50 μsec after the laser pulse. At lower pressures we failed to observe the same phenomenon, though the sensitivity was quite sufficient. This shows that the decrease of dissociation selectivity is mainly connected with thermal dissociation.

REFERENCES

1. R. V. Ambartzumian, V. S. Letokhov, E. A. Ryabov and N. V. Chekalin, JETP Lett. 20, 597 (1974).

2. R. V. Ambartzumian, Yu. A. Gorokhov, V. S. Letokhov, and G. N. Makarov, JETP Lett. 21, 375 (1975).

3. R. V. Ambartzumian, Yu. A. Gorokhov, V. S. Letokhov, G. N. Makarov, E. A. Ryabov and N. V. Chekalin. Proceedings of the Second International Conference on Laser Spectroscopy, 23-27 June 1975, Megève, France, (Springer-Verlag, 1975).

4. R. V. Ambartzumian, Yu. A. Gorokhov, V. S. Letokhov, G. N. Makarov and A. A. Puretzki, JETP 71, N8, (1976).

5. R. V. Ambartzumian, Yu. A. Gorokhov, V. S. Letokhov, G. N. Makarov and A. A. Puretzki, JETP Lett. 23, 26 (1976); ibid 22, 374 (1975).

6. R. V. Ambartzumian, Yu. A. Gorokhov, V. S. Letokhov and G. N. Makarov, JETP 69, 1956 (1975).

7. R. V. Ambartzumian, N. P. Furzikov, Yu. A. Gorokhov, V. S. Letokhov, G. N. Makarov and A. A. Puretzky, JETP Lett. 23, 217 (1976).

8. R. V. Ambartzumian, N. P. Furzikov, Yu. A. Gorokhov, V. S. Letokhov, G. N. Makarov and A. A. Puretzki, Opt. Commun. (to be published).

9. R. V. Ambartzumian, Yu. A. Gorokhov, V. S. Letokhov and G. N. Makarov, JETP Lett. 22, 96 (1975).

10. R. V. Ambartzumian, Yu. A. Gorokhov, G. N. Makarov, N. P. Furzikov and A. A. Puretzki, Chem. Phys. Lett. (to be published).

COLLISIONLESS DISSOCIATION OF POLYATOMIC MOLECULES
BY MULTIPHOTON INFRARED ABSORPTION

N. Bloembergen
Harvard University, Cambridge, MA 02138, USA
C.D.Cantrell
Los Alamos Scientific Laboratory, Los Alamos, NM 87545, USA
and
D.M. Larsen
M.I.T. Lincoln Laboratory, Lexington, MA 02173, USA

1. Brief Review of Experimental Characteristics

During the past year numerous experimental and theoretical investigations have been carried out on the prompt dissociation of several polyatomic molecules by intense CO_2 laser pulses. The fact that this dissociation process is isotopically selective [1,2] undoubtedly provided the major impetus for this activity. In this presentation attention will not be focused on the important application of laser isotope separation, but on the physical processes by which a single molecule, without the benefit of collisions, may by exposure to an infrared laser pulse acquire sufficient energy for dissociation to occur.

Historically fluorescence from dissociation products was first detected as a consequence of heating of the gas by absorption of c.w. CO_2 radiation. Gas-kinetic collisions obviously play an important role in this case [3].

The discovery by ISENOR et al. [4] of a fast dissociation process occurring during the short duration of a TEA laser pulse with $t_p < 10^{-7}$ sec, provided the first indication of a collisionless mechanism, and these authors already pointed out the importance for dissociation of a quasi-continuum of energy levels for highly excited polyatomic molecules. Since AMBARTZUMIAN et al. demonstrated [1] that this fast dissociation process is isotope-selective, a large amount of experimental data has been reported. While the early data were sometimes conflicting and caused some to hold onto the belief that collisions are necessary for dissociation to occur, it is now well established that collisionless dissociation can take place. The following observations for dissociation of SF_6 molecules are indicative of this fact:
1. The dissociation of SF_6 molecules in a high vacuum molecular beam apparatus has been reported at a recent conference.
2. Dissociation probabilities close to unity occur for molecules irradiated with a single pulse with a duration less than 0.5×10^{-9} sec at pressures less than 100 millitorr [5]. The gas-kinetic collision time is 50 ns/torr, so that each molecule has a chance of less than 10^{-3} to undergo a collision during the pulse. Intermolecular V-V transfer can be excluded. The energy in the pulse was about 0.1 joule and the peak power density exceeded 5×10^8 watts/cm^2.
3. The dissociation probability per molecule and isotopic selectivity both decrease monotonically with increasing pressure [6].

The collisionless dissociation of SF_6 has been studied [7] as a function of wavelength, power flux density and energy of the CO_2 pulses and as a function of partial pressures of SF_6 and some buffer or scavenging gases (H_2).

These data have been reviewed by AMBARTZUMIAN [8], LETOKHOV [9] and ROCKWOOD [10], and the salient features for our purposes may be summarized in terms of the following observations:

4. For a pulse duration $t_p \sim 8 \times 10^{-8}$ sec, the dissociation rate has a relatively abrupt threshold at 23×10^6 watts/cm^2. Above this threshold the dissociation rate increases roughly as the cube of the intensity, but it saturates and reaches unity at power levels of about 0.5×10^9 watts/cm^2.

5. Above threshold level the dissociation rate at room temperature as a function of frequency (for $S^{32}F_6$) shows a maximum at 940 cm^{-1} and is an order of magnitude smaller at 930 and 950 cm^{-1}. The frequency response is a bell-shaped curve with a width (FWHM) of about 8 cm^{-1}. The dissociation rate for $S^{34}F_6$ is given by a similar curve displaced 17 cm^{-1} towards lower frequencies. At low temperature ($\sim 190°$K) the peak dissociation rate appears to be shifted only about 4 cm^{-1} from the Q-branch band head [9].

6. Dissociation may occur at power levels as low as $\sim 10^3$ watts/cm^2, provided higher power fluxes are provided at a second frequency, which need not correspond to an absorption band of the SF_6 molecule at room temperature. The radiation at the second frequency may, for example, be provided by a second CO_2 laser in the 9.2 or 9.6 μm band.

7. Dissociation may also occur by single frequency irradiation near combination bands. The effective dissociation rate/molecule appears to be proportional to the dipole matrix element of the combination transition. Again, the frequency response is maximum about 8 cm^{-1} below the Q-branch head of the transition involved at room temperature.

8. The same process is effective in the dissociation of many other polyatomic molecules. Attempts to dissociate diatomic molecules by this method have always met with failure.

This brief review of experimental findings is concluded with a table of pertinent physical quantities for the SF_6 molecule. The spectroscopy of the

Table I. Relevant energies for the collisionless dissociation of SF_6 by infrared radiation

Physical Quantity	Symbol	Energy in cm^{-1}
Dissociation energy	W_{diss}	40,000
Fundamental frequency of ν_3 mode	$h\nu_3$	948
Thermal energy	kT	200
Rotational width of P and R branch	$2B <J>$	5
Coriolis splitting	------	0.1-10
Anharmonic defect	$\lvert 2X_{33}h\nu_3 \rvert$	2.88
Electric dipole matrix element	$\hbar\,\omega_R = \frac{1}{2}\lvert\mu\rvert\,\lvert E_{th}\rvert$	0.6
Rotational constant	B	0.0906
Effective rotational constant	$\tilde{B} = B(1-\zeta_3)$	0.027
Experimental spectral width	$\hbar\,\Delta\,\omega_L$	0.03
Pulse-limited width	$\hbar\,t_p^{-1}$	10^{-4}
Collisional width at one millitorr	$\hbar\,\tau_C^{-1}$	10^{-6}
Radiative width	$\hbar\,\tau_{IR}^{-1}$	10^{-8}

SF$_6$ ν_3-band has been studied extensively [11], and recently high resolution studies at low pressures with semiconductor lasers have been reported [12]. Our primary interest will be in the relative magnitude of the various spectroscopic quantities involved. The electric dipole $|\mu| = 0.3$ Debye for the fundamental transition $v_3 = 0 \rightarrow 1$ is consistent with an absorption cross-section of $\sigma = 0.5 \times 10^{-17}$ cm^2 observed [11] in the unresolved band at high temperature with a width (FWHM) of 20 cm^{-1}. A similar value for $|\mu|$ has been obtained by ALIMPIEV and KARLOV [13]. The strength of the dipole matrix element can be expressed in terms of frequency (cm^{-1}) by introducing the Rabi precession frequency at resonance in a field with an amplitude $|E_{th}| \sim 1.3 \times 10^5$ volts/cm. This field strength corresponds to a power flux density of 23×10^6 watts/cm, which is the threshold value for dissociation in the experiments of AMBARTZUMIAN [8]. Thus, the Rabi frequency $\omega_R = (3^{-\frac{1}{2}}/2)\hbar^{-1}|\vec{\mu}||E_{th}| = 0.6$ cm^{-1} has been entered in the table. The factor $3^{-\frac{1}{2}}$ arises because only one projection of $\vec{\mu}$ is important for each of the P, Q and R bands. This value is smaller than the anharmonic defect [11], $\Delta W_{anh} = -2 X_{33}\nu_3 = +2.8 \pm 0.4$ cm^{-1}. The energy levels of the vibrational levels relative to the ground state $v_3 = 0$ of one anharmonic ν_3-mode can be conveniently written,

$$W(v_3) = v_3 h\nu_3 + h\nu_3 X_{33} v_3(v_3 - 1) . \tag{1}$$

In this expression $\nu_3 = 948$ cm^{-1} is the fundamental resonant frequency. The photon energy for an n-photon transition from the ground state to the level $v_3 = n$ is

$$h\nu_n = h\nu_3 - (n-1)|X_{33}|h\nu_3 . \tag{2}$$

The value of the rotational constant $B = 0.0906$ cm^{-1} in the ground state is small, so that at room temperature the statistical distribution over rotational states has a maximum at $J \approx 55$. About three-quarters of the SF$_6$ molecules may be found in states with J in the range 30-80. High resolution data have led to a complete assignment of the complex band structure [12], associated with the ν_3 fundamental modes. The three-fold degeneracy of the ν_3 vibration leads to important Coriolis splittings, which may exceed the rotational spacing. In fact, the rotational constant requires an important correction from the ν_3-mode degeneracy, and assumes an effective value $\tilde{B} = B(1-\zeta_3) = 0.027$ cm^{-1}.

Furthermore, there are many hot bands, because other normal modes may be thermally excited. Coriolis interactions and anharmonic coupling with overtones and nearly degenerate combination states rapidly increase in importance with increasing vibrational energy. Even for $v_3 = 1$, the Coriolis interactions have a magnitude of about 0.1 cm^{-1}, while for $v_3 = 2$ there appears already a cluster of states separated by a few wave numbers which are a superposition of states belonging to rotational manifolds of several different vibrational states [12].

The density of vibrational states has been computed by JACKSON [14]. The number of vibrational states per cm^{-1} amounts to 30 at a total vibrational excitation of 3000 cm^{-1}, and increases to 10^3 at 5000 cm^{-1} excitational energy. Each of these states has a rotational manifold associated with it.

The quantities at the bottom of the energy scale in the table are the homogeneous linewidths due to finite lifetimes. The uncertainty principle sets a width $\Delta\nu = t_p^{-1}/2\pi \sim 10^{-4}$ cm^{-1} for a pulse duration of $t_p = 5 \times 10^{-8}$ sec. The linewidth of the CO$_2$ laser radiation from a typical TEA laser is much larger and amounts to about 0.03 cm^{-1}. The infrared radiative lifetime is about $\tau_{IR} = 10^{-3}$ sec, corresponding to 10^{-8} cm^{-1}. Since this paper is

concerned with collisionless dissociation, the pressure is assumed sufficiently small that even phase-perturbing collisions and rotational interchange are negligible. Since no phase-interrupting processes take place,
the coherence of the excitation of the individual molecule and the radiation
field must be taken into account.

2. Quantummechanical Model for Collisionless Dissociation

For dissociation by monochromatic radiation one looks, of course, for a ladder of equally spaced energy levels. It is clear that a harmonic oscillator
would readily acquire a level of excitation larger than the dissociation energy. The coupling of the mechanical harmonic oscillator with the coherently
excited harmonic oscillator of the radiation field leads to coherent mechanical excitation, a Glauber coherent wave packet [15], with a Poisson distri-
bution over the energy eigenstates $|n_v\rangle$. The mean excitation energy is equal
to that acquired by a classical harmonic oscillator. When the oscillator is
in the ground state at t = 0, the expectation value for the energy $\langle n_v \rangle \, h\nu_0$
is determined by,

$$< n_v(t)> = \left(\frac{\omega_R}{\omega - \omega_0}\right)^2 \sin^2 \frac{1}{2}(\omega - \omega_0)t \tag{3}$$

The maximum excitation, corresponding to a quantum number $(\omega_R/\omega - \omega_0)^2$ is
reached in a time $\pi(\omega - \omega_0)^{-1}$. Exactly at resonance the expectation value
for the vibrational quantum number at the end of a laser pulse of duration
t_p is given by $\frac{1}{4}(\omega_R t_p)^2$. For a pulse of 8 x 10^{-8} sec at the threshold value,
the data in the table would yield $< n_v(t_p)> \sim 3 \times 10^4$, well above the dissociation energy, $< n_{diss}> \geq 42$.

The simple equation (3) assumes that the excitation energy of the electromagnetic field oscillator is very large compared to the mechanical oscillator.
The pulse energy in a single EM mode is about 10^{-3} joules, very much larger
than the dissociation energy of one molecule, so that (3) should give an adequate description. Exact solutions of two coupled coherent quantummechanical
oscillators are available in the literature [16], if needed.

It is clear that a harmonic vibrator would be excited to a multiple of
the dissociation energy in small fraction of the pulse duration t_p at field
amplitudes considerably below $|E_{th}|$, but it is also evident that in practice
the anharmonicity of the vibrator plays a dominant role. This is true even
at the lowest levels of excitation, v_3 = 2 or 3, if $\hbar \omega_R < |X_{33}| h\nu_3$.

2.1 Anharmonic Vibrator

The unequal spacing of the energy levels of a nondegenerate slightly anharmonic vibrator is described by (1). For "weak" infrared radiation fields,
$\omega_R << 2\pi X_{33}\nu_3$, perturbation theory is applicable. Eq.(2) shows that on the
low frequency side of the fundamental resonance, a set of equally spaced
multi-photon resonances will occur with a spacing of 1.4 cm^{-1}. The effective
matrix element for the two-photon transition in the rotating wave approximation is given by the well-known expression,

$$\hbar \, \omega_R^{eff}(2) = < 0|\mu E|2 >_{eff} = \frac{< 0|\frac{1}{2} \mu E|1 > < 1|\frac{1}{2} \mu E|2 >}{X_{33} h\nu_3} \tag{4}$$

and similarly for the higher order processes. Since $|X_{33}| << 1$, the zero-
order wave functions for the harmonic oscillator may be used to evaluate the

matrix element. In this manner one derives an effective Rabi frequency for the n-photon transition, from the ground state to the level $v_3 = n$,

$$\omega_R^{eff}(n) = \frac{\omega_R^n \, (n!)^{\frac{1}{2}}}{(\omega_3|X_{33}|)^{n-1} \prod_{n'=1}^{n-1} n'(n-n')} \tag{5}$$

For a given applied frequency ω, the problem is essentially reduced to a two-level problem, and the probability for the system to be in the state $|v_3 = n >$ at time t is given by,

$$P_R(n,t) = \frac{\{\omega_R^{eff}(n)\}^2}{4\pi^2(\nu-\nu_n)^2 + \{\omega_R^{eff}(n)\}^2} \sin^2 \frac{1}{2}\left(4\pi^2(\nu-\nu_n)^2 + \{\omega_R^{eff}(n)\}^2\right)^{\frac{1}{2}}t \tag{6}$$

where ν_n is given by (2) and $\omega_R^{eff}(n)$ by (5). For $\omega_R^{eff}(n)t_p >> 1$ and $\nu = \nu_n$, the system essentially oscillates between the ground state and the state $|v_3 = n >$. The time-averaged probability to be in this state is plotted schematically as a function of frequency for n = 1,2,3,4 in Fig. 1. The width of each response curve is, of course, proportional to $\omega_R^{eff}(n)$. This simple perturbation approach is valid if the overlap between the individual two-level resonances is negligible, $\omega_R << 2\pi|X_{33}|\nu_3$.

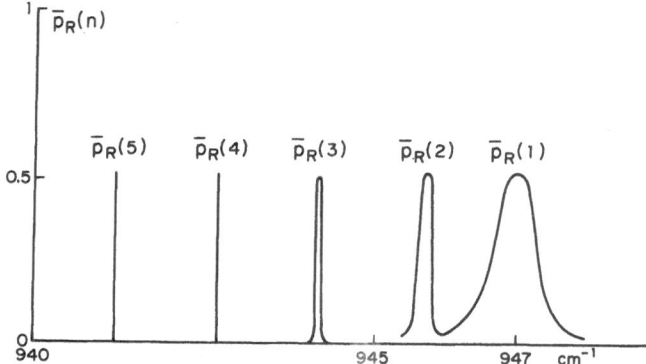

Fig. 1 Time-averaged probability $\bar{p}_R(n)$ to occupy the nth excited level of an isolated anharmonic vibrator by an n-photon process. The horizontal frequency scale is for the ν_3 vibration of SF_6. Two-level perturbation theory is used with $\hbar \omega_R \sim \frac{1}{4} h\nu_3 X_{33}$.

More complete analyses [17-19] have been made by considering the first n+1 levels together in a (n+1) x (n+1) matrix problem. The coupling by the harmonic field is still between adjacent levels, but $\omega_R \gtrsim 2\pi|X_{33}|\nu_3$ is now permissible. A numerical solution obtained by LARSEN [17] is presented in

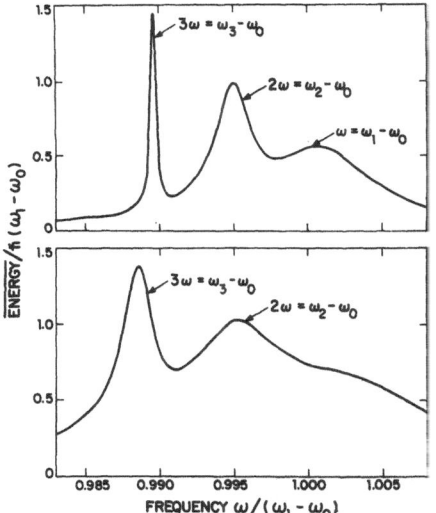

AVERAGE ENERGY STORED IN 4-LEVEL OSCILLATOR

Figure 2

Energy of excitation of anharmonic vibrator as a function of frequency for an exact four-level calculation, with $\hbar\,\omega_R = \frac{1}{2}\,h\nu_3 X_{33}$ (top) and $\hbar\,\omega_R = h\nu_3 X_{33}$ (bottom).

Fig. 2. Note that on the vertical energy is plotted, or $p_R(n)$ $nh\nu_3$, rather than $\overline{p_R(n)}$ in Fig. 1. As expected, the individual n-photon resonances broaden and start to overlap. For $\omega_R \gg |X_{33}|\nu_3$ the quantum-mechanical expectation value for the excitation of the anharmonic oscillator would approach the values derived from the classical anharmonic oscillator model [20], if a sufficient number of quantum levels is taken into account in the quantum calculation.

Just above the experimental threshold intensity for dissociation, $\omega_R \ll |X_{33}|\nu_3$, (5) yields $\omega_R^{eff}(4) = 2.7 \times 10^{-3}$ cm^{-1} and $\omega_R^{eff}(5) \approx 10^{-5}$ cm^{-1}. Thus, for $t_p \sim 8 \times 10^{-8}$ sec, the level $v_3 = 4$ could ideally still be reached, but the tuning would have to be extremely selective as the width of the four-photon resonance would be 2.7×10^{-3} cm^{-1}. The levels $v_3 \geq 5$ could not be reached at all. Thus, this model needs modification and extensions in order to obtain satisfactory agreement with experiment.

2.2 Rotating Anharmonic Vibrator

Considerable improvement can be obtained by taking into account the rotational band structure. The compensation of anharmonic defects by rotational energy differences has been pointed out by several workers [17,21,22]. The expression for the effective Rabi frequency for the n-photon transition between the state $|v_3 = 0, J_0 >$ and the state $|v_3 = n, J_n >$ is

$$\omega_R^{eff}(n,J) = \frac{< 0,J_0|\frac{1}{2}\,\mu E|\,1,J_1>\,<1,J_1|\frac{1}{2}\,\mu E|\,2,J_2> \ldots}{\hbar^n(\omega_3 + \Delta\omega_{J_0} - \omega)(2\omega_3 - X_{33}\omega_3 + \Delta\omega_{J_1} - 2\omega) \ldots}$$

$$\frac{< n-1,J_{n-1}|\frac{1}{2}\,\mu E|n,J_n >}{\{(n-1)\omega_3 - (n-2)X_{33}\omega_3 + \Delta\omega_{J_{n-2}} - (n-1)\omega\}} \qquad (7)$$

Here $\Delta\omega_{Ji}$ is the difference in rotational energy of states with J_i and J_{i+1}. We have, of course, the selection rule $|J_n - J_0| \leq n$. Since the width of the P and R branches is much larger than the anharmonic defect, it is possible to reach the vibrational level $v_3 = 3$ with very small energy defects in the intermediate steps. If the simple energy spectrum $\hbar\,\omega_n + \tilde{B}J(J+1)$ is assumed for the rotational-vibrational spectrum of SF_6, the sequence $|v_3 = 0, J_0 = 53> \rightarrow |1,52> \rightarrow |2,52> \rightarrow |3,53>$ would have essentially equally spaced levels. There will be a considerable range of initial J_0 values, where the mismatch in the energy denominators is smaller than the corresponding dipole matrix elements in the numerator. In these cases the perturbation expression (7) is not valid. If one wishes to avoid extensive numerical calculations for a coupled four-level problem, a crude estimate [23] is obtained by assigning a maximum value of unity to each quotient of dipole matrix element to energy mismatch in (7). The maximum value of $\omega_R^{eff}(n)$ is obviously $\omega_R(1)$ for a one-photon transition. It is clear that the state $|v_3 = 3>$ can be populated with an average probability of one-half for those molecules which have initial J_0 values in the range $52 \pm \Delta J_0$, with $2\tilde{B}\Delta J_0 \sim \hbar\,\omega_R(1)$.

At the threshold field amplitude $|E_{th}|$, the probability to reach the states $v_3 = 4$, or even 5, during the pulse duration is still considerable for a selected range of values J_0. Fig. 3 illustrates a P P R R sequence to the level $v_3 = 4$, in which one has a small energy mismatch of 1.4 cm^{-1} in only two denominators. Eq.(7) with the cut-off rule for a maximum quotient of unity for successive ratios of $\omega_R^{eff}(n)/\omega_R^{eff}(n-1)$ permits an estimate of what fraction of molecules can reach a certain vibrational state during the laser pulse. One must expect that at 10^8 watts/cm^2 a considerable fraction of the molecules will reach levels with $v_3 = 4$ if the laser frequency is tuned from 948 cm^{-1} to 940 cm^{-1}. A smaller fraction will reach a level $|v_3 = 5>$ via P P Q R R or other sequences. Perhaps some molecules may even reach the state $|v_3 = 6>$ although $\omega_R^{eff}(n)$ drops rapidly for $n \geq 5$, because of the increasing anharmonicity. The model of the rotating anharmonic vibrator, which should be a good representation for diatomic molecules, does not permit excitation to the dissociative levels. There is no way to reach levels with $v_3 \sim 10$, let alone $v_3 > 40$. This result is in agreement with the negative attempts to induce collisionless dissociation in diatomic molecules. Our model must be extended to include the effects of anharmonic and Coriolis coupling with other vibrational modes.

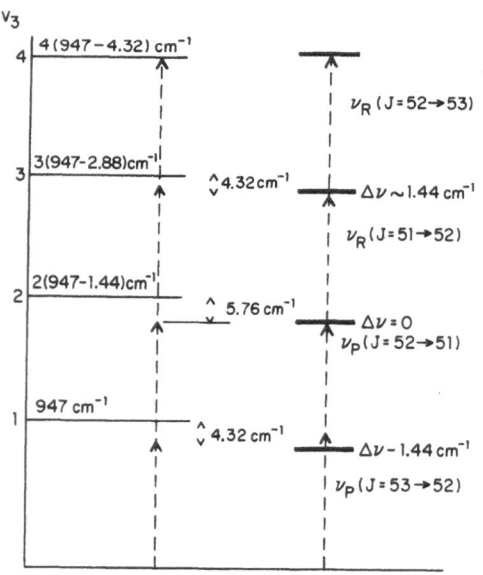

$4\nu_3$ TRANSITION IN SF_6

Figure 3

A four-photon process with partial compensation of the anharmonic energy defect with rotational energy differences.

2.3 Coupling to the Rotational-Vibrational Quasi-Continuum

The molecule SF_6 has 15 normal modes, the totally symmetric ν_1 mode, the two-fold degenerate ν_2 mode and four three-fold degenerate modes. Only the ν_3 and ν_4 are IR-active and have dipole matrix elements connecting neighboring oscillator states. The coupling of the three degenerate ν_3 modes should be considered first. The degeneracy is partially lifted by Coriolis and anharmonic interactions. These may push some states up relative to the center of gravity of the zero-order degenerate state [12]. This effect may be even more important in creating a ladder of states with nearly equal spacing than the compensation of the anharmonic down shift of the center of gravity by rotational splittings.

Coupling of the ν_3 modes with the other normal modes is especially important when overtone or combination states of the latter are nearly degenerate. This situation will always occur for high vibrational excitation energy where the density of vibrational states is high. This coupling with the quasi-continuum has been considered by many authors [4,17,19,20,24,26].

The coupling of a selected "zero-order" pure ν_3 state to this quasi-continuum of other vibrational states takes place through a Coriolis term in the Hamiltonian. A typical form of this term is [25]

$$\mathcal{H}_{Cor} = J^2 \sum_{i \neq 3} C_i (p_i q_3 - p_3 q_i) \qquad (8)$$

This term will take on values large compared to the spacing between the "zero-order" vibrational levels, and perturbation theory would have to be carried out to very high order. Additional coupling to the quasi-continuum is provided by higher order terms in the anharmonic potential with a representative form,

$$\mathcal{H}_{anh} = V_{anh} f(q_3) \underset{i \neq 3}{F} (q_1 \ldots q_i \ldots) \qquad (9)$$

where f is a polynomial in the "selected" normal coordinate q_3 and F is a polynomial in the other normal coordinates.

It is not possible to carry out detailed quantitative calculations in the quasi-continuum, but obviously some ν_3 character will be admixed to many states in this continuum. Conversely, the higher steps in the ν_3 ladder will acquire a progressively higher admixture of the other states and will eventually become indistinguishable from the continuum, as shown on the left in Fig. 4. At the excitation $\nu_3 = 6$ the density of states $\rho_v(\nu)$ in the quasi-

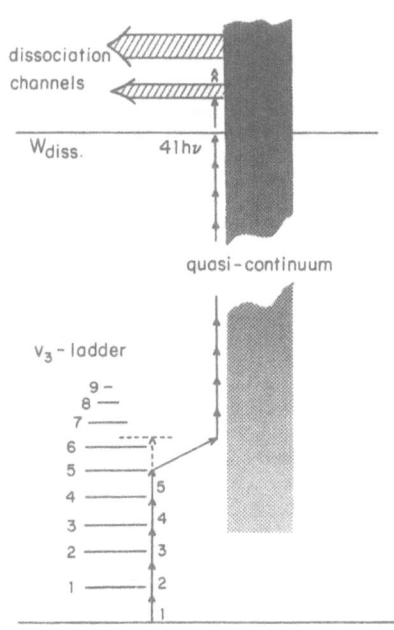

dissociation
channels

$W_{diss.}$ 41hν

quasi-continuum

ν_3 - ladder

Figure 4

The ladder of the selected "zero-order" normal mode $|v_3\rangle$ and the quasi-continuum of vibrational states. A possible path of multiphoton dissociation is indicated.

continuum exceeds $10^3/cm^{-1}$. Since $\rho_v(\nu) > t_p$, the quasi-continuum becomes a true continuum in the sense of the uncertainty principle.

The effect of the coupling is twofold. The states in the ν_3 ladder with an oscillator strength $|\mu(\nu)|^2\delta\{\nu(v_3+1)-\nu(v_3)\}$ will be displaced in energy. Some states may be pushed up in energy so that the anharmonic defect is partially compensated by Coriolis interaction. Simultaneously, part of the oscillator strength will be distributed over a continuum of frequencies. Suppose that the ν_3 admixture by the interactions (8) and (9) is spread out over a frequency interval $\Delta\nu$ in the continuum near the zero-order pure ν_3 state. Its order of magnitude is given by $h\Delta\nu \sim |\mathcal{H}_{Cor}| \sim 10 \ cm^{-1}$. Denote the total amount of admixture to the continuum by $|C_3(\nu_3)|^2\Delta\nu$. A reasonable value for this admixture may lie in the range of 1-10 percent for $v_3 \sim 6$.

In the preceding section it was shown that for a suitably chosen frequency and intensity of the laser pulse, states with $v_3 = 5$ might still be reached, but that the probability to reach the zero-order state $v_3 = 6$ is negligible. For $v_3 \geq 3$, it becomes, however, possible to make a transition to the continuum with the borrowed oscillator strength from the ν_3 vibration, while the energy can be conserved by the spread of this oscillator strength over a continuous range. Consider for definiteness the six-photon process, in which for the first five steps the large discrete oscillator strength of the pure "zero-order" ν_3 ladder is used, while for the final energy conserving step the reduced borrowed oscillator strength is used with a density $|C_3(\nu)|^2$ $|\frac{1}{2} \mu E|^2$. The transition rate to this continuum may be calculated with Fermi's Golden Rule, as coherence effects are washed out by an integration of the Rabi-type probability $p_R(n+1)$ over a distribution of frequencies. The time proportional probability to make an (n+1)-photon transition to the continuum is consequently

$$d \ \bar{p}_{cont}(n+1)/dt = \bar{p}_R(n) \ \hbar^{-2}|\frac{1}{2} \mu E|^2|C_3(\nu)|^2 \tag{10}$$

where $\bar{p}_R$ may be calculated with the aid of (6) and (7). The bar denotes not only a time average, but also a statistical average over the initial distribution of rotational states and a summation over possible final rotational states. It is, of course, not entirely correct to divide the process into two separate factors, one for transitions in the discrete ν_3 ladder, and one for the continuum. The advantage of this procedure is that it affords insight as to how the dipole oscillator strength is distributed in the frequency domain. Formally a time proportional transition probability by using (6) $p(n+1,t)$ with $\omega_R^{eff}(n+1)$ given by (7), and summing over all possible final vibrational-rotational states is equivalent. The last integration leads again to a constant transition rate. Eq.(10) shows clearly that even for $|C_3(\nu)|^2$ as low as $10^{-2}/cm^{-1}$ the probability to reach the bottom of the continuum during the early part of the laser pulse is near unity for molecules in the proper range of initial values of J_0, since even for field strengths well below the threshold field the condition $|\frac{1}{2} \mu E|^2|C_3(\nu)|^2 t_p > 1$ can be satisfied.

2.4 Continuum Absorption and Dissociation

The argument may be extended to further absorption of photons by the distributed dipole oscillator strength in the continuum. As the excitation energy increases, all vibrational and rotational degrees of freedom become completely scrambled, and the normal coordinates are no more privileged than the Cartesian coordinates of the individual atoms. The dipole oscillator strength associated at low excitation with the ν_3 and ν_4 normal modes is now perhaps spread into a continuum of several hundred wave numbers. It can best be expressed in terms of an absorption cross-section for the continuum [20] $\sigma_{cont}(\nu)$. Its order of magnitude may be estimated to be about one hundred times smaller than the maximum absorption cross-section in the P branch of the ν_3 vibration at room temperature, since the oscillator strength is spread out over a hundred times wider frequency range. Thus, $\sigma_{cont}(\nu) \approx 10^{-19}$ cm^2 per molecule should be considered an average value. Especially for the lower energy part of the continuum, $\sigma(\nu)$ will display structure.

Several authors have discussed the ergodic nature of the vibrational continuum [19,26,27,28] in connection with monomolecular reactions. The models are necessarily simplified. Most often harmonic oscillator levels are used with damping and/or anharmonic coupling terms added in an ad hoc fashion. The elaborate numerical calculations cannot be trusted to give realistic results for the complex real situation. It appears, nevertheless, likely that the molecular state does not cover the whole ergodic surface uniformly until dissociation channels open up. Fortunately, the details of the continuum interactions and the question of ergodicity are immaterial for our purpose. If the molecule by continuum absorption has acquired sufficient energy to open up one or more dissociation channels, it will dissociate essentially instantaneously on the time scale of t_p. It does not matter how this energy is distributed until the dissociation limit is exceeded.

The amount of energy absorbed by the molecule is obtained by adding the continuum absorption during the later part of the laser pulse to the energy of the photons absorbed in the first few steps on the discrete ladder during the early part of the pulse. Within the accuracy of these calculations, the total amount of energy is approximated by

$$W_{abs} = \sigma_{cont}(\nu) \, I(\nu) \, t_p \tag{11}$$

where $I(\nu)$ is the intensity (power flux density) of the laser pulse. When this amount exceeds the dissociation energy W_{diss} by an activation energy ΔW_{act} dissociation will take place. The probability for a molecule to be dissociated is consequently,

$$P_{diss} \approx \bar{p}_R(n) \quad \text{for} \quad W_{abs} > W_{diss} + \Delta W_{act}$$
$$P_{diss} \approx 0 \quad \text{for} \quad W_{abs} < W_{diss} + \Delta W_{act} \tag{12}$$

Here $\bar{p}_R(n)$ is the average probability occurring in (10) to reach a vibrational level with $v_R = n$, given by (6) and (7). The choice for n in the case of SF$_6$ lies probably in the range 3 to 5. Eq.(11) is, of course, an oversimplification which ignores both the time interval and the energy involved in the initial n-photon selective process. For $\Delta W_{act} \ll W_{diss}$, this simple model leads to an energy threshold per pulse,

$$\sigma_{cont}(\nu) \, I_{th}(\nu) \, t_p \approx W_{diss} \tag{13}$$

which is determined by the continuum absorption cross-section.

3. Comparison with Experimental Results and Discussion

The model developed in Section 2 can account for most experimental features enumerated in Section 1, which are here taken up in the same order:

1. Dissociation of isolated polyatomic molecules is clearly theoretically possible. Collisions are not necessary; neither rotational nor vibrational energy needs to be provided.

2. For power densities well above threshold the dissociation probability per molecule will approach unity. At a power level of 10^9 watts/cm^2 with a pulse duration of 10^{-7} sec, nearly all molecules regardless of initial rotational quantum number J_0 may be excited up to and through the continuum to the dissociation level.

3. Collisions with non-excited molecules will lead to a loss of excitation energy and thus to a lowering of the dissociation rate. Collisions are not necessary to reach the dissociation limit. Thus, the dissociation rate per molecule is highest at the lowest pressure. It is assumed that the dissociation products cannot recombine chemically in the interval between pulses. If recombination reactions occur, a scavenger gas at a sufficient partial pressure may be beneficial.

4. The experimental threshold of 23×10^6 watts/cm^2 for $t_p = 8 \times 10^{-8}$ sec is in agreement with (13), if $\sigma_{cont}(\nu)$ is chosen to be 4×10^{-19} cm^2, which is a reasonable value. No data are available to verify the prediction that the process has an energy threshold, i.e. the power threshold is inversely proportional to the pulse duration.

Above the threshold the intensity dependence of the dissociation probability per molecule is determined by $\bar{p}_R(n)$ given by a suitable average of (6) and (7). As the intensity increases, a larger fraction of the molecules is lifted into the quasi-continuum. If the quotients in the perturbation expression (7) are small compared to unity, the initial range of rotational quantumnumbers for molecules which could be raised to $v_3 = n$, would be given by $2\tilde{B}\Delta J_0 \sim \hbar \omega_R^{eff}(n)$. This expression would predict that the fraction of molecules dissociated would increase as $|\mu E|^n$ or proportional to $I^{n/2}$. As explained in Section 2, the perturbation approach has very limited validity, and several of the quotients in (7) would be near their maximum value of unity. This would lower the power dependence of $\bar{p}_R(n)$ on the radiation intensity I. Without detailed computer calculations, all that can be said is that the experimentally observed power law I^m, where m ranges between 1.5 and 3 above threshold, is not inconsistent with the requirement to raise the molecules in a n-photon process to the $v_3 = n$ with n lying in a range between 3 and 6.

5. The downward frequency shift S from the Q band head, at which maximum dissociation occurs should also be indicative of the number of the photons involved in the selective initial excitation process. With the simple energy level spectrum used in Sections 1 and 2, (2) would predict a coherent n-photon process, for a frequency detuning S, given by

$$n-1 = |S|/|X_{33}|\nu_3$$

Unfortunately the experimental values of $|S|$ have been variously reported as 8 cm^{-1} (single laser beam experiments with maximum dissociation at 940 cm^{-1}) and as 4 cm^{-1} (two experiments with a second laser beam at a different frequency). The peak intensities of the first beam, at which these experiments were performed should also be taken into account. Furthermore, $|X_{33}|$ is not known to better than 25 percent. One may again conclude that these results are consistent with a value of n lying in the range between 3 and 6. The dissociation response curve has a width that is larger than that of the individual P and R branches, and this is not inconsistent with the selection rule $|\Delta J| \leq n$.

6. Continuum absorption is not selective in frequency. Once a molecule is excited to the level $v_3 = n$, it may be dissociated by an intensity exceeding the threshold value at an arbitrary frequency in a relatively wide frequency range, because $\sigma_{cont}(\nu)$ has no sharp features. The frequency selective absorption in the first stages of the ladder may be provided by radiation at much lower intensities. At power levels as low as 10^4 watts/cm^2 molecules in selected rotational states will still be excited to levels with $v_3 = 3$ and 4 by P Q R and P P R R sequences. This feature was not appreciated in early double resonance experiments [29,30]. There is a need for more systematic double resonance experiments at low pressures, in which the frequencies and power levels of two simultaneous irradiation fields are varied independently.

7. When the probability $\bar{p}_R(n)$ can still be appreciable at such low power levels, it is also clear from the form of (7) that it can be appreciable for smaller dipole matrix elements. Thus, irradiation near weak combination bands, e.g. $\nu_2 + \nu_4$, will also lead to dissociation, with the same threshold set by continuum absorption. The dissociation probability per molecule has been reported [8] to be proportional to the strength of the dipole matrix element $|\mu|$. This last feature is not understood within the context of this paper. For a three-photon transition to the state $3\nu_2 + 3\nu_4$, one would expect that molecules in a selected range of values ΔJ_0 given by $2\bar{B}\Delta J_0 = |\mu E|^3$ would be able to find a P Q R ladder of nearly equally spaced energy levels. The experimental observation suggests that only one step is made on the combination band transition, or alternatively that the higher combination states derive an increased oscillator strength from ν_3 and/or ν_4 admixtures.

8. Since diatomic molecules have no vibrational continuum, they cannot be dissociated by monochromatic infrared radiation. The mechanism should, however, be effective in a variety of polyatomic molecules. Although the general features are similar, the distinct characteristics of individual molecular species will undoubtedly be apparent in detailed parametric studies of the power and frequency dependence for each species. Much further experimental and theoretical work is needed.

As the intensity is increased to high values so that $\hbar \omega_R$ is larger than the anharmonic defect necessary to reach the levels $v_3 = 3$ or 4, the quantummechanical model yields results similar to the classical anharmonic oscillator, as shown by comparing the energy of excitation of the quantummechanical model in Fig. 2 with the classical response curve [20]. This classical anharmonic oscillator must also be coupled to a continuum to reach the dissociation energy. The dynamics of the continuum can be modeled in a classical manner, as proposed by LAMB [31]. The continuum dynamics may be described in terms of Cartesian coordinates of the individual atoms. They are coupled to the IR field via effective charges. They interact with each other via effective induced polarizabilities and short range repulsive forces. The periodic shaking of the classical system will eventually lead to the escape of one or more fluorine atoms from the molecule SF$_6$. A type of Monte-Carlo program for different initial conditions is probably within the capabilities of present-day calculating machines. In our opinion no classical program can be used to describe the first steps up the anharmonic ladder for $\hbar \omega_R \ll h\nu_3 |X_{33}|$ where the distinct resonances of Fig. 1 occur.

The isotopic selectivity in the collisionless regime is a straightforward consequence of the model. The frequency response $\bar{p}_R(n)$ of the excitation of the first n-steps of the anharmonic ladder is simply displaced in frequency for a second isotopic species, as the resonant frequency is inversely proportional to the square root of the reduced mass of the normal vibration. For $S^{34}F_6$ the shift is 17 cm^{-1} toward lower frequencies compared to $S^{32}F_6$ for the ν_3 vibration. Isotopic separation can be obtained with a probability of unity [32] for very modest powers of the selective excitation field, provided an off-resonance continuum pumping field is present to satisfy the threshold condition given by (13).

If the isotopic shift is smaller than the width of the average response function $\bar{p}_R(n)$, which is on the order of the width of the rotational bands, one may still obtain an adequate selectivity factor by operating on the slope of the response curves $\bar{p}_R(n,\omega_A) - \bar{p}_R(n,\omega_B)$ for the two isotopic species A and B.

In practice, requirements of high throughput and efficient use of available infrared power will necessitate operation in a regime of pressures where collisions, even during the short laser pulse, cannot be ignored. Scavenger gas pressures may also be necessary to take away the dissociation products. The problem of collisions with V-V transfer and chemical reactions between excited species of molecules adds a new dimension of complexity to the problem. Many papers have stressed the role played by collisions in chemical dissociation reactions [32]. This paper does not deal with these vast, and sometimes bewildering, aspects of laser chemistry. The discussion has been restricted to the physics of one polyatomic molecule in an infrared radiation field. This problem already offers plenty of complexity, but it is believed that the essential features of collisionless dissociation have been identified.

References

1. R.V. Ambartzumian, V.S. Letokhov, E.A. Ryabov, N.V. Chekalin: JETP Lett. 20, 273 (1974)

2. J.L. Lyman, R.J. Jensen, J. Rink, C.P. Robinson, S.D. Rockwood: Appl. Phys. Lett. 27, 87 (1975)

3. M.C. Borde, A. Henry, M.L. Henry: C.R. Acad. Sc. (Paris) B262, 1389 (1966) and B263, 619 (1967); N.V. Karlov, Yu. Petrov, A.M. Prokhorov, D.M. Stel'Markh: JETP Lett. 11, 135 (1970)

4. N.K. Isenor, M.C. Richardson: Appl. Phys. Lett. 18, 225 (1971); N.K. Isenor, V. Merchant, R.S. Hallsworth, M.C. Richardson: Can. J. Phys. 51, 281 (1973)

5. E. Yablonovitch: (to be published)

6. R.V. Ambartzumian, Yu.A. Gorokhov, V.S. Letokhov, G.N. Makarov: ZhETF Pis. Red. 21, 375 (1975) [JETP Lett. 21, 171 (1975)]

7. R.V. Ambartzumian, Yu.A. Gorokhov, V.S. Letokhov, G.N. Makarov: ZhETF Pis. Red. 22, 96 and 374 (1975)

8. R.V. Ambartzumian: paper presented at the Nordfjord Conference on Tunable Lasers and Applications, June (1976)

9. L.S. Letokhov: paper presented at the Laser Chemistry Conference, Steamboat Springs, Colorado, February (1976)

10. S.D. Rockwood: papers presented at Nordfjord Conference on Tunable Lasers and Applications, June (1976) and at the Laser Chemistry Conference, Steamboat Springs, Colorado, February (1976)

11. A.V. Nowak, J.L. Lyman: J. Quant. Spectrosc. Rad. Transfer 15, 945 (1975)

12. J.P. Aldridge, H. Filip, H. Flicker, R.F. Holland, R.S. McDowell, N.G. Nereson, K. Fox: J. Mol. Spectrosc. 58, 165 (1975); R.S. McDowell, H.W. Galbraith, B.J. Krohn, C.D. Cantrell, E.D. Hinkley: Opt. Comm. (to be published); C.D. Cantrell, H.W. Galbraith: J. Mol. Spectrosc. 58, 158 (1975)

13. S.A. Alimpiev, N.V. Karlov: ZhETF 66, 542 (1974) [Sov. Phys.- JETP 39, 260 (1974)]; see also, I. Burak, J.I. Steinfeld, D.G. Sutton: J. Quant. Spectrosc. Rad. Transfer 9, 959 (1969)

14. C.D. Jackson: Los Alamos Scientific Laboratory Report LA-6025-MS-1975

15. R.J. Glauber: Phus. Rev. 130, 2529 and 131, 2766 (1963)

16. W.H. Louisell: Quantum Statistical Properties of Radiation (John Wiley, New York 1973) p. 205

17. D.M. Larsen, N. Bloembergen: Opt. Comm. to be published (1976)

18. C.J. Elliott, B.J. Feldman: Bull. Am. Phys. Soc. 20, 1282 (1975)

19. S. Mukamel, J. Jortner: (to be published); receipt of a preprint, "A model for isotope separation via molecular multiphoton dissociation", is acknowledged

20. N. Bloembergen: Opt. Comm. 15, 416 (1975)

21. R.V. Ambartzumian, Yu.A. Gorokhov, V.S. Letokhov, G.B. Makarov, A.A. Puretzky: ZhETF Pis. Red. 23, January (1976)

22. Private communications by P.L. Kelley, R.V. Khokhlov, V. Platonenko

23. F.H.M. Faisal: (to be published); receipt of a preprint, "A model for dissociation of polyatomic molecules by multiple absorption of photons", is acknowledged

24. V.S. Letokhov, A.A. Makarov: Opt. Comm. (to be published)

25. G. Herzberg: Molecular Spectra and Molecular Structure, Vol. 2, 1st ed. (Van Nostrand, New York 1945) pp 375 and 447

26. M.F. Goodman, J. Stone: Phys. Rev. A5, 1355 (1972); J. Stone, E. Thiele, M.F. Goodman: J. Chem. Phys. 59, 2909 (1973); M.F. Goodman, J. Stone, E. Thiele: J. Chem. Phys. 63, 2929 (1975); J. Stone, E. Thiele, M.F. Goodman: J. Chem. Phys. 63, 2936 (1975); J. Stone, M.F. Goodman: Phys. Rev. A (in press)

27. E.J. Heller, S.A. Rice: J. Chem. Phys. 61, 936 (1974)

28. S. Nordholm, S.A. Rice: J. Chem. Phys. 61, 203 and 768 (1974)

29. D.S. Frankel, J.I. Steinfeld: J. Chem. Phys. 62, 3358 (1975)

30. J.I. Steinfeld, I. Burak, D.G. Sutton, A.V. Nowak: J. Chem. Phys. 52, 5421 (1970)

31. W.E. Lamb, Jr.: paper presented at the Conference on Laser Chemistry, Steamboat Springs, Colorado, February (1976)

32. See, for example, the following reviews: V.S. Letokhov: Science 180, 451 (1973); C.B. Moore: Accounts Chem. Res. 6, 323 (1973); N.V. Karlov: Appl. Opt. 13, 301 (1974); J.P. Aldridge, J.H. Birely, C.D. Cantrell, D.C. Cartwright: submitted to Physics of Quantum Electronics, Vol. 4 S.F. Jacobs, M. Sargent III, M.O. Scully, C.T. Walker, eds. (Addison-Wesley, Reading); V.S. Letokhov, C.B. Moore: Sov. J. Quant. Electronics 3, 247 and 485 (1976).

LASER EXCITATION OF MOLECULES TO HIGH STATES OF VIBRATION

K.L. Kompa

Max-Planck-Gesellschaft zur Förderung der Wissenschaften e.V.
Projektgruppe für Laserforschung, 8046 Garching, Germany

I. Introduction

High vibrational levels of molecules have gained interest in connection with
some new concepts of isotope separation. Specific vibrational excitation of
molecules is also of very basic interest to the reaction dynamicist who wishes
to study the reactive behaviour of molecules prepared in a well-defined set
of quantum states. It is the general context of laser chemistry at which the
following report is primarily oriented. It reflects the early phase of a systematic study of the reactive behaviour of vibrationally excited molecules.
Obviously these are old problems in the sense that the detailed energy dependence of chemical reaction rates has always been of central importance in
our understanding of chemical processes. The new approach to these old problems is made promising by a new tool, the laser. This has indeed yielded
some unexpected results. It is practical to distinguish from an experimental
point of view work that has been done with single level excitation of low
lying vibrational levels from distributed high level excitation. While the
study of elementary chemical processes by specific excitation of low vibrational levels has been the subject of a number of papers within the last
few years [1], the possibility to bring molecules into states of very high
excitation has only opened up very recently. One method by which this can be
accomplished is now termed "Infrared molecular superexcitation". This involves
the direct - often dissociative - excitation of high molecular levels
through purely radiative processes (fig. 1). This new phenomenon which appears
to have been demonstrated by now for a number of polyatomic molecules [2] -
e.g. SF_6, BCl_3, OsO_4, SiF_4, CF_2Cl_2, CH_2Cl_2, CCl_4 - is to be looked upon as
one out of several different approaches having the same goal. One of these
schemes uses electronic excitation to generate high vibrational levels in the
electronic ground state either following visible or UV fluorescence ("Franck-
Condon pumping") or via internal conversion. These two possibilities are ex-
emplified in figs. 2 and 3. A third approach which may be referred to as
"Treanor pumping" [4] rests on the probability of V-V-up-pumping in collisions between excited molecules as indicated in fig. 4. In a practical experimental situation it is likely that a combination of these schemes can be
employed. In particular the interaction of superexcitation (fig. 1) and collisional excitation transfer (fig. 4) is of relevance here. It is indeed this

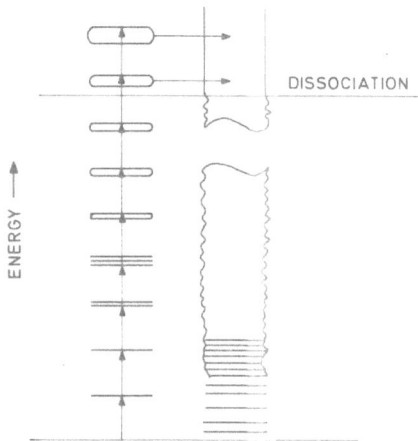

Fig.1: Direct mechanism of dissociative
superexcitation (after COTTER [3])

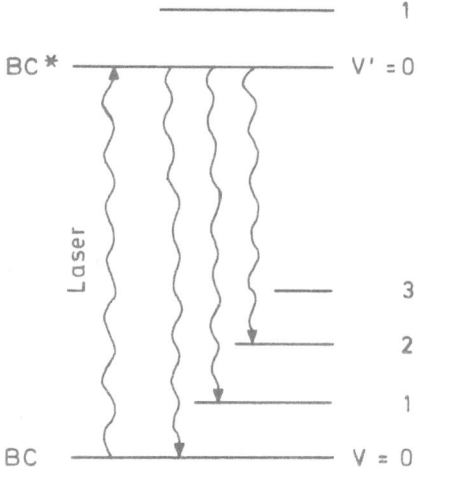

Fig.2: Pumping of vibrational levels
of the ground state molecule
BC via laser-induced
fluorescence of BC*

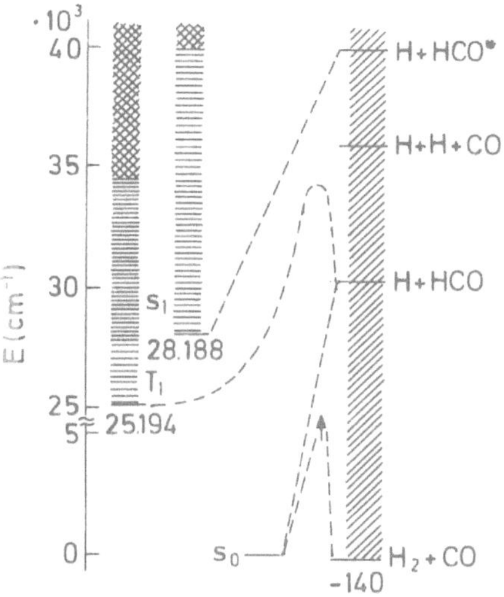

Fig.3: Internal conversion as discussed for the laser-induced
dissociation of formaldehyde
in [5]

combination which provides most interesting perspectives not only for basic kinetic and spectroscopic studies but also for future chemical syntheses.

II. Experiments in Vibrational Superexcitation

The first experiments that were done by American and Russian workers to selectively dissociate certain isotopic constituents of $^{32,34}SF_6$ or $^{10,11}BCl_3$ by a pulsed CO_2 laser [2] used very simple arrangements of the type shown in fig. 5. To avoid the inherent difficulties of measurements with focussed beams FUSS and COTTER [7] chose to use for their measurements cavity reflection cells as indicated in fig. 6 which permit high radiation levels together with a rather homogeneous intensity distribution over relatively large volume areas. Typical results are given in fig. 7. It is clearly demonstrated here that SF_6 can indeed be dissociated with high isotopic selectivity and high probability under the conditions chosen. With these simple table-top experiments in the case of SF_6 chemical conversions nearly in the milligram range are attained with a single laser shot. Without much further optimization it is likely that these laboratory set-ups could be scaled up to production rates of grams per day of, for instance, $^{32}SF_6$. It was shown in separate experiments by SCHMID et al. [8] using mass spectrometry and optical absorption measurements in the UV that two main products of SF_6 fragmentation under these conditions are SF_4 and F_2. F_2 was observable at laser energies $\geqslant 2$ J/cm². This result is surprising as the dissociation of SF_6 into SF_4 and F_2 does not constitute the lowest energy dissociation channel. Details connected with the chemistry of the process and the role of scavengers for the dissociation products are omitted from this discussion for the sake of brevity.

The above mentioned measurements of the dissociation yield were supplemented by measurements of the laser energy deposition for various CO_2 laser lines by STAFAST [9]. In fig. 8 some representative results are reproduced which show the variation of the energy absorption cross section with varying laser energy.

Various models have been proposed to understand the mechanisms of molecular superexcitation and dissociation [10]. We briefly summarize here the approach of COTTER [3] which subsequently has been applied to the interpretation of the absorption and dissociation measurements described above. The general characteristics of this model are the following. The absorbing molecules are described as an ensemble of one-dimensional, forced anharmonic oscillators which are subject to near resonant coherent excitation. The excitation depends on the frequency ω and amplitude E of the applied electromagnetic field. The parameters of the oscillator which enter into the discussion are the effective dipole transition moment μ, the small amplitude resonance frequency ω_o, the anharmonic shift coefficient x, and the average damping rate γ. In the lowest order of approximation the expectation value of the average vibrational excitation $\bar{\upsilon}$ is given (with $\omega_R = \mu E/\hbar$)

$$\bar{\upsilon} = \omega_R^2 / (\omega_o + x\bar{\upsilon} - \omega)^2 + \gamma^2$$

It is shown by COTTER [3] how the excitation function of the oscillator $\bar{\upsilon}(\omega_o-\omega)$ may be derived from this expression by choice of the relevant root. The excitation function has Lorentzian wings, and a peaked asymmetric central region around ω_o of frequency width $\sim (\mu E/\hbar)^{2/3} x^{1/3}$ and maximum excitation $\sim (\mu E/\hbar)^{2/3}$. A qualitative correlation is provided by this model for the absorption measurements reproduced in fig. 8. These measurements were performed at four frequencies within the ν_3 band of SF_6 with laser intensities resulting in excitation of up to $\bar{\upsilon} \cong 40$. These data can be represented by a theoretical fit with $\mu \cong 3$ debye and $x \cong 1$ cm^{-1}. It is assumed that the molecule dissociates rapidly when a certain level of excitation $\bar{\upsilon}_{Diss}$ is

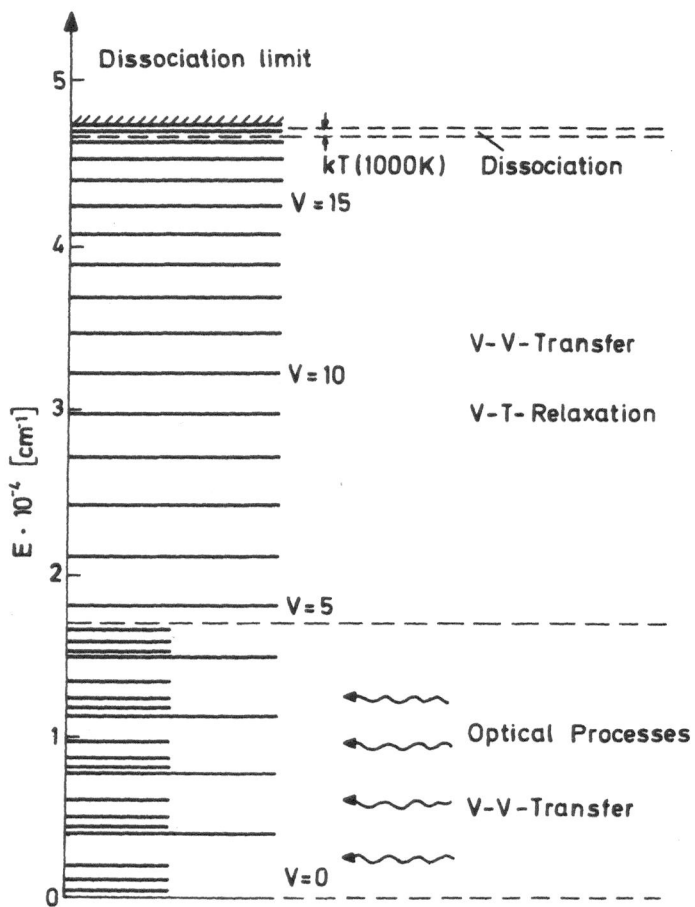

Fig.4:. Collisional excitation transfer in the case of hydrogen fluoride
(after PUMMER [6]). Following the optical excitation of the vibrational levels V ≤ 4 by a high power pulsed HF laser V-V-up-
pumping can provide for energy transport up to the dissociation
limit

reached. The curves in fig. 7 can also be derived theoretically with the
adjustable parameters μ and x to show the dependence of the dissociation
probability P on the laser energy from a threshold near $P \simeq 10^{-3}$, to saturation at $P \simeq 1$.

Previous investigations of molecular dissociation via IMS suffered from
the fact that time-integrated measurements had to be used. In such experiments chemical changes that occur after the laser pulse due to collisional
and wall effects can sometimes not fully be taken into account. Therefore a
molecular beam study was initiated by BRUNNER et al. to investigate in an

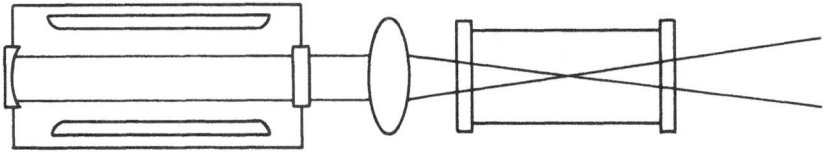

Pulsed CO₂ Laser, Lens, Absorption Cell

Fig.5: Experimental arrangement to dissociate polyatomic molecules by the focussed pulse of a high power CO_2 laser

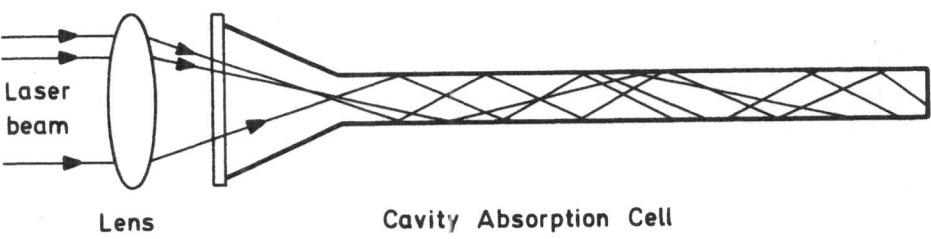

Laser beam

Lens   Cavity Absorption Cell

Fig.6: Cavity absorption cell used in [7] in the study of the superexcitation of SF_6

unambiguous fashion the dissociation probability of SF_6 under irradiation with a range of CO_2 laser lines and energies. To this end a molecular beam apparatus has been built which permits the study of any reduction in beam intensity as a result of the laser interaction as well as the study of possible fragmentation products with regard to their chemical nature and primary kinetic energy. The design characteristics of this experiment are similar to the photofragment spectrometer first described by K. WILSON [11]. Details of the experiment are shown in fig. 9. Two cryo pumps (> 10^4 1/sec) produce oil-free UHV conditions necessary for low noise mass spectrometric measurements. The beam geometry is variable in such a way that the beam points either directly towards the ion source if beam reactive recoil loss effects are to be observed or that it is directed perpendicular to the line of sight of the mass spectrometer for the study of fragments scattered from the beam. The CO_2 laser is a UV-preionized, line selectable oscillator with a maximum pulse energy of 30 J. A cylindrical lens condenses the laser beam in such a way that complete overlap with the molecular beam over the region of interest is assured. A quadrupole mass spectrometer together with a single-ion pulse counting technique is used as the detector. Experimental control and data processing is performed by a PDP 11/10 laboratory computer. The set-up has been used up to now to study the superexcitation of sulfur hexafluoride irradiating the molecule in its ν_3 mode by the P(20) line of the 001 - 100

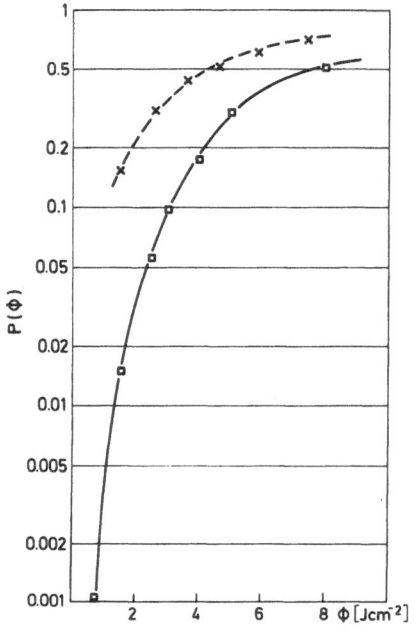

Fig.7: Dissociation probability P(ϕ) of SF_6 versus laser energy ϕ measured in the arrangement of fig. 6 by FUSS and COTTER. For a detailed discussion see [3] and [7]. For the upper curve (dashed line) which has been determined by BRUNNER et al. [12] see below

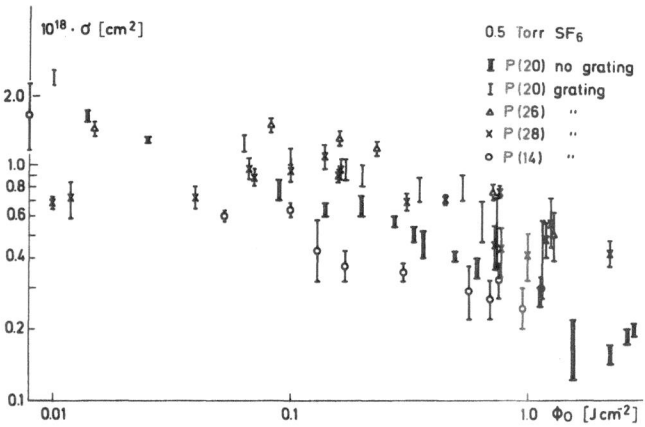

Fig.8: Absorption cross section versus laser energy for four CO_2 laser lines across the ν_3 band of SF_6 (after STAFAST et al. [9]). The characteristic structure of these plots is discussed in [3]

CO_2 laser band. The molecular beam was formed by a glass capillary array at a temperature of 120 K. The dissociative recoil loss of molecules from the beam by the laser interaction manifests itself as a change in the countrate for any of the ions that are generated from SF_6 in the mass spectrometer. The sum of these single ion counts can also be determined if the quadrupole mass filter is switched off. Knowing the temperature and the flight path of the molecules and assuming a Boltzmann velocity distribution in the beam the loss effect can also be simulated numerically with the dissociation probability as the only adjustable parameter. Measurements were made with laser energies ranging from 0.8 - 8 J cm^{-2}. The experimental results are reproduced in fig. 10. It is seen by the integral measurement of the total reduction of the particle flow (quadrupole mass filter switched off) that the density of molecules in the beam is reduced to ~ 30 % of its initial value at the highest laser energies used. We believe that this is a true measure of the dissociation of SF_6 molecules by the laser. This assumption is justified because the ionization energy in the mass spectrometer is with 70 eV sufficiently high that the total ionization probability of a molecule will not depend on its state of vibrational excitation. The variation in the fragmentation

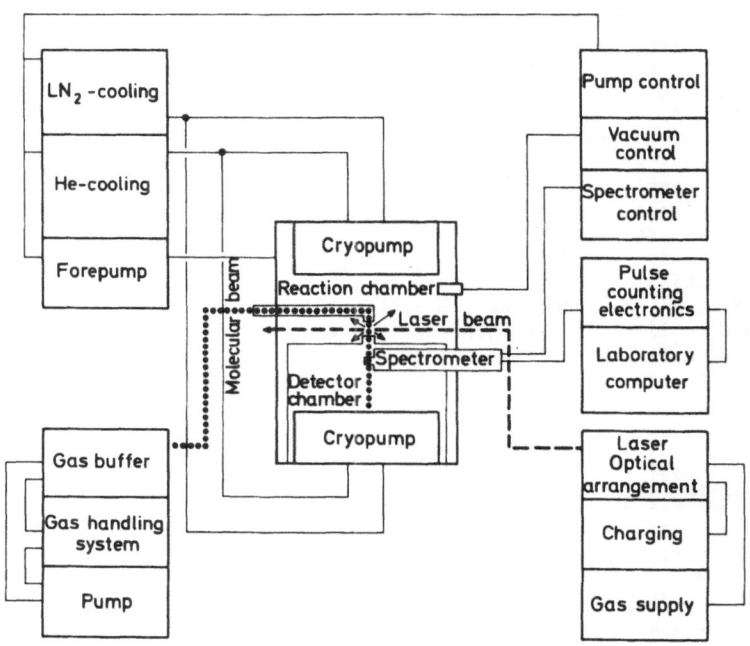

Fig.9: Functional diagram of molecular beam experiment to study vibrational excitation and fragmentation of polyatomic molecules by infrared lasers [12]

ion concentration as function of the laser energy, however, seems to indicate that the <u>fragmentation pattern</u> in the mass spectrometer does indeed depend on the <u>vibrational</u> energy content of the molecules. Thus different secondary ions are formed at different laser energies. Correspondingly the appearance probability for SF_5^+, SF_4^+, SF_3^+ is given by two contributions, the dissociation probability $P(\phi)$ and the probability that a certain ion channel is used $W(\phi)$. Calculated on the assumption that the integral measurement gives $P(\phi)$ the various probabilities $W(\phi)$ (normalized to the probability $W(o) = 1$ for the thermal gas at the temperature chosen) are shown in fig. 11. While the formation of SF_5^+, the dominant ion at thermal energies is increasingly less probable as the laser energy increases the ions SF_4^+ and SF_3^+ are formed in higher concentrations. The decrease in their concentrations at the highest energies used indicates that even smaller fragment ions are entering into the picture then.

The data reported here are preliminary in the sense that their presicion is estimated to be 20 % and that radiational losses that may affect the vibrational energy of the molecules on the flight path prior to detection have not been taken into account.

The conclusions to be drawn from these molecular beam measurements are the following:

1) To our knowledge these measurements represent the first direct observation of collisionless dissociation via infrared laser superexcitation.

2) The dissociation probability of SF_6 under irradiation with the P(20) line of the CO_2 laser is found to be $P(\phi) \simeq 0.5$ at an energy of $\phi = 5$ J/cm².

3) It is believed that the characteristic mass spectrometric fragmentation pattern of SF_6 which is found for different laser energies also provides a diagnostic for the state of excitation of the molecules prior to dissociation. It is hoped that results like the ones shown in fig. 11 can be directly linked to the vibrational energy content of the molecules.

As is shown in fig. 7 there remains still a difference to be explained for the dissociation yield measurements between the molecular beam results of BRUNNER et al. [12] and the time-integrated data of FUSS and COTTER [7]. As has been mentioned above, however, the two experiments started with SF_6 molecules of different thermal temperatures. Besides, it is hoped that a detailed discussion of the experimental parameters may resolve this discrepancy.

III. Experiments Involving Collisional Excitation Transfer

Unfortunately the direct combination of superexcitation and V-V-transfer collisions which is involved in some of the SF_6 experiments is still hard to interpret and will therefore not be discussed in detail in this paper. Instead, we present here the study of vibrational excitation and dissociation of HF as a powerful example of V-V-up-pumping in a simple diatomic molecule. The characteristics of the pumping scheme according to PUMMER [6] are displayed in fig. 4. While the lowest vibrational levels are radiatively coupled to the HF laser pump the energy transport into the higher vibrational manifold is accomplished by collisional vibrational-vibrational energy exchange. The vibration is partly decoupled from the other degrees of freedom in a way first proposed by TREANOR et al [4]. A computer model has beed developed to describe the system using the rate equation approximation for the first 20

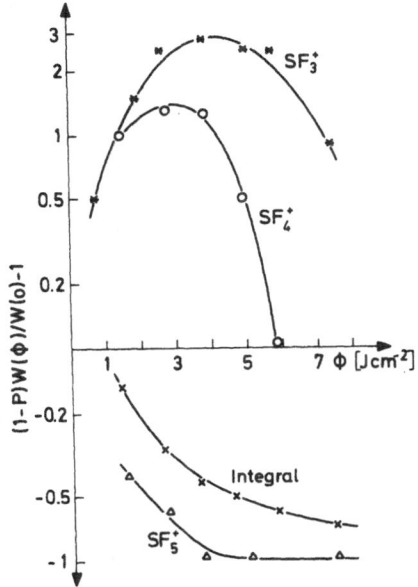

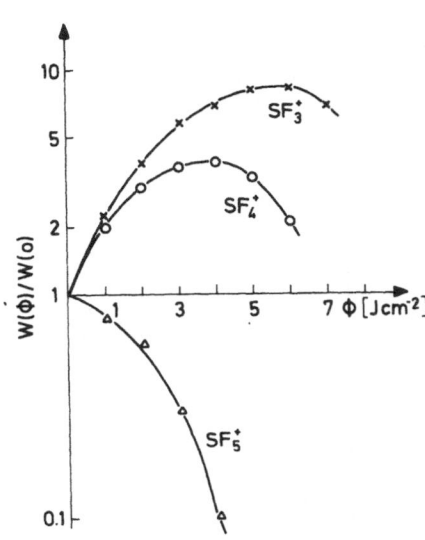

Fig.10: Apparent dissociation probability of SF_6 as function
of laser energy according to
BRUNNER et al. [12]. For
discussion see text

Fig.11: Fragment ions of vibrationally
excited SF_6 as function of the
laser energy [12] showing the
increasing probability for the
formation of lower mass fragments as the vibrational energy
content of SF_6 increases

vibrational levels, level 20 being connected to the dissociation continuum.
The laser pulses used in the calculations and experiments are in the range
of 0.1 to 100 MW/cm² with τ fwhm = 200 nsec and consist of 14 lines in the
P branch of the V = 4 - 3, 3 - 2, 2 - 1, and 1 - 0 bands. Gas pressures are
several 10 Torrs to guarantee an appreciable number of collisions during the
laser pulse.
    The model predicts
    a) a strong intensity dependent absorption which increases with increas-
       ing intensities,
    b) population inversion for some high rotational states of the V = 3 - 2,
       2 - 1, 1 - 0 bands, and
    c) considerable dissociation of the HF molecules.
These three points were verified experimentally. Compared to the necessary
simplifications made in the computer model, the agreement between theory and
experiment is remarkably good. It was demonstrated experimentally that the
vibrotational inversion created in this way can be used to operate an HF laser pumped HF laser (fig. 12).
    The results can be understood in terms of an energy reservoir formed by
the highly vibrationally excited molecules. These molecules are extremely

186

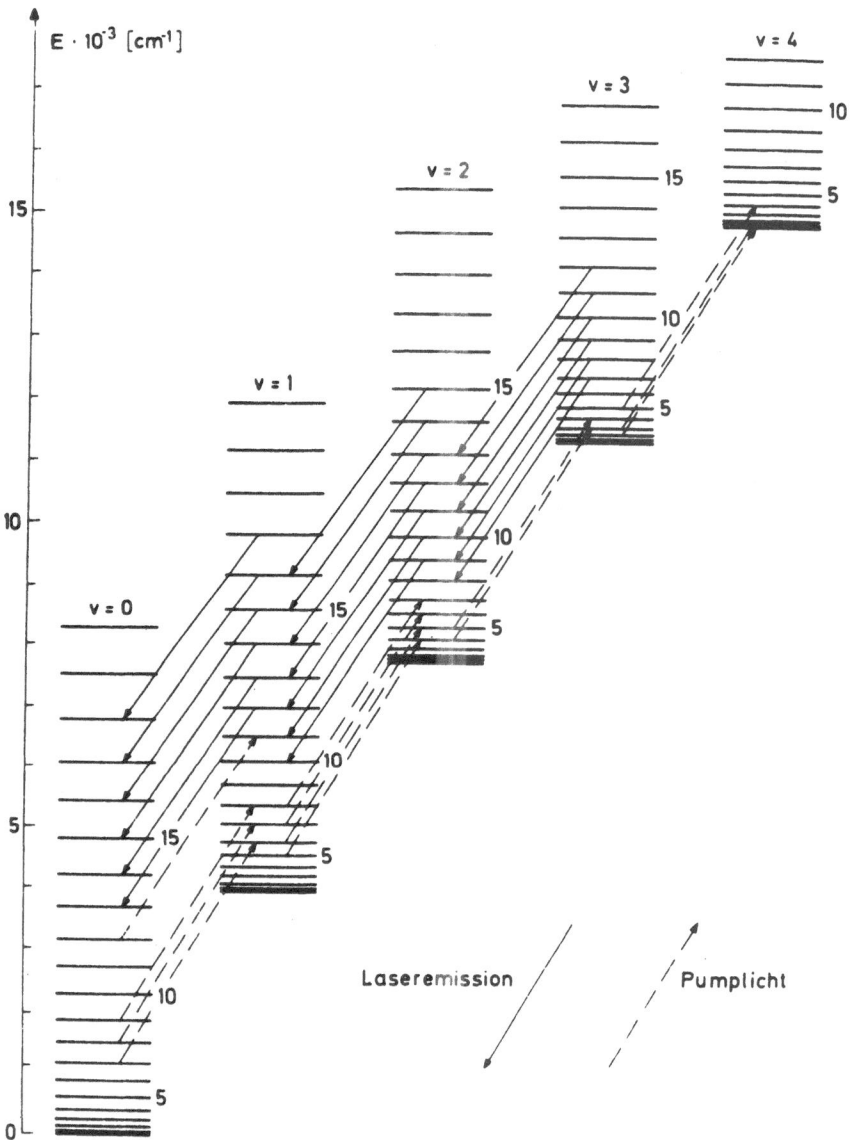

Fig.12: Pump and stimulated emission lines of the HF laser-pumped HF laser.
The emission lines indicated have been observed experimentally by
PUMMER et al. [6]

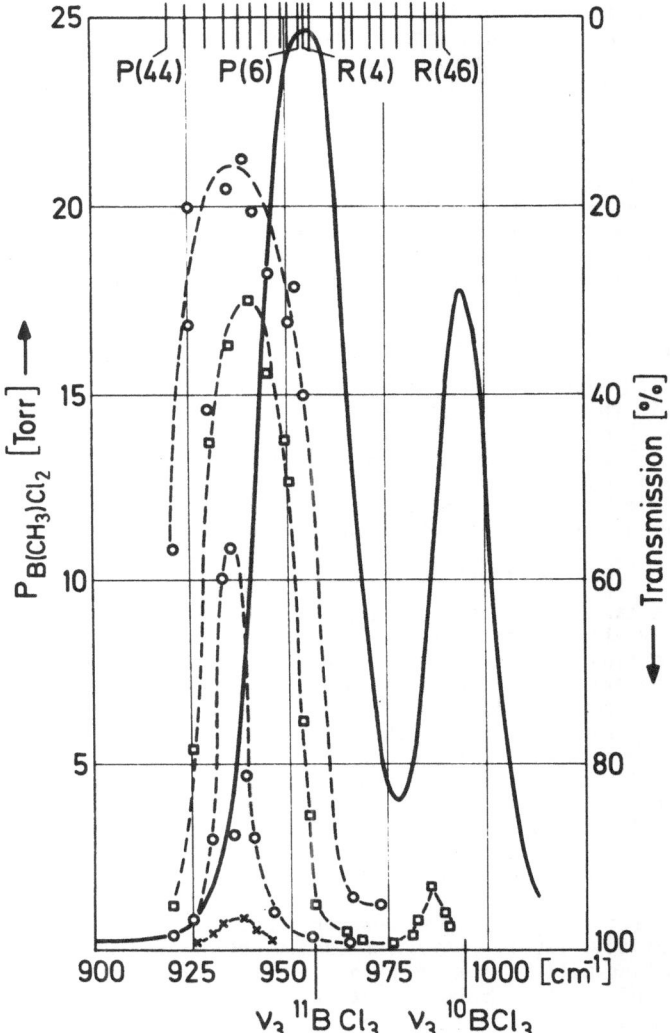

Fig.13: Infrared absorption of BCl₃ in comparison to the wavelength dependence of the product yield in the exchange reaction of BCl₃ with
B(CH₃)₃. Indicated at the top is the range of CO₂ laser lines which
can be used to initiate the reaction. For details see [14]

efficient in removing vibrational energy out of the optically pumped states by fast V-V collisions, in the same time restoring one collision partner to a state from which it can absorb again. A build-up of such a reservoir can thus enhance dissociation and absorption in a self-accelerating way. The increase of absorption with increasing pulse intensities which cannot be caused by cascade effects can be explained by this model.

The authors stress the general nature of this scheme for the excitation of molecules to high states of vibration.

In addition, a build-up of a reservoir containing an appreciable number of highly excited molecules causes very fast energy removal out of low vibrational states. In high power HF lasers operating at some 100 Torr this could be a major loss mechanism which limit the energy extraction from the laser medium.

IV. Outlook to the Future: Synthetic Application in Chemistry

Parallel to the work mentioned above a series of experiments was conducted to demonstrate the usefulness of specific laser excitation for synthetic chemistry. Many of these experiments were done with CW CO_2 lasers and the molecules involved often contained boron [13]. Without any systematic discussion we will only present one example of this type of laser chemistry here which comes from the work of RINCK et al. [14]. This example concerns the CO_2 laser-driven reaction of boron trichloride BCl_3 with boron trimethyl $B(CH_3)_3$.

$$2BCl_3 + B(CH_3)_3 \longrightarrow 3B(CH_3)Cl_2$$

This reaction can be initiated by many CO_2 laser emission lines and substantial product yields are possible. It is worth mentioning, however, that the highest conversion rates are found not for those laser lines for which maximum absorption exists but for lines which are shifted to the long-wavelength side from the absorption maximum. Fig. 13 shows the "yield spectrum" versus the absorption spectrum of BCl_3. The effect is interpreted by RINCK et al. [14] by the key role of an absorbing species that is only generated in the process and which absorbs at wavelengths longer than the small amplitude resonance absorption of BCl_3. It is assumed that this species is vibrationally excited BCl_3. In view of an estimated negative anharmonicity coefficient of BCl_3, $x \approx 1.6$ cm^{-1} the observed shift of about 15 cm^{-1} would be indicative of an activation energy roughly corresponding to 9 - 10 CO_2 laser quanta or $E_A \approx 25$ kcal/mole. A detailed discussion of these results would be beyond the scope of this report. We believe, however, that the data show the relevance of high vibrational levels even under conditions of CW irradiation. Future work of this kind will show to which extent laser chemistry is not only an academic area of reaction kinetics but also provides fruitful routes for industrial chemistry.

References

1 There are numerous discussions on the dynamic behaviour of state selected reactants in the literature. For a recent reference see for instance M.J. Berry, Laser Studies of Gas Phase Chemical Reaction Dynamics in Ann. Rev. Phys. Chem. 26 (1975)

2 R.V. Ambartzumian, V.S. Letokhov et al., JETP Lett. 20, 273 (1974), 21, 171 (1975), ZhETF Pis. Red. 22, 96, 374 (1975), J.L. Lyman et al., Appl. Phys. Lett. 27, 87 (1975), J.L. Lyman, S.D. Rockwood, J. Appl. Phys. in print, A. Yogev, R.M.J. Benmair, J. Am. Chem. Soc. 97, 4430 (1975)

3 T.P. Cotter, to be published

4 C.E. Treanor, J.W. Rich, R.G. Rehn, J. Chem. Phys. <u>48</u>, 1798 (1968)

5 E.S. Yeung, C.B. Moore, Appl. Phys. Lett. <u>21</u>, 109 (1972), J. Chem. Phys. <u>58</u>, 3988 (1973)

6 H. Pummer, Max-Planck-Institut für Plasmaphysik Laboratory Report IPP IV/87, March 1976, H. Pummer, D. Proch, U. Schmailzl, K.L. Kompa, to be published

7 W. Fuss, T.P. Cotter, to be published

8 W.E. Schmid, H. Stafast, K.L. Kompa, to be published

9 H. Stafast, T.P. Cotter, W.E. Schmid, to be published

10 For a general reference see Laser Spectroscopy, Proc. of the Second International Conference, Mêgeve, June 23 - 27, 1975

11 G.E. Busch. J.F. Cornelius, R.T. Mahoney, R.I. Morse, D.W. Schlosser, K. R. Wilson, Rev. Sci. Instr. <u>41</u>, 1066 (1970)

12 F. Brunner, T.P. Cotter, D. Proch, K.L. Kompa, to be published

13 H.R. Bachmann, K.L. Kompa, H. Nöth, R. Rinck, Chem. Phys. Lett. <u>29</u>, 627 (1974, <u>33</u>, 261 (1975), Chem. Phys. Lett., to be published, J. Phys. Chem., in print, H.R. Bachmann, F. Bachmann, K.L. Kompa, H. Nöth, R. Rinck, Chem. Ber., in print

14 R. Rinck, F. Bachmann, H. Nöth, K.L. Kompa, to be published Chem. Phys. Lett.

DOUBLE RESONANCE AND ENERGY TRANSFER
IN SULFUR HEXAFLUORIDE

J.I. Steinfeld and C.C. Jensen
Department of Chemistry, Massachusetts Institute of Technology
Cambridge, Massachusetts 02139, USA.

Since the discovery of multiple infrared photon dissociation of SF_6 and other molecules, by AMBARTZUMYAN [1] and other investigators [2], a great deal of effort has gone into an attempt to understand the mechanism of this process. Theoretical models have been proposed by BLOEMBERGEN [3], LARSEN [4], and others [5]. A principal question in these models is, how the absorption proceeds up through the vibrational manifold--that is, the time- and frequency-dependent distribution of dipole strength in laser-pumped SF_6 or other molecules. Another question of interest is the effect of collisional redistribution of vibrational energy on the dissociation process. Although it seems clear that the process occurs most efficiently under collision-free conditions, knowledge of the effect of relaxations would be helpful in scaling potentially useful multiphoton laser-induced reactions to higher pressures.

Existing data on the nature of the absorption transients do not yet present a consistent picture. LETOKHOV [6] has shown that dissociation can be produced by a second high-intensity field at 9.6 μm following initial excitation at 10.6 μm. However, COTTER [7] finds no absorption of 9.6-μm pulses in SF_6 pumped at 10.6 μm. FRANKEL [8] finds that absorption transients at 10.6 μm appear with a time constant far exceeding the gas-kinetic collision rate, thus indicating some sort of rapid intramolecular equilibrium.

The infrared double-resonance method [9] is the ideal technique for studying these absorption transients. Our initial experiments used a line-tunable CO_2 laser to probe SF_6 pumped by a fixed-frequency CO_2 laser. The signals observed in such an experiment are shown in Fig. 1. From these experiments, we can obtain the absorption spectrum of SF_6 molecules with an average vibrational excitation of one quantum of ν_3 (dashed curve), and the V-T relaxation time in the presence of added buffer gases (from the decay of the time-resolved signal). The observed shift of the ν_3 absorption band would correspond to a vibrational temperature of about 200°C, using the shock-tube results of LYMAN and NOWAK [10]; when fully equilibrated, the total rise in system temperature is about +25°C. The rise time of the absorption transients gives the V-V equilibration time. It is this part of the signal that has been studied by FRANKEL [8], using a picosecond-pulse mode-locked CO_2 laser. In those experiments, the density of vibrational levels reached at the power levels employed is far too low to permit collisionless coupling of the levels; thus, the levels must be power-broadened sufficiently during the pulse to induce a coupling between vibrational levels that is absent in the truly isolated (i.e., absence of radiation field) molecule.

We have obtained some preliminary results using a tunable PbSSe diode laser to probe the detailed dipole-strength distribution in laser-pumped SF_6. The width in frequency of the power-broadened "hole" burnt into the SF_6 absorption profile should be equal to the Rabi frequency,

$$\Delta \nu = \mu |\varepsilon| / \hbar. \tag{1}$$

The field amplitude $|\varepsilon|$ is related to the pumping laser intensity,

$$I = c |\varepsilon|^2 / 2\pi. \tag{2}$$

At a relatively low power, $I = 1.4 \times 10^4$ W/cm^2, we observe a power-broadened line width of approximately 0.9 GHz. The theoretical value, using Eq. (1) with $\mu = 3 \times 10^{-19}$ esu-cm [11], is 1.5 GHz. This linewidth encompasses a number of fine-structure components of the P(56) through P(60) transitions in the ν_3 band of SF_6 [12]; thus, even at low pumping power levels, a multiplicity of rotational levels is excited. At the nominal threshold for multiple

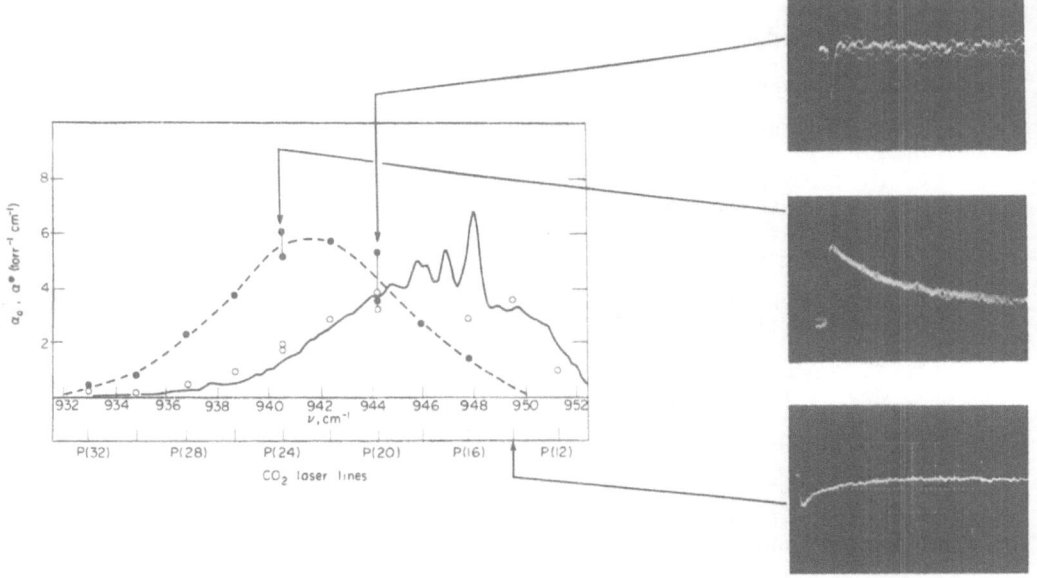

Figure 1. Infrared double resonance at CO_2 laser frequencies in SF_6. Left-hand panel shows normal room-temperature absorption spectrum of SF_6 (—o—), along with excited state spectrum deduced from double resonance experiment (---•---). Right-hand panel shows time-resolved absorption transients at (a) P(20) [pump frequency], (b) P(24), and (c) P(14) laser transitions. Adapted from [9].

infrared photon dissociation, 2.3×10^7 W/cm^2, the power-broadened linewidth would be 62 GHz, or nearly 2 cm^{-1}.

These experiments are being continued, in order to measure the hole-burning width in more detail, to locate, assign, and measure formation and decay kinetics of hot-band transitions, and to determine off-resonant absorption at high pump power levels.

Acknowledgements

This work was supported by the U.S. Energy Research and Development Administration.

References

1. R. V. Ambartzumyan, Yu. A. Gorokhov, V. S. Letokhov, and G. N. Makarov, J.E.T.P. Letts. 21, 375 (1975).

2. J. L. Lyman, R. J. Jensen, J. Rink, C. P. Robinson, and S. D. Rockwood, Appl. Phys. Letts. 27, 87 (1975).

3. N. Bloembergen, Opt. Commun. 15, 416 (1975).

4. D. M. Larsen and N. Bloembergen, Opt. Commun. (in press).

5. B. Kivel, AVCO-Everett Research Report No. 426 (April 1976).

6. V. S. Letokhov, at 6th Winter Colloquium on Laser-Induced Chemistry (Steamboat Springs, Colorado, February 1976).

7. T. Cotter, idem.

8. D. Frankel, Paper D.5, 9th International Quantum Electronics Conference (Amsterdam, June 1976).

9. J. I. Steinfeld, I. Burak, D. G. Sutton, and A. V. Nowak, J. Chem. Phys. 52, 5421 (1970).

10. A. V. Nowak and J. L. Lyman, J. Quant. Spectroscopy and Rad. Transfer 15, 945 (1975).

11. I. Burak, J. I. Steinfeld, and D. G. Sutton, J. Quant. Spectroscopy and Rad. Transfer 9, 959 (1969).

12. R. S. McDowell, H. W. Galbraith, B. J. Krohn, C. D. Cantrell, and E. D. Hinkley, Opt. Commun. (in press).

V. Laser Photokinetics

LASER INDUCED COLLISIONS*

S.E. Harris, R.W. Falcone, W.R. Green, D.B. Lidow
J.C. White and J.F. Young
Edward L. Ginzton Laboratory
Stanford University, Stanford, California 94305, USA

We describe processes where one or more photons are utilized to
conserve energy between the initial and final states of colliding
atoms. Energy transfer is thus initiated, and directed to particular states, by the optical radiation. We report new experiments
which we believe demonstrate a laser induced collision.

Introduction

If the energy defect of an atomic process ΔE is large with respect to

kT , then the cross section for collision or chemical reaction will be

quite small. In this paper we consider collision processes where one or

more photons are utilized to conserve energy, i.e., $n\hbar\omega \cong \Delta E$. A proto-

type system is shown in Fig. 1. Energy is first stored in the designated

s state of the A atom. During collision of the A and B atoms an

electromagnetic field at frequency $\hbar\omega$ causes the A atom to make a vir-

tual transition. Long range dipole–dipole coupling between the two atoms

causes this excitation to be transferred to the B atom to complete the

excitation. Energy transfer is thus initiated or "switched" by the pres-

ence of the optical radiation. The possibility of collision processes of

this type have been predicted by GUDZENKO and YAKOVLENKO [1] and by HARRIS

and LIDOW [2]. Recent theoretical work is given by PAYNE and NAYFEH [3],

GELTMAN [4], and GEORGE, et al. [5]. Though an apparently successful ex-

periment recently reported by LIDOW, et al. [6] is now in question, we re-

port additional experimental results which we believe demonstrate a laser

induced collision.

*   This work was jointly supported by the U.S. Office of Naval Research,
    the Advanced Research Projects Agency, and the Energy Research and
    Development Administration.

Optical processes involving the near simultaneous collision and absorption of photons may be considerably more general than that shown in Fig. 1. They may involve the absorption of several photons instead of one photon. They may take place between states involving dipole-quadrupole or quadrupole-quadrupole coupling. Processes involving charge exchange collisions, spin exchange collisions, and even free-bound reactions are possible.

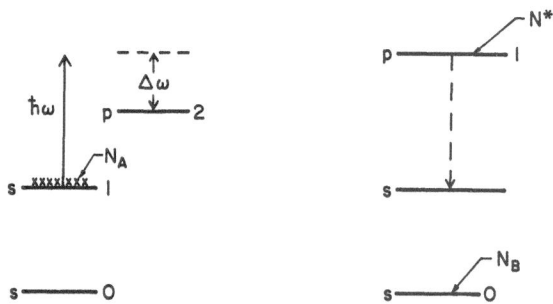

Fig. 1 Schematic of laser induced collision process. Energy is stored in the s state of atom A and may be rapidly transferred to the upper p state of atom B.

Collision processes of this type may have application to the construction of short wavelength lasers, to the mapping out of interatomic potential surfaces, to the development of radiative collision and Raman lasers and frequency converters, and to the initiation of selected chemical reactions.

In the following sections of this paper I will describe these processes, summarize the theory as it is known to date, and describe recent experiments wherein we observed a laser induced collision in a mixture of Sr and Ca.

Theory

I will first describe these effects from three related viewpoints. The first of these is the virtual transition viewpoint shown in Fig. 2. Assume first, that energy is stored in an excited s state of the first atom. In the presence of an electromagnetic field this atom makes a virtual transition. Excitation is then transferred to the second atom via dipole-dipole coupling. The process is thus described as a virtual electromagnetic transition followed by a real collision. Now consider the reverse process

where energy is stored in the upper p level of the second atom, and radiation at the same frequency $\hbar\omega$ is again present. As the atoms approach
each other they each make virtual transitions in the opposite direction to
the arrows denoted by 2 . The process is completed by the emission of
a photon of frequency $\hbar\omega$. The overall process is thus described as a
virtual collision followed by a real emission. For collision velocities
high enough that a straight-line trajectory is a good approximation, the
collision cross sections for the forward and reverse processes are equal.

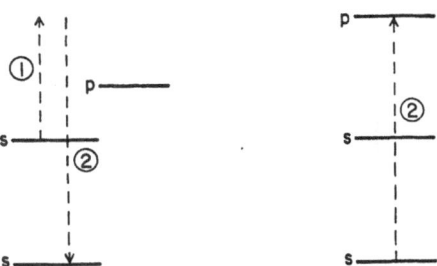

Fig. 2 Virtual transition viewpoint of laser induced collisions. For flow
of energy from the left- to the right-hand atom, the process is best
considered as a virtual electromagnetic transition followed by a
real collision. For the reverse process, we view the process as a
virtual collision followed by a real transition.

We next consider the process from the viewpoint of an electromagnetic
transition between the states of the quasi-molecule which is formed during
the time of a collision between an A and B atom. Fig. 3 shows the
energy of the quasi-molecular state as a function of the distance between
the colliding atoms. The lower state represents the sum energy of an excited A atom and a ground state B atom, while the upper state is the
final state, where B is excited and A is at ground. This viewpoint
leaves out the dynamics of the intermediate state and is thus not always
correct. In any case it is somewhat misleading. Using stationary perturbation theory one calculates a matrix element between the initial and
final states of the quasi-molecule. Since the square of this matrix element varies as $1/R^6$, one might at first expect the electromagnetic absorption to increase in strength as the atoms approach each other, and
thus expect a broad linewidth (at least several hundred wavenumbers) as a
function of the energy of the incident photon $\hbar\omega$. More exact analysis
shows that the square of the effective interaction time between the col-

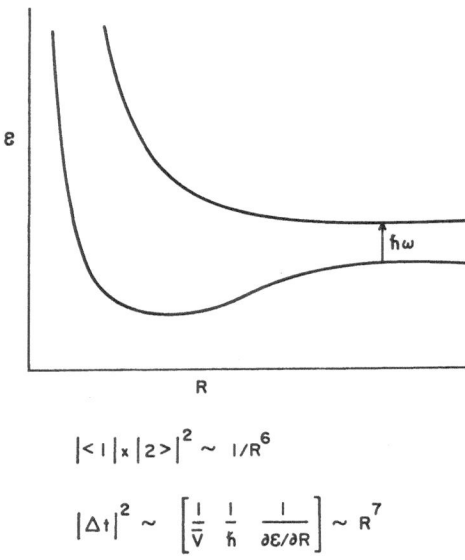

$$|<1|x|2>|^2 \sim 1/R^6$$

$$|\Delta t|^2 \sim \left[\frac{1}{\bar{V}} \quad \frac{1}{\hbar} \quad \frac{1}{\partial \mathcal{E}/\partial R} \right] \sim R^7$$

Fig. 3 Quasi-molecular viewpoint of laser induced collisions. A photon of frequency $\hbar\omega$ causes an electromagnetic transition between the initial and final state of the quasi-molecule.

liding species varies inversely as the slope of the relative energy difference of the initial and final states and thus increases as R^7 . An additional R^2 dependence is introduced when integrating over impact parameter. We will see that the linewidth for the dipole-dipole process is in fact quite narrow and peaks at the R = ∞ energy separation of the colliding atoms.

Following the reasoning of Fig. 3, one expects the collision cross section to peak at the R = ∞ energy separation for dipole-dipole and dipole-quadrupole processes. For quadrupole-quadrupole, spin exchange, charge exchange, and free-bound processes a peak at the R = ∞ frequency is not expected.

The third viewpoint is ad hoc. In this viewpoint the electromagnetic field applied to atom A causes an effective per atom dipole moment at the sum frequency of the energy stored in atom A and the applied field. Each A atom possesses a near electric field often many orders of magnitude larger than the macroscopic electric field. As an A atom passes a B atom an electromagnetic transition of the B atom is induced. The dephasing event allowing the transition is simply the passage of the A atom by the B atom.

I now give a brief description of the theory. The overall procedure is to apply perturbation theory at fixed interatomic separation R ; to integrate over $R(t)$ for fixed impact parameter ρ and velocity $\overline{V}$; and to then integrate over ρ and $\overline{V}$. The principal assumptions are that the collisions are slow compared to the orbiting velocity of an electron, and that the net energy defect is small compared to the incident energy of the particles, thus implying that the trajectories are unchanged. We work in a basis set of product eigenfunctions of the separated atoms. Thus

$$\psi = \sum_n c_n(t) \, u_n \, \exp - jE_n t/\hbar \tag{1}$$

The u_n are the product eigenfunctions of the atomic states and the E_n are the sum energies.

The classical interaction hamiltonian is

$$H' = ex_1 E + ex_2 E + \frac{e^2 x_1 x_2}{R^3} \tag{2}$$

where x_1 and x_2 are the local coordinates of the electrons of each atom, E is the applied electromagnetic field, and R is the distance between atoms. The first two terms represent the interaction of the electromagnetic field with each atom separately while the latter term gives the dipole-dipole coupling between the two atoms. Following the notation of Fig. 1 we assume three pertinent product states. c_1 is the amplitude of the product state whose energy is the sum of the storage s state of the first atom and the ground state of the second atom. c_2 is the amplitude of the p state of atom A and the ground state of atom B , while c_3 is the amplitude of the upper p state of atom B and the ground state of atom A . We consider the on line center case where $\hbar\omega$ = the energy defect ΔE . Substituting into Schrödinger's equation we obtain

$$\frac{\partial c_1}{\partial t} = \frac{1}{j\hbar} \frac{\mu^{A1} E}{2} c_2 \exp j\Delta\omega t \tag{3a}$$

$$\frac{\partial c_2}{\partial t} = \frac{1}{j\hbar} \frac{\mu^{A1} E}{2} c_1 \exp - j\Delta\omega t + \frac{1}{j\hbar} \frac{\mu^{A2} \mu^{B}}{R^3(t)} c_3 \exp - j\Delta\omega t \tag{3b}$$

$$\frac{\partial c_3}{\partial t} = \frac{1}{j\hbar} \frac{\mu^{A2} \mu^{B}}{R^3(t)} c_2 \exp j\Delta\omega t \tag{3c}$$

$\Delta\omega$ is the energy separation of the upper p levels, as shown in Fig. 1.
μ^{A1} , μ^{A2} , and μ^{B} are defined as: $\mu^{A1} = \langle 1|x_1|2\rangle^A$, $\mu^{A2} = \langle 2|x_1|0\rangle^A$,
and $\mu^{B} = \langle 1|x_2|0\rangle^B$.

If the relative rate of change of c_1 , c_3 , and $1/R^3(t)$ are slow
compared to $\Delta\omega$ then (3b) may be integrated and combined into (3a) and
(3c) to yield coupled equations between the initial and final states.
These two coupled states have an effective interaction hamiltonian

$$H_{13} = H_{31} = \left(\frac{\mu^{A1}E}{2\hbar\Delta\omega}\right)\frac{\mu^{A2}\mu^{B}}{R^3(t)}$$

By varying the strength of the electromagnetic field E , the strength of
the interaction hamiltonian may be varied. By varying the frequency of
the electromagnetic field an effective curve crossing may be created at
arbitrary R .

By a change of variable of the form

$$c_i = a_i \exp\left[-\frac{j}{\hbar}\int_{-\infty}^{t}\frac{c_6}{R^6(t)}\,dt\right]$$

(3) become

$$\frac{\partial a_1}{\partial t} = \frac{H_{13}}{j\hbar}a_3 \exp\left[-\frac{j}{\hbar}\int_{-\infty}^{t}\frac{c_6}{R^6(t)}\,dt\right] \tag{4a}$$

$$\frac{\partial a_3}{\partial t} = \frac{H_{13}}{j\hbar}a_1 \exp\left[\frac{j}{\hbar}\int_{-\infty}^{t}\frac{c_6}{R^6(t)}\,dt\right] \tag{4b}$$

In these equations I have neglected a small ac Stark shift which for the
power densities which we will consider will be (at most) a few tenths of
a cm^{-1}. The (relative) van der Waals constant c_6 includes not only terms
which come directly from (3), but also all other contributions pertinent to
the difference of the energy shifts between the initial and final states.
This shift is quite important and sets the minimum impact parameter which
contributes to the interaction.

We now examine the weak-field case where for all impact parameters we
may neglect the depletion of a_1 . We take $R(t) = [\rho^2 + \bar{v}^2 t^2]^{\frac{1}{2}}$.

We assume a constant impact parameter ρ and integrate (4b) over
$t = -\infty$ to $t = +\infty$. For impact parameters such that the exponent in

(4b) is less than about 1 radian, the transition probability as a function of impact parameter is

$$
|a(\rho)|^2 = \frac{1}{\hbar^4} \left(\frac{\mu^{A1} \mu^{A2} \mu^{B}}{\Delta\omega \overline{v} \rho^2} \right)^2
\tag{5}
$$

The cross section for collision is now obtained by integrating over all impact parameters from ρ_0 to infinity, or

$$
\sigma_c = 2\pi \int_{\rho_0}^{\infty} |a(\rho)|^2 \, \rho d\rho
$$

Combining the above we obtain the result

$$
\sigma_c = \frac{\pi}{\hbar^4} \left(\frac{\mu^{A1} \mu^{A2} \mu^{B} E}{\overline{v} \Delta\omega \rho_0} \right)^2
\tag{6a}
$$

where

$$
\frac{1}{\hbar} \int_{-\infty}^{+\infty} \frac{c_6}{R^6(t)} dt \cong \frac{3\pi}{8} \frac{|\mu^{A2}|^2 |\mu^{B}|^2}{\hbar^2 \overline{v} \Delta\omega \rho_0^5} = 1
\tag{6b}
$$

The second part of (6b) determines the minimum impact parameter ρ_0 which is to be used in (6a).

If we assume that all involved transitions have relatively large oscillator strength, then at thermal velocities, ρ_0 will typically be greater than 10 Å and the details of the atomic potentials for small interaction distances need not be considered. For $\Delta\omega \cong 5000$ cm^{-1}, typical predicted collision cross sections are about $\sigma_c \cong 5 \times 10^{-23} \frac{P}{A}$ (W/cm^2) cm^2. Thus an incident power density of about 1 MW/cm^2 is required to obtain cross sections of atomic dimension. Though I will not go into detail here, it is expected that the cross section for collision will continue to increase linearly with power density until reaching approximately $\pi\rho_0^2$. We believe that collision cross sections as large as 10^{-13} cm^2 should be obtainable, and experiments to demonstrate such cross sections are underway.

To this point we have considered the cross section for collision induced by a laser tuned to the $R = \infty$ frequency of the separated atoms. The expected line shape as a function of the frequency of the transfer laser is of considerable interest. If it were not for the dependence of the eigenenergies of the quasi-molecular states on the interatomic spacing then the line shape for transfer would be a first order modified Bessel function of the third kind with argument $(2\pi\Delta f\rho_0/\bar{V})$, where Δf is the frequency defect. For a situation such as that in Fig. 3 where the energy level of the lower quasi-molecular state falls faster than the upper state, a quite rapid fall-off is expected on the low frequency side of the $R = \infty$ frequency. On the high frequency side, the tunable laser in effect causes a curve crossing at arbitrary R . One might intuitively expect a sharp fall off on the red side, and a slowly varying tail on the blue side. Further work — both theoretical and experimental — is necessary.

Experimental Results

In our presentation at Loen, we described a key difficulty with a pre-viously reported experiment. The collision process studied in this earlier experiment is described by

$$Sr(5p^1P^0) + Ca(4s^2\ ^1S) + \hbar\omega(6409\ \text{Å}) = Sr(5s^2\ ^1S) + Ca(4d^1D) \qquad (7)$$

Energy was first stored in the radiatively trapped $5p^1P^0$ level of Sr I. The level was populated by two-photon pumping of the $5d^1D$ Sr level, fol-lowed by radiative decay. A second laser at 6409 Å was used to induce in-elastic collision to the $4d^1D$ level of Ca I. The overall process is best viewed as a virtual collisional excitation, followed by a real absorption. For this experiment, $\hbar\Delta\omega = 1954$ cm^{-1} , and (6) predicts $\rho_0 = 16.7$ Å , and $\sigma_c = 2.9 \times 10^{-23}\ \frac{P}{A}$ (W/cm^2) cm^2 .

As shown in Fig. 4, the difficulty with this experiment arises due to the presence of a transition at 6408.5 Å within the triplet series of Sr. This transition differs by only .1 Å from the interatomic frequency of 6408.6 Å, and is beneath our resolution. A possible artifact path now consists of collisional transfer from the Sr 5p to the Sr $4d^3D$ levels, followed by laser excitation at 6408.5 Å and collisional transfer to the Ca $4d^1D$ level. The fact that signals of comparable magnitude were ob-served when the transfer laser was tuned to other lines in the same triplet series, as well as the narrow symmetrical linewidth of the observed signal, indicates that the artifact rather than the true effect was observed.

Following the conference at Loen we succeeded in demonstrating laser induced collisions in the Sr-Ca system, but using two different target

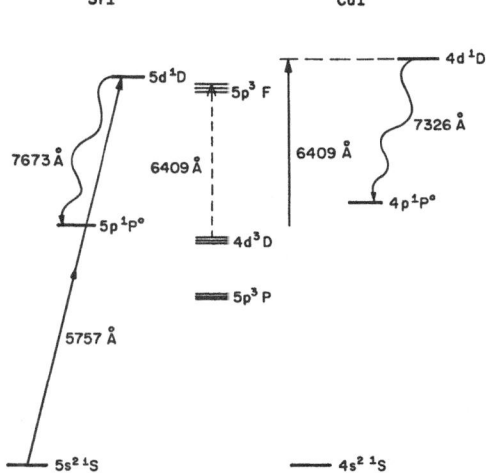

Fig. 4 Energy level diagram for Sr-Ca induced-collision experiment.

states in Ca, i.e., the Ca($6s^1$S) and Ca($5d^1$D) states respectively. These experiments are described by

$$Sr(5p^1P^0) + Ca(4s^2\ ^1S) + \hbar\omega(4977\ \text{Å}) = Sr(5s^2\ ^1S) + Ca(6s^1S) \qquad (8a)$$

$$Sr(5p^1P^0) + Ca(4s^2\ ^1S) + \hbar\omega(4711\ \text{Å}) = Sr(5s^2\ ^1S) + Ca(5d^1D) \qquad (8b)$$

In both of these experiments pumping of the $Sr(5p^1P^0)$ storage level was accomplished by direct single photon pumping of the resonance line. Typically this pumping laser was on the red side of the line and was detuned by about 50 cm^{-1}. An energy level diagram for the Ca($6s$) target state experiment is shown in Fig. 5. The power densities of the pumping and transfer lasers were about 10^5 W/cm^2 and 10^6 W/cm^2 respectively. The number density of Sr and Ca was $\sim 10^{15}$ atoms/cm^3.

Collisional transfer was monitored by examining the fluorescence from the Ca($6s^1$S) to the Ca($4p^1P^0$) state. The amplitude of this fluorescence signal, and thus the magnitude of the cross section for laser induced collision, as a function of the wavelength of the transfer laser is shown in Fig. 6. We note that the peak of the laser induced collision cross section occurs at about the R = ∞ wavelength of 4976.8 Å (wavelengths are given

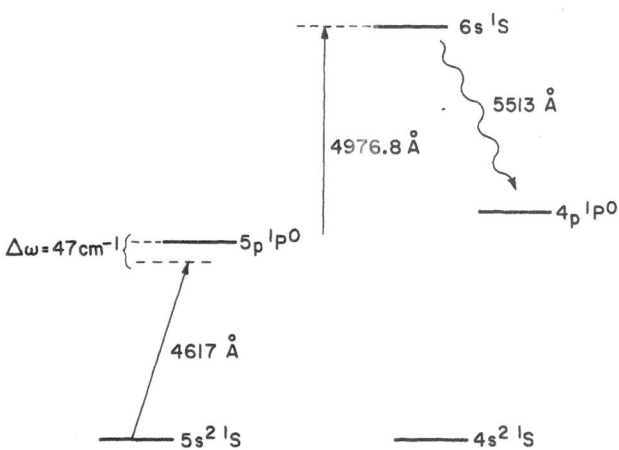

Fig. 5 Energy level diagram for the recent Sr-Ca induced collision experiment.

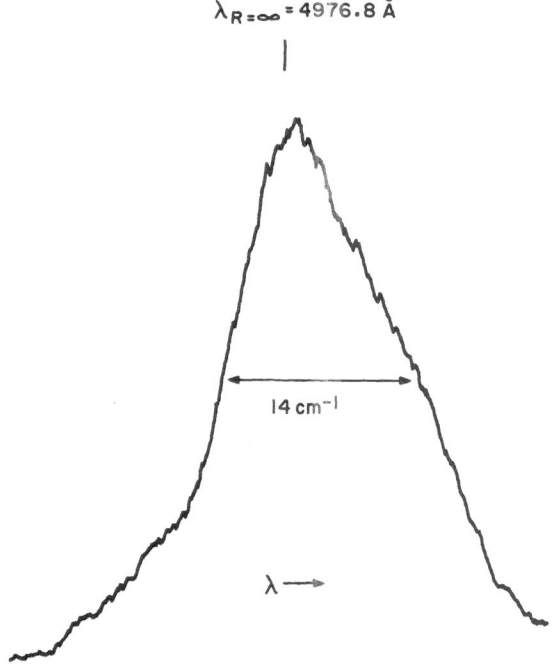

Fig. 6 Experimental results showing collision cross section as a function of transfer laser wavelength. The position of the R = ∞ wavelength is uncertain to 0.2 Å.

in air). The half-power linewidth for the process is 14 cm^{-1} and the line has a slight asymmetry to the red. The linewidth of the transfer laser used in these experiments was 2 cm^{-1}.

The slight asymmetry to the red indicates that the van der Waals constant C_6 for the Ca(6s^{1}S) target state is comparable in magnitude to that of the Sr(5p^{1}P^0) storage state. This is consistent with a rough calculation. The linewidth of the process associated with the Ca(5d^{1}D) target state is about 30% narrower than that for the Ca(6s) state, and is also slightly asymmetrical to the red.

We have not yet properly calibrated the magnitude of these effects. The amplitude of the fluorescent signals obtained by the laser induced collisional process is about 1/500th of that obtained by tuning the transfer laser to the two-photon absorption frequency of the Ca(6s) level. The signal to background ratio is about five. If we assume theory to be correct, then at the power density of the transfer laser, the observed signal corresponds to a storage density in the Sr(5p) level of about 10^{13} atoms/cm^3. This is roughly consistent with our estimates of population obtained by absorption probing. Over a total scan range of about 400 cm^{-1}, the peak shown in Fig. 6 is the only one which is not explained as either a two-photon summing process or as a direct atomic line excitation, followed by collision.

A possible alternate mechanism for the observed signal is two-photon absorption to the Ca(6s) state, where one of the photons result from spontaneous emission from the Sr(5p) level, and the second is furnished by the transfer laser. To eliminate this possibility we varied the density of an argon buffer gas from about 10^{17} to 10^{18} atoms/cm^3. Over this range, no observable change was noted in the cross section vs. transfer laser linewidth (Fig. 6). Over this same range of argon pressure, the 50% transmission width of the resonance line varied by $\sqrt{10}$. Also, calculations indicate that such a two-photon absorption should be about 10^4 smaller than the observed signal.

Applications

I will now briefly describe a number of potential applications and devices which may result from collisional processes of this type. We first consider a radiative collision laser [1]. A schematic of a radiative colli-

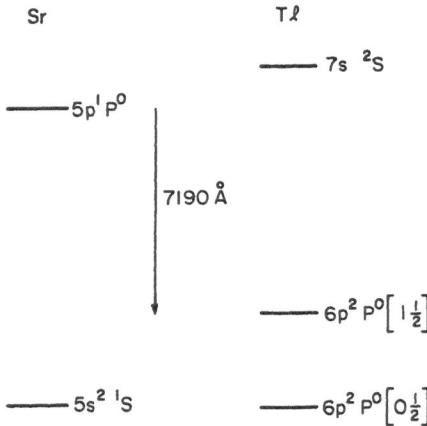

Fig. 7 Energy level diagram of a possible radiative collision laser.

sion laser is shown in Fig. 7. In a laser of this type energy would be
stored in the radiatively trapped resonance line of Sr. During collision
with a $T\ell$ atom, a Sr atom would be de-excited while a $T\ell$ would be excited
from ground to the $T\ell$ $6p^2P^0$ [$1\frac{1}{2}$] level. A photon at 7190 Å would experience gain rather than loss. The gain coefficient g is related to the
collision cross section σ_c according to

$$g = \frac{N_A N_B \bar{V} \hbar \omega \sigma_c}{P/A} \tag{9}$$

where N_A and N_B are the number density of the two species. If we take
$\sigma_c = 10^{-22}$ P/A , $N_A = 10^{16}$ atoms/cm^3 , $N_B = 10^{19}$ atoms/cm^3 , $\omega = 10^{15}$
radians/sec , and $\bar{V} = 10^5$ cm/sec , we find a gain cross section of
$g = 0.1$ cm^{-1} . This number is somewhat optimistic for a typical system.
A laser of this type may be of practical interest for two reasons: First,
is the attractive possibility of radiatively trapped resonance line storage.
Second, is the variable gain cross section provided by varying the number
density of the second (ground state) species.

Though most of the discussion of this paper has been concerned with a
single-photon transfer process, all ideas continue to hold for a multi-
photon process. By storing energy in high lying states of atoms or ions,

and then using multi-photon processes with the intense VUV light sources now available, it should be possible to reach states to at least the 100 Å spectral region.

Coherent collisionally induced Raman processes such as that shown in Fig. 8 may be of interest for up-conversion of long wavelength radiation. This process is alternately described as a collisionally aided Stokes process which allows an output frequency higher than the input frequency, or alternately as an anti-Stokes process in a quasi-molecular system. The ability to demonstrate processes of this type will be dependent on achieving high number densities of the ground state species.

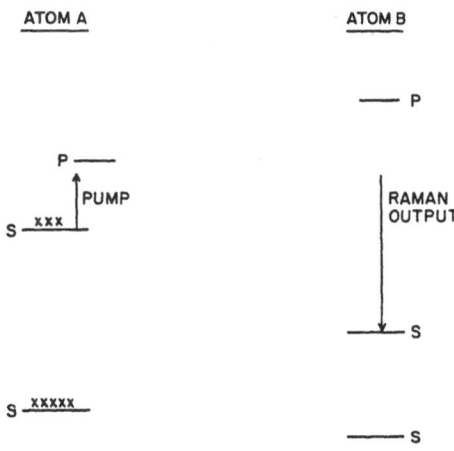

Fig. 8 In a coherent collisional Raman process gain is obtained at the Raman output frequency. Note that atom A need not be inverted, and that the frequency of the Raman output may be higher than that of the pumping frequency.

Other areas of application of these processes include the selective ionization of innershell electrons, laser induced charge transfer collisions, and perhaps radiative charge transfer lasers.

Laser induced collisions may someday allow one to direct the flow of energy and thus to influence reaction rates of general gas phase kinetic processes.

The authors acknowledge helpful discussions with David Bloom and Alan Gallagher. We wish to thank Mr. Ben Yoshizumi for help with the experiments reported here.

References

1. L. I. Gudzenko and S. I. Yakovlenko, Zh. Eksp. Teor. Fiz. 62, 1686 (1972) [Sov. Phys. JETP 35, 877 (1972)].

2. S. E. Harris and D. B. Lidow, Phys. Rev. Lett. 33, 674 (1974), and 34, 172(E) (1975).

3. M. G. Payne and M. H. Nayfeh, "Laser Enhanced Collisional Energy' Transfer" (to be published).

4. Sydney Geltman, "Theory of Laser-Stimulated Collisional Excitation Transfer" (to be published).

5. Thomas F. George, Jian-Min Yuan, I. Harold Zimmerman, and John R. Laing, "Radiative Transitions for Molecular Collisions in an Intense Laser Field," Disc. Faraday Soc. No. 62.

6. D. B. Lidow, R. W. Falcone, J. F. Young, and S. E. Harris, Phys. Rev. Lett. 36, 462 (March 1976).

APPLICATION OF PICOSECOND LASER PULSES TO THE DETERMINATION OF VIBRATIONAL TIME CONSTANTS OF POLYATOMIC MOLECULES IN LIQUIDS

W. Kaiser and A. Laubereau

Physik-Department, Technische Universität, München, Germany

During the past several years picosecond light pulses have found extensive application in physics and chemistry for the investigation of ultrafast processes(1). In this short report we focus our attention on studies of the dynamic behavior of vibrational modes of polyatomic molecules (in the electronic ground state). In particular we are interested in finding pertinent population lifetimes and dephasing times of well-defined vibrational states.

In liquids our knowledge of vibrational relaxation processes is rather scarce. Ultrasonic waves interact predominantly with the lower vibrational modes and do not provide specific relaxation values of the (many) higher normal modes of poly-atomic molecules. Line width measurement of infrared and Raman bands are the main source of "vibrational" relaxation times. In general, line width data are difficult to interpret, because a number of physical processes contribute to the observed line shape. Line broadening factors are: phase relaxation, energy relaxation, rotational motion, isotope splitting and inhomogeneous broadening due to a distribution of vibrational frequencies. Under certain assumptions it is possible to separate the rotational contribution. The rest, which is frequently called the intrinsic vibrational part of the band conture contains the other line-broadening factors. At present, there exists no possibility to isolate the different contributions by spectroscopic methods. For instance, the important question concerning the population lifetime of an exicted vibrational state was unknown up to very recently for any vibrational mode in the liquid state. Similarly, the inhomogeneous contribution of a spectroscopic line is not known in most cases and, as a result, a dynamic time constant cannot be deduced from the width of the vibrational spectral line.

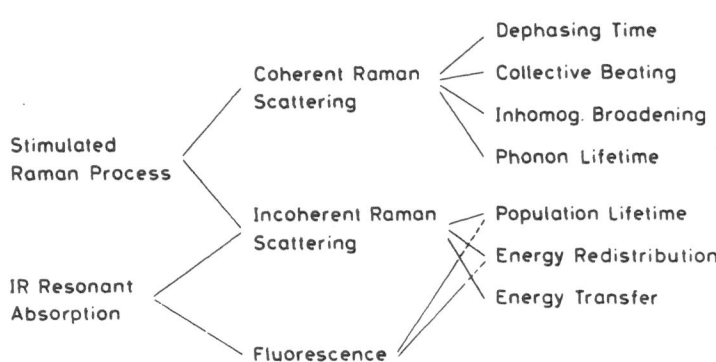

Excitation	Probing	Information

Fig. 1   Physical processes for the excitation and the probing
of molecular vibrations in liquids. The dynamical
information is listed on the right.

In this paper we briefly discuss new experimental techniques
to study directly ultrafast processes related to vibrational
modes in liquids. A number of recent experimental results
should illustrate the state of the art and the potential of
the new methods. Fig. 1 outlines schematically the experimental techniques used and various results obtained. The excitation of molecules or lattice vibrations is achieved by an intense laser pulse via stimulated Raman scattering or directly
by a resonantly absorbed infrared pulse. After the passage of
the first pulse the excitation process rapidly terminates and
free relaxation and/or energy transfer of the excited mode occurs. A second weak probe pulse properly delayed with respect
to the first pulse monitors the instantaneous state of the vibrationally excited system. Three types of probing methods
were used in order to obtain the dynamical information listed
on the right side of Fig. 1.

(i) Coherent Raman scattering results from the interaction
of the probe pulse with the coherently excited system. For
molecular modes with essentially one vibrational frequency the
dephasing time of the excited mode is deduced from the loss of
phase correlation within the excited volume i.e. from the decay of the probing signal (2).The situation is more complex for
molecular vibrations with different neighboring frequencies
(e.g. several isotope components) or with a distribution of
vibrational frequencies (inhomogeneous lines). For these
cases we have developed a highly selective phase matching geometry in order to isolate one frequency component for the study
of the phase relaxation time. In this way we are able, for
the first time, to measure a dynamic time constant of inhomogeneously broadened spectral lines. With less selective
phase matching of the probe pulse we observe an interesting
collective beating of neighboring frequency components. Finally,

with coherent probe scattering we have studied the lifetime of optical phonons in three well-known crystals.

(ii) Spontaneous anti-Stokes Raman scattering, an incoherent scattering process, allows us to study directly the momentary degree of occupation of a vibrational energy state. With this technique is was possible, for the first time, to observe population lifetimes, energy transfer and energy redistribution (3-5).Time constants between 1.0 and approximately 100 psec were measured for different dynamical processes in a number of polyatomic molecules.

(iii) We have developed a double resonance probing technique where a first infrared pulse excites a well-defined vibrational mode and a second probe pulse promotes' the molecules close to the bottom of the fluorescent singlet state. In this way the degree of fluorescence is a direct measure of the occupation of certain vibrational quantum states. This technique was found very useful for the investigation of highly diluted systems (6).

In the following part of this report we wish to discuss in some detail (i) the dephasing time of a homogeneously broadened transition; (ii) the vibratioanl excitation of an inhomogeneously broadened line; (iii) the dephasing time of one vibrational component of complex vibrational modes and the collective beating due to different vibrational species; (iv) the population lifetime and the energy redustribution after direct infrared excitation.

(i) In a previous publication we have reported on the population lifetime T_1 of the CH_3-symmetrical stretching mode of CH_3CCl_3 (5). A value of $T_1 = 5.2$ psec was found for the first excited vibrational state in the neat liquid. In Fig. 2 we present the results on the dephasing time $\tau = T_2/2$ of the same vibrational mode. The Raman scattering signal is plotted as function of delay time of the probe pulse. From the exponential decay of the signal curve we deduce a dephasing time of $\tau = T_2/2 = 1$ psec. Since the decay rates are related by the equation: $2/T_2 = 1/T + 1/\tau_{ph}$ we conclude that pure dephasing τ_{ph} determines the loss of phase relaxation in Fig. 2. It should be noted that the time constant $\tau = (2\pi c \delta \tilde{\nu})^{-1} = 1.0$ psec calculated from the line width $\delta \tilde{\nu}$ of spontaneous Raman line agrees will with our dephasing time, i.e. for the mode discussed here, the dephasing time is able to fully account for the Raman line width. We shall see in the following example that in other cases (inhomogeneously broadened bands) the latter statement does not hold.

(ii) We have made a detailed theoretical study of the vibrational excitation of modes consisting of neighboring vibrational states or a continuous distribution of frequencies. It can be shown that transient stimulated Raman scattering allows to excite molecules coherently over a certain frequency range depending upon the ratio t_p/T_2, pulse duration to dephasing time. During the excitation process molecules of the center

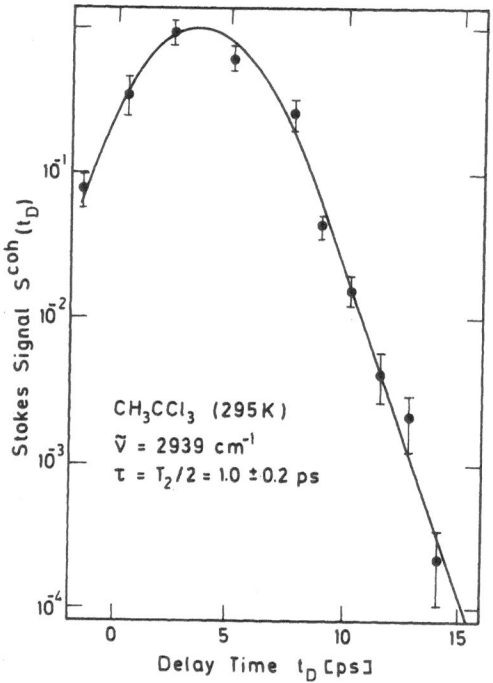

Fig. 2. Coherent Stokes probe scattering signal $S^{coh}(t_D)$ versus delay time between pump and probe pulse for the symmetric CH_3-stretching mode of CH_3CCl_3.

of the vibrational distribution are locked together and driven with the frequency $\omega_L - \omega_S$, where ω_L and ω_S are the laser and Stokes frequencies respectively. After the excitation has terminated the molecules relax freely with their individual resonance frequencies. The following spectra should illustrate the forced excitation and free relaxation process (7). The spectra 3b, 3c and 3d are obtained on the symmetric CH_3-stretching mode of ethylene glycol $(CH_2OH)_2$ at 2970 cm^{-1}. The spontaneous Raman band of this mode shows a large band width (FWHH) of $\delta\tilde{\nu}$ spon $\simeq$ 60 cm^{-1}. In Fig.3 four spectra are presented: (a) The spectrum of the incident (or transmitted) laser pulse at $\tilde{\nu}_L = 18,910$ cm^{-1} shows a line width of $\delta\tilde{\nu}_L \simeq 5$ cm^{-1} which is the Fourier transform of the pulse duration of 3-4 psec. (b) The spectrum of the stimulated Stokes pulse has a band width of $\delta\tilde{\nu}_S \simeq 8$ cm^{-1}. We know that the Stokes pulse is shorter than the pump pulse by approximately a factor of two, since the transient build-up the material excitation scatters the incoming probe pulse for approximately half of the pulse duration. The reduced pulse duration gives a stimulated Stokes spectrum $\delta\tilde{\nu}_S \simeq 8$ cm^{-1}. (c) The spectrum of the Stokes scattered probe light, $S^{coh}(t_D=0)$ is taken at the same time as the previous spectrum (b), but with a perpendicular polarization direction. As expected from the isotopic part of the Raman tensor of the normal mode investigated here, the spectrum of $S^{coh}(t_D=0)$ is the same as the previous one in (b).

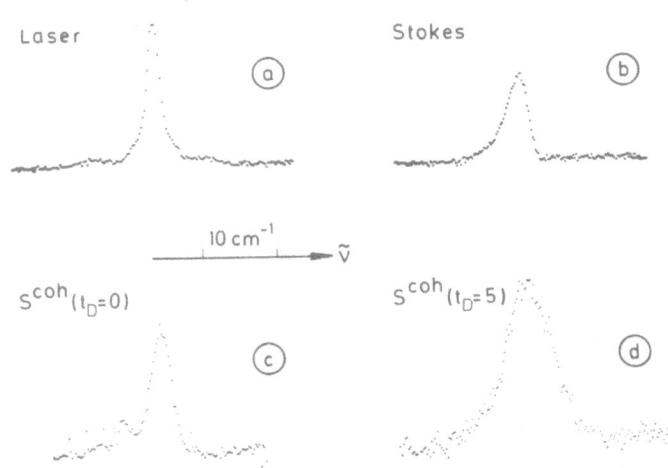

Fig. 3   Spectral band shapes; a) incident laser pulse; b) generated stimulated Stokes pulse; c) and d): coherent
Stokes scattering S^{coh} of the probe pulse with time
dealy of $t_D = 0$ and $t_D = 5$ psec, respectively.

d) The Stokes scattered pulse S^{coh} ($t_D=5$ psec) sees a completely different situation. For times $t \gtrsim t_p$ the molecules relax
freely with their resonance frequencies. From the spectrum
of Fig. 3d we duduce a coherently excited band width of
~ 18 cm^{-1}. This spectral observation indicates that the molecules which were driven off resonance during the transient
stimulated Raman process return to their resonance frequencies
during the free relaxation process.

(iii) Many liquids are composed of several isotope species.
The resulting multiplicity of the vibrational modes may or may
not be resolved by spontaneous Raman spectroscopy. It will be
shown here that coherent excitation and coherent probing allow
the determination of the dephasing time of one isotope component. In addition a new beating phenomenon between the different isotope components is observed.

Two well-known molecules, the tetrahalides CCl_4 and $SnBr_4$
were chosen for demonstration. The various isotopes of Cl and
Br give rise to vibrational multiplicity of the totally symmetric tetrahedron vibration around 460 cm^{-1} and 220 cm^{-1} respectively. The spontaneous Raman bands are depicted in
Fig. 4. In CCl_4 we see a distinct line splitting of approximately 3 cm^{-1} due to the two chlorine isotopes ^{35}Cl and ^{37}Cl.
The molecular components $C^{35}Cl_4$ (1), $C^{37}Cl^{35}Cl_3$ (2), $C^{37}Cl_2{}^{35}Cl_2$
(3) and $C^{37}Cl_3{}^{35}Cl$ (4) have concentration ratios of
0.77 : 1 : 0.48 : 0.10 in agreement with the respective Raman
intensities. The overlap of the lines does not allow an accur-

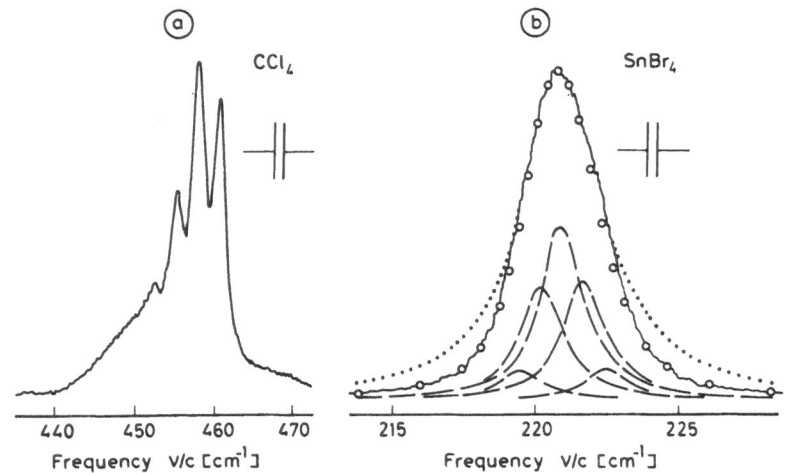

Fig. 4   a) Spontaneous Raman spectrum of the totally symmetric tetrahedron mode of CCl₄; the isotopic line
splitting is clearly resolved. b) the same for SnBr₄
at 305 K (solid curve). The band shape differs notably from a Lorentzian line (dotted curve). The
broken curves represent the Raman lines of the individual isotope species and were calculated with the
help of our picosecond data. Summation of the five
isotope components (open circles) account well for
the measured spontaneous Raman band (see text).

ate measurement of the line width of a single isotope component.
The spectrum of SnBr₄ is considerably different (see
Fig. 4b). The smooth band shows no structural features and a
band width of $\delta\tilde{\nu} \simeq 3.2$ cm^{-1}. The isotope splitting is expected
to be smaller in SnBr₄ on account of the lower vibrational fre-
quency and the smaller relative mass effect $\Delta m/m$ of the ^{79}Br
and ^{81}Br isotopes. We estimate a frequency spacing $\Delta\omega \simeq$
$(\omega/8)$ $(\Delta m/m)$ $\{1+(\Delta m/m)\}^{-1} \simeq 0.67$ cm^{-1}. This value is just one-
fifth of the total Raman line width. The isotope distribution
is masked here by the homogeneous broadening of the closely
spaced isotope components. The wings of the observed band differ notably from a Lorentzian line (dotted curve). The broken
lines in Fig. 4 will be explained below.

We have measured coherent probe scattering signals with various phase matching geometries. a) Under selective phase
matching conditions we study only one isotope component; b)
Without phase matching selectivity we observe the interference
of serveral scattered probe beams.

To a): Phase matching of a coherent anti-Stokes scattering signal requires an off-axis geometry for CCl_4. Wave vector calculations indicate that the scattered probe emission of the different isotope components of CCl_4 occurs under slightly different angles of $\Delta\gamma \simeq 6$ mrad. Selective phase matching was experimentally adjusted with a small aperture which selects a probe scattered beam of ~2 mrad. The coherent scattering signal S^{coh} was measured as a function of delay time t_D between the excitation and the interrogating pulse. In Fig. 5a and 6a we present experimental data of one isotope component of CCl_4 and $SnBr_4$ respectively (8).

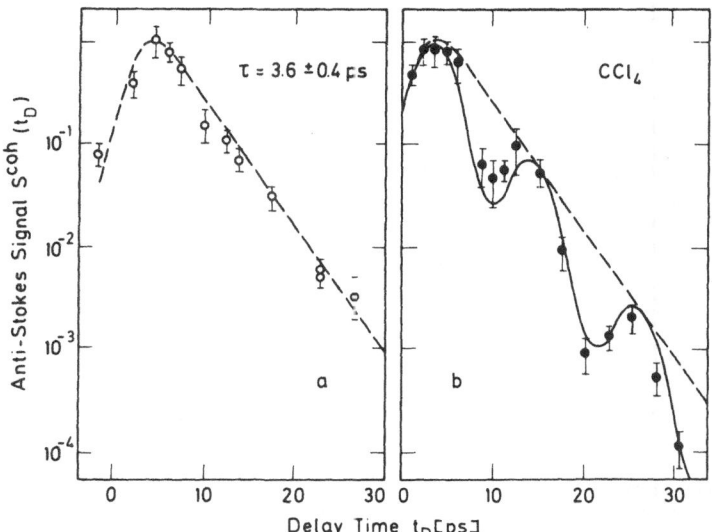

Fig. 5   Coherent probe scattering versus delay time t_D of CCl_4 with natural isotope abundance; a) selective k-matching observing a single isotope component; b) non-selective k-matching with coherent superposition of vibrational states; curves are calculated.

From the exponential slopes of the signal curves we obtain dephasing times $\tau=T_2/2=3.6$ psec for CCl_4 and $\tau=T_2/2=3.0$ psec for $SnBr_4$. We note that in CCl_4 our measured dephasing time accounts fully for the spontaneous Raman line width of one isotope component. In $SnBr_4$ our dephasing time corresponds to a homogeneous line width of 1.8 cm^{-1}, a value is significantly smaller than the band width of 3.2 cm^{-1} observed in the spontaneous spectrum (see Fig. 4b).

To b): Different results are observed for a non-selective k-geometry. For the investigation of CCl_4, the aperature in the off-axis geometry is removed; now the solid angle of acceptance is considerably larger (~10 mrad) and probe scattering beams of slightly different directions can be detected. The observed total signal is expected to show the interference of the different scattered beams. Fig 5 b and 6b give the experimental data for CCl_4 and $SnBr_4$ respectively (8). There is a drastic difference to the corresponding results of Fig. 5a and 6a. In Fig. 5b, two minima and additional maxima of the signal curve are clearly indicated by the experimental data.

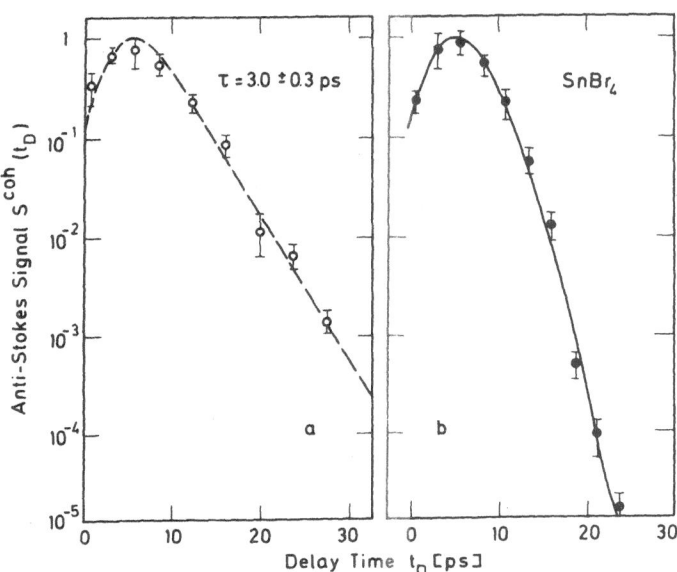

Fig. 6   Same investigations as in Fig. 5 with $SnBr_4$ at 305 K.
a) Selective k-matching, observing the homogeneous dephasing time of a single isotope component.
b) Non-selective k-matching with coherent superposition of vibrational states. Comparison of a) and b) directly shows the isotopic substructure of the Raman band, not resolved by spontaneous Raman spectroscopy.

These results are caused by collective beating of the isotope species of CCl_4. The solid curve is calculated taking into account the scattering contributions of the four most abundant isotope species of CCl_4. Since the dephasing time is determined in an independent experiment (see Fig. 5a) the only fitting parameter in the calculation is the frequency difference $\Delta\omega$ between neighboring vibrational components. A value of $\Delta\omega/2\pi c = 2.9+0.15$ cm^{-1} is found which favorably compares with the spontaneous Raman data of Fig. 4a. Considerably different are the experimental data in $SnBr_4$. Coherent superposition of the closely spaced isotope components gives rise to a rapid decay of the signal points over five orders of ten during a time interval of $\sim$18 psec. For $SnBr_4$ we estimate a long beat period of 50 psec. As a result the signal does not reach the first minimum and subsequent maximum within our measuring range. In fact, the beat maximum around t_D = 50 psec is expected 9 orders of ten below the peak scattering signal. The solid curve of Fig. 6b is calculated for the non-selective phase matching situation using the beat frequency $\Delta\omega$ as a fitting parameter and the measured vaule of the dephasing time τ=3 psec which was determined in Fig. 6a. Good agreement of the theoretical curve with the experimental data (full points) in the Figure is obtained for a value of $\Delta\omega/2\pi c = 0.7\pm0.15$ cm^{-1} which favorably compares with the theoretical estimates of $\Delta\omega_{th}\approx0.67$ cm^{-1}.

The apparent decay time of the solid curve in Fig. 6b of 1.2 psec does not represent a relaxation time of the vibrational system. The observed time dependence results from the destructive interference of the coherently exicted vibrational modes. A relaxation time of 1.2 psec would correspond to a spectral linewidth of 4.4 cm^{-1} which is notably larger than the observed spontaneous bandwidth of 3.2 cm^{-1}.

Knowing the dephasing time τ, the frequency spacing $\Delta\omega$ and the abundance of the SnBr$_4$ species, the composed spontaneous Raman band was evaluated. The natural abundance of the two Br isotopes leads to a relative concentration of the five components Sn^{81}Br$_x$^{79}Br$_{4-x}$ (x=0 to 4) of 0.173, 0.685, 1.0, 0.653 and 0.160. In Fig. 4b the five isotope components are depicted by broken curves. The sum of the five Raman lines gives the open circles which agree completely with the experimental Raman band of common SnBr$_4$. This result demonstrates that our picosecond data of τ and $\Delta\omega$ are fully consistent with the integral information supplied by spontaneous Raman spectroscopy.

Concluding this section the following remarks should be made. The beating phenomenon originates from the coherent superposition of the vibratioanl amplitudes $<q>_j$ of the excited quantum states. $<q>_j$ is equal to the expectation value of the displacement operator. There are differences to the quantum beats previously observed in gases: (a) The different excited states belong to various molecular species; (b) the vibrational states have no optically allowed transitions i.e. there is no emission to be observed. (c) The beat frequency is very high, of the order of 10^{11} sec^{-1}, and the dephasing times are very short, a few 10^{-12} sec. Ultrashort pump and probe pulses are required to study the beating process in the condensed phase.

We now turn to the discussion of the population lifetime T_1 and of vibrational redistribution in small molecules. Data on the CH$_3$-stretching vibrations of CH$_3$I are reported here. Solutions of one to five mol per cent of CH$_3$I in CCl$_4$ were investigated. In Fig. 7a and b the incoherent anti-Stokes Raman signal which is proportional to the occupation of the first vibrational state is plotted as a function of delay time between an infrared excitation pulse and a green (0.53 μm) interrogating pulse (9). The observed spontaneous anti-Stokes Raman signal was measured at a frequency shift of 2950 cm^{-1} which corresponds to the symmetric CH$_3$-stretching vibration. There is an important difference in the experimental conditions of Fig. 7a and 7b:

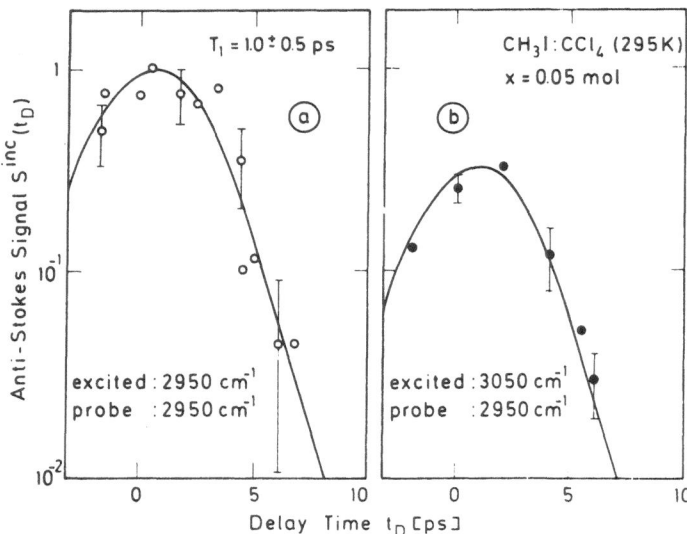

Fig. 7   Incoherent anti-Stokes probe scattering S^{inc} (t_D) versus delay time of the symmetric CH_3-stretching mode
($\tilde{\nu}$ = 2950 cm^{-1}) of CH_3I dissolved in CCl_4 (mol fraction 0.05). a) The mode at 2950 cm^{-1} is directly populated by the resonant infrared excitation pulse.
b) The asymmetric CH_3-stretching mode at 3050 cm^{-1} is
excited by the tunable pump pulse; the observed excess population of the vibration at 2950 cm^{-1} indicates
rapid energy redistribution between the neighboring
modes.

In one experiment we excited the symmetric CH_3-mode by an infrared pulse at a frequency of 2950 $^{-1}$ (Fig. 7a) and observed
the population lifetime T_1 of the same vibrational mode. A value of T_1 = 1.0±0.5 psec is directly indicated by the data. In
the second experiment we tuned our infrared pulse to 3050 cm^{-1}
in order to excite the asymmetric CH_3-mode. We observed again
the build-up and decay of the symmetric lower lying CH_3-vibra-
tion (Fig. 7b). In both experiments we find very fast time
dependence of the generated vibrational excess population and
a similar signal decay in the first and second experiment respectively. This investigation indicates that even in relatively small molecules the population relaxation can proceed
very fast and that there exists a very rapid energy exchange
between the two neighboring CH_3-stretching vibrations with a
time constant of approximately 1.5 psec. Recent model calculations by S. FISCHER suggest fast population decay via overtone
and combination states of the CH_3 bending modes.

In summary we wish to say that the investigations discussed
here allow us to measure a series of molecular parameters which
advance our understanding of ultrafast dynamical processes in
the liquid state.

References

1. For a reviw see A Laubereau and W. Kaiser, Ann. Rev. Phys. Chem. <u>26</u>, 83 (1975)

2. D. von der Linde, A. Laubereau and W. Kaiser, Phys. Rev. Lett. <u>26</u>, 955 (1971)

3. A. Laubereau, D. von der Linde and W. Kaiser, Phys. Rev. Lett. <u>28</u>, 1162 (1972)

4. R. R. Alfano and L. L. Shapiro, Phys. Rev. Lett. <u>29</u>, 1655 (1972).

5. A. Laubereau, L. Kirschner and W. Kaiser, Opt. Commun. <u>9</u>, 182 (1973)

   A. Laubereau, G. Kehl and W. Kaiser, Opt. Commun. <u>11</u>, 74 (1974).

6. A. Laubereau, A. Seilmeier and W. Kaiser, Chem. Phy. Lett. <u>36</u>, 232 (1975).

7. G. Wochner, A. Laubereau and W. Kaiser, to be published.

8. A. Laubereau, G. Wochner and W. Kaiser, Optics Commun. <u>17</u>, 91 (1976) and Phys. Rev., in press.

9. K. Spanner, A. Laubereau and W. Kaiser, to be published.

OPTICAL COHERENT TRANSIENTS BY LASER FREQUENCY-SWITCHING*

R. G. Brewer and A.Z. Genack
IBM Research Laboratory, San Jose, California 95193

ABSTRACT: Coherence phenomena such as photon echoes, free induction and nutation effects are easily detected in molecular iodine using a frequency-switched tunable cw dye laser. This technique is generally applicable to atoms, molecules, and solids and offers unique ways for probing dynamic interactions in a selective manner. Elastic and inelastic collision mechanisms for I_2 are examined independently and compared with theory.

We present a simple and versatile laser technique which should prove useful in observing coherent optical transient phenomena in atoms, molecules and solids. In concept, the method is analogous to pulsed nuclear magnetic resonance techniques (1-3), but in practice, it more closely resembles the recently introduced Stark-switching method (4-7).

To illustrate its use, we demonstrate photon echoes (8), free induction decay (FID) (5) and nutation effects (4,9) in numerous lines of the visible electronic transition of I_2 (Fig. 1). Similar observations have been made also on the sodium D lines. Our work contrasts with classical optical line broadening measurements where the various broadening mechanisms invariably remain hidden within the optical lineshape. Coherent transient methods, on the other hand, allow one to distinguish individual dephasing processes. As an initial example, we inspect the collisional properties of I_2.

In the earlier Stark-switching experiments, a prescribed sequence of low voltage Stark pulses switched a molecular sample into or out of resonance with an infrared beam from a fixed frequency cw CO_2 laser (4-6). Coherent emission or absorption transients were detected in the transmitted beam. A variant of this idea was later realized by Hall and Kramer (10) who frequency-switched the laser instead of the sample. Optical transients arose due to a methane sample located inside the cavity of a frequency modulated 3.39 micron He-Ne laser.

---

* Work supported in part by the U.S. Office of Naval Research

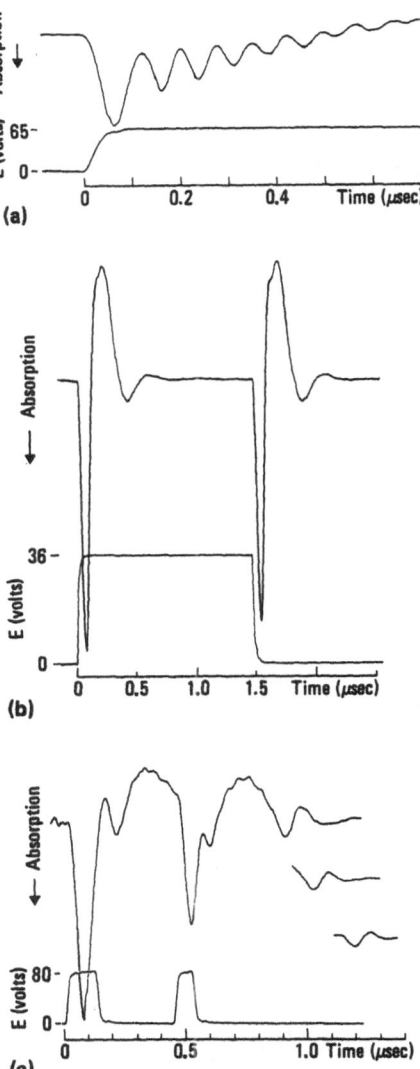

Figure 1   Coherent optical transient phenomena in I_2 vapor.   (a) Free induction decay where the emission and laser produce a 13 MHz
beat; (b) Two optical nutation patterns; and (c) Photon echoes
occurring at ~1 μs where the successive echoes decay with increasing pulse delay time.   The frequency-switching pulse pattern is displayed in the lower trace of each figure.

In our configuration, Fig. 2, a stable tunable cw dye laser (11) is frequency-switched while the sample's transition frequency remains constant. The sample is now external to the laser system so as to not affect its performance. Coherent transient signals again appear in the forward beam. Frequency-switching is achieved with an electro-optic crystal of ammonium dihydrogen phosphate (ADP), which is inside the dye laser cavity and is driven by a sequence of low voltage pulses. The laser frequency follows the refractive index variations induced in the ADP crystal. Hence, the experiment is controlled electronically and in such a way that the advantages inherent in the Stark technique are preserved here as well. We find, therefore, that (1) the only transient observed is the desired coherent transient itself; this is not the case with pulsed laser sources as the small coherent transient signal often rides on top of the laser pulse and the two are not easily separated; (2) heterodyne detection is possible because the coherently radiated light propagates with the laser beam in the forward direction and is shifted from it in frequency; this increases the signal amplitude several orders of magnitude and facilitates measuring the decay of emission signals; (3) a further improvement in signal to noise results with signal averaging, which is possible because the pulse sequency is repetitive; and (4) the entire class of coherent optical transient effects can be monitored since the electronic pulse sequence can be tailored to the particular experiment of interest. Moreover, when these features are combined with the broad tuning range available in a dye laser, it is apparent that coherent transient phenomena can now be observed with ease in a large number of optical transitions in various atomic, molecular and solid state systems.

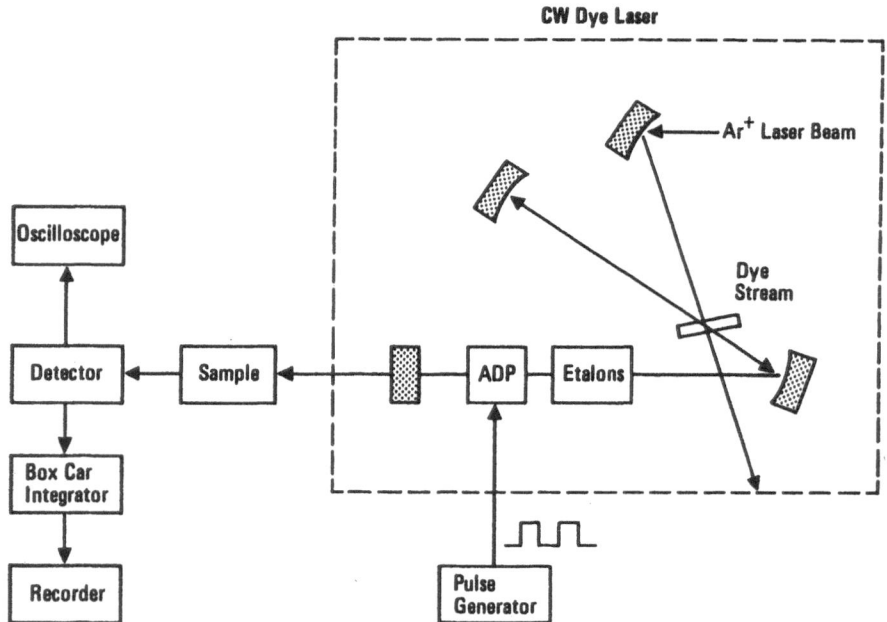

Figure 2   Schematic of the apparatus for observing coherent optical transients using a frequency-switched cw dye laser

A Spectra-Physics 580A cw dye laser is utilized but modified to include the ADP modulator. The dye is Rhodamine 6G. The output beam is single mode, linearly polarized and has a power up to 100 mW in a beam diameter of 0.5 mm. The collimated beam irradiates in single pass an evacuated and sealed off cell, of 20 cm length, containing I_2 at a vapor pressure (3 - 150 mTorr) determined by a refrigerated cold finger. Laser tuning by means of an intracavity etalon allows selecting a particular I_2 line where the overlapping Doppler-broadened I_2 hyperfine components span ~1 GHz (Doppler width: 395 MHz FWHM). Coherent transients in I_2 are seen even at a fraction of 1 mW laser power.

The intracavity ADP crystal is driven by an HP 1900A or 214A puls generator with a single or double pulse sequence and at a 25 kHz repetition rate. A PIN photo-diode monitors the forward beam, and transients are observed with a Tektronix 7704 sampling oscilloscope or a box car integrator. From the observed FID beat frequency, we find that the ADP electro-optic frequency shift parameter is 0.2 MHz/volt. Hence ~30 volt pulses are adequate for nonadiabatically switching the laser frequency outside an I_2 homogeneous linewidth of ~1 MHz. On the other hand, the laser does not emit a transient signal itself, which otherwise would obscure observations in the sample, because switching occurs inside the dye's homogeneous linewidth of $\gtrsim$ 200 MHz.

From the multitude of $^{127}I_2$ lines accessible, we selected in these initial studies only one line, $(v, J) = 2, 59 \to 15, 60$ of the electronic transition $X\,^1\Sigma_g^+ \to B\,^3\pi_{ou}+$. It falls at 16,956.43 cm^{-1}, 7.6 GHz to the high frequency side of the sodium D line. The vibration-rotation assignment was verified from the calculated line position and the fluorescence spectrum using a 150,000 resolving power spectrometer.

The three coherent transient effects shown in Fig. 1 are (a) FID of an I_2 velocity group that is prepared under steady-state conditions and where the laser frequency is abruptly switched by a step-function voltage pulse; (b) optical nutation patterns arising from an I_2 velocity group that is suddenly excited at the beginning of the switching pulse and another at the end; and (c) the photon echo pulse which follows two short switching pulses. The theory (5,6,12) of these processes parallels the molecular infrared case for vibration-rotation transitions where Stark-switching was employed. For electronic transitions, however, we must generalize these density matrix calculations to allow the upper (level a) and lower (level b) transition levels to depopulate at different rates, $\gamma_a \neq \gamma_b$, where γ_a and γ_b are the total decay rates, radiative and nonradiative (elastic and inelastic) of the diagonal density matrix elements (13). Furthermore, to agree with our experiments, it will be necessary to consider that during elastic collisions upper and lower transition states shift by significantly different amounts, so that the off-diagonal element exhibits quantum mechanical phase interruptions rather than classical velocity changes. This subtle point has emerged recently in certain linebroadening theories (14) and is consistent with the results obtained here. It follows that the normalized echo field amplitude will decay with pulse delay time τ as

$$E_c(t=2\tau) = e^{-\gamma t} \tag{1}$$

where $\gamma = (\gamma_a + \gamma_b)/2 + \gamma_\phi$ (15) is the rate the optically induced dipole dephases and γ_ϕ is the elastic collision rate for phase interruptions caused by perturber-induced energy level shifts. We note that the infrared echo results (6) represent the other limiting case where elastic collisions are dominated by velocity changes and the echo decay law is not a simple exponential. Our results, therefore, support the Berman-Lamb theory (14) for these limiting cases.

The two-pulse sequence of Fig. 1(c) also allows a measurement of the rate of population recovery, which results from inelastic collisions and radiative decay. The first pulse causes the upper level(a) to gain in population at the expense of the lower level(b) while the second pulse nutation signal is a measure of the extent that the population difference of the transition levels has recovered between the two pulses. The method is described elsewhere (6). Using the density matrix equations of motion, we find that the normalized second nutation amplitude grows with pulse delay time τ as

$$S_\infty - S(t=\tau) = e^{-\gamma_a t}(1 + \frac{\gamma_1}{\gamma_a - \gamma_b}) + e^{-\gamma_b t}(1 - \frac{\gamma_1}{\gamma_a - \gamma_b}) \tag{2}$$

where S_∞ is the value at $\tau = \infty$, which we identify with the first pulse nutation amplitude. The quantity γ_1 is the decay rate for the single channel (a) $\to$ (b); from the fluorescence intensities of the lines originating in (a), we estimate that $\gamma_1 \sim 0.1 \gamma_a$. The rate γ_b is, of course, restricted to collisional processes. Equation (2) suggests that lower and upper state decay rates can be determined independently.

The two pulse nutation measurements are characterized by essentially a single exponential decay in the pressure range 17 - 130 mTorr. This implies that upper and lower states depopulate by collisions at essentially the same rate so that $\gamma_b \sim \gamma_a$. It follows from (2) that the short time decay rate is $(\gamma_a + \gamma_b + \gamma_1)/2$. The observed value is

$$(\gamma_a + \gamma_b + \gamma_1)/2 = (0.71 + 0.029\ p)\ \mu s^{-1} \tag{3}$$

where the I_2 pressure p is in mTorr. From the pressure-independent part, we obtain an upper state radiative lifetime of $1.41\ \mu s$. The value is in reasonable agreement with the previous literature (16) and also agrees with our direct fluorescence decay measurements, giving $1.32\ \mu s$ at zero pressure. The pressure-dependent part of (3) yields a total inelastic collision cross section $\sigma_I = 530\ \text{Å}^2$. This result is about one order of magnitude larger than previous fluorescence measurements (16), which often are insensitive to upper state vibration-rotation quantum jumps of the emitter. In the pressure regime below 17 mTorr, the lower state seems to decay more slowly than the upper state, in accord with (2), but laser jitter and drift prevent quantitative measurements at present. Frequency-locking the dye laser to the I_2 line of interest should remove this difficulty in the future.

The echo measurements reveal a different aspect of the problem, namely, the degree to which coherence is preserved following collisions. We find that the echo decays exponentially as predicted by (1), and no evidence is found for an e^{-Kt^3} decay law at short times (~ 100 ns) which would be symptomatic of velocity-changing collisions (6). The echo decay rate is

$$\gamma = (0.79 + 0.071\ p)\ \mu s^{-1} \tag{4}$$

with p in mTorr of I_2. Utilizing the pressure-dependent parts of (3) and (4) and the relation $\gamma = (\gamma_a + \gamma_b)/2 + \gamma_\phi$, we obtain the <u>elastic collision cross section</u> σ_E = 780 Å^2 associated with the phase interruption rate γ_ϕ. We believe this to be the first optical coherence measurement of phase interrupting collisions. While this information is contained in the optical linewidth, it cannot generally be separated from other causes such as power, Doppler and inelastic collision broadening.

Although several papers (16) have dealt with the I_2 relaxation problem in the past, the measurements have been restricted almost exclusively to the upper state and to inelastic collisions, primarily those that terminate spontaneous emission such as predissociation. Some evidence for quasi-elastic I_2 collisions has appeared just recently also (17).

These preliminary results may obviously be extended in several different directions - to other optically excited systems and to other coherent transient phenomena, in a manner resembling the elegant methods of pulsed nmr.

We express our gratitude to A. Schenzle, P. R. Berman, S. Grossman, K. L. Foster, and D. E. Horne for aid and encouragement.

References

1. E. L. Hahn, Phys. Rev. <u>80</u>, 580 (1950)
2. H. C. Torrey, Phys. Rev. <u>76</u>, 1059 (1949)
3. A. Abragam, The Principles of Nuclear Magnetism, (Oxford University Press, London, 1961)
4. R. G. Brewer and R. L. Shoemaker, Phys. Rev. Lett. <u>27</u>, 631 (1971)
5. R. G. Brewer and R. L. Shoemaker, Phys Rev. <u>A6</u>, 2001 (1972)
6. J. Schmidt, P. R. Berman and R. G. Brewer, Phys. Rev. Letters <u>31</u>, 1103 (1973); P. R. Berman, J. M. Levy and R. G. Brewer, Phys. Rev. <u>A11</u>, 1668 (1975) and references therein
7. M. M. T. Loy, Phys. Rev Lett. <u>32</u>, 814 (1974)
8. N. A. Kurnit, I. D. Abella and S. R. Hartmann, Phys. Rev. Lett. <u>13</u>, 567 (1964); ibid Phys. Rev. <u>141</u>, 391 (1966)
9. G. L. Tang and B. D. Silverman, Physics of Quantum Electronics, edited by P. Kelley, B. Lax and P. E. Tannenwald (McGraw Hill, New York, 1966), p. 280
10. J. L. Hall, Proceedings of the Third International Conference on Atomic Physics, edited by S. J. Smith, G. K. Walters and L. H. Volsky (Plenum Press, N. Y., 1973), p. 615
11. P. P. Sorokin and J. P. Lankard, IBM J. Res. Dev. <u>10</u>, 162 (1966); F. P. Schäfer, W. Schmidt and J. Volze, Appl. Phys. Lett. <u>9</u>, 306 (1966)
12. F. A. Hopf, R. F. Shea and M. O. Scully, Phys. Rev. <u>A7</u>, 2105 (1973)
13. A. Schenzle, A. Genack and R. G. Brewer (to be published)
14. P. R. Berman and W. E. Lamb, Jr., Phys. Rev. <u>A6</u>, 2435 (1970); P. R. Berman, Phys. Rev. <u>A5</u>, 927 (1972) and references therein
15. M. Sargent III, M. O. Scully and W. E. Lamb, Jr., Laser Physics (Addison-Wesley, Reading, Mass., 1974), p. 85-87
16. A. Chutzian, J. K. Link and L. Brewer, J. Chem. Phys. <u>46</u>, 2666 (1967); G. A. Capelle and H. P. Broida, J. Chem. Phys. <u>58</u>, 4212 (1973); J. I. Steinfeld and A. N. Schweid, J. Chem. Phys. <u>53</u>, 3304 (1970) and references therein
17. D. L. Rousseau, G. D. Patterson and P. F. Williams, Phys. Rev. Lett. <u>34</u>, 1306 (1975)

RELAXATION IN MACROSCOPIC SYSTEM: AN INFORMATION THEORETIC APPROACH*

R.D. Levine
Department of Physical Chemistry
The Hebrew University, Jerusalem, Israel

1. Introduction

There are essentially two reasons why a simple yet quantitative approach to relaxation of a system in disequilibrium is relevant to laser research and applications. The direct and obvious one is that the operation of any atomic or molecular laser is dependent on maintaining the active medium in a state of disequilibrium ('population inversion'). Non-radiative relaxation processes drive the system towards equilibrium and thus tend to reduce the efficiency. A realistic modelling of the laser must allow for such loss processes. In addition however, lasers offer a particularly concenient way of selectivity driving a system away from equilibrium. This selective excitation offers many potential applications (1-3). Here again, the relaxation processes tend to diminish the selectivity of the initial excitation, degrading it ultimately into heat. As a concrete example we shall consider vibrational relaxation (4,5). Diatomic molecules, in an excess of buffer gas are displaced from vibrational equilibrium (See, e.g. (6)). Collisional (vibration-to-translation) energy transfer will then act to restore equilibrium. The problem is to determine the fraction of molecules in any given vibrational state at any instant during the relaxation. Even when the (very many) rate constants (or cross sections) for collisional energy transfer (from all possible initial states to all possible final states) are available one still requires the numerical solution of the kinetic (so called 'master') equations. Even under such optimal circumstances, an interpretation of the solution is then required. Typically however, many of the relevant rate constants

---

*Work supported by the Stiftung Volkswagenwerk.

are not known (see however (7-9)). The technical point made
here is that for the purpose of determining the population time
evolution knowledge of the very many rate constants can be
replaced by more accessible information. Near equilibrium,
the mean (over all molecules) vibrational energy provides all
the required input. At earlier times it may be necessary to
know also the vibrational specific heat, etc. We shall refer
to the variables whose mean values suffice to determine the
distributions as the 'independent' (or 'informative') constraints. Internal checks are inherently present in the
formalism and these provide the signal as to when further input
is required. The present approach (4,5,13) has been motivated
by previous applications of the surprisal analysis (and synthesis) to the problems of energy disposal and energy requirements
in isolated molecular collisions (so-called 'micro disequilibrium' (7-11)).

2. The Surprisal

As a measure of the deviation of the population of the state v
at time t from its equilibrium value we use the surprisal, $I(v,t)$,

$$I(v,t) = \ln\{P(v,t)/P^o(v)\} \tag{1}$$

Here $P(v,t)$ is the propability (i.e. the fraction) of molecules
in the state v at the time t and $P^o(v)$ is the magnitude of
$P(v,t)$ at equilibrium (i.e. as $t \to \infty$). For the example of
isothermal vibrational relaxation, $P^o(v)=\exp(-E_v/kT)/\sum_v\exp(-E_v/kT)$
where E_v is the vibrational energy and T is the temperature of
buffer gas.

As in our studies (9,10) of micro disequilibrium, we shall
find for macroscopic disequilibrium as well that the v (and t)
dependence of the surprisal is simpler (and more readily interpreted) than the similar dependencies of the actual probabilities. In fact, we shall find that the dependence of the
surprisal on v is essentially synonymous with the 'independent'
constraints that are required to uniquely specify the disequilibrium state of the system. This finding can then be used
in either of two ways. If the populations are known, one com-
putes the surprisal, via eq. (1), and thereby identifies the
nature of the variables whose average values suffice to
characterize the system, at the time t. (See eq. (9) or eq. (14)
below). Typically, the number of such variables (the number of
independent constraints) is well below the number of states.
Moreover, we shall argue that the number of independent constraints diminishes as time increases. The procedure of examining
the surprisal, the so called surprisal analysis, accomplishes
our first aim, that of offering a compact representation of the
distribution.

As an illustration of the power of surprisal analysis,
figure 1 shows the surprisal, at several time instances, vs.
E_v using probabilities computed via an exact numerical solution
of the kinetic (or 'master') equation. The results shown
correspond to a (singular) initial distribution P(v=9,t=0) = 1,
P(v≠9; t=0) = 0. Even so, it is evident from the drawing
that well before equilibrium has been reached the surprisal is
at most a polynomial of the second degree in E_v. Since normalization of the probabilities serves to define one parameter in
the surprisal, it follows that at most two parameters (say the
coefficients of E_v and E_v^2) are required to represent the surprisal and hence the entire distribution. At later times,
only one parameter is required and, as t → ∞, I(v,t) → 0.

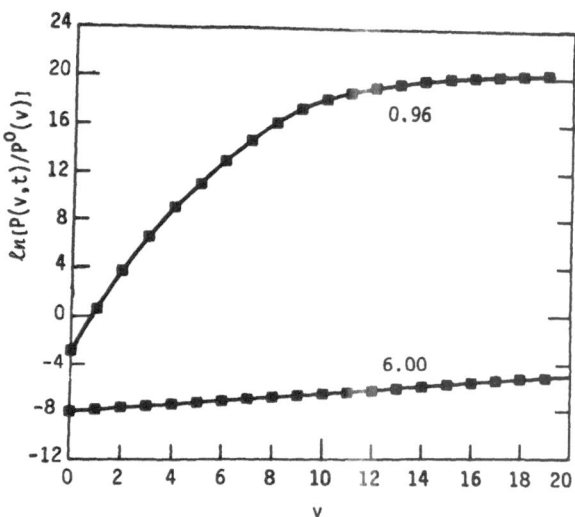

Figure 1.   Surprisal plot of the vibrational population distribution at two times (in units of the relaxation time
of the vibrational energy) when at t=0 all molecules
are in the v=9 state. The populations, P(v,t), were
determined via a numerical solution of the master
equation with a realistic set of rate constants (5).
The ordinate refers to t=0.96. The t=6 curve was
shifted down by 6 ordinate units. At t=0.96 two
terms are required in the surprisal expansion, eq. (14),
At t=6 a linear approximation, eq. (15), is clearly
sufficient. As t → ∞, the surprisal plot is a horizontal line. (Adapted from Ref.5).

When a smooth initial distribution is employed, the surprisal plot, figure 2, is far simpler. The coefficient of E_v^2 is much smaller and declines to zero much faster.

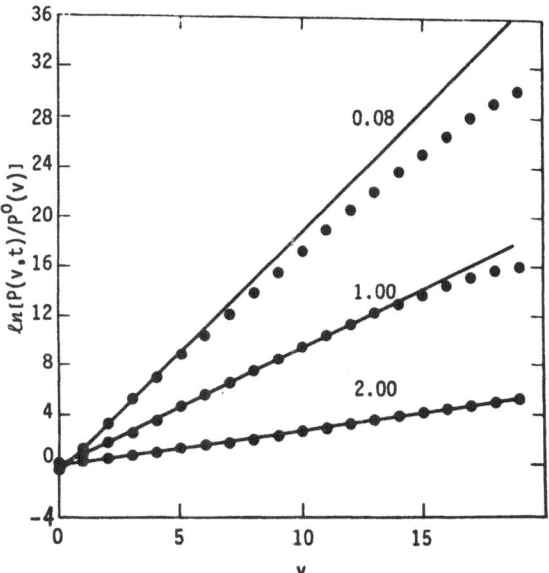

Figure 2. Surprisal plot of the vibrational population distribution when at t=0 the vibrational temperature is 1200°K (for Cl_2) and the buffer gas is at 300°K. The straight lines are the linear approximation to the surprisal which for the present case (cf. fig. 1) is valid even at the early stages of the relaxation. (Adapted from Ref. 5).

The observation that fewer moments appear in the surprisal as time progresses is a central one for the simplicity of the approach.

The utility of the surprisal analysis as a means of offering a compact representation for disequilibrium distributions has been extensively reviewed for the micro-level case (9,10). The surprisal offers more than just a convenient starting point for an analysis of the distribution. Given the constraints, one can find the surprisal and hence predict the distribution. The method, the so called surprisal synthesis, will be offered below as an alternative to the numerical solution of the master equation.

3. The Entropy Deficiency

The average deviation between the population distribution at time t and the equilibrium one is the entropy deficiency (9,10)

$$\Delta S = \sum_v P(v,t) \ln\{P(v,t)/P^o(v)\}. \tag{2}$$

In actual computations it is customary to measure ΔS in entropy units (1 e.u. = 1 cal/deg.mole). It is then necessary to multiply the dimensionless sum in eq. (2) by the gas constant R. The inequality $\ln x \geqslant 1 - 1/x$ insures that ΔS is non negative (9). During the relaxation, the magnitude of ΔS will change with time, decaying to zero as $t \to \infty$. In fact, the decline of ΔS towards zero is monotonic $-d\Delta S/dt \geqslant 0$, where the null value obtains only if $P(v,t) = P^o(v)$ for all v.

To predict ('synthesise') the population time evolution we employ the following principle: Given (complete or, typically, incomplete) data on the system at a given time t consider the set of all distributions that are consistent with the data and select the one that corresponds to the minimal value of ΔS. The reasons for this procedure have been discussed before (9, 10,12), and only the aspect unique to the time-dependent macro-relaxation case is considered in the sequel section.

4. Work

Work was required to drive the vibrational state distribution to its disequilibrium state. During the irreversible relaxation this work is dissipated. How much work can, in fact, be extracted if the disequilibrium distribution is harnessed to do useful work? To simplify the discussion we limit attention to isothermal relaxation and exclude concentration gradients. It can then be easily shown that the maximal available work (per mole) when the buffer gas is at the temperature T is $RT\Delta S$, (13) where R is the gas constant. The proof is immediate. The (vibrational) free energy at equilibrium, A^o is defined by

$$\exp(-A^o/RT) = \sum_v \exp(-E_v/RT) \tag{3}$$

Hence $P^o(v) = \exp\{(A^o-E_v)/RT\}$ and substituting in eq. (2)

$$\Delta S = \sum_v P(v,t) \ln\{P(v,t)/P^o(v)\} = \sum_v P(v,t)\{\ln P(v,t)-(A^o-E_v)/RT\}$$

$$= \{<E_v> - E_v^o - T(S - S^o)\}/RT \tag{4}$$

Here $<E_v> = \sum_v E_v P(v,t)$, $A^o = E_v^o-TS^o$ and $S = R\sum_v P(v,t) \ln P(v,t)$. We recognize $<E_v> - E_v^o$ as the amount of energy lost by the system during the relaxation from the distribution $P(v,t)$ to equilibrium. This energy can appear either as work, ΔW or as heat ΔQ, $\Delta E = <E_v> - E_v^o = \Delta Q + \Delta W$ (the descrease in energy equals the heat removed plus the work done). Now the decrease in vibrational entropy $(S-S^o)$ must be at least as high as the increase in the entropy of the buffer gas $(\Delta Q/T)$ hence

$$\Delta W \leqslant RT\Delta S \tag{5}$$

The equality sign applies to the limit where the work is extracted in a reversible process.

Our basic conceptual point of view is that a system in disequilibrium is a system capable of doing work. In practical applications this capacity can be manifested in different ways. We may be interested in actually extracting this work (e.g. as the coherent radiation emitted during lasing (14)) or we may be interested in consuming this work towards some particular goal (e.g. as the free energy necessary to 'unmix' different isotopes). Relaxation processes diminish the capacity of the system to do useful work, and bring it ultimately to a state equilibrium. The purpose of this review is to show that this abstract point of view can be very readily converted into a practical algorithm which enables one to determine both the populations of the different energy states and the available work as a function of time. We obtain the required algorithm by a slight generalization of the very conventional thermodynamic recipe: the most probable state of the system is the one of lowest free energy. All we do here is to compute the 'free energy', i.e. the maximal available work (RTΔS) when the system is not necessarily in thermal equilibrium and find the distribution for which ΔS is minimal, subject to the condition that the distribution be consistent with the available data on the system.

5. The Distribution of Minimal ΔS

We assume that the data on the system represents the results of the measurements of the mean value of one or more observables. Hence, as in other applications of the information theoretic procedure (9,12) data can be given as bulk expectation values of some n observables

$$<A_r(t)> = \sum_v A_r(v) P(v,t), \quad r = 1, \ldots, n . \tag{6}$$

We also define the r=0 observable by $A_0(v) = 1$ so that

$$<A_0(t)> = \sum_v P(v,t) = 1 \tag{7}$$

is the normalization condition which always obtains. In general, we thus have n+1, n $\geqslant$ 0, pieces of data.

Seeking the distribution that minimizes ΔS subject to the n+1 conditions , eqs. (6) and (7) proceed as usual (9,12). Introducing n+1 parameters, λ_r, r = 0,1 ., .., n one seeks the unconstrained minimum of the Lagrangian $\mathscr{L}$

$$\mathscr{L} = \Delta S - \sum_r \lambda_r <A_r(t)> - <A_0(t)> . \tag{8}$$

This gives through standard manipulations (9,12)

$$P(v,t) = P^o(v) \exp\{- \sum_{r=0}^{n} \lambda_r A_r(v)\} = P^o(v) \exp\{-\lambda_0 - \sum_{r=1}^{n} \lambda_r A_r(v)\},$$

$$\tag{9}$$

where the $n+1$ 'Lagrange multipliers' λ_r need be determined by the requirement that the distribution $\hat{P}(v,t)$ is to be consistent with the data, i.e. that the $n+1$ conditions, eqs. (6) and (7) obtain. A set of n coupled implicit equations

$$\langle A_r(t) \rangle = \sum_v A_r(v) P^O(v) \exp\{- \sum_{r=1}^{n} \lambda_r A_r(v)\} / \sum_v P^O(v) \exp\{-\sum_{r=1}^{n} \lambda_r A_r(v)\} \tag{10}$$

need be solved for λ_r's, $r=1,\ldots,n$. Subsequently, λ_0 can be determined by the normalization condition, i.e.

$$\exp(\lambda_0) = \sum_v P^O(v) \exp\{- \sum_{r=1}^{n} \lambda_r A_r(v)\} . \tag{11}$$

Alternatively, one can solve the $n+1$ simultaneous equations

$$\langle A_r(t) \rangle = \sum_v A_r(v) P^O(v) \exp\{- \sum_{r=0}^{n} \lambda_r A_r(v)\} \tag{12}$$

where $r= 0,1,\ldots,n$, $\langle A_0(t)\rangle =1$ and $A_0(v) = 1$. The standard proof (9) that among all possible distributions that are consistent with the data, eq. (9) has the lowest possible value of ΔS, i.e.

$$\Delta S\{P\} = - \lambda_0 - \sum_{r=1}^{n} \lambda_r \langle A_r(t) \rangle \tag{13}$$

applies here as well. The only non-standard feature is that the Lagrange multiers are now time-dependent. The surprisal corresponding to the data is $I(v,t)$

$$I(v,t) = - \ln\{P(v,t)/P^O(v)\} = \sum_{r=0}^{n} \lambda_r A_r(v) . \tag{14}$$

One can prove (5) that in the surprisal expansion eq. (14) (or in eq. (13)) the only non-vanishing multipliers (the only finite λ_r's) are those that correspond to independent observables.

6. Applications

The surprisal analysis in figs. 1 and 2 show that near equilibrium there is only one independent moment (i.e. only one nontrivial term in the expansion eq. (14)). Indeed, using the magnitude of $\langle E_v \rangle$ (as a function of time) one can reproduce (via equations (9) $-$ (13)) the near equilibrium $P(v,t)$ distribution with considerable (better than 1%) accuracy. A single function of time suffices to characterize the population time evolution.

For shorter times it is evident from fig. 1 and 2 that there are additional independent moments. In practice however, we will now know the exact population distribution. How can we monitor the need for additional data? Figure 3 shows the procedure. Given both $\langle E_v \rangle$ and $\langle E_v^2 \rangle = \sum_v E_v^2 P(v,t)$, we employ the magnitude of $\langle E_v \rangle$ only, predict the distribution and hence are in a position to compute $\langle E_v^2 \rangle$. We now compare, fig. 3, the actual magnitude of $\langle E_v^2 \rangle$ with the magnitude predicted us-

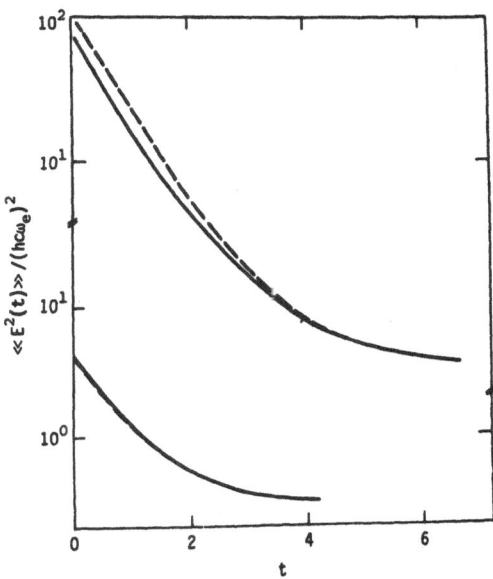

Figure 3. $<E_v^2(t)>$ vs. t computed using a numerical solution of
a master equation (solid line) and predicted using
only the magnitude of $<E_v(t)>$ as input. Top: System
of figure 1; Bottom: System of figure 2. As is
evident from the surprisal analysis, $<E_v(t)>$ is the
only independent moment over much of the relaxation
stage for system 2 but not so for the early stages
of the time evolution in system 1.

ing $<E_v>$ as the only independent constraint. As long as both
magnitudes are in accord, knowledge of $<E_v>$ alone will suffice.
At earlier times, it will not and we need to determine the
distribution which is consistent with the magnitudes of both
$<E_v>$ and $<E_v^2>$.

An important variational property of ΔS, equation (13), is
shown in figure 4. If we fail to include all the independent
constraints, the magnitude of ΔS predicted by eq. (13) will always
be below the actual value. When we include all the independent
moments we reach the exact value. Moreover, adding extra,
dependent, constraints will not cause the ΔS computed via the
procedure of section 5 to exceed its exact value (computed from
eq. (2) using the exact distribution). The exact value is an
upper bound that can be realised using all the n independent
constraints as input to the procedure of section 5.

7. Vibrational Temperature

A particularly interesting aspect of vibrational relaxation is
that an initial induction period is often followed by a time

232

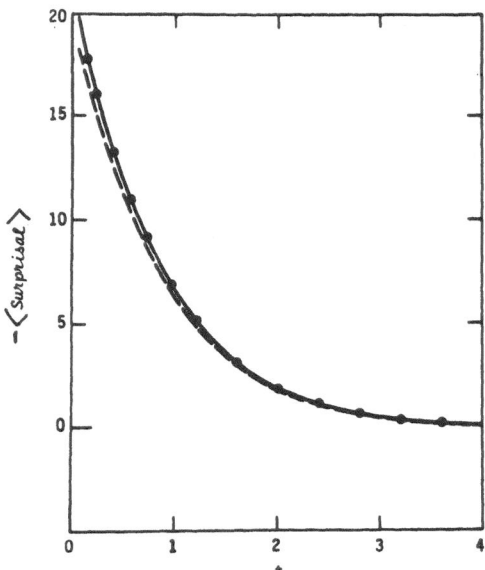

Figure 4: Entropy deficiency (ΔS in units of the gas constant, R),
for the surprisals shown in figure 1 vs. time.
Solid line: ΔS computed from the populations obtained
by an exact solution of the master equation. Broken
line: ΔS predicted using as input the time dependence
of the lowest moment of the vibrational energy,
$<E_v(t)>$. Dots: ΔS predicted using the three lowest
moments as input. (Adapted from Ref. 5).

interval where the higher vibrational states decay faster (6,15).
We now show that this is the time regime where the mean vibrational energy is sufficient to describe the populations. This
time regime will be particularly pronounced for diatomic molecules of high vibrational frequencies or, in general, when the
characteristic vibrational temperature $\theta = hc\omega_e/k$ is high compared to the buffer gas temperature T, i.e. when $\tilde{u} = hc\omega_e/kT > 1$.
(The opposite limit has also been considered (5,16)).

Given $<E(t)>$ as the only constraint we obtain as the distribution consistent with it and of minimal ΔS (cf. eq. (9))

$$P(v,t) = P^o(v)\exp\{-\lambda_0(t) - \lambda_1(t)E_v\}. \qquad (15)$$

For any given time t the Lagrange parameters $\lambda_0(t)$ and $\lambda_1(t)$ are
obtained by the condition that $<E(t)>$ has the given value and that
the distribution is normalized. The first approximation is
already remarkably accurate, as is evident from figures 1-4.

The concept of a 'vibrational temperature' is thus valid whenever a single independent constraint suffices to characterize the distribution. It offers however a good approximation even when it is not quantitatively accurate.

When the buffer gas temperature is not too high (i.e. when $hc\omega_e/kT > 1$), the concept of vibrational temperature does imply that the higher vibrational states decay faster (4,5,15). Thus, from eq. (15)

$$dP(v,t)/dt = -\{(d\lambda_0/dt) + (d\lambda_1/dt)E_v\}P(v,t) . \qquad (16)$$

Summing both sides of eq. (16) over v and noting eq. (7), we have that

$$- (d\lambda_0/dt) = (d\lambda_1/dt)<E_v(t)> \qquad (17)$$

and hence

$$dP(v,t)/dt = (d\lambda_1/dt)\{<E_v(t)> - E_v\}P(v,t) \qquad (18)$$

Assume now that $hc\omega_e/kT >> 1$ then, at equilibrium, most of the molecules will be in the ground, v=0 state. Also, $<E_v(\infty)> \leq kT < hc\omega_e \leq E_v$ for v > 0. Hence, for any v > 0, $dP(v,t)/dt \cong -E_vP(v,t)$. Similarly, if one is sufficiently far from equilibrium, $P(v,t)/P^o(v) >> 1$, for any v > 0. Hence $P(v)/\{P(v)-P^o(v)\} \simeq 1$ and, from eq. (18)

$$-d\ell n\{P(v,t) - P^o(v)\}/dt \propto E_v. \qquad (19)$$

The 'fan like' behavior eq. (19) of $\ell n\{P(v,t) - P^o(v)\}$ vs. t is evident in figure 5.

8. Summary

A practical method for using bulk averaged values of observables for the characterization and prediction of the molecular population time evolution during isothermal relaxation has been presented. In practical applications to realistic examples of vibrational relaxation very few bulk averages are required to accurately predict the population distribution even when the initial population was very strongly inverted. The bulk average values are used as constraints in a maximal entropy procedure for the determination of the population distribution. It is shown that the procedure is of a variational type. Monotonic convergence of the information theoretic predicted distribution to the exact one is guaranteed upon inclusion of additional macroscopic input. The concept of 'independent moments' is introduced for this purpose. Only independent observables are informative, i.e. provide independent data which is required for convergence. The number of informative observables decreases with time and is typically very much smaller than the number of significantly populated molecular energy states.

234

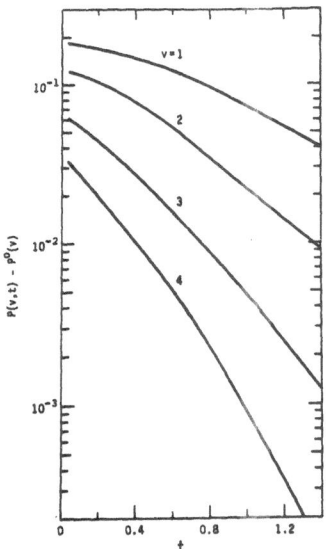

Figure 5: The populations obtained via the solution of the
master equation vs. t for the time regime where
$P(v,t)/P^0(v) > 1$. Initial population as in figure 2.
(Adapted from Ref. 5).

The method was illustrated by comparing its predictions
to the results of a numerical solution of the master equation,
which a realistic set of rate constants and for different
initial conditions. The application of the surprisal analysis
to the interpretation, characterization and compaction of the
population distribution has been demonstrated. Turning to
predictions ('surprisal synthesis'), only strongly inverted
initial populations required three independent moments, during
the initial stages. Over much of the relaxation a single moment
('vibrational temperature') sufficed for an accurate prediction.

An important practical problem which has not been explicitly
discussed is the adaptation of the method to the case when there
are concentration gradients. Now the probabilities of the
different states are functions of not only time t but also of
the position in space r, $P(v,rt)$. The principle to be employed
is again the same except that now the Lagrange parameters
λ_r will be functions of both space and time coordinates.
Concrete applications of this, more general approach, are in
progress.

Acknowledgment

The work here reviewed was originally carried out and published
in collaboration with Drs. I. Procaccia and Y. Shimoni.

References

1. R.B. Bernstein, Israel J. Chem. $\underline{9}$, 615 (1971)
2. R.N. Zare and P.J. Dagdigian, Science, $\underline{185}$, 739 (1974)
3. V.S. Letokhov and C.B. Moore, Soviet J. Quant. Elec. (in press)
4. I. Procaccia and R.D. Levine, J. Chem. Phys. $\underline{62}$, 3819 (1975)
5. I. Procaccia, Y. Shimoni and R.D. Levine, J. Chem. Phys., in press
6. See for example: (a) J.F. Bott and N. Cohen, J. Chem, Phys., $\underline{59}$ 447 (1973); (b) I. Burak, Y. Noter, A.M. Ronn and A. Szøke, Chem. Phys. Letts., $\underline{16}$, 306 (1972); ibid, $\underline{17}$, 345 (1972); (c) B. Hopkins and H. Chen. J.Chem Phys., $\underline{57}$, 3816 (1972); (d) J.L. Ahl and T.A. Cool, J. Chem. Phys., $\underline{58}$, 5540 (1973); (e) R.M. Osgood, Jr., P.B. Sackett and A. Javan, J. Chem. Phys. $\underline{60}$, 1464 (1974); (f) F. Menard-Bourcin, J. Menard and L. Henry, J. Chem. Phys. $\underline{63}$, 1479 (1975)
7. I. Procaccia and R.D. Levine, J. Chem. Phys. $\underline{62}$,2496; $\underline{63}$,4261 (1975)
8. I. Procaccia and R.D. Levine, J. Chem. Phys. $\underline{64}$, 808 (1976)
9. R.D. Levine and R.B. Bernstein in Dynamics of Molecular Collisions, ed. W.H. Miller (Plenum, N.Y. 1976)
10. R.D. Levine and R.B. Bernstein, Accts. Chem. Res. $\underline{7}$, 393 (1974)
11. M. Rubinson and J.I. Steinfeld, Chem. Phys. $\underline{4}$, 467 (1974)
12. E.T. Jaynes, Phys. Rev. $\underline{106}$, 620 (1957)
13. R.D. Levine, J. Chem. Phys., in press; I. Procaccia and R.D. Levine, to be published
14. (a) R.D. Levine and O. Kafri, Chem. Phys. $\underline{8}$, 426 (1975); (b) Chem. Phys. Letters $\underline{27}$, 175 (1974); A. Ben-Shaul, O. Kafri and R.D. Levine, Chem. Phys. $\underline{10}$, 367 (1975)
15. W.D. Breshears, Chem. Phys. Letters, $\underline{20}$, 429 (1973)
16. I. Procaccia, Y. Shimoni and R.D. Levine, J. Chem.Phys. $\underline{63}$, 3181 (1975)

VI. Atmospheric Photochemistry and Diagnostics

TROPOSPHERIC PHOTOCHEMICAL AND PHOTOPHYSICAL PROCESSES

J.N. Pitts, Jr.
Department of Chemistry and Statewide Air Pollution Research Center
University of California, Riverside, CA 92502
and
B.J. Finlayson-Pitts
Department of Chemistry, California State University
Fullerton, CA 92634, USA

I. Introduction

Our goal in this paper is to illustrate some areas where tunable lasers--IR, UV, and vacuum UV--might play a definitive role in achieving a significantly better understanding of the complex chemical and physical transformations occurring in the natural and polluted troposphere. Such an advancement in knowledge would not only be of fundamental scientific interest per se, but would also serve to greatly improve the quality and breadth of the technical data base upon which the air pollution control strategies of all of our countries should be founded.

The sources, transport and transformation, effects, and control of air pollution may be treated logically in terms of an integrated air pollution system (Fig. 1). Pollutants emitted directly into the air (by definition, "primary" pollutants) are transformed into other compounds (termed "secondary" pollutants) as they are transported downwind from their sources. As these secondary pollutants are formed and subsequently decay, they impact on various receptors, including man, plants, materials, and even our climate. Monitoring of the concentration of these species in ambient air and establishing the nature and extent of their effects form the bases for the implementation of rational control measures through legislative action, thus completing the loop of the system.

We shall focus on the "transformation" portion of the system, starting with a brief discussion of the "classical" and current mechanisms of photochemical air pollution. Emphasis will be placed on possible applications of tunable lasers to identification and monitoring of:

● Pollutants capable of absorbing solar radiation and producing reactive species in air.

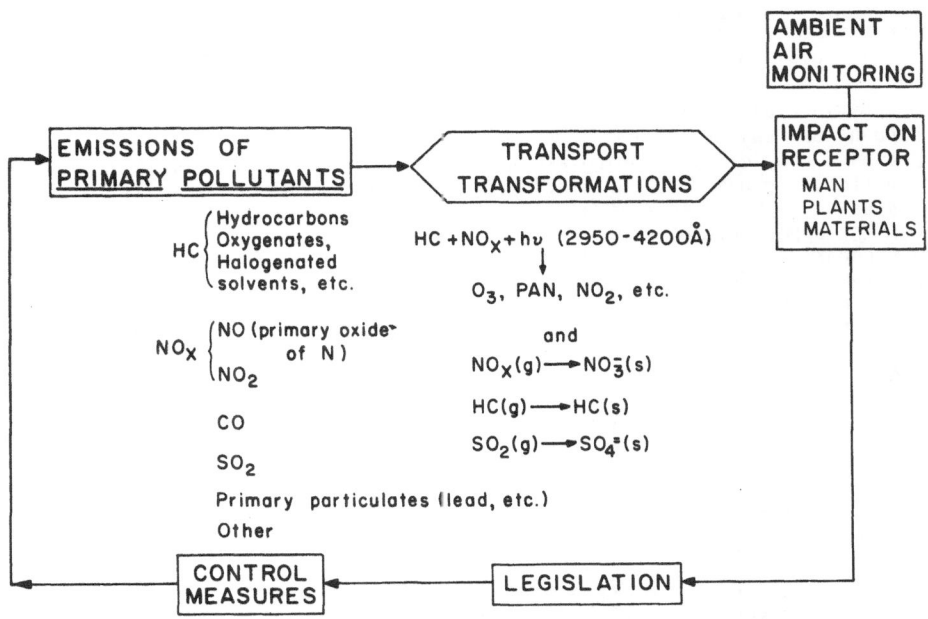

Fig. 1.   Schematic diagram of an air pollution system

- Atomic and free radical intermediates formed in primary and secondary processes.

- The ultimate products of these photooxidations both in simulated and real atmospheres.

We shall not review the use of lasers to measure gaseous pollutants, such as carbon monoxide, ozone, etc., as this is discussed in some detail both in other papers at this conference and elsewhere [1]. Rather, we shall discuss those species known or suspected to be present in ambient air which may have significant chemical, physical, or biological effects and for which no satisfactory conventional monitoring techniques now exist. In addition, the importance of secondary aerosols (particulates) produced by gas-to-particle conversion processes in the polluted troposphere will be discussed, again with special emphasis on the problems to which tunable lasers might be applied. Because of space limitations, no attempt can or will be made to cite extensively the original literature; this has been done in three recent reviews [2,3]. Furthermore, the very active area of the chemistry of fluorocarbons in the troposphere, as well as in the stratosphere, is beyond the scope of this article.

II. Early Theories of the Mechanisms of Formation of Air Pollution

A. Historical

Two distinct air pollution conditions have been recognized. The first, termed "London" or "sulfurous" smog, has been known for centuries and is characterized by a chemically reducing atmosphere containing sulfur dioxide and particulates, including sulfuric acid aerosols and sulfates. The second, "Los Angeles" or "photochemical" smog, was first recognized about thirty years ago from the appearance of crop damage, eye irritation, and haze. It was shown subsequently, starting with HAAGEN-SMIT and co-workers in the early 1950's [2], that these symptoms were due to the formation of oxidants, including ozone and peroxyacetyl nitrate (PAN), in addition to other gaseous species and particulates, from hydrocarbons (HC) and oxides of nitrogen (NO_x):

$$NO_x + HC + h\nu \text{ (sunlight; } \sim 290 < \lambda < \sim 430 \text{ nm)} \rightarrow$$

$$O_3 + NO_2 + CH_3C\underset{OONO_2}{\overset{O}{\diagdown}} + \text{particulates} + ---- \tag{1}$$
$$\text{(PAN)}$$

Solar ultraviolet irradiation was found to be essential; in the troposphere, the effective short wavelength cutoff of this light source is about 290 nm [4].

Today, however, especially with the increased use of high-sulfur fuels due to the energy shortage, oxides of sulfur (SO_x) and particulate sulfate are also often found at significant concentrations in photochemically oxidizing atmospheres. Thus, the distinction between the two types of air pollution is now sufficiently blurred that it is unrealistic to treat them separately.

Not only has the distinction as to the "type" of smog one is studying disappeared, but the earlier idea that photochemical smog is generally confined to southern California has been discarded during the past few years. For example, in May 1976, The Netherlands experienced an air pollution episode in which the ambient oxidant concentration reached 0.3 ppm, well over California's first-stage alert level of 0.20 ppm. Sydney, Australia, which has recorded oxidant concentrations as high as 0.35 ppm, now exceeds the World Health Goal of 0.06 ppm ozone (O_3) more than 100 days per year.

Indeed, modeling studies [5] in our laboratories have shown that the potential for photochemical smog formation, in terms of solar UV flux at latitudes as high as Fairbanks, Alaska, and Leningrad, Russia, is significant in the summer and early months of fall; indeed, on June 21, it virtually equals that of Los Angeles, and, thus, an oxidant episode with the accompanying rapid gas $\rightarrow$ particle conversion processes may arise if there are significant emissions of HC and NO_x and the appropriate meteorology (e.g., stagnant air and clear days).

B. "Classical" Mechanism

Until the late 1960's, attention was generally focused on nitrogen dioxide as being the most important light-absorbing species in photochemical smog. Thus, while the predominant form of NO_x emitted into the atmosphere is the colorless and relatively nontoxic nitric oxide, in the presence of sunlight and hydrocarbons and at concentrations in the ppm-pphm region, this NO is rapidly converted to toxic brown NO_2, both in real and simulated atmospheres.

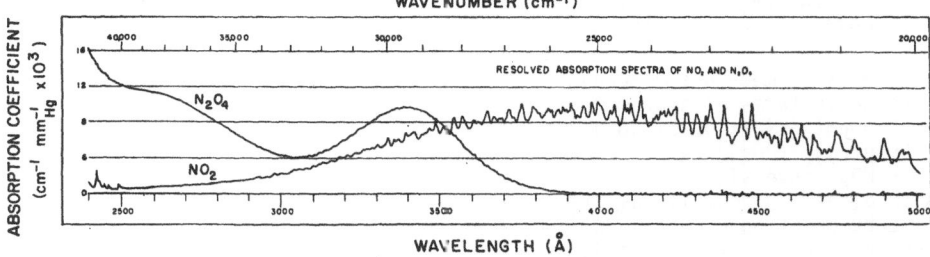

Fig. 2. Absorption spectra of NO_2 and N_2O_4 from 250 to 500 nm [4]

It was well known, at that time, that whereas NO did not absorb actinic UV radiation (e.g., $\lambda > 290$ nm), NO_2 absorbed strongly throughout the entire region and well into the visible (Fig. 2) [4].

On the basis of several experimental studies, it was postulated that the major mode of photodecomposition of NO_2 in the actinic UV was direct photo-dissociation into NO and a ground state oxygen atom [4]:

$$NO_2 + h\nu \ (\lambda < 430 \text{ nm}) \rightarrow NO + O(^3P) \tag{2}$$

This was subsequently confirmed [6], and it is now accepted that the primary quantum yield of photodissociation rises rapidly from zero at ~430 nm to approximately unity at wavelengths less than 398 nm [6].

Under atmospheric conditions, it was recognized that over 99% of the $O(^3P)$ atoms formed in (2) must react with molecular oxygen to form ozone:

$$O(^3P) + O_2 \overset{M}{\rightarrow} O_3 \tag{3}$$

However, if there are significant amounts of NO present, O_3 is rapidly destroyed:

$$O_3 + NO \rightarrow NO_2 + O_2 \tag{4}$$

Thus, O_3 and NO cannot simultaneously coexist at high concentrations, and at the steady state in a strictly inorganic system (i.e., NO, NO_2, O_2, and UV) only small amounts of O_3 are present [4].

When traces of hydrocarbons or other reactive organics were added to the inorganic $NO-NO_2-O_2-UV$ system, the chemistry drastically changed, as shown in the first smog chamber studies [4]. A typical set of time-concentration profiles for such conditions (i.e., a simulated polluted atmosphere) are shown in Fig. 3. They were generated recently by irradiation of 0.53 ppm propylene and 0.59 ppm NO_x in one atmosphere of purified air in the SAPRC 5800-ℓ evacuable smog chamber [7]; similar diurnal variations of primary and secondary pollutants have long been observed in ambient air. It is seen that both of the primary reactants, NO and propylene, disappear rapidly, while NO_2 increases to a maximum at about three hours of irradiation and then drops

240

off.[1] When the NO concentration drops to low levels, significant and increasing amounts of O_3 and PAN start to form.

The early mechanisms accounted for the loss of HC and the NO → NO_2 conversion in terms of a primary attack on the HC by ozone and oxygen atoms and subsequent oxidation of NO via the secondary radicals produced therein [e.g., alkylperoxy (RO_2)] [4]. However, the observed rates of loss of HC and of the NO → NO_2 conversion were too great to be accounted for by this general mechanism alone. Hence, it was clear that an unspecified intermediate(s) must also be reacting with the HC (propylene in Fig. 3) and generating additional radicals capable of oxidizing NO → NO_2.

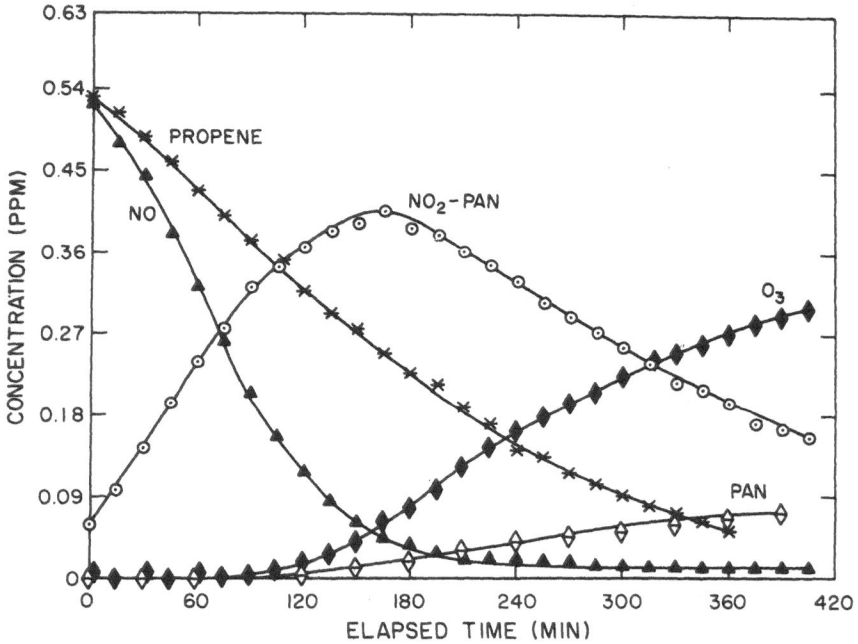

Fig. 3. Concentration-time profiles of the major primary and secondary pollutants during irradiation of 0.53 ppm propylene and 0.59 ppm NO_x in one atmosphere of air in an evacuable smog chamber [7]

---

[1]The curve (NO_2-PAN) is the *net* NO_2, that is the total NO_2 as measured by a chemiluminescent technique *minus* the PAN present which, in commercial NO_2 chemiluminescent instruments, can give an interferring response as though it were NO_2 [45].

III. "Current" Theory--Role of the Hydroxyl Radical

The first accurate determinations of the absolute room temperature rate constants for the reactions of the hydroxyl radical, OH, with carbon monoxide [8] and with a series of simple hydrocarbons [9] revealed that these rate constants were much greater than hitherto believed. Thus, it was suggested that the hydroxyl radical might play a major role in the oxidation of hydrocarbons as well as in the oxidation of NO $\rightarrow$ NO$_2$ [10]. A series of chain reactions were then postulated in which OH was primarily responsible for the HC oxidation, particularly in the early stages of the reaction, and in which these OH radicals were regenerated via reactions of RO$_2$ and hydroperoxyl (HO$_2$) free radicals, both of which convert NO $\rightarrow$ NO$_2$. Typical reactions in such a chain sequence involving simple alkyl radicals include:

$$\cdot OH + R\text{-}CH_3 \rightarrow H_2O + RCH_2\cdot \quad (R = H, CH_3, C_2H_5) \tag{5}$$

$$RCH_2\cdot + O_2 \rightarrow RCH_2O_2\cdot \tag{6}$$

$$RCH_2O_2\cdot + NO \rightarrow RCH_2O\cdot + NO_2 \tag{7}$$

$$RCH_2O\cdot + O_2 \rightarrow RCHO + HO_2\cdot \tag{8}$$

$$HO_2\cdot + NO \rightarrow \cdot OH + NO_2 \tag{9}$$

$$\cdot OH + CO \rightarrow H + CO_2 \tag{10}$$

$$H + O_2 \overset{M}{\rightarrow} HO_2\cdot \tag{11}$$

For larger hydrocarbons, other reactions--e.g., alkoxy radical isomerizations--must also be included [48].

The rates and mechanisms of many of the reactions believed to be important have now been determined. Their application in computer modeling of the chemistry of these systems, sometimes including over two hundred elementary reactions for a one hydrocarbon-NO$_x$ system, leads to concentration-time profiles of the major pollutants which are generally in good agreement with the experimentally determined curves [3]. In addition, the hydroxyl radical was detected for the first time in ambient air in 1975 (see below) [11]. Thus, it is now accepted that OH and HO$_2$ (as well as RO$_2$) play a central role in the formation of photochemical air pollution, and probably in the enhanced rate of oxidation of SO$_2$(g) to SO$_4^=$ particulates as well (see below) [47].

IV. Photochemical Reactions Leading to the Production of OH, HO$_2$, RO$_2$, and Other Free Radicals

A. Sources of OH

Direct sources of ambient OH include the photolysis of nitrous acid (HONO), and, probably to a lesser extent, hydrogen peroxide (H$_2$O$_2$). Indirect sources include the oxidation of NO by HO$_2$ (9), and the attack of O(^{1}D) atoms, formed from O$_3$ photolysis below ~318 nm, upon water vapor [2,3].

Ozone.--The absorption spectrum of O$_3$ from 200 to 330 nm is shown in Fig. 4 [4]. At wavelengths less than approximately 318 nm (there is still some uncertainty as to the exact wavelength of the onset of this reaction),

242

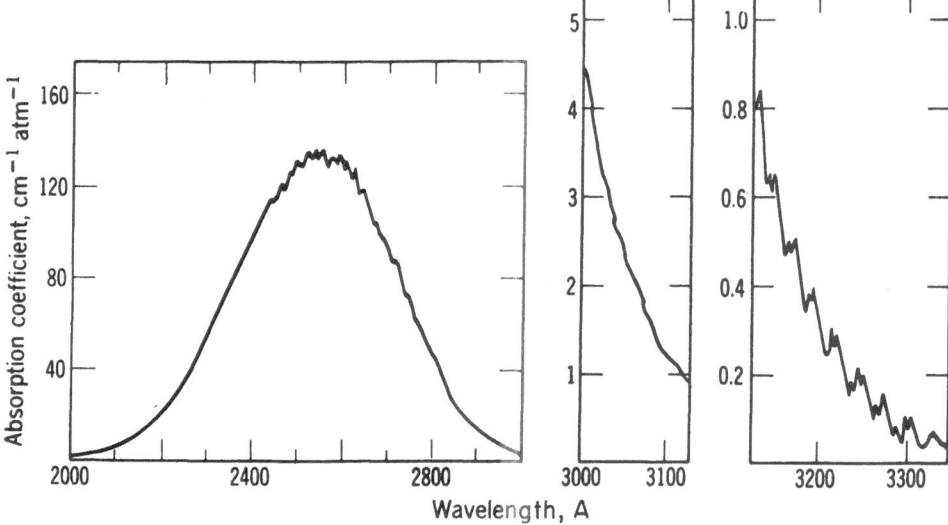

Fig. 4. Absorption spectrum of O_3 from 200 to 330 nm [4]

the products are oxygen atoms and molecular oxygen, both in electronically excited states [12]:

$$O_3 \xrightarrow{\lambda < 318 \text{ nm}} O(^1D) + O_2(^1\Delta_g) \tag{12}$$

The predominant fate of the $O(^1D)$ atoms is quenching to the ground state, $O(^3P)$. However, in the troposphere, a small but significant fraction of the $O(^1D)$ atoms react with water to produce OH radicals [13].

$$O(^1D) + H_2O \rightarrow 2OH \tag{13}$$

While this mechanism of OH production is relatively unimportant in the early stages of smog production, compared to HONO or formaldehyde photolysis, it can become significant as ambient levels of O_3 build up.

Singlet molecular oxygen.--While discussing O_3 photolysis, it is relevant to digress briefly and consider the atmospheric implications of the other excited species formed in reaction (12). Thus, prior to the measurement of any gas phase absolute rate constants for the reactions of singlet molecular oxygen--$O_2(^1\Delta_g)$--with olefins, this species was postulated as a possible oxidant to account for the excess hydrocarbon consumption, at least in part [2]; it has since been shown in our laboratory [2] and others [2,14] that its rates of reaction with olefins are too slow to lead to any significant consumption of the organic. However, the *products* of its reactions with various organic species, *in vivo* as well as *in vitro*, may well warrant further consideration for their health implications.

Thus, the lifetime of $O_2(^1\Delta_g)$ produced in a typical photochemically polluted atmosphere, after removal from the atmosphere by inhalation, is calculated to be sufficiently long for a small fraction of that present in the ambient air to be available for possible reactions with lung tissue [15]. Furthermore, the reaction products of $O_2(^1\Delta_g)$ with various organics include hydroperoxides, endoperoxides, and dioxetanes, depending on the particular reactant [2,16]:

$$O_2(^1\Delta_g) + \begin{array}{c} CH_3 \\ CH_3 \end{array}C=C\begin{array}{c} CH_3 \\ CH_3 \end{array} \rightarrow \begin{array}{c} CH_3 \\ CH_3 \end{array}\overset{OOH}{\underset{}{C}}-C\begin{array}{c} CH_2 \\ CH_3 \end{array} \tag{14}$$

$$O_2(^1\Delta_g) + \text{[anthracene with } \phi \text{]} \rightarrow \text{[endoperoxide product]} \quad (\phi = C_6H_5) \tag{15}$$

$$O_2(^1\Delta_g) + (CH_3O)_2C=C(OCH_3)_2 \rightarrow \begin{array}{c} CH_3O \\ CH_3O \end{array}\overset{O-O}{\underset{}{C-C}}\begin{array}{c} OCH_3 \\ OCH_3 \end{array} \tag{16}$$

Recent biochemical studies of α-ketoperoxides, for example, indicate they may be toxic to humans [17], while the effects of the other oxidation products and their metabolic fates are generally unknown.

A number of mechanisms have been postulated for the formation of $O_2(^1\Delta_g)$ in polluted ambient air, including direct absorption from the ground state (a highly forbidden transition), energy transfer from organic and inorganic molecules in their excited triplet states, photolysis of ozone, exothermic chemical reactions which produce O_2, oxygen-enhanced absorption by organic species, and the hydrolysis of PAN [2]. However, $O_2(^1\Delta_g)$ has never been detected directly or indirectly in ambient air, although its concentration has been estimated to be quite low, $\approx 10^{-5}$ ppm [2]. Such a determination would be of particular interest because of the potential health effects (direct and indirect) and because all of its sources, and hence its antici-pated concentration in polluted atmospheres, have not been quantitatively established.

In addition, the possibility of a photosensitized formation of $O_2(^1\Delta_g)$ on surfaces, for example on atmospheric particulates, by energy transfer from a triplet excited species, is an area which, though first proposed in 1967 [2], still awaits detailed investigation. The feasibility of such a process is suggested by the observation of $O_2(^1\Delta_g)$ generation upon photolysis of oxygen-saturated heterogeneous liquid solutions containing a dye sensitizer bound to copolymer beads [18]. Atmospheric particulates contain a variety of organics, for example polycyclic aromatics, which might act as sensitizers to form $O_2(^1\Delta_g)$, which could then in turn oxidize the aromatic. Since it is actually the oxidized forms of many so-called "carcinogens" which are toxic, the products of such surface reactions may also represent a health problem.

Nitrous acid.--HONO absorbs light at wavelengths below 400 nm (Fig. 5) [19], forming OH and NO with a quantum yield of approximately unity [20]:

$$HONO + h\nu \ (\lambda < 400 \text{ nm}) \rightarrow HO + NO \tag{17}$$

Sensitive and specific techniques for the detection and measurement of ambient HONO are vitally necessary because it is believed to be a major source of OH radicals in polluted atmospheres. It has been identified in a synthetic air pollution mixture by Fourier transform infrared spectroscopy [21] and once in ambient air using a wet chemical method [22] (the latter are, however, notoriously subject to interferences).

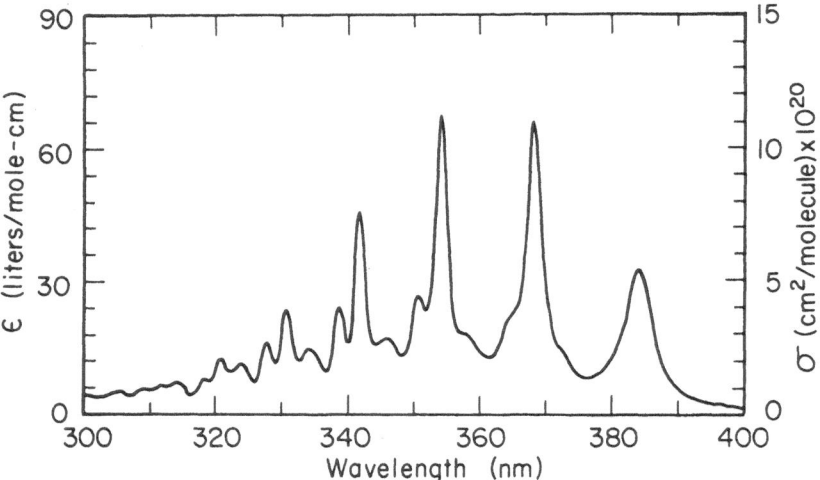

Fig. 5. Absorption spectrum of HONO from 300 to 400 nm [9]

Modes of formation of HONO in the troposphere and in smog chambers are still not known with certainty; indeed, they may be homogeneous, heterogeneous, or both. Some proposed sources include the following [2,3]:

$$OH + NO \xrightarrow{M} HONO \tag{18}$$

$$HO_2 + NO_2 \rightarrow HONO + O_2 \tag{19}$$

$$NO_2 + NO + H_2O \rightleftharpoons 2\ HONO \tag{20}$$

However, even these processes are controversial at the present time. For example, reaction (19) has just been shown to proceed, at least in part, by addition to form peroxynitric acid, $HOONO_2$, which was directly identified in the gas phase, using Fourier transform IR spectroscopy [49].

In the past, reaction (20) was believed to be predominantly heterogeneous in the forward direction; the results of recent studies [23], however, suggest that the homogeneous component of this forward reaction may be sufficiently fast that, before dilution, it might be a significant source of HONO in power plant plumes, auto exhaust, etc. This, as well as the rate of the reverse reaction, is currently the subject of debate.

Hydrogen peroxide.--Hydrogen peroxide absorbs weakly in the actinic ultraviolet and appears to photodissociate into two hydroxyl radicals, although the quantum yield in air is not established.

$$H_2O_2 + h\nu\ (\lambda < \sim 370\ nm) \rightarrow 2\ OH \tag{18}$$

Here again, detection and measurement techniques are needed as ambient H_2O_2 measurements have only been reported in one study in which a wet chemical method was employed [24].

B. *Sources of HO₂*

As we have seen, the hydroperoxyl radical not only oxidizes NO → NO₂, but also regenerates OH in the process. Sources of HO₂ [2,3] include the reaction of H atoms with O₂ (11), as well as H atom abstraction from alkoxy radicals (8).

Hydrogen atoms can be produced by the photolysis of formaldehyde (produced both as a primary pollutant--e.g., auto exhaust--and as a secondary pollutant-- e.g., ozone-olefin reactions), which absorbs in the actinic ultraviolet (Fig. 6) [25].

$$HCHO + h\nu \ (\lambda < 370 \ nm) \begin{cases} \overset{(a)}{\longrightarrow} H + HCO \\ \overset{(b)}{\longrightarrow} H_2 + CO \end{cases} \qquad (19)$$

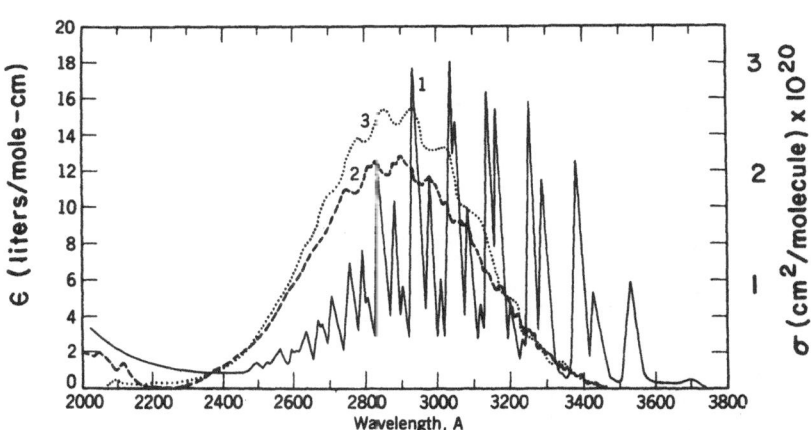

Fig. 6. Absorption spectra in the region 200-300 nm for 1) HCHO (≈75°C), 2) CH₃CHO (25°C), and 3) C₂H₅CHO (25°C) [25]

Unfortunately, even after many studies using a variety of techniques, the absolute, or indeed the relative, quantum yields of the two paths as a function of wavelength or the effect of molecular oxygen on them are not known accurately [2,3,26]. Such quantum yield determinations must be carried out under atmospheric as well as laboratory conditions, since total pressure and the presence of excess oxygen may well affect the relative quantum yields of (19a) and (19b) and, indeed, the overall mechanism of the photooxidation.

The formyl radical produced in (19a) may also be an HO₂ source via (20), which is known to occur rapidly at low total pressures:

$$HCO + O_2 \rightarrow HO_2 + CO \qquad (20)$$

However, there is some question as to whether addition to form the peroxy-formyl radical, HCO₃, is the predominant reaction mode at atmospheric pressure [3].

As expected from the preceding discussion, formaldehyde has a dramatic effect when introduced into a synthetic air pollutant mixture. The synthesis of such mixtures containing known pollutant concentrations in the ppb-ppm concentration range and their irradiation and analysis under reproducible, simulated atmospheric conditions require experimental apparatus and techniques not commonly used in conventional laboratory studies. Figure 7 is a schematic diagram of a 5800-ℓ evacuable environmental chamber and accompanying 25-KW solar simulator designed for such studies and currently in operation in our laboratory. The apparatus is described in detail elsewhere [2,27].

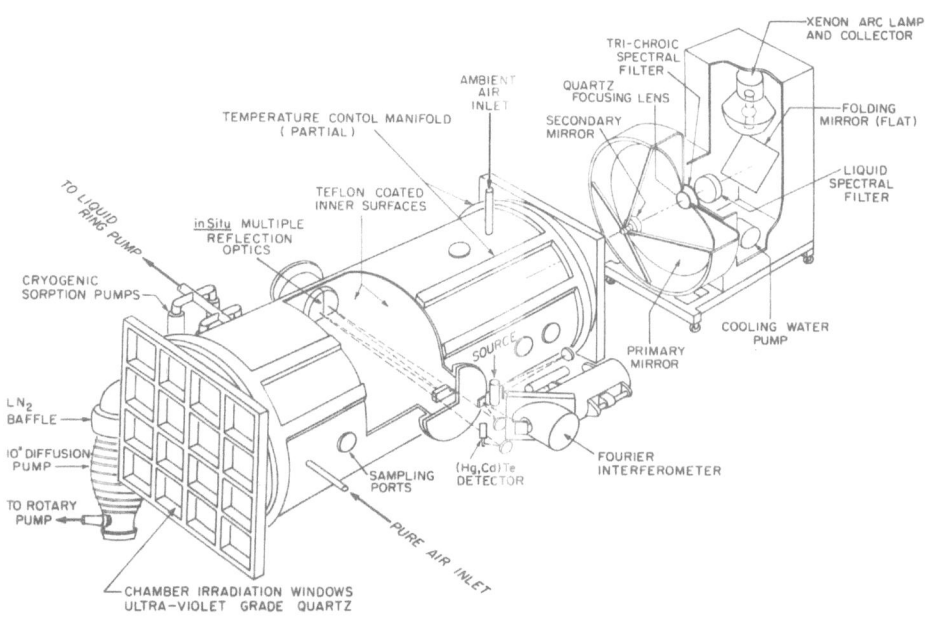

Fig. 7. Schematic of the evacuable smog chamber solar simulator facility at the Statewide Air Pollution Research Center, University of California, Riverside, California

The results of adding 0.16 ppm HCHO to a toluene-NO_x-air mixture [7] are shown in Fig. 8. Clearly, the rates of oxidation of NO to NO_2 and of O_3 formation increase significantly in the presence of HCHO. Such acceleration undoubtedly occurs in "aged smog," where, under stagnant air conditions, the secondary pollutants produced prior to sunrise include aldehydes either formed the previous day or by dark reactions during the night.

C. Sources of RO_2

RO_2 is produced predominantly by the reaction (6) of O_2 with alkyl radicals R. The latter result from a variety of hydrocarbon oxidations by OH, O_3, $O(^3P)$, etc. [2,3], as well as by the photolysis of aliphatic aldehydes (e.g.,

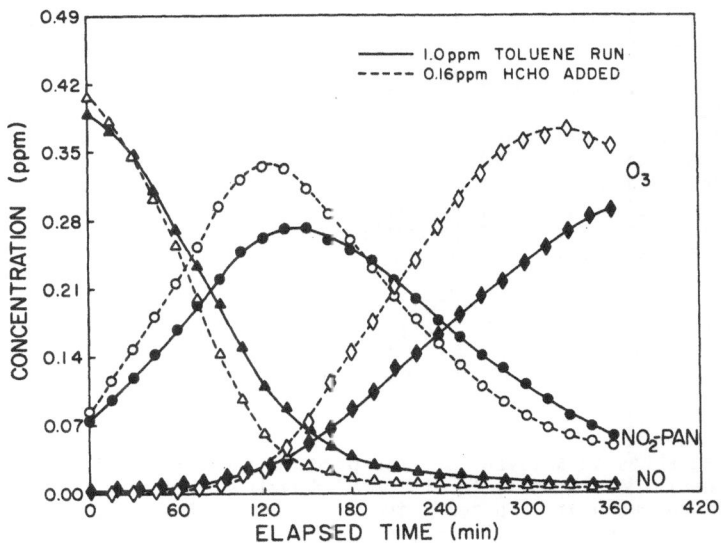

Fig. 8. Concentration-time profiles of O_3, NO, and NO_2 + PAN during the irradiation of toluene and NO_x in one atmosphere of purified air in an evacuable smog chamber [7]

one mode is RCHO + $h\nu$ → R + CHO) and ketones. For example, methyl ethyl ketone is believed to photodissociate by two paths in the actinic UV [25]:

$$CH_3COC_2H_5 \xrightarrow{h\nu} \begin{array}{l} \text{(a)} \longrightarrow CH_3CO + C_2H_5 \\ \\ \text{(b)} \longrightarrow CH_3 + C_2H_5CO \end{array} \qquad (21)$$

As in the case of formaldehyde and the aliphatic aldehydes, the absolute and relative quantum yields under atmospheric conditions are not known with any degree of accuracy, if at all. However, recent computer modeling studies, using the best available data (in which 21a is highly favored over 21b), suggest that methyl ethyl ketone, for example, while less important than HONO or HCHO at the start of the irradiation of a synthetic pollutant mixture, may become equally important in generating O_3 at later reaction times [28].

Not only are accurate determinations of primary photochemical and photophysical quantum yields under atmospheric conditions essential, but also the development of simple and accurate measurement techniques is needed for aliphatic aldehydes and ketones in real and simulated atmospheres. Thus, while gas chromatography and combined gas chromatography-mass spectrometry may be applied to the determination of specific compounds, simultaneous measurement of a wide variety of these oxygenates by these techniques is complex and time-consuming. As a result, there are relatively few unambiguous reports of their ambient concentrations in the literature.

D. Other Possible Primary Processes Leading to the Formation of Free Radicals

In addition to the photodissociative processes described above, there may be others which contribute, although to a lesser extent, to the production of free radicals. For example, alkyl peroxides (ROOR') and hydroperoxides (ROOH) absorb weakly at ~300 nm [25]. Such species may be produced in various hydrocarbon oxidations, but are difficult to detect, even in laboratory systems, because of their labile nature and sensitivity to surface reactions. For example, the reaction of O_3 with olefins has been postulated to lead to the formation of α-ketohydroperoxides [29]:

$$O_3 + \underset{H}{\overset{CH_3}{\diagup}}C=C\underset{H}{\overset{CH_3}{\diagdown}} \rightarrow \underset{H}{\overset{CH_3}{\diagup}}C-C\underset{H}{\overset{O}{\diagdown}}CH_3 \rightarrow \underset{H}{\overset{CH_3}{\diagup}}\overset{O\cdot}{C}-\overset{OO\cdot}{C}\underset{H}{\overset{CH_3}{\diagdown}} \rightarrow CH_3\overset{O}{\underset{}{C}}-\overset{OOH}{\underset{H}{C}}-CH_3 \quad (22)$$

α-ketohydroperoxide

However, attempts to analyze for this product using GC and GC-MS have not been successful due largely to its rapid decomposition on contact with surfaces. Hence, accurate and sensitive *in-situ* techniques are needed, not only to clarify if they are formed in such reactions, but also to determine whether they are present in ambient air.

V. Identification of Trace but Significant Molecular Species

As mentioned earlier, there have been several reports of the use of wet chemical methods to measure trace molecules, such as H_2O_2 and HONO. Similar determinations for HNO_3 have also been reported recently [30]. However, such wet techniques are subject to numerous interferences. As a result, it is important to develop *in-situ* physical monitoring techniques for the detection of these and other trace pollutants.

Long-path infrared absorption spectroscopy is one such analytical method that is currently being applied to these problems with some success. For example, the highly irritating and toxic compound, ketene, is known to be a product of several ozone-olefin reactions, including that of *cis*-2-butene [2,3]:

$$O_3 + \underset{H}{\overset{CH_3}{\diagup}}C=C\underset{H}{\overset{CH_3}{\diagdown}} \rightarrow CH_3CHO + CH_3\overset{\cdot}{C}HOO\cdot \rightarrow \text{other products} \quad (23)$$

Criegee
biradical

$$CH_2=C=O + H_2O$$
ketene

Figure 9 shows its detection in our laboratory by Fourier transform infrared spectroscopy (FTIR) in this system at a total pressure of 2 torr [31]. In spite of the strong absorptions in the same region by CO, as seen, for example, in the C_2H_4 reaction in which ketene is not a product (Fig. 9), it can still be detected at low pressures if the spectral resolution is sufficiently great. In the runs with propylene and *cis*-2-butene at atmospheric pressure, its presence might be inferred by the overall distribution in the band shape. Ketene has not been detected in ambient air during air pollution episodes and, hence, an in-depth application of such monitoring techniques to this problem is needed in the near future.

One such study has been reported in which various major and trace pollutants were detected and monitored during a severe photochemical air pollution episode in Pasadena, California, using FTIR [32]. With an optical path

50 ppm Olefin + 10 ppm O_3
Pathlength 53–64 meters
Spectral resolution 0.13 cm^{-1}

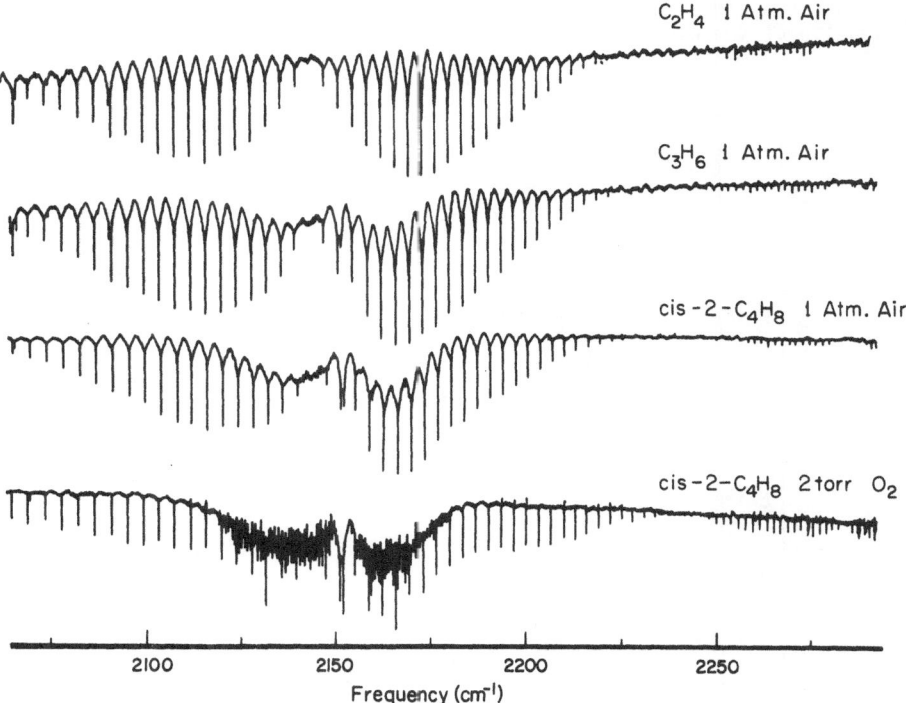

Fig. 9. Fourier transform infrared spectrum in the region of 2100 cm^{-1} for the reaction of O_3 with several olefins; strong, well-spaced lines are due to CO, while the distortion in the second and third curves and the fine structure in the fourth is due to $CH_2=C=O$ [31]

length of 417 m, several trace species, including formic acid (HCOOH) and methanol (CH_3OH), were identified.

In addition to the above species, there are a variety of trace contaminants of interest from a health viewpoint. For example, N-nitrosodimethylamine, a notorious carcinogen, has recently been detected in ambient air [33]. Other

250

known carcinogens, including some epoxides, are suspected to be present in polluted atmospheres as products of hydrocarbon oxidations, although they have not yet been detected. Additionally, secondary ozonides $(R_1R_2C{<}^{O-O}_O{>}CR_3R_4)$, and likely peroxides and hydroperoxides, are produced in ozone oxidations in the laboratory [2,3], yet remain to be identified under atmospheric conditions.

VI. *Identification of Free Radicals by Spectroscopic Techniques*

The calculated importance of $O(^3P)$, HO_2, O_3, and OH with respect to consumption of an olefin during a typical environmental chamber experiment is shown in Fig. 10 [34]. It is seen that OH is responsible for the majority of the hydrocarbon consumption, particularly in the early stages of the reaction, and that HO_2 and especially $O(^3P)$ play relatively minor roles in the direct oxidation of the olefin.

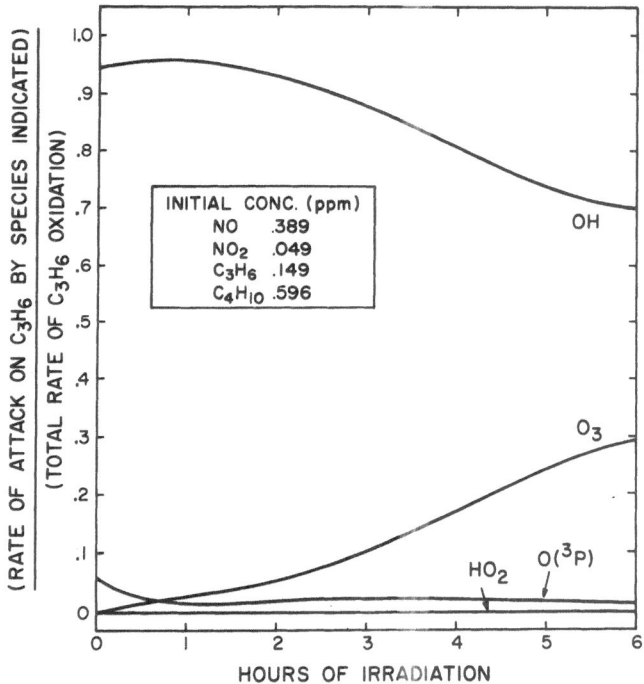

Fig. 10. Predicted relative importance of several reactive intermediates during photooxidation of propene-n-butane mixture under simulated atmospheric conditions [34]

The hydroxyl radical has in fact been detected recently at ground level in ambient air at concentrations of $\sim 10^5$-10^7 radicals/cc using laser excited resonance fluorescence [11]; these concentrations are in agreement with those predicted by computer modeling studies and are consistent with observed rates of hydrocarbon decay in the atmosphere [46]. Airborne studies of OH concentrations have very recently been reported for the troposphere and stratosphere [35]. Additionally, the resonance fluorescence technique has just been applied to smog chamber studies [36]. This laser technique shows great promise, although some interferences have been noted, including the two-photon absorption of water and photolysis of O_3 to produce $O(^1D)$, followed by its reaction with water, both of which produce OH [11].

Ground-state oxygen atoms, $O(^3P)$, have also been measured in the stratosphere recently, using a resonance fluorescence technique [50].

The application of such laser methods to the determination of such atoms and free radicals, as for example $O(^3P)$, HO_2, RO_2, and HCO, is also sorely needed for studies of the elementary chemical processes occurring in photochemical smog and for their determination in real and simulated atmospheres. Progress has been made in this direction with the establishment of the HO_2 ultraviolet (peak at ~ 210 nm) [37] and infrared absorption bands (1095, 1390, 3410 cm^{-1}) [37] (6649, 7018, and 7968 cm^{-1}) [38]. In addition, several peroxyl radicals, including CH_3O_2, $C_2H_5O_2$, and $CH_3 \cdot CO \cdot O_2$, have also been detected in laboratory studies, using their UV and near IR absorption bands [51]. Techniques such as laser magnetic resonance [39], which has been applied to OH and HO_2, may also prove helpful in meeting this goal.

VII. *Secondary Aerosol Formation*

Secondary aerosols produced in gas-to-particle conversion processes in a photochemically oxidizing atmosphere often exceed particulates emitted directly. Because of their light-scattering properties and potential effects on health, visibility, and climate, especially during long-range transport, a knowledge of their chemical and physical properties is essential.

A. *Physical Properties*

Examination of the mass, size, and volume distributions of "aged aerosols" as a function of particle diameter generally shows a bimodal distribution with a minimum at ~ 1-3 µm [3,40]. The larger or "coarse" particles--e.g., dust, sand, etc.--generally originate from mechanical processes, while the "fine" particles less than ~ 1-3 µm are produced by condensation on Aitken nuclei of the type produced in combustion processes. The "fine" particulate range, in fact, may itself contain two distinct size regions, as it does in "fresh" smog near strong automobile sources. The very small (≤ 0.1 µm) particles arise from primary emissions from combustion sources, while the larger ones (~ 0.1-1 µm) arise from the coagulation of the small nuclei or from their growth by condensation of photochemical reaction products, etc. While dependent on the particular conditions, on the average approximately one-third to one-half of the total aerosol mass is found in the fine particle range below ~ 1-3 µm [40].

The size range in which secondary aerosols from gas-to-particle conversion processes are expected to accumulate is from ~ 0.1 to ~ 2 µm. This is confirmed by studies of chemical composition as a function of particle size. Figure 11, for example, shows the size distribution of total suspended particulate (TSP),

252

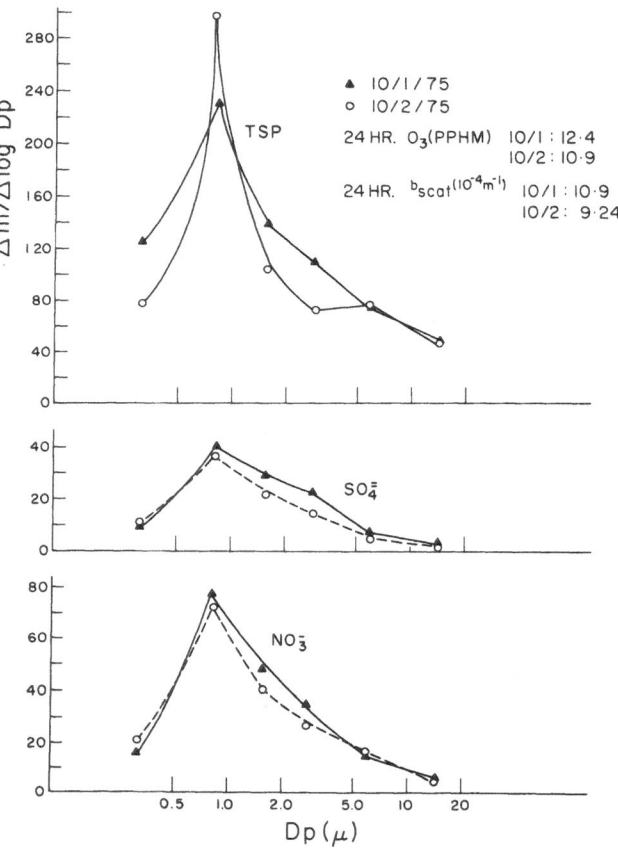

Fig. 11. Size distribution of total suspended particulates (TSP), nitrate (NO₃⁻), and sulfate (SO₄⁼) of 24-hour samples taken in Riverside, California, October 1 and 2, 1975 [41]

sulfate, and nitrate during a photochemical air pollution episode [41]. It is seen that all three peak at ~0.8 μm. This is particularly significant, since particles in this size range cause maximum light scattering and also have the potential for maximum health impact [3].

B. *Chemical Composition*

The composition of secondary aerosols is highly complex and includes sulfates, nitrates, and organic compounds produced by the oxidation of SO_2, NO_x, and hydrocarbons, respectively. These gas-to-particle conversions are of particular concern because they increase in rate substantially in a photochemically oxidizing atmosphere [3]. For example, the rate of photooxidation of SO_2 in pure air is ≤0.1% per hour, while in ambient air during photochemical air pollution episodes, it may be as high as 10% per hour.

Such enhanced rates of secondary aerosol formation are of particular concern because of the potential transport of pollutants many hundreds of kilometers downwind. It has been shown, for example, that for at least two episodes elevated ozone levels at Harwell, England, and southern Ireland result primarily from the transport of pollutants from the continent [42]. Such long-range transport during which gas-to-particle oxidations are occurring often results in 1) the appearance of haze in locales which are relatively free of local pollutant sources, and 2) possible widespread climatic effects, etc. [43].

Nitrate and sulfate anions likely exist, to a large extent, as ammonium salts in aerosols. Thus, recent studies of the aerosol in the Los Angeles air basin indicate that sufficient ammonium ion is present that, on the average, all of the nitrate and half of the measured sulfate could exist in this form [41].

The actual chemical form of particulate nitrate and sulfate *in situ*, however, is uncertain. One study, using the relationship between light scattering and relative humidity to detect the reaction of NH_3 with particulates, has shown that sulfate particles in the St. Louis, Missouri, U.S.A., region are in the form of H_2SO_4, NH_4HSO_4, and $(NH_4)_2SO_4$, the relative amounts depending on the sampling site. The development of further specific and sensitive *in-situ* techniques for the determination of the chemical form of sulfates and nitrates in aerosols is sorely needed, particularly in view of the uncertainties in the effects of the measurement techniques currently in use. For example, it is possible that at least a portion of the measured ammonium nitrate may not exist *in situ*, but rather is formed during sampling by the reaction of gaseous NH_3 with HNO_3 on the filter.

Organic compounds in an oxidized state also comprise a significant portion of the aerosol mass [3]. Thus, the application of a variety of analytical techniques, including GC, GC-MS, high resolution MS, and IR, to the problem of particulate composition has revealed the presence of a variety of oxygenated multifunctional organics, including those which may be represented by the formulae

$(CH_2)_a COOH$

$\bigcirc$

$HOOC(CH_2)_b COOH$

$X-(CH_2)_c-Y$

(a = 0-3)   (b = 0-5)   (c = 1-5)

where X and Y = -COOH, CH_2OH, -CHO, $-CH_2ONO$, $-CH_2ONO_2$, $-\overset{O}{\overset{\|}{C}}ONO$, $-\overset{O}{\overset{\|}{C}}-ONO_2$

Again, *in-situ* analysis of such compounds is a most desirable goal.

VIII. *Summary*

From this necessarily cursory examination of the chemistry of air pollution, it is obvious that there are many important problems to which tunable lasers might be profitably applied. These include:

• The detection and measurement of molecular species of interest in the real and polluted troposphere, particularly those which are difficult to determine by conventional techniques, for example, hydroperoxides and carcinogens, such as nitrosamines.

- Development or extension of techniques for determining free radical and electronically excited intermediates [e.g., $O_2(^1\Delta_g)$].

- Application of tunable lasers to the problem of elucidating the mechanisms of the elementary processes and overall chemical reactions involved in tropospheric pollution, including those which may be occurring on surfaces.

- Development of *in-situ* analytical techniques for better establishing the chemistry of particulates, especially those falling in the size range of 0.1-2 μm.

It is hoped that solution of these problems, in part through the use of tunable lasers, will be forthcoming in the near future.

Acknowledgments

We wish to thank our colleagues at UCR in the Department of Chemistry and at the Statewide Air Pollution Research Center for their contributions to the research results cited herein, as well as to those in the Departments of Chemistry and Physics, California State University, Fullerton, for helpful and thought-provoking discussions.

We would also like to express our appreciation to our conference hosts and to the distinguished organizations sponsoring this International Conference for a travel grant and for the opportunity to be present at such a significant and stimulating scientific event.

Finally, we gratefully acknowledge the generous financial assistance of the following agencies in support of the research cited in this paper; California Air Resources Board (Grant No. ARB-4-214); Environmental Protection Agency (EPA Grant No. R-800649); National Science Foundation (Grant Nos. GP-35424 and MPS73-08638-A02); and National Science Foundation--Research Applied to National Needs (NSF-RANN Grant No. AEN73-02904-A02); Petroleum Research Fund of the American Chemical Society (Type G Grant No. 3080-G2).

References

1. See, for example, Proc. NATO Expert Conf. on Laser Spectroscopy of the Atmosphere, Opt. Quantum Electr. 8 (1976); E. D. Hinkley, K. W. Nill, F. A. Blum: Laser Focus, pp. 47-51 (April 1976); E. D. Hinkley, R. T. Ku, K. W. Nill, J. F. Butler: Appl. Opt., in press; T. H. Maugh, II: Science 177, 1090 (1972); J. A. Hodgeson, W. A. McClenny, P. L. Hanst: Science 182, 248 (1973); Proc. Int. Conf. on Environmental Sensing and Assessment, Las Vegas, Nevada (September 14-19, 1976)

2. J. N. Pitts, Jr., B. J. Finlayson: Ang. Chem. (Int. Ed.) 14, 1 (1975), and references therein.

3. B. J. Finlayson, J. N. Pitts, Jr.: Science 192, 111 (1976); B. J. Finlayson, J. N. Pitts, Jr.: In Advances in Environmental Science and Technology, ed. by J. N. Pitts, Jr., R. L. Metcalf, Vol. 7 (Wiley-Interscience, New York, in press), Ch. 9 and references therein

4. P. A. Leighton: Photochemistry of Air Pollution (Academic Press, New York, 1961) and references therein

5. H. Nieboer, W. P. L. Carter, A. C. Lloyd, J. N. Pitts, Jr.: Atmos. Environ. (1976), in press

6. J. N. Pitts, Jr., J. H. Sharp, S. I. Chan: J. Chem. Phys. 40, 3655 (1964); I. T. N. Jones, K. D. Bayes: J. Chem. Phys. 59, 4836 (1973); H. Gaedtke, J. Troe: Ber. Bunsenges. Physik. 79, 184 (1975)

7. J. N. Pitts, Jr., A. M. Winer, K. R. Darnall: "Mechanisms of photochemical reactions in urban air-smog chamber studies," Final Report, U. S. Environmental Protection Agency, Grant No. R-800649, Vol. II, (1976), in preparation

8. J. T. Herron: J. Chem. Phys. 45, 1854 (1966); G. Dixon-Lewis, W. E. Wilson, Jr., A. A. Westenberg: J. Chem. Phys. 44, 2877 (1966)

9. N. R. Greiner: J. Chem. Phys. 46, 2795, 3389 (1967); J. Chem. Phys. 53, 1070, 1284 (1970)

10. J. Heicklen, K. Westberg, N. Cohen: Center for Air Environmental Studies, Report No. 115-69 (1969); D. H. Stedman, E. D. Morris, Jr., E. E. Daby, H. Niki, B. Weinstock: "The role of OH radicals in photochemical smog reactions," presented at the 160th Natl. Meeting, Am. Chem. Soc., Chicago, Illinois (September 14-18, 1970)

11. C. C. Wang, L. I. Davis, Jr: Phys. Rev. Lett. 32, 349 (1974); C. C. Wang, L. I. Davis, Jr., C. H. Wu, S. Japar, H. Niki, B. Weinstock: Science 189, 797 (1975); C. C. Wang, L. I. Davis, Jr., C. H. Wu, S. Japar: Appl. Phys. Lett. 28, 14 (1976)

12. G. K. Moortgat, P. Warneck: Z. Naturforsch. 30a, 835 (1975), and references therein; D. L. Philen, R. T. Watson, D. D. Davis: "A quantum yield determination for $O(^1D)$ production from ozone via laser flash photolysis," presented at the 12th Informal Conf. on Photochemistry, Gaithersburg, Maryland (June 28-July 1, 1976)

13. R. Simonaitis, J. Heicklen: Int. J. Chem. Kinet. 5, 231 (1973); R. Simonaitis, J. Heicklen: J. Phys. Chem. 77, 1096 (1973)

14. R. E. Huie, J. T. Herron: Int. J. Chem. Kinet. 5, 197 (1973)

15. J. G. Calvert, "Interactions of air pollutants," Proc. Conf. on Health Effects of Air Pollutants, Assembly of Life Sciences, National Academy of Sciences--National Research Council, October 3-5, 1973, Serial No. 93-15 (U. S. Government Printing Office, Washington, DC, 1973) pp. 19-101

16. C. S. Foote: Accts. Chem. Res. 1, 104 (1968); D. R. Kearns: Chem. Rev. 71, 395 (1971)

17. D. B. Menzel: Personal communication (1975)

18. A. P. Schaap, A. L. Thayer, E. C. Blossey, D. C. Neckers: J. Am. Chem. Soc. 97, 3741 (1975)

19. R. A. Graham: Ph.D. Dissertation, University of California, Berkeley, California (1975) (unpublished)

20. R. A. Cox, R. G. Derwent: J. Photochem. $\underline{4}$, 139 (1975); R. A. Cox; J. Photochem. $\underline{3}$, 175 (1974)

21. J. M. McAfee, J. N. Pitts, Jr., A. M. Winer: "In-situ long-path infrared spectroscopy of photochemical air pollutants in an environmental chamber," presented at the Pacific Conf. on Chemistry and Spectroscopy, San Francisco, California (October 16-18, 1974)

22. T. Nash: Tellus $\underline{26}$, 175 (1974)

23. W. H. Chan, R. J. Nordstrom, J. G. Calvert, J. H. Shaw: Chem. Phys. Lett. $\underline{37}$, 441 (1976)

24. J. J. Bufalini, B. W. Gay, Jr., K. L. Brubaker: Environ. Sci. Technol. $\underline{6}$, 816 (1972)

25. J. G. Calvert, J. N. Pitts, Jr.: Photochemistry (John Wiley & Sons, Inc., New York, 1966) and references therein

26. C. B. Moore: "Photochemistry and radiationless processes in formaldehyde," presented at the Int. Conf. on Tunable Lasers and Applications, Loen, Nordfjord, Norway (June 7-11, 1976); E. K. C. Lee, R. S. Lewis, R. G. Miller: "Photochemistry of formaldehydes: past and present," presented at the 12th Informal Conf. on Photochemistry, Gaithersburg, Maryland (June 28-July 1, 1976)

27. J. N. Pitts, Jr., P. J. Bekowies, A. M. Winer, J. M. McAfee, G. J. Doyle: To be published in Environ. Sci. Technol. (1976)

28. A. C. Lloyd, W. P. L. Carter, J. N. Pitts, Jr.: To be submitted for publication (1976)

29. H. E. O'Neal, C. Blumstein: Int. J. Chem. Kinet. $\underline{5}$, 397 (1973)

30. D. F. Miller, C. W. Spicer: J. Air Pollut. Contr. Assoc. $\underline{25}$, 940 (1975); C. W. Spicer: In Advances in Environmental Science and Technology, ed. by J. N. Pitts, Jr., R. L. Metcalf, Vol. 7 (Wiley-Interscience, New York, in press), Ch. 7

31. J. M. McAfee, A. M. Winer, J. N. Pitts, Jr.: "An infrared product analysis of ozone-olefin reactions by Fourier interferometry," presented at the CODATA Symp., Chemical Kinetics Data for the Lower and Upper Atmosphere, Warrenton, Virginia (September 16-18, 1974)

32. P. L. Hanst, W. E. Wilson, R. K. Patterson, B. W. Gay, Jr., L. W. Chaney, C. S. Burton: "A spectroscopic study of California smog," U. S. Environmental Protection Agency, Research Triangle Park, North Carolina (February 1975); P. L. Hanst: Opt. Quantum Electr. $\underline{8}$, 87 (1976)

33. D. H. Fine, D. D. Rovenbehler, N. M. Belcher, S. S. Epstein: "N-nitroso compounds in the environment," Proc. Int. Conf. on Environmental Sensing and Assessment, Las Vegas, Nevada (September 14-19, 1975); D. Shapley: Science $\underline{191}$, 268 (1976)

34. A. C. Lloyd, K. R. Darnall, A. M. Winer, J. N. Pitts, Jr.: J. Phys. Chem. 80, 789 (1976)

35. D. D. Davis, A. Moriarty, W. Heaps, R. Schiff, T. J. McGee: "In-situ measurements of the OH radical in the lower stratosphere," presented at the 1st Chemical Congress of the North American Continent, Mexico City (November 30-December 5, 1975); D. D. Davis, W. Heaps, T. McGee: "Direct measurements of natural tropospheric levels of OH via an aircraft-borne tunable dye laser," to be published; J. G. Anderson: Geophys. Res. Lett. 3, 165 (1976)

36. B. Weinstock, H. Niki, and co-workers: Personal communication (1976)

37. T. T. Paukert, H. S. Johnston: J. Chem. Phys. 56, 2824 (1972); C. J. Hochanadel, J. A. Ghormley, P. J. Ogren: J. Chem. Phys. 56, 4426 (1972)

38. H. E. Hunziker, H. R. Wendt: J. Chem. Phys. 60, 4622 (1974)

39. H. E. Radford, K. M. Evenson, C. J. Howard: J. Chem. Phys. 60, 3178 (1974); C. J. Howard, K. M. Evenson: J. Chem. Phys. 61, 1943 (1974)

40. K. T. Whitby, B. Cantrell: "Atmospheric aerosols--characteristics and measurements," Proc. Int. Conf. on Environmental Sensing and Assessment, Las Vegas, Nevada (September 14-19, 1975)

41. D. M. Grosjean, G. J. Doyle, T. M. Mischke, M. P. Poe, D. R. Fitz, J. P. Smith, J. N. Pitts, Jr.: "The concentration, size distribution, and modes of formation of particulate nitrate, sulfate and ammonium compounds in the eastern part of the Los Angeles Basin," presented at the Air Pollution Control Assoc. Meeting, Portland, Oregon (June 28-July 2, 1976)

42. R. A. Cox, A. E. J. Eggleton, R. G. Derwent, J. E. Lovelock, D. H. Pack: Nature 255, 118 (1975); R. A. Cox: Summaries Workshop: Photochemical Air Pollution, TNO, The Netherlands (May 18, 1976)

43. B. Bolin, R. J. Charlson: Ambio 5, 47 (1976)

44. R. J. Charlson, A. H. Vanderpol, D. S. Covert, A. P. Waggoner, N. C. Ahlquist: Atmos. Environ. 8, 1257 (1974)

45. A. M. Winer, J. W. Peters, J. P. Smith, J. N. Pitts, Jr.: Environ. Sci. Technol. 8, 1118 (1974); C. W. Spicer, D. F. Miller: J. Air Pollut. Contr. Assoc. 26, 45 (1976)

46. J. G. Calvert: Environ. Sci. Technol. 10, 257 (1976)

47. D. D. Davis. G. Klauber: Int. J. Chem. Kinet. Symp. 1, 543 (1975); J. G. Calvert, R. D. McQuigg: Int. J. Chem. Kinet. Symp. 1, 113 (1975)

48. W. P. L. Carter, K. R. Darnall, A. C. Lloyd, A. M. Winer, J. N. Pitts, Jr.: "Evidence for alkoxy radical isomerization in C_4-C_6 alkanes in NO_x-air systems," presented at the 12th Informal Conf. on Photochemistry, Gaithersburg, Maryland (June 28-July 1, 1976); W. P. L. Carter, K. R. Darnall, A. C. Lloyd, A. M. Winer, J. N. Pitts, Jr.: Chem. Phys. Lett. (1976), accepted for publication; K. R. Darnall, W. P. L. Carter, A. M. Winer, A. C. Lloyd, J. N. Pitts, Jr.: J. Phys. Chem. (1976), accepted for publication

49. H. Niki, P. Maker, C. Savage, L. Breitenbach: "IR Fourier transform spectroscopic studies of atmospheric reactions," presented at the 12th Informal Conf. on Photochemistry, Gaithersburg, Maryland (June 28-July 1, 1976)

50. J. G. Anderson: Geophys. Res. Lett. $\underline{2}$, 231 (1975)

51. H. E. Hunziker: "Near IR detection of peroxyl radicals in mercury-photo-sensitized reactions," presented at the 12th Informal Conf. on Photochemistry, Gaithersburg, Maryland (June 28-July 1, 1976); M. R. Whitbeck, J. W. Bottenheim, S. Z. Levine, J. G. Calvert: "A kinetic study of CH_3O_2 and $(CH_3)_3CO_2$ radical reactions by kinetic flash spectroscopy," presented at the 12th Informal Conf. on Photochemistry, Gaithersburg, Maryland (June 28-July 1, 1976)

PHOTOCHEMISTRY IN THE STRATOSPHERE

H.S. Johnston
Department of Chemistry, University of California
Berkeley, California 94720, USA

ABSTRACT

The current understanding of stratospheric photochemistry and dynamics is based partly on direct observations and partly on model calculations. At present it is desirable to increase the ratio of observations to calculations in this subject, although some components of the problem are now well established.

Ozone is formed from far ultraviolet solar radiation and oxygen in the middle and upper stratosphere. The local and the global rates of ozone formation can be reliably evaluated. Of the ozone formed below 45 km, 20 to 30 percent is destroyed by ozone itself, 10 to 20 percent is destroyed by free radicals derived from water, and 50 to 70 percent is catalytically destroyed by natural stratospheric oxides of nitrogen. The natural rate of production of stratospheric nitrogen oxides is known, and this source strength provides a simple, general measure of how large man-made pollution of the stratosphere would have to be in order to cause a large reduction of ozone. Current civilian aircraft, current military aircraft, and the proposed 50 space shuttles per year add ozone-destroying catalysts to the stratosphere at a rate much less than their natural formation. A fleet of supersonic, passenger aircraft that would be large enough to pay for its development costs or continuous usage of chlorofluorocarbons at the current rate would each add ozone-destroying catalysts to the stratosphere at about the same rate that nature now produces stratospheric nitrogen oxides. To estimate the magnitude of ozone reduction caused by this approximately doubling of ozone-destroying catalysts, one must turn to the detailed model calculations involving two or three dimensions of atmospheric motions; such computations are quoted here.

Introduction

This article does not attempt to give a complete account of the photochemistry in the stratosphere, both because such a complete account would require much more space than allotted here and there have been recent reviews of the subject [1-8]. The full problem is indeed complicated, involving atmospheric motions in three dimensions, solar radiation with diurnal and seasonal variations, 20 to 30 chemical species, and 60 to 100 chemical and photochemical reactions. However, if one is willing to sacrifice some technical completeness for general comprehension, the main subject can be presented in a simple form, which is the goal of this article.

Much of the complexity of stratospheric photochemistry arises from the need to calculate the concentrations of various trace species, whereas measurements of these species in the stratosphere would greatly simplify the problem. Of 27 trace species discussed in this article, 15 have been observed in the stratosphere, and the average, observed concentrations to one significant figure are entered in Table 1. Twelve species have not been observed (in journals available at the time of writing this article) and are listed at the bottom of the figure. On the basis of atmospheric observations, laboratory data on the spectroscopic and chemical properties, field observations of solar intensities, and a theory of atmospheric motions, various modelers have calcu- lated the concentrations of trace species in the stratosphere. Table 2 gives typical values for 19 molecules.

Table 1

Average concentrations of trace species observed in the sunlit stratosphere

km	15	20	25	30	35	40	45	ref
T°K	217	217	222	227	237	250	264	[5]
M	4(18)	2(18)	8(17)	4(17)	2(17)	8(16)	4(16)	[5]
O_3	3(12)	5(12)	4(12)	3(12)	2(12)	7(11)	2(11)	[5]
O			5(7)	2(8)	4(8)	2(9)		[10]
H_2O	1(13)	3(12)	2(12)	1(12)	8(11)	5(11)	2(11)	[8]
CH_4	6(12)	2(12)	1(12)	3(11)	1(11)	3(10)	1(10)	[8]
H_2	3(12)	1(12)	6(11)	3(11)	1(11)	3(10)	1(10)	[8]
CO	2(11)	1(11)	1(11)	5(10)				[8]
HO							8(6)	[8]
N_2O	3(11)	1(11)	3(10)	6(9)			< (8)	[8]
NO, max		1(9)	1(9)	2(9)	2(9)			[11]
min		1(8)	1(8)	1(8)	3(8)			[11]
NO_2, max	6(9)	4(9)	3(9)	2(9)	8(8)			[11]
min	1(9)	1(9)	1(9)	1(9)	5(8)			[11]
HNO_3, max	1(10)	2(10)	8(9)	1(9)	3(8)			[11]
min	1(9)	2(9)	1(9)	5(8)	1(8)			[11]
CF_2Cl_2	7(8)	3(8)	7(7)	2(7)	6(6)			[12,13]
$CFCl_3$	5(8)	1(8)	2(7)	4(6)	5(5)			[12,13]
CCl_4	4(8)	2(8)						[14]
HCl	1(9)	6(8)						[15]

Not measured: $O(^1D)$; H_2O_2, H_2CO, HOO, H; N_2O_5, NO_3, N; $ClNO_3$; ClO, Cl

Concentrations: molecules cm^{-3}, $3(12) = 3 \times 10^{12}$

In some cases, laboratory data for making model calculations of strato- spheric photochemistry are incomplete or have never been obtained. Tunable lasers promise to solve some of these problems. There has been some use of lasers, for example [9], to measure key components in the stratosphere, and it is hoped that more stratospheric measurements will be made. The goal of stratospheric measurements is not to eliminate all model calculations, but rather it is to provide a large number of calibration points for the theories and thus greatly to reduce their uncertainties.

Table 2

Average concentrations of trace species in the sunlit stratosphere as calculated by modelers

A. Cases reviewed by [8]

km	15	20	25	30	35	40	45
O(^{3}P)	1(5)	1(6)	1(7)	4(7)	3(8)	1(9)	5(9)
O(^{1}D)	0.1	0.5	2 -	10	60	100	200
N$_2$O	8(11)	3(11)	1(11)	3(10)	6(9)	1(9)	1(8)
NO$_2$	2(9)	2(9)	3(9)	2(9)	1(9)	3(8)	8(7)
NO	2(9)	1(9)	1(9)	1(9)	1(9)	6(8)	3(8)
HNO$_3$	7(9)	7(9)	2(9)	1(9)	1(8)	1(7)	
H$_2$O	2(13)	8(12)	3(12)	2(12)	7(11)	3(11)	2(11)
CH$_4$	4(12)	2(12)	1(12)	3(11)	1(11)	3(10)	1(10)
H$_2$	2(12)	1(12)	6(11)	2(11)	8(10)	2(10)	1(10)
CO	2(11)	6(10)	1(10)	8(9)	4(9)	3(9)	2(9)
H$_2$O$_2$	(6 to 9)	(7 to 9)		(7 to 9)			(6 to 7)
H$_2$CO	6(6)	5(6)	6(6)	7(6)	6(6)	4(6)	1(6)
H		(-1)		(0)	(2)	(4)	(5)
HO	3(5)	4(5)	7(5)	2(6)	3(6)	6(6)	1(7)
HOO	1(7)	1(7)	2(7)	2(7)	2(7)	2(7)	1(7)

B. Cases reviewed by [16]

HCl	2(8)	8(8)	8(8)	4(8)	2(8)	1(8)	5(7)
ClO		1(7)	4(7)	5(7)	3(7)	9(6)	1(6)
Cl						8(4)	8(4)

C. Cases in B recalculated to include ClNO$_3$

HCl		5(8)	3(8)	2(8)	2(8)	1(8)	5(7)
ClO		7(6)	2(7)	2(7)	2(7)	9(6)	1(6)
ClNO$_3$		2(8)	5(8)	3(8)	4(7)	8(5)	3(3)

The Temperature Structure of the Troposphere and Stratosphere [1]

The surface of the earth is heated by solar radiation, and the warmed air rises, expands, and thereby cools. From simple thermodynamics it can be shown that dry air cools 10°K per kilometer as it is adiabatically lifted in the atmosphere. The decrease of temperature with increasing elevation is a familiar phenomenon, but in the troposphere the rate of cooling is substantially less than 10°K per kilometer. The substance <u>water</u> is responsible for this lesser rate of cooling with elevation. As water vapor condenses to cloud droplets in the atmosphere, the latent heat of vaporization is injected into the air, partially off-setting the "dry adiabatic lapse rate."

At an elevation of about 10 to 15 kilometers the air has cooled so much from adiabatic expension that there is very little water vapor left, only a few parts per million, which is not enough for significant thermal effects from further condensation. One might expect air above 15 kilometers to cool with the dry adiabatic lapse rate of 10°K per kilometer, and at an elevation of about 30 kilometers the sky should rain liquid air. Of course, this sort of rain does not occur. Instead the air temperature ceases to decrease with increasing elevation at about 10 to 15 kilometers, and the temperature increases with elevation up to about 50 kilometers, which is defined as the top of the stratosphere. Such an increase of temperature with elevation is called a "temperature inversion," and it confers great stability to air against vertical mixing. An inert gaseous tracer placed at about 20 kilometers in the stratosphere has a residence time (1/e decay time) of about three years [17], but water-soluble tracers are removed from the turbulent, rainy troposphere within several days or a few weeks.

The souce of heat in the stratosphere is primarily the photochemistry of <u>ozone</u>, which is described in general terms in the next two sections. The absorption spectrum of ozone between 150 to 725 nm is given by Fig. 1. Except for a narrow gap between 350 and 450 nm, ozone absorbs over this entire range. The absorption of visible radiation is weak, and ozone removes only 10^{-2} to 10^{-3} of the visible radiation between 450 and 750 nm. Even so, this absorption converts a significant amount of solar energy to heat, and it gives rise to photochemical reactions wherever there is ozone in the atmosphere. Below about 300 nm, ozone acts as a cut-off filter for solar radiation, and a substantial amount of heat is injected into the stratosphere by this band of sunlight.

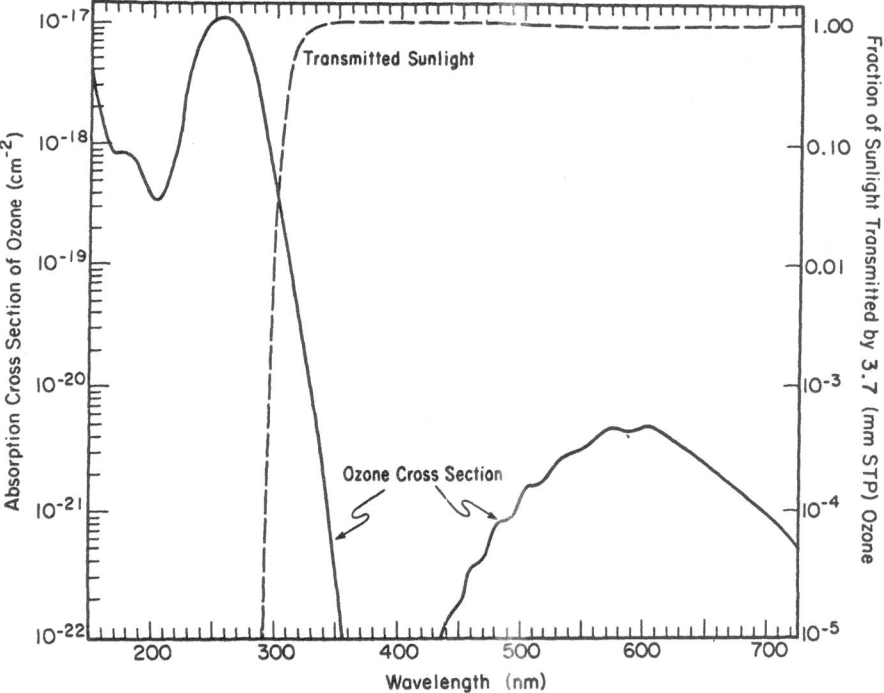

Figure 1. Absorption cross section for ozone, $I = I_0 \exp(-\sigma NL)$, where N is molecules cm^{-3}, L is path length cm, and σ is cm^2. The dashed line gives the fraction of solar radiation transmitted by 9.9×10^{18} molecules ozone cm^{-2}.

Screening of Solar Radiation

The radiation leaving the sun varies from X-rays of very short wave lengths, ultraviolet radiation with wave lengths up to 400 nm, visible radiation, and radiation of the infrared, microwave, and radiowave regions. The principal components of air, nitrogen and oxygen, strongly absorb the X-rays and very short wave length ultraviolet radiation to form ions and atoms well above the stratosphere. Ultraviolet radiation up to about 180 nm is absorbed above the stratosphere by the Schumann-Runge bands of molecular oxygen. Little radiation below 190 nm enters the stratosphere.

Between 190 and 242 nm, solar radiation is removed more or less equally by stratospheric ozone and by molecular oxygen through its weak, continuous, Herzberg, absorption continuum. By the time solar radiation has penetrated to 20 km, virtually all radiation in this spectral band has been absorbed by either ozone or oxygen. When oxygen absorbs this radiation, it is converted to ozone. When ozone absorbs this radiation, the radiation is converted to heat.

The minimum in the ozone absorption spectrum at 205 nm, Fig. 1, is quite important in stratospheric photochemistry. This minimum acts as a "window" allowing some energetic ultraviolet radiation to penetrate as low as 20 km. This radiation is absorbed by several relatively stable molecules that enter the stratosphere from the troposphere, in particular, nitrous oxide and chlorinated hydrocarbons.

Stratospheric ozone is the only effective absorber of solar radiation between 250 and 300 nm. Some biologists define the B-band of ultraviolet radiation as that abosrbed by DNA [4]. With this definition, ozone screens out about 99 percent of solar UV-B from the surface of the earth. A reduction of ozone by 10 percent would increase the incidence of UV-B at the surface by about 30 percent [4]. The fraction of sunlight transmitted by an overhead sun for a typical column of ozone is included in Fig. 1, and the sharp cut-off near 300 nm is noteworthy.

Near ultraviolet radiation (300-400 nm) and visible radiation (400-750 nm) are transmitted by the atmosphere to the surface of the earth. Although there are no strong molecular absorbers of this radiation, it is subject to Rayleigh scattering and to scattering by clouds and particulate matter.

By virtue of its role in determining the temperature structure of the stratosphere [1] and in screening the surface of the earth against biologically damaging radiation [4], ozone is the most important trace component of the stratosphere.

Ozone Formation and Destruction [1]

As oxygen absorbs solar ultraviolet radiation, ozone is formed

$$O_2 + h\nu \ (\lambda < 242 \ nm) \rightarrow O + O \tag{1}$$
$$\underline{O + O_2 + M \rightarrow O_3 + M, \ twice} \tag{2}$$
$$net: \ 3O_2 + h\nu \rightarrow 2O_3$$

It is a straightforward matter in three dimensional geometry to calculate the instantaneous, global rate of this reaction from the intensity of solar radiation above the atmosphere, the observed distribution of ozone and oxygen, and the absorption spectra of ozone and oxygen [7]. This rate is 5.0×10^{31} molecules sec^{-1} for the global stratosphere below 45 km. The total global inventory of ozone is about 4.4×10^{37} molecules. The total inventory divided by the instantaneous rate of formation gives the global "ozone replacement time," which is about 10 days. This simple calculation shows that ozone is in a dynamic steady state, being rapidly formed by solar radiation and equally rapidly being destroyed by some process or processes.

Below about 310 nm ozone strongly absorbs solar radiation to form singlet oxygen atoms

$$O_3 + h\nu \ (\lambda < 310 \text{ nm}) \rightarrow O_2 + O(^1D) \tag{3}$$

$$\downarrow M$$

$$O_3 + h\nu \ (\lambda > 310 \text{ nm}) \rightarrow O_2 + O(^3P) \tag{4}$$

and above 310 nm, including the broad weak absorption of visible light, ozone is photolyzed to form triplet oxygen atoms (4). Singlet oxygen atoms are rapidly quenched to ground state atoms by collision with any molecule M. However, this photolysis of ozone does not constitute destruction of ozone, because it is usually followed by re-formation (2)

$$
\begin{array}{ll}
O_3 + h\nu \rightarrow O_2 + O & (3,4) \\
\underline{O + O_2 + M \rightarrow O_3 + M} & (2) \\
\text{net: null reaction}
\end{array}
$$

This null reaction injects heat into the stratosphere, screens the surface of the earth against UV-B radiation, and in less than one second set ups a steady state concentration of oxygen atoms, Tables 1 and 2.

In the pure oxygen system (O_X stands for O and O_3), ozone is destroyed by the couple of reactions

$$
\begin{array}{ll}
O_3 + h\nu \rightarrow O_2 + O & (3,4) \\
\underline{O + O_3 \rightarrow O_2 + O_2} & (5) \\
\text{net: } 2O_3 + h\nu \ (\text{VIS, UV}) \rightarrow 3O_2
\end{array}
$$

Since (3,4) is usually followed by (2), (5) is the "rate determining step," and the rate of ozone destruction is twice the rate of the elementary reaction (5). Ozone is formed by the couple of reactions (1,2), and in this case (1) is the rate determining step. The rate of formation of ozone is twice the rate of photolysis of oxygen, as an excellent approximation. Of course, it is not necessary to write these reactions as sets (1,2), (3-4,2), and (3-4,5) with "rate determining steps." However, it is very instructive to do so, and with this approach one can understand more about stratospheric ozone than if one simply writes out the long list of photochemical reactions that occur. This approach is followed throughout this article.

The rate of (5) integrated over the global, three-dimensional stratosphere below 45 km is equal to only about 20 percent of the rate of (1) similarly integrated over the globe [7]. Something else is very important in the global ozone balance.

The free radicals derived from water, H; HO, HOO, destroy ozone [18,1]. In the lower stratosphere there is a homogeneous catalytic cycle that destroys ozone but not HO_X

$$HO + O_3 \rightarrow HOO + O_2 \tag{6}$$
$$HOO + O_3 \rightarrow HO + O_2 + O_2 \tag{7}$$

net: $2O_3 \rightarrow 3O_2$

Step (6) is not always followed by (7); there may be a cycle neutral with respect to ozone formation or destruction

$$HO + O_3 \rightarrow HOO + O_2 \tag{6}$$
$$HOO + NO \rightarrow HO + NO_2 \tag{8}$$
$$NO_2 + h\nu \ (300\text{-}400 \ nm) \rightarrow NO + O \tag{9}$$
$$O + O_2 + M \rightarrow O_3 + M \tag{2}$$

net: null reaction

Thus (7) is the rate determining step in this catalytic cycle. Neutral cycles such as this are important in determining the ratio of various free-radical species even though they do not form or destroy ozone. In the upper stratosphere, another catalytic cycle destroys ozone

$$HO + O_3 \rightarrow HOO + O_2 \tag{6}$$
$$HOO + O \rightarrow HO + O_2 \tag{10}$$

net: $O_3 + O \rightarrow O_2 + O_2$

Here (10), again involving HOO, is the rate determining step. Mostly above the stratosphere, a third catalytic cycle occurs

$$H + O_3 \rightarrow HO + O_2 \tag{11}$$
$$HO + O \rightarrow H + O_2 \tag{12}$$

net: $O_3 + O \rightarrow O_2 + O_2$

In the natural stratosphere, the active oxides of nitrogen (NO_x = NO and NO_2) destroy ozone [19,1]. In the middle stratosphere, which is the primary ozone formation region, the oxides of nitrogen destroy ozone by a homogeneous catalytic cycle

$$NO + O_3 \rightarrow NO_2 + O_2 \tag{13}$$
$$NO_2 + O \rightarrow NO + O_2 \tag{14}$$

net: $O_3 + O \rightarrow O_2 + O_2$

Step (13) is not always followed by (14); there is a competing neutral cycle

$$NO + O_3 \rightarrow NO_2 + O_2 \tag{13}$$
$$NO_2 + h\nu \ (300\text{-}400) \rightarrow NO + O \tag{9}$$
$$O + O_2 + M \rightarrow O_3 + M \tag{2}$$

net: null reaction

Thus (14) is the rate-determining step in the NO_x catalytic cycle. The rate constant for (14) has been repeatedly observed in the laboratory, at temperatures found in the stratosphere [1]. The reactants in the rate-determining step (14), namely O and NO_2, have been directly observed in the stratosphere, Table 1. Thus, reaction (14) has been demonstrated to occur in the stratosphere. There is another relatively unimportant ozone-destroying cycle in the lower stratosphere [20]

$$NO_2 + O_3 \rightarrow NO_3 + O_2 \tag{15}$$
$$NO_3 + h\nu \text{ (VIS)} \rightarrow NO + O_2 \tag{16}$$
$$\underline{NO + O_3 \rightarrow NO_2 + O_2} \tag{13}$$
$$\text{net: } 2O_3 + h\nu \text{ (VIS)} \rightarrow 3O_2$$

Photolysis of NO_3 to give $NO_2 + O$ is neutral with respect to ozone formation and destruction, and thus (16) is the rate-determining step in this catalytic cycle.

Free radicals based on chlorine, Cl and ClO, are capable of destroying ozone in a catalytic cycle that is parallel to the NO_x cycle [21-26,6]

$$Cl + O_3 \rightarrow ClO + O_2 \tag{17}$$
$$\underline{ClO + O \rightarrow Cl + O_2} \tag{18}$$
$$\text{net: } O_3 + O \rightarrow O_2 + O_2$$

Reaction (17) is not always followed by (18); there is a parallel neutral cycle

$$Cl + O_3 \rightarrow ClO + O_2 \tag{17}$$
$$ClO + NO \rightarrow Cl + NO_2 \tag{19}$$
$$NO_2 + h\nu \rightarrow NO + O \tag{9}$$
$$\underline{O + O_2 + M \rightarrow O_3 + M} \tag{2}$$
$$\text{net: null reaction}$$

Thus (18) is the rate-determining step in this catalytic cycle. The rate constant for (18) has been observed in the laboratory, oxygen atoms have been observed in the stratosphere, but observations of ClO in the stratosphere have not been reported (this is expected to change at any time).

Air motions play a vital role in determining the distribution of stratospheric ozone. The relative role of photochemistry and air motions in shaping the distribution of ozone can be inferred by comparing the photochemical replacement times of ozone (local concentration divided by local gross formation rate) with transport times. The time for stratospheric east-west winds to move one-quarter way around the globe is typically a few days. The time for north-south eddy diffusion and net winds to transport air one-half the way between equator and pole is typically a few months. The time for air motions to mix air over a vertical distance of about 10 kilometers in the lower and middle stratosphere is a few years. (The loose term "few" is used here to communicate the general idea without getting involved in a detailed quantitative discussion, for which see [1].) At 45° latitude in the summer, the average photochemical ozone replacement times are about one day at 40 km, about one month at 27 kilometers, and about two years at 20 km. Thus the seasonal north-south air motions are faster than local ozone photochemistry below the 25 to 30 km vertical band, and the slow vertical mixing of air is faster than local ozone photochemistry below the 15 to 20 km vertical band (the range is function of latitude and season).

The amount of ozone transported into the troposphere and destroyed at the ground has been evaluated [27], and this rate of ozone destruction is less than one percent of the global rate of ozone production below 45 km.

For an average, observed, three-dimensional distribution of ozone, the global rate of ozone formation has been found [7], and for elevations below 45 km this can be stated as ± 100 percent. The global rate of ozone destruction by ozone (5) is 20 to 30 percent, depending on whether one does not consider or does consider Rayleigh scattering and albedo effects in the calculation. The rate of ozone destruction by the HO_x family of reactions (7,10, 12) was found to be between 10 and 20 percent, depending on the values taken for uncertain rate constants. The rate of destruction of ozone by the chlorine now in the stratosphere is less than one percent [16]. Thus O_x, Cl_x, and HO_x reactions can account for 30 to 45 percent of the ozone produced globally. The rate of ozone destruction derived from the average value of observed NO_2 and O, Table 1, is more than enough to account for 70 percent of the ozone produced over the same range of elevation. If the so-far observed NO_2 [11] is typical of the global average, it appears that natural NO_x destroys 55 to 70 percent of the ozone formed below 45 km. The NO_x family of reactions is the most important mechanism in the natural destruction of stratospheric ozone.

Sources, Sinks, and Reservoirs of Ozone-Destroying Catalysts

Each coupled reaction that catalytically destroy ozone involves a pair of free radicals, that is, molecules with an odd number of electrons: HO, HOO; NO, NO_2; Cl, ClO. Most molecules have an even number of electrons, and from the simplest properties of numbers one sees that free radicals must be created and destroyed in pairs. It is instructive to classify stratospheric chemical reactions in terms of "free radical arithmatic." In terms of molecules with an even number of electrons, 0, and an odd number, 1, the sources and sinks of ozone-destroying catalysts may be represented as:

$$\text{Source:} \quad 0 \rightarrow 1 + 1 \tag{20}$$
$$\text{Sink:} \quad 1 + 1 \rightarrow 0 \tag{21}$$

A large number of stratospheric reactions act to convert one free radical into another

$$0 + 1 \rightarrow 1 + 0 \tag{22}$$

Some compounds formed by combination of two radicals (21) are unstable in the stratospheric environment and after a time short compared to stratospheric residence times they are again broken down to free radicals (20). These substances are called "reservoirs" of ozone-destroying catalysts.

HO_x family of reactions. Singlet oxygen atoms from (3) react with H_2O, CH_4, and H_2 to form HO_x free radicals, for example

$$O(^1D) + H_2O \rightarrow 2HO \tag{23}$$

The primary destruction of HO_x radicals is

$$HO + HOO \rightarrow H_2O + O_2 \tag{24}$$

The reaction of two hydroperoxyl radicals

$$HOO + HOO \rightarrow H_2O_2 + O_2 \tag{25}$$

does not really destroy HO_x because the hydrogen peroxide so formed is readily photolyzed by solar radiation to produce two hydroxyl radicals. Thus hydrogen peroxide is a stratospheric reservoir of HO_x radicals. In the photochemical oxidation of methane, formaldehyde H_2CO is an intermediate product; it too is a reservoir of HO_x radicals in that it is, in part, photolyzed to give H and HCO radicals, both of which react with oxygen to form HOO. When hydroxyl radicals react with nitric acid an HO_x radical is destroyed and an NO_x radical is created from the nitric acid "reservoir." In this case the HO_x sink is

$$\begin{array}{ll} & \quad\quad M \\ & HO + NO_2 \rightarrow HNO_3 & (26) \\ & \underline{HO + HNO_3 \rightarrow H_2O + NO_3} & (27) \\ \text{net:} & 2HO + NO_2 \rightarrow H_2O + NO_3 & \end{array}$$

NO_x family of reactions. Nitrous oxide N_2O is an inert compound. It is formed by bacteria in the soil and in fresh and ocean waters. These bacteria reduce nitrates to N_2 and N_2O. The troposphere contains about 0.3 parts per million by volume (ppm) of nitrous oxide. The tropospheric residence time of nitrous oxide is a matter of decades. The air motions that mix stratospheric and tropospheric air, on the average, move a material in the direction of its decreasing mole fraction or mixing ratio by volume. Nitrous oxide is slowly mixed into the stratosphere where most of it is photolyzed to form molecular nitrogen and oxygen and some of it reacts with singlet oxygen atoms to form nitric oxide [28-30]

$$N_2O + O(^1D) \rightarrow 2NO \tag{28}$$

The maximum observed mixing ratio of NO_y ($NO + NO_2 + HNO_3$) in the stratosphere is about 10 parts per billion (ppb), far higher than observed in the unpolluted troposphere [31]. Thus the same air motions that mix N_2O upward into the stratosphere mix NO_y downward into the troposphere where NO_y is rapidly removed by rain. The tropospheric rain-out of NO_y is the principal sink of stratospheric NO_x. Another, minor sink occurs in the top-most stratosphere. Nitric oxide is photolyzed by narrow bands of solar radiation that slip through the fine-structure of the Schumann-Runge oxygen bands.

$$\begin{array}{ll} & NO + h\nu \rightarrow N + O & (29) \\ & \underline{N + NO \rightarrow N_2 + O} & (30) \\ \text{net:} & 2NO + h\nu \rightarrow N_2 + 2O & \end{array}$$

Nitric oxide is regenerated if the nitrogen atom reacts with O_2 or O_3 but destroyed if it reacts with NO (30).

A reservoir of NO_x free radicals is nitrogen pentoxide, which is produced at night from nitrogen dioxide and ozone

$$\begin{array}{ll} & NO_2 + O_3 \rightarrow NO_3 + O_2 & (15) \\ & \quad\quad M \\ & \underline{NO_2 + NO_3 \rightarrow N_2O_5} & (31) \\ \text{net:} & 2NO_2 + O_3 \rightarrow N_2O_5 + O_2 & \end{array}$$

During the day N_2O_5 is largely photolyzed back to nitrogen dioxide. Table 3 gives self-consistent calculated values of stratospheric species with emphasis on the diurnal variation of N_2O_5.

The global rate of formation of nitric oxide from nitrous oxide by (28) was calculated by ISAKSEN [32] and found to be (4 to 5) x 10^{26} molecules sec^{-1}. From the measured distribution of stratospheric nitric acid between northern Alaska and southern Argentina, LAZRUS and GANDRUD [33] deduced the global loss of stratospheric nitric acid to be (4 ± 2) x 10^{26} molecules sec^{-1}. Thus the approximate magnitude of the global NO_x source and sink is known

$$\text{Source} \approx 5 \times 10^{26} \text{ molecules sec}^{-1} \approx \text{Sink} \tag{32}$$

This quantity is used in a subsequent section to compare various possible man-made sources of stratospheric NO_x with the natural source.

Table 3

Examples of trace species calculated with diurnally varying solar radiation

A. Self-consistent model with diurnal variation, spring equinox, noon, 60°N

km	18	23	28	33	38	43	48
M	2.5(18)	1.1(18)	5.2(17)	2.4(17)	1.1(17)	5.4(16)	2.7(16)
T°K	217	220	225	231	245	259	271
O_3	4.7(12)	4.7(12)	3.1(12)	1.9(12)	9.9(11)	3.9(11)	1.3(11)
$O(^3P)$	1.4(6)	7.8(6)	3.2(7)	1.3(8)	6.2(8)	2.5(9)	6.8(9)
$O(^1D)$	1	4.6	16	60	142	316	435
NO_2	2.3(9)	1.7(9)	3.6(9)	1.8(9)	7.3(8)	1.6(8)	7.1(6)
NO	1.2(9)	8.5(8)	2.6(9)	2.1(9)	1.6(9)	1.5(9)	3.6(8)
N_2O_5	6.6(8)	4.2(8)	6.7(8)	1.2(8)	9.6(6)	6.4(3)	6 (-1)
NO_3	4.0(5)	4.0(5)	5.0(5)	2.3(5)	9.9(4)	1.5(4)	4.5(2)
HNO_3	2.1(9)	2.9(9)	1.4(9)	1.4(8)	1.9(7)	1.9(6)	3.2(4)
H_2O_2	1.1(8)	1.1(8)	1.1(8)	9.2(7)	4.6(7)	2.0(7)	9.6(6)
H	(-4)	(-2)	(-1)	17	6.3(2)	1.5(4)	1.5(5)
HO	7.1(5)	1.1(6)	1.8(6)	2.8(6)	4.9(6)	7.0(6)	6.6(6)
HOO	2.6(7)	3.8(7)	4.0(7)	3.7(7)	2.8(7)	2.2(7)	1.6(7)

B. One minute before sunrise and one hour before sunset

N_2O_5 6 AM	1.3(9)	9.6(8)	1.9(9)	2.5(9)	4.2(8)	1.5(7)	1.4(6)
N_2O_5 5 PM	4.4(8)	2.7(8)	3.7(8)	4.3(7)	1.2(6)	2.0(3)	1 (0)

The ClX family of reactions. Methyl chloride CH_3Cl is produced naturally in the ocean and released into the atmosphere. Most of it is destroyed by hydroxyl radicals in the troposphere to produce water-soluble chlorine compounds that are rapidly removed by rain. Some methyl chloride survives to be transported by air motions into the stratosphere. Ultraviolet radiation in the "window" near 205 nm, Fig. 1, ejects the chlorine atom from methyl chloride. This photolysis is a general feature for organic chlorine-containing molecules, including man-made molecules such as CF_2Cl_2, $CFCl_3$, and CCl_4. The HCl observed in the stratosphere, Table 1, largely originated from photolysis of CH_3Cl and CF_2Cl_2 [16].

Chlorine atoms react with many hydrogen-containing species to produce hydrogen chloride, for example

$$Cl + CH_4 \rightarrow HCl + CH_3 \tag{33}$$

Hydrogen chloride is an inert reservoir of stratospheric chlorine; it is a reservoir in that it is reconverted to active ozone-destroying catalysts by hydroxyl radicals

$$HO + HCl \rightarrow H_2O + Cl \tag{34}$$

In the lower stratosphere HCl is the dominant ClX species, and when it is mixed down into the troposphere it is rapidly rained out. This mechanism of physical removal of HCl from the stratosphere and rain-out in the troposphere is the only known sink for stratospheric chlorine.

Other reservoirs. An interaction between the HO_x and NO_x systems of re-actions produces a major stratospheric reservoir for each, nitric acid

$$HO + NO_2 \xrightarrow{M} HNO_3 \tag{26}$$
$$HNO_3 + h\nu \ (\lambda < 330 \ nm) \rightarrow HO + NO_2 \tag{35}$$

The distribution of nitric acid and of the NO_x radicals is indicated by the observed data in Table 1 and by the model calculations in Tables 2 and 3.

An interaction between the ClX and NO_x systems of reactions produces a potentially important stratospheric reservoir, chlorine nitrate [34]

$$ClO + NO_2 \xrightarrow{M} ClNO_3 \tag{36}$$
$$ClNO_3 + h\nu \ (\lambda < 350 \ nm) \rightarrow ClO + NO_2 \tag{37}$$

Chlorine nitrate is an especially interesting compound in that it simultaneous ties up the key substance in the rate-determining step for destruction of ozone in both the ClX and NO_x systems, (18) and (14).

The relationships between the ozone-destroying catalysts, ClO and NO_2, and the stratospheric reservoirs HCl, $ClNO_3$, N_2O_5, and HNO_3 are given in Fig. 2. The first-order rate constant for HCl destruction is k_{34} [HO], compare (34); and the time for (34) to destroy the fraction $1/e$ of HCl is $1/k_{34}$ [HO]. At 30 km elevation with a 60° zenith solar angle, this time is 240 hours. The reservoir compounds, $ClNO_3$, N_2O_5, and HNO_3, are primarily destroyed by photolysis. The first-order photolysis constants are conventionally abbreviated as j. The lifetime (1/e) of each species with respect to photolysis is $1/j$. At 30 km elevation and with a 60° zenith solar angle, the lifetimes of $ClNO_3$, N_2O_5, and HNO_3 are respectively 5, 5, and 69 hours. These times are entered on Fig. 2. The photochemical relaxation time for NO_2, NO interconversion is about two minutes, and for the ClO, Cl interconversion it is about 0.5 minute. Some stratospheric, photochemical relaxation times are short compared to the 24 hour diurnal cycle and some are long compared to 24 hours.

This wide spread in photochemical relaxation times introduces an important complication. Most model calculations of stratospheric photochemistry uses a constant, average solar intensity. Should one use a 24 hour average or a 12 hour average during the day and zero at night? If a process of interest varies linearly with solar intensity, it does not matter which average one uses. If a process is non-linear, it does matter, and the relation between photochemical relaxation time and the length of a day is the key to which average one should use. If the photochemical relaxation time is short compared to the length of a day (for example, ClO, NO_2, N_2O_5, $ClNO_3$ in Fig. 2), one should use full solar intensity half the time. If the photochemical relaxation time is long compared to day time (for example, HCl and HNO_3 in Fig. 2), one should use the 24 hour average solar intensity. Since some species have short relaxation times and others long relaxation times, it is clear that there is no single average of solar intensity that is suitable for all species. A realistic treatment, then, requires explicit consideration of the diurnal cycle.

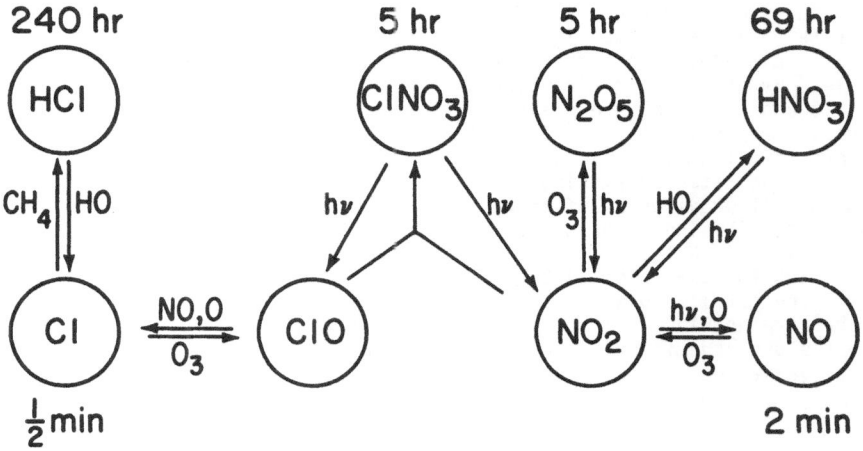

Figure 2. Relationships and photochemical relaxation times for ozone destroying catalysts, (Cl, ClO) and (NO, NO₂), and for stratospheric reservoirs of
these catalysts

Diurnal Variations

Ozone is photochemically formed by day and photochemically destroyed by day.
In the stratosphere very little ozone is formed or destroyed at night, and
thus ozone shows little diurnal variation. The photochemical lifetime of
nitric acid at 30 km elevation and with a 60° zenith solar angle is 69 hours,
Fig. 2; and the corresponding chemical relaxation time for HCl is 240 hours.
These two reservoirs of ozone-destroying catalysts would show only minor diurnal variation.

In a few minutes after ultraviolet sunset, virtually all nitric oxide in
the stratosphere is converted to nitrogen dioxide by (13), and a few minutes
after ultraviolet sunrise a steady-state balance between NO and NO₂ is attained through (9) and (13). The response of NO and NO₂ to sunlight is fast,
and these species are in phase with the diurnal solar cycle.

Between sunset and sunrise there is a slow build-up of nitrogen pentoxide
from (15) and (31). Between sunrise and sunset, the nitrogen pentoxide is
almost completely photolyzed. Nitrogen pentoxide is very nearly 90° out of
phase with sunlight. There is no average intensity of sunlight that is suitable for this problem. It is necessary to consider the full diurnal variation
of solar intensity, and the results of one such study is given in Table 3.
Table 3 A. gives the concentration of various species at noon, spring equinox,
60°N latitude. Table 3 B. gives N₂O₅ just before sunrise and an hour before
sunset. As N₂O₅ is photolyzed during the day, both NO₂ and NO slowly build
up, compare [9,35].

After sunset, virtually all ClO will be picked up by NO_2 to form chlorine nitrate (34). After sunrise, the chlorine nitrate will be dissociated by solar radiation. The time constant is about five hours at 30 km and 60° solar angle. By noon a large fraction of the chlorine nitrate would be dissociated to ClO and NO_2. The steady-state fraction of all ClX that is $ClNO_3$ is inversely proportional to the square of solar intensity. Chlorine nitrate is interesting from two points of view: (a) It ties up the key species (ClO and NO_2) in the rate-determining step of ozone for both the ClX and NO_x system, and thus it provides a buffering action against an increase in stratospheric chlorine. (b) The chlorine nitrate derives most of its chlorine from the slow reservoir HCl and much of its nitrogen from the slow reservoir HNO_3, and at a high sun the molecule breaks down to release both ClO and NO_2. At such times the total amount of ClO and NO_2 is greater than it would have been if there were no $ClNO_3$. There is no average solar intensity suitable for treating the photochemistry of chlorine nitrate; it must be followed over the diurnal variation of solar intensity.

Jobs That Tunable Lasers May Be Able To Do

Singlet oxygen atoms are the source of HO_x free radicals (23) and of nitric oxide (28) in the natural stratosphere. The critical energy for formation of $O(^1D)$ from ozone occurs at a wave length of 310 nm (3). However, the quantum yield for production of $O(^1D)$ is not a step function at 310 nm, but it is somewhat rounded off below and above that wave length. The shape of this quantum-yield curve depends on temperature. So far, only low-resolution measurements have been made of this action spectrum [36]. The rate of production of hydroxyl radicals in the lower stratosphere and in the troposphere is strongly dependent on the exact shape of this quantum-yield curve. A well-designed experiment with a laser tunable over the region 300 to 325 nm should be able to solve this important stratospheric problem.

The photolysis of NO_3 involves a similar threshold problem. Only below 580 nm is there sufficient energy to produce NO_2 + O from the photolysis of NO_3. It is possible to produce NO + O_2 at any wave length where NO_3 absorbs light. There appears to be a competition between fluorescence and photolysis. The quantum yields for NO and for NO_2 as a function of wave length are needed over the entire visible spectrum of NO_3 [37].

The photolysis of NO_2 (9) has a threshold at 400 nm, with fluorescence of NO_2 primarily occurring above 400 nm. Although the quantum yield of this photolysis has been studied [38], a high resolution study as a function of temperature and wave length is still needed.

Eqs. (5,7,10,12,14,16,18) give the rate-determining steps in the catalytic cycles that destroy ozone in the O_x, HO_x, NO_x, and ClX systems. The species involved are O_3, O, HO, HOO, ClO, NO_2, and NO_3. Of these, O_3, O, HO, and NO_2 have been observed in the stratosphere, and HOO, ClO, and NO_3 have not been observed. Methods are needed to observe HOO and ClO, and additional observations are needed for HO and O.

Observations or additional observations are needed for the reservoir species H_2O_2, H_2CO, N_2O_5, and $ClNO_3$.

To confirm, modify, or disprove components of the theory of stratospheric ozone, one needs to make simultaneous measurements of several sets of compounds. The following sets of stratospheric species should be observed

HOO, HO, O_3, H_2O
NO_2, NO, O_3
ClO, $ClNO_3$, NO_2, HCl
O, O_3
N_2O_5, NO_2, O_3
HNO_3, O_3, H_2O, NO_2

Good ideas and new methods to make these measurements are needed.

Pollution of the Stratosphere

Several products of human activity have been asserted to reduce stratospheric ozone [39,20,21-26,6,2-4]. One way to test these assertions is to carry out model calculations, preferably including at least two dimensions of atmospheric motions and at least 50 chemical reactions. Such calculations can be and have been carried out [40-42, 2-4], and they have been checked and verified by other experts in this activity. These model calculations bridge so many scientific disciplines and are so complicated that non-experts, even scientists in closely related fields, find it difficult to appreciate the strengths and the weaknesses of these calculations. There is a continuing need for simpler, more generally comprehensible treatments to demonstrate which threats to stratospheric ozone are plausible and which are implausible. A simple "meter stick" to assess the plausibility of proposed threats to stratospheric ozone is given here.

It was shown (32) that the global natural source and sink of stratospheric NO_x is about 5×10^{26} molecules sec^{-1} and that this quantity of nitrogen oxides destroys more than half of the ozone formed from sunlight below 45 km. Global geophysical quantities of this sort are not sharp numbers, and factor of two uncertainties are not unexpected. Even so, one may conclude that a steady input of man-made NO_x of this magnitude would constitute a large source of stratospheric pollution. Such an input of artificial NO_x would have to be taken seriously. It would require some special mechanism to be operative if such an input did not seriously perturb stratospheric ozone. On the other hand, if an artificial source of ozone-destroying catalysts was very small compared to the natural source of NO_x, it would require the operation of some very special mechanism for such an input to have a serious affect.

Supersonic transports are said to reduce stratospheric ozone through nitrogen oxides in the exhaust gases [20]. In considering the plausibility of this proposal, one should consider the number of supersonic transports that would pay off the development costs, not just the first few to fly. Environmental impact calculations should be made on the number of aircraft proposed for economic justification. In 1971 the U.S. National Planning Association [3] set a target figure of 800 Boeing SST; the development costs were expected to be recovered by 400; and the second 400 would return a 100 percent profit on the development costs. Even so, environmental impact calculations have usually been made on the basis of 500 SSTs. With the properties they were expected to have in 1971, 500 Boeing SSTs would constitute a global input of NO_x in the stratosphere of 7×10^{26} molecules sec^{-1} [20,4]. With current properties, 500 Concorde SSTs would constitute a global input of about 2×10^{26} molecules sec^{-1} [4]. These inputs of NO_x into the stratosphere are comparable to the natural source of NO_x, (32). Even on the basis of this simple general argument, they must be regarded as probable threats to stratospheric ozone.

The Climatic Impact Assessment Program [1-3] devoted three years and \$22 million to detailed study of this problem. It was judged that if stratospheric aviation should be a success, the amount of fuel burned in the stratosphere would be at least 10^{11} kg yr^{-1}. If future SST engines have the same NO_x emission characteristics as present SST engines, the amount of NO_x emitted per year would be 7.5 x 10^{26} molecules sec^{-1}. It turns out that this NO_x estimate for a future SST fleet that would pay off its development costs is very nearly the same as that expected in 1971 from 500 Boeing SSTs, and thus this figure has been used as a standard case. For the standard case of an input of 7.5 x 10^{26} molecules sec^{-1} of NO_x at 20 km, the study at MIT that included three-dimensional atmospheric motions [4] found a 16 percent ozone reduction in the northern hemisphere, an eight percent ozone reduction in the southern hemisphere, and a maximum reduction of the vertical ozone column of 25 percent near the flight corridor. This three-dimensional model study and numerous other less detailed analyses [1-4] confirmed that NO_x injections of this magnitude seriously reduce stratospheric ozone.

In considering man-made ozone-destroying catalysts in the stratosphere, one should consider the elevation of insertion as one of the major variables in the problem. The lower in the stratosphere that the insertion occurs, the faster it will be mixed into the troposphere, and the lower will be the effect on stratospheric ozone. The MIT three-dimensional model was solved for only one case, 7.5 x 10^{26} molecules NO sec^{-1} inserted at 20 km. With an extensive, two-dimensional model, WIDHOPF [41] carried out calculations for the natural stratosphere and the stratosphere perturbed by insertion of 6.9 x 10^{26} molecules NO sec^{-1} at 20 km and at 17 km and between 45° and 55°N latitude. These two elevations simulate the cruise heights of the formerly proposed Boeing supersonic transport at 20 km and the Concorde at 17 km. The results obtained are given in Table 4. The reduction of ozone caused by insertion of nitric oxide at 17 km is only 2/3 as great as that caused by insertion at 20 km, according to this study. CIAP concluded that current jet aircraft add only a very small amount of nitrogen oxides to the stratosphere.

Table 4

Ozone reduction by supersonic transports as a function of elevation of NO_x insertion as calculated by WIDHOPF [41] with a two-dimensional model of atmospheric motions

Cruise height, km	20	17
Added NO, molecules sec^{-1}	6.9 x 10^{26}	6.9 x 10^{26}
Ozone column reduction, %		
Global ave.	15	9
Local max.	41	27
N. hemisphere	23	15
S. hemisphere	6	4

With this simple method calibrated by the CIAP [1-3], it can be used to assess other proposed sources of stratospheric pollution. U.S. military aircraft inject NO_x in the stratosphere at an average rate of 1.5 x 10^{23} molecules sec^{-1} [43,4] which is a factor of 3000 less than the natural source of NO_x. This source of stratospheric pollution appears to be very small. Fifty space shuttles per year would inject an average of 2 x 10^{24} molecules HCl sec^{-1} into the stratosphere [4]. The total catalytic effect of NO_x and ClX on ozone is about the same, and on this basis 50 space shuttles per year is a factor of 200 less in terms of ozone-destroying catalysts than the natural oxides of nitrogen. Although this source is small compared to natural NO_x, there are special circumstances about the mode of introduction of HCl from the space shuttle that require careful consideration [44].

The 1974 ground-level source of chlorine atoms in the form of $CFCl_3$ and CF_2Cl_2 was 3×10^{26} molecules sec^{-1} [6]. This material is building up at a linear rate in the troposphere, and within experimental error (which is still large) the total amount ever produced is equal to the total amount now in the atmosphere. These compounds have been observed in the stratosphere in amounts consistent with theoretical calculations [14]. Two "ifs" apply here. If there is no as-yet-unidentified tropospheric sink for the chlorofluorocarbons and if there is no as-yet-unidentified stratospheric sink or reservoir for chlorine, then in a few decades (after the troposphere is saturated with respect to this source and the stratospheric sink) 3×10^{26} molecules Cl sec^{-1} from chlorofluorocarbons released at the earth's surface would be a large source of stratospheric pollution. Direct observations of ClO in the stratosphere is needed to answer the question about stratospheric reservoirs. Improved analytical chemistry in observing chlorofluorocarbons in the troposphere is needed to ascertain if there is or is not a tropospheric sink.

Summary

Although acknowledging the great complexity of the full three-dimensional time-dependent problem, one can focus on the dominant factors in stratospheric photochemistry and achieve a general understanding of the natural and possibly perturbed situation. Ozone is formed in the upper and middle stratosphere by the photolysis of molecular oxygen

$$O_2 + h\nu \ (\lambda < 242 \text{ nm}) \rightarrow 0 + 0 \tag{1}$$

followed by addition of 0 to O_2 to make O_3. In terms of the rate-determining steps in coupled, usually catalytic, reactions, ozone is destroyed in the upper (U), middle (M), or lower (L) stratosphere by the following chemical reactions

(U)	$O_3 + 0 \rightarrow O_2 + O_2$	(5)
(U)	$HO + 0 \rightarrow H + O_2$	(12)
(U)	$HOO + 0 \rightarrow HO + O_2$	(10)
(U)	$ClO + 0 \rightarrow Cl + O_2$	(18)
(M)	$NO_2 + 0 \rightarrow NO + O_2$	(14)
(L)	$NO_3 + h\nu \rightarrow NO + O_2$	(16)
(L)	$HOO + O_3 \rightarrow HO + O_2 + O_2$	(7)

A major complexity of stratospheric photochemistry concerns the rates of formation, destruction, interconversion, and storage in molecular reservoirs of the free-radical ozone-destroying catalysts; this complexity has been relieved to some extent and can be relieved to a greater extent by observations of trace species in the stratosphere. Air motions transport ozone faster than it is formed as follows:

(U,M,L)	east-west winds
(M,L)	north-south mixing
(L)	vertical mixing

Some computations have been carried through that include three-dimensional or two-dimensional atmospheric motions and ozone photochemistry; numerous computations have been made considering only vertical components of air motion.

According to current observations and computations, the various chemical families destroy ozone below 45 km in the natural stratosphere as follows:

NO_x 50 to 70 percent
O_x 20 to 30 percent
HO_x 10 to 20 percent
ClX < 1 percent

On the basis of limited observations, the natural global source of stratospheric NO_x has been found to be the order of magnitude of 10^{27} molecules sec^{-1}. This known source strength of natural NO_x and its known strong catalytic effect in destroying natural stratospheric ozone provide a simple reference point to judge which man-made pollutions of the stratosphere would probably have a large effect and which would probably have a small effect. Such comparisons show that man-made ozone-destroying catalysts would be injected into the stratosphere at rates small compared to nature by:

(a)   current military aircraft
(b)   current civilian aircraft
(c)   50 space shuttles per year

Such comparisons show that ozone-destroying catalysts would be introduced into the stratosphere at the same order of magnitude as the natural rate by either

(e)   an economically successful fleet of supersonic transports
(f)   chlorofluorocarbons if the current rate of release is continued for several decades.

On general grounds one can say that unless special, unknown mechanisms are operative, the pollution of the stratosphere according to (e) or (f) would cause a large reduction of stratospheric ozone. Detailed computations including two and three dimensions of atmospheric motions have given explicit estimates for (e), and large reductions of ozone are indicated.

References

1.   The Natural Stratosphere of 1974, CIAP monograph 1 (Department of Transportation. DOT-TST-75-51, National Technical Information Service, Springfield, Virginia 22151. Sept. 1975) 1350 pp.

2.   The Stratosphere Perturbed by Propulsion Effluents, CIAP Monograph 3 (Department of Transportation. DOT-TST-75-53, National Technical Information Service, Springfield, Virginia 22151. Sept. 1975) 765 pp.

3.   A. J. Grobecker, S. C. Coroniti, and R. H. Cannon, Jr. The Effects of Stratospheric Pollution by Aircraft, CIAP Report of Findings (Department of Transportation. DOT-TST-75-50, National Technical Information Service, Springfield, Virginia 22151. March 1975) 850 pp.

4.   Environmental Impact of Stratospheric Flight (National Academy of Sciences — National Research Council, Washington, D.C. April 1975) 348 pp.

5. M. Nicolet, Rev. Geophys. Space Phys. 13, 593 (1975).

6. F. S. Rowland and M. J. Molina, Rev. Geophys. Space Phys. 13, 1 (1975).

7. H. S. Johnston, Rev. Geophys. Space Phys. 13, 637 (1975).

8. T. Shimazaki and R. C. Whitten, Rev. Geophys. Space Phys. 14, 1 (1976).

9. C. K. N. Patel, E. G. Burkhardt, and C. A. Lambert, Science 184, 1173 (1974).

10. J. G. Anderson, Geophys. Res. Letts. 2, 231 (1975).

11. M. Ackerman, J. Atmos. Sci. 32, 1649 (1975).

12. L. E. Heidt, R. Lueb, W. Pollock, and D. H. Ehhalt, Geophys. Res. Letts. 2, 445 (1975).

13. A. L. Schmeltekopf, P. D. Goldan, W. R. Henderson, W. J. Harrop, T. L. Thompson, F. C. Fehsenfeld, H. I. Schiff, P. J. Crutzen, I. S. A. Isaksen and E. E. Ferguson, Geophys. Res. Letts. 2, 393 (1975).

14. W. J. Williams, J. J. Kosters, A. Goldman, and D. G. Murcray, submitted to Geophys. Res. Letts. 1976.

15. A. L. Lazrus, B. W. Gandrud, R. N. Woodward, and W. A. Sedlacek, Geophys. Res. Letts. 2, 439 (1975).

16. P. J. Crutzen and I. S. A. Isaksen, preprint 1975.

17. H. S. Johnston, D. Kattenhorn, and G. Whitten, J. Geophys. Res. 81, 368 (1976).

18. B. G. Hunt, J. Geophys. Res. 71, 1385 (1966).

19. P. J. Crutzen, Quart. J. Roy. Meteorol. Soc. 96, 320 (1970).

20. H. S. Johnston, Science 173, 517 (1971).

21. R. S. Stolarski and R. J. Cicerone, Can. J. Chem. 52, 1610 (1974).

22. P. J. Crutzen, Can. J. Chem. 52, 1569 (1974).

23. P. J. Crutzen, Geophys. Res. Letts. 1, 205 (1974).

24. S. C. Wofsy and M. B. McElroy, Can. J. Chem. 52, 1582 (1974).

25. S. C. Wofsy, M. B. McElroy, and N. D. Sze, Science 187, 535 (1975).

26. R. J. Cicerone, R. S. Stolarski, and S. Walters, Science 185, 1165 (1974).

27. P. Fabian and C. Junge, Arch. Meteorol. Geophys. Bioklimatol. Ser. 19, 161 (1970).

28. P. J. Crutzen, J. Geophys. Res. 76, 7311 (1971).

29. M. Nicolet and W. Peetermans, Ann. Géophys. 28, 751 (1972).

30. M. B. McElroy and J. C. McConnell, J. Atmos. Sci. 28, 1095 (1971).

31. J. F. Noxon, Science 189, 547 (1975).

32. I. A. Isaken, Pure Appl. Geophys. 106-108, 1438 (1973).

33. A. L. Lazrus and B. W. Gandrud, J. Atmos. Sci. 31, 1102 (1974).

34. F. S. Rowland, J. E. Spencer, and M. J. Molina, Paper presented at 12th International Symposium on Free Radicals, Laguna Beach, Calif., January 1976.

35. A. W. Brewer, C. T. McElroy, and J. B. Kerr, Nature 246, 129 (1973).

36. G. K. Moortgat and P. Warneck, Z. Naturforsch. 30, 835 (1975).

37. R. A. Graham, The Photochemistry of NO₃ and the Kinetics of the N₂O₅-O₃ System (Ph.D. Dissertation, University of California, Berkeley, November 1975) 174 pp.

38. I. T. N. Jones and K. D. Bayes, J. Chem. Phys. 59, 4836 (1973).

39. H. Harrison, Science 170, 734 (1970).

40. E. Hesstvedt, Can. J. Chem. 52, 1592 (1974).

41. G. F. Widhopf, Paper presented at Am. Geophys. Union, San Francisco, Calif. Dec. 1974.

42. F. N. Alyea, D. M. Cunnold, and R. G. Prinn, Science 188, 117 (1975).

43. D. Elliott, National Aeronautical and Space Council, personal communication, 1971.

44. R. C. Whitten, W. B. Borucki, I. G. Poppoff, and R. P. Turco, J. Atmos. Sci. 32, 613 (1975).

REMOTE SENSING USING TUNABLE LASERS

K.W. Rothe and H. Walther
Sektion Physik der Universitet München
Garching, Federal Republic of Germany

Introduction, Survey on the Methods for Remote Sensing

Shortly after the invention of the laser it became obvious
that it is also a useful tool for remote measurements of
atmospheric parameters via backscattering processes. The
properties of lasers such as their large spectral brightness, small divergence and frequency tunability are key
factors in the development of various remote sensing techniques. In this paper a brief survey on measurements performed so far will be given; hereby the pollution studies
will be emphasized. The paper also reports some so far unpublished results obtained in our laboratory.

The scattering of laser light in the lower atmosphere
is governed by Mie scattering [1] which is mainly caused by
aerosols dust or other small particles. The Rayleigh scattering by the molecular constituents of the air is in general two orders of magnitude smaller. However, at altitudes
higher than 30 km where the aerosol concentration is negligible, the Rayleigh scattering is dominant. In this case it
is therefore possible to determine from the Rayleigh data
the molecular density variation and from that the atmospheric
pressure and temperature [2-7] .

In order to detect trace constituents in the lower
atmosphere fluorescence e.g. [8-12] or Raman scattering
e.g. [13-21] can be observed. In addition absorption measurements are possible e.g.[22-24] . The absorption studies
are the most straightforward and most sensitive ones for
pollution monitoring. The measurements give the integrated
pollution concentration over the light path. In those investigations a closed path, double ended long path or single

---

*
Also Max-Planck-Gesellschaft zur Förderung der Wissenschaften e.V., Projektgruppe für Laserforschung

ended arrangements with retro-reflectors or topographical
reflectors are used. The sensitivity achieved in pollution
monitoring is usually in the ppb range. The low transmitted
power required in those studies also allows the use of frequency tunable continuous wave infra-red diode laser sources
[24] or Raman-spin-flip lasers [24]. The sensitivity for the
detection of weak return signals in the absorption arrangements can still be significantly increased by using heterodyne detection [26] with the appropriate laser as the local
oscillator and an infra-red photodetector as the mixer. The
heterodyne detector is several orders of magnitude more sensitive than direct photodetectors therefore the efficiency
of the reflective surface can be much lower, or the path
length can be made much longer. In connection with heterodyne detection fast room temperature detectors such as e.g.
pyroelectric detectors become competitive in sensitivity
with the cooled infra-red detectors.

Heterodyne detection can also be used for the remote
sensing of pollutants when the temperature or emissivity
of the gases differs from the temperature or emissivity of
the background as e.g. in the exhaust of a smoke stack [25].
In this case the thermal radiation from the pollutant and the
output of a tunable laser are incident on the detector. By
scanning the laser frequency a beat signal is obtained that
can be used to determine the emission spectrum and thus detect the presence of particular pollutants. Another advantage of heterodyne detection is the strong spatial filtering
which is diffraction limited [27]. This is a big advantage
when a strong incident background radiation has to be suppressed.

Although long path absorption provides the most sensitivity of any remote detection method, it has the disadvantage of being double ended and lacking range resolution.
The atmospheric studies using scattering phenomena do not show
this drawback. When pulsed laser sources are used, a RADAR-like
measurement is possible. Since light replaces the radio waves
the techniques are usually summarized under the term LIDAR. A
survey on the atmospheric properties which can be studied by
LIDAR is given in Table 1. Reviews on the subject have been
published by e.g. FIOCCO [38] (Rayleigh and Mie scattering)
and by BYER [37] (pollution measurements).The relative size
of the backscattered intensity for the various processes is
given in Table 2. These data show that Raman scattering is not
suitable to detect pollutants over larger ranges. The highest
detection sensitivity was obtained so far by HIRSCHFELD et al.
[20]. For this investigation a frequency doubled ruby laser
of 260 mJ was used; the diameter of the receiving telescope
was 3 m. It was possible to detect 1.7 ppm H_2O, 30 ppm CO_2,
and SO_2 over a distance of 200 m. An advantage of the observation of Raman scattering is that no frequency tunable lasers
are required; furthermore no pumping beam depletion by the
pollutant itself arises.However, the lack of sensitivity has
so far severely limited the application of Raman scattering
to ambient pollutant monitoring. Some suggestions have been

Table 1: Atmospheric Studies by Means of Lasers (for further
references see text)

Process	Information	References
refractive index fluctuation	scintillation	28
	clear air turbulence	29,30
	wind velocity	31
Rayleigh-scattering	total density of molecular constituents of the atmosphere, temperature and pressure distribution	6
Mie-scattering	aerosol density and size	32
	clouds	33
	dust, smog	34
Raman-scattering	molecular species	19
	temperature	21
resonance absorption and fluorescence	molecular and atomic species	35,36

Table 2: Relative Size of Backscattered Intensity.
For comparison see Ref. [36]

Scattering Process	Relative Signal Size	Remarks
Rayleigh-scattering	1	
Mie-scattering	80	visibility 5km
Raman-scattering	10^{-3} . Np/Na	Np,Na number densities of the pollutant and the air
fluorescence-scattering	80	cross-section $10^{-16}cm^2$ and quenching factor 10^{-2}; the concentration was assumed to be 0.3 ppm

made to improve the sensitivity as e.g. using resonance Raman
effect or to apply four wave Raman mixing to generate a coherent anti-Stokes signal with high efficiency [39] . There is not
yet enough experience with the latter method to make a final

statement. In the case of the resonance enhancement of Raman scattering, however, it is better to tune the used laser exactly on resonance and excite resonance fluorescence directly; since the fluorescence scattering is still about two orders of magnitude larger than the resonance Raman signal. There have also been proposals to use the rotational Raman scattering instead of the vibrational-rotational Raman effect since the corresponding cross sections are larger. The rotational lines of most molecules, however, are too close to the exciting line; this is the reason why the expected improvement could not be realized.

The detection of fluorescence backscattering seems to be more favourable. For molecules, however, the backscattered fluorescent light for electronic transitions is distributed over a rather large spectral region due to the rotational and vibrational splitting of the fundamental state. In order to detect sufficient fluorescent light the receiver system has to be rather broad-banded. This, of course, lowers the ratio between fluorescent and background signal. An additional disadvantage is that a large amount of excitation energy gets lost by quenching processes under atmospheric pressure conditions. This was considered by a factor 10^{-2} in Table 2. (The cross section $10^{-16} cm^2$ assumed in Table 2 is valid for electronic transitions; for vibrational fluorescence the corresponding value is smaller. For details see [36]). Furthermore there may be a local variation of the quenching factor; therefore the fluorescence cross section is never precisely known. The quenching effects play no role in the upper atmosphere; therefore sodium and potassium atoms could be detected in concentrations as low as 10^3 atoms/cm^3 over a distance of 90 km [8-11].

The quenching processes are also important when infra-red transitions of pollutants are investigated. In this case an additional difficulty arises by the lack of depth resolution due to the relatively long lifetimes of the vibrational levels. The sensitivity of fluorescence observation is surpassed by that of the differential absorption method. This latter technique will be discussed in the following.

Differential Absorption Method and Measurements

As mentioned above the observation of absorption seems to preclude the possibility of a RADAR-like measurement. This disadvantage can be overcome by using the ubiquious Mie scattering as a reflector. The position of the "Mie reflector" is determined by the time delay between the emission of the laser pulse and the observation of the backscattered radiation. Mie scattering can be considered to be monotonic at least over the spectral region of a molecular absorption line. Therefore the selective absorption by the pollutant can be separated by tuning the transmitted wavelength on and off an absorption line. This method which is called "differential absorption method" was first suggested by SCHOTLAND in 1964 [40] using a searchlight as a light source. Later the method was compared to the other LIDAR techniques in theoretical studies which show that differential absorption is the most sensitive method [35, 41]. It combines in an optimum way the depth resolution with the large absorption cross sections and resultant high sensitivity of the absorption method. The first application to pollution studies over larger distances was performed by ROTHE et al. [45, 46] who measured ambient NO_2. The LIDAR system was equipped with a tunable dye laser. The energy of the laser pulse was about 1 mJ. The backscattered light was collected by a simple Cassegrain optics of which the large mirror (60 cm diameter) consisted of a military searchlight reflector. With this set-up a NO_2 concentration of 0.2 ppm at distances up to 4 km could be measured. As an extension of these measurements the study of strong localized emission sources has also been performed by the same autors [47] With a spatial resolution of 100 m the concentration of NO_2 in a plume of a smokestack and over the area of a chemical factory was measured. In the latter case a two-dimensional NO_2 density plot was obtained. (For other publications on the method see e.g. Ref. [36, 42-44, 48-50, 53])

The extension of the method to other pollutants requires laser in the ultra-violet (electronic transitions) and infra-red (vibrational transitions) spectral region. As the transmission of the atmosphere is drastically reduced at wavelengths shorter than 2500 Å, due to the O_2 absorption, the measurements in the ultra-violet region are essentially restricted to SO_2 and O_3. The near infra-red spectral region is more interesting for a general application of the method since all important pollutants have absorption lines in this region. The required laser pulse energy is in the range of several mJ. Therefore under the available tunable lasers for the infra-red spectral region primarily the parametric oscillator seems to be the most suitable one [52]. However, the complexity of this system still opposes a general use. A rather good substitute are the TEA lasers [51]. They are operating primarily with CO, CO_2, HF, DF and N_2O in the infra-red region and have many emission lines the wavelengths of which coincide reasonably well with vibrational transitions of many important pollutants. These coincidences have been extensively investigated in connection with absorption studies in the atmosphere and the data obtained there are also very useful for the measurements by means of the differential absorption method. A survey on the absorption studies performed by means of fixed frequency infra-red lasers is given in Table 3. The emission spectra of the DF, CO_2, N_2O lasers and partially also that of the CO laser coincide with transmission windows of the atmosphere. This is not the case for the HF laser lines; therefore those can only be used for measurements over relatively short distances.

Figures 1 to 3 demonstrate some coincidences between laser lines and the absorption lines of different gases and vapors. The lower trace of the figures shows the original laser spectrum which is obtained when the spectrum of the TEA laser is recorded by means of a monochromator. The upper traces show the

Table 3: Infra-Red Absorption Measurements with Fixed Frequency Lasers

Pollutant	Laser	References
Ammonia	CO	57
	CO_2	26,59,60,61
Acetonitrile	CO_2	63
Benzene	CO_2	57,63
1,3-Butadiene	CO	57
	CO_2	57
1-Butene	CO	57
	CO_2	57
n-Butane	DF	62
Iso-Butane	DF	62
t-Butanol	CO_2	63
Carbon Dioxide	DF	58
Cyclohexane	CO_2	63
1,2-Dichloroethane	CO_2	63
Ethylene	CO_2	26,57,59,60,61
Ethylacetate	CO_2	63
Freon	CO_2	59
Freon 12	CO_2	63
Freon 113	CO_2	63
Furane	CO_2	63
Hydrogen Chloride	DF	50*
Iodo Propane	CO_2	63
Methyl Chloroform	CO_2	63
Methyl Ethyl Cetone	CO_2	63
Methane	DF	50*,54,58
	He-Ne	55
Methanol	CO_2	57
Nitrous Oxide	DF	50*,54,58
Nitric Oxide	CO	26,57
Nitrogen Dioxide	CO	57
Ozone	CO_2	26,59,60,64,65,66
Propylene	CO	57
Iso-Propanol	CO_2	63
Perchloroethylene	CO_2	59
Propane	DF	62
Sulfur Dioxide	CO_2	26
Trichloroethylene	CO_2	57,59,61
Vinyl Chloride	CO	57
	CO_2	57,63
Water (H_2O)	CO	57
	CO_2	53,59
Water (HDO)	DF	58

* Field measurements

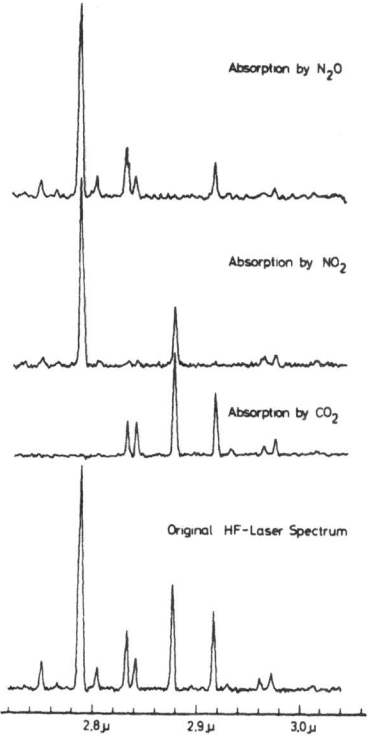

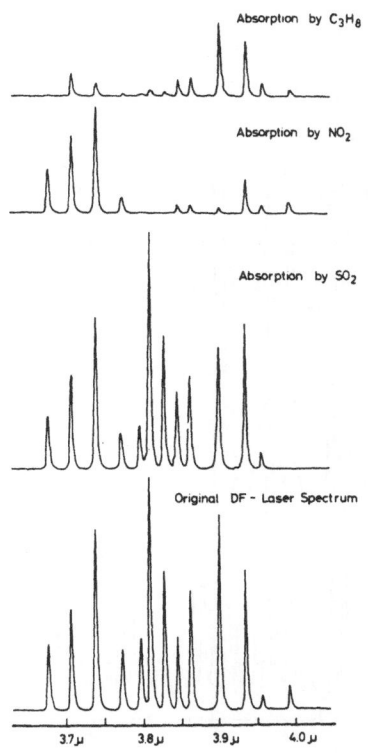

Figure 1: Absorption of the HF laser lines by different gases. The lowest spectrum is the original laser spectrum for multi-line operation of the TEA laser. The three strongest lines correspond to the $P_1(8)$, $P_2(7)$, and $P_2(8)$ transitions.

Figure 2: Absorption of the DF laser lines by different gases. The lowest spectrum is the original laser spectrum for multi-line operation of the TEA laser. The strongest lines correspond to the $P_2(6)$, $P_2(8)$, $P_3(7)$, and $P_3(8)$ transitions at the wavelenths 3.73μ, 3.80μ, 3.89μ and 3.93μ respectively.

respective spectra with absorption by various gases. For these measurements the laser light was passed through cells of a length of 1 m containing a partial pressure of the respective gas or vapor in the range of 100 Torr. The gases have been mixed with air to a total pressure of 1 atm to obtain the same pressure broadening conditions as in the free atmosphere.

From the Fig. 1-3 it can be seen that the observation of the total spectrum at the same time gives the possiblity to study several pollutants simultaneously [51] . For this pur-

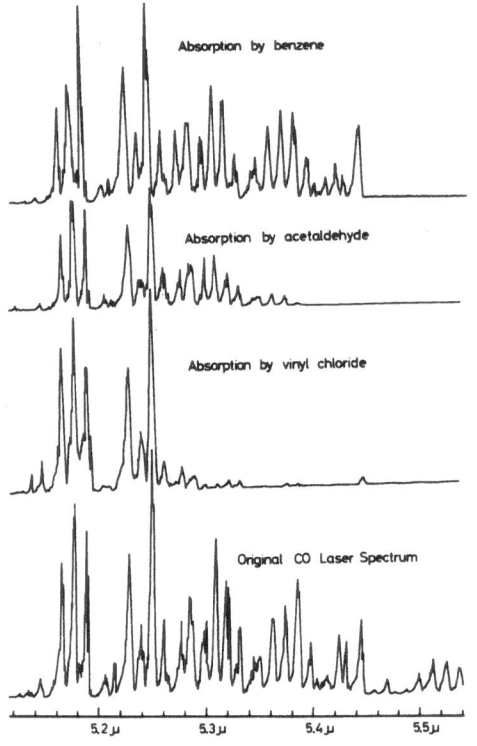

Figure 3: Absorption of the CO laser lines by different gases and vapors. The lowest spectrum is the original laser spectrum for multi-line operation of the TEA laser. The strong groups of transitions at 5.19µ, 5.22µ, 5.3µ, 5.39µ, and 5.42µ correspond to the P-transitions between the 7-6, 8-7, 9-8, 10-9, 11-10 vibrational levels respectively.

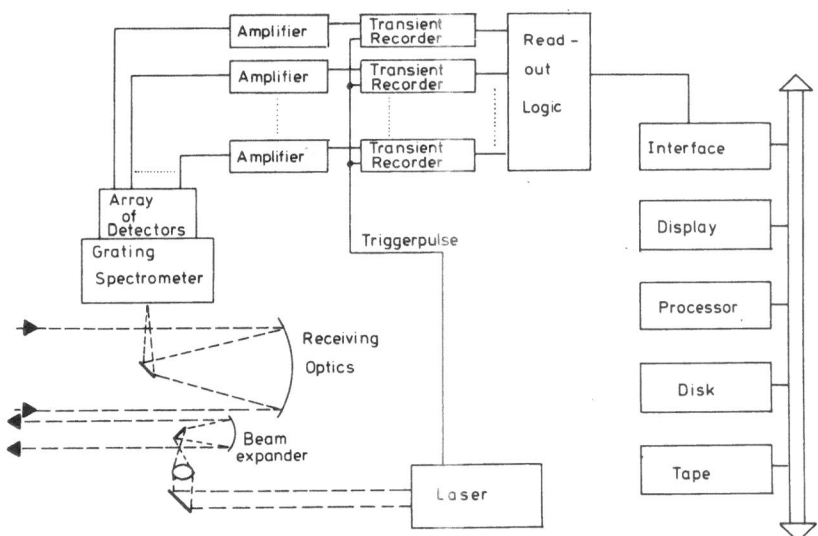

Figure 4: LIDAR set-up with multi-line detection.

Table 4: Parameters of the LIDAR System

Laser	Lumonics TEA 801 A
Repetition rate	up to 100 pulses/sec
Transmitted energy per	
pulse CO_2 - Laser	1000 mJ (typically, single line)
CO - Laser	50 mJ (all lines)
HF - Laser	600 mJ (all lines)
DF - Laser	300 mJ (all lines)
Pulsewidth	0.2 µs to 1.0 µs ⎫ (depending on the
Beam divergence	1 mrad to 3 mrad ⎭ gases and optics)
Diameter of the expand-	
ing telescope	20 cm
Receiving Optics	
Telescope parameters	
Diameter	60 cm
Focal length	240 cm
Field of view	0.04 mrad
Detector parameters	
Type	Ge:Hg, 20-element array
Size	0.1 x 0.1 mm for each element
Detectivity	$2 \cdot 10^{10}$ cm $Hz^{1/2}$ W^{-1}
Time constant	1 ns (with 50 Ω load)

pose our new LIDAR set-up has an array of 20 infra-red detectors
built into the grating spectrograph of the receiver system (see
Fig. 4). Each channel of the diode array has its own amplifier
and transient recorder. In this way the backscattered signal of
all lines can be measured as function of time for the individual
laser shots. The relevant parameters of our new LIDAR set-up are
given in Table 4.

The laser does not work as nicely in multi-line operation
for CO_2 as it does for DF, HF or CO. Therefore the multi-chan-
nel observation is not possible. The laser must be tuned over
the different CO_2 laser lines by means of an inner cavity grating. A result obtained with the CO_2 laser is shown in Fig. 5.
Plotted is the backscattered light intensity versus the range
from which the backscattering is observed. The two signals are
obtained from the scattering of the R(18) and R(20) lines respectively. The plotted signal intensity is normalized by the total emitted laser energy. The stronger decrease of the R(20) line is due to the strong water absorption. The peak at 3 km for
the R(18) signal results from the backscattering by a forest.

Due to the weak overlap between the emitted laser beam and the observation region of the receiving optics the backscattered signal from distances shorter than about 500 m is not in scale. To obtain this result the signal of about 100 laser shots for each line has been averaged. For the R(20) line no signal reflected by the forest is observed due to the water vapor absorption. The slow decay of the signal at 3 km is due to the time constant of our infra-red detector. The signal to noise ratio obtained in the measurement shows that the return signal is well above noise for a distance of 2 km. The laser energy used was about 1 J. This is essentially the same result which was also obtained by MURRAY et al. [53] in a recent publication.

For the evaluation of the pollution concentration from the differential absorption data, it is necessary to have sufficiently accurate absorption cross sections. Many of the relevant values are published in the literature (Table 3). In our laboratory also a series of measurements has been performed [51] . For the determination of the absorption coefficients absorption cells of a length between 1 and 10 m have been used. The longer absorption cells are necessary in order to keep the partial pressure of the investigated gas small. Otherwise the data are too strongly influenced by self broadening and the extrapolation to small

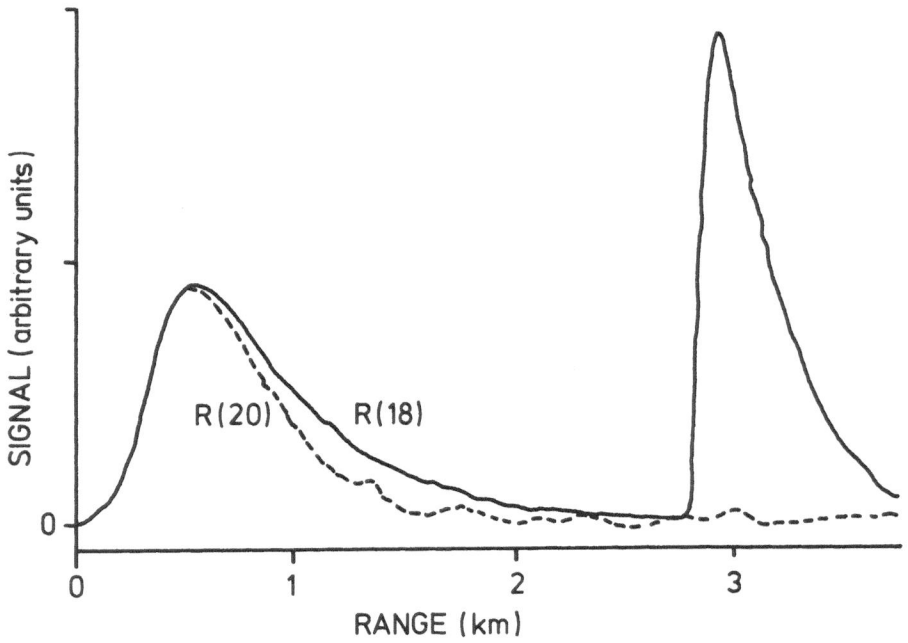

Figure 5: Atmospheric backscatter signals for the R(20) and R(18) lines of the CO_2 laser. The R(20) line is strongly absorbed by H_2O. For details see text.

partial pressures is not accurate enough. All measurements were
performed in mixtures with air (total pressure: 1 atm) to consider the foreign gas pressure broadening. Some of our results
are compiled in Table 5. The Table gives typical absorption
cross sections which are measured for the respective laser lines. The last column gives the values for the concentrations for
which a 1% absorption is observed for a path length of 1 km.
The ultimate sensitivity which can be obtained with the differential absorption method dependes of course on the respective
laser power and on the size of the receiving optics.The smallest
detectable concentrations are in general smaller than those given in Table 5.

In our measurements the influence of self pressure broadening was carefully studied. Since the absorption cross sections are small, most of the absorption measurements described
in the literature have been performed at large partial pressures. The coincidence between the laser line and the molecular
absorption line is generally not perfect, therefore a rather
large influence of line broadening on the respective molecular

Pollutant	Laser	Cross-Section [cm ^{2}]	Sensitivity [ppm] (1km,1% Absorption)
Sulfur Dioxide	DF	$2.0 \cdot 10^{-21}$	0.9
Nitrogen Dioxide	DF	$3.8 \cdot 10^{-21}$	0.5
Nitric Oxide	DF	$3.8 \cdot 10^{-21}$	0.5
Nitrous Oxide	DF	$3.5 \cdot 10^{-21}$	0.5
Carbon Dioxide	HF	$3.2 \cdot 10^{-21}$	0.6
Carbon Monoxide	HF	$5.0 \cdot 10^{-22}$	3.1
Hydrogen Fluoride	HF	$3.0 \cdot 10^{-20}$	0.06
Propylene	CO	$2.7 \cdot 10^{-21}$	0.7
Propane	DF	$2.4 \cdot 10^{-21}$	0.8
Vinyl Chloride	CO	$3.0 \cdot 10^{-21}$	0.6
Ethylene	CO	$2.0 \cdot 10^{-20}$	0.09
Toluene	CO	$6.8 \cdot 10^{-21}$	0.3
Benzene	CO	$2.5 \cdot 10^{-20}$	0.07
Acetaldehyde	CO	$6.0 \cdot 10^{-21}$	0.3
Propylene Oxide	CO	$1.3 \cdot 10^{-21}$	1.4

Table 5: Absorption Cross Sections for Different Gases and
Vapors

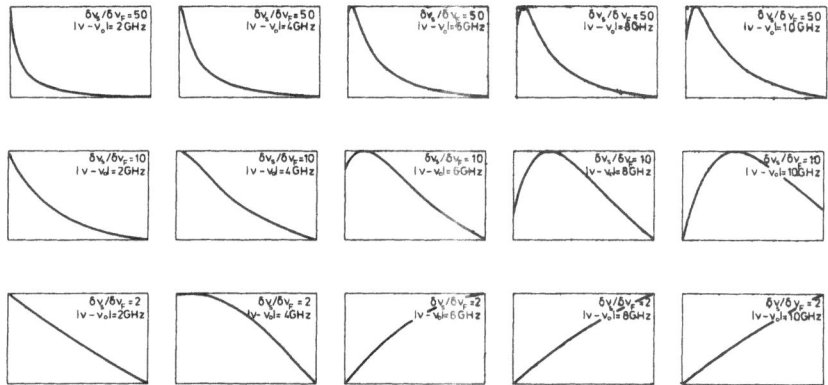

Figure 6: Dependence of the absorption coefficient on the self pressure. $\nu - \nu_0$ is the distance between the absorption and the laser line. $\delta\nu_S$ and $\delta\nu_F$ are the self broadening and foreign broadening coefficients. For details see text.

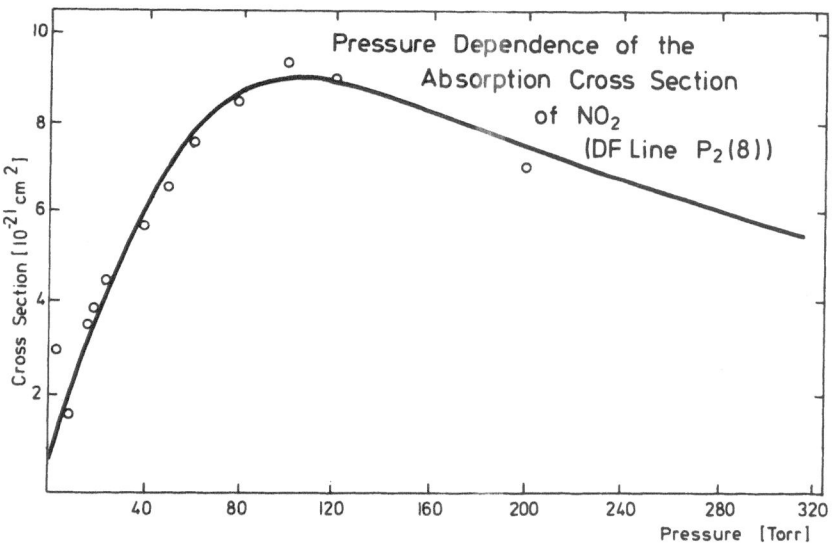

Figure 7: Dependence of the absorption coefficient on the self pressure for NO_2. For details see text.

absorption can be observed. The latter behaviour is shown on
Fig. 6. There the dependence of the absorption coefficient versus self pressure is plotted for different values of the distance between laser and absorption line $\nu - \nu_0$ and for different ratios of the self broadening $\delta\nu_s$ and foreign broadening $\delta\nu_F$
coefficients. For the calculations a Lorentzian line shape was
assumed. The results show that an extrapolation to small partial pressures may be quite problematic when the absorption measurements were not performed over a pressure range which is sufficiently large. Fig. 7 shows the experimental pressure dependence of the absorption cross section of NO_2 measured for a DF
laser line. The circles are the experimental results and the
solid curve represents at fit of the theoretical pressure dependence to the experimental data. There is a good agreement between experiment and theory.

Conclusion

Summarizing the present situation of pollution monitoring by
means of LIDAR it can be said that it is proven that the differential absorption method is the most sensitive technique known
at present. TEA lasers are suitable light sources for a general
application in the infra-red spectral region. The sensitivity
which can be expected with those lasers allows to probe the atmosphere in the vicinity of chemical factories or other pollution sources over distances of about 3 km. Further improvements
are possible e.g. by the use of heterodyne detection or by the
use of frequency up-conversion for signal detection. A considerable step forward can still be expected when continuously
tunable lasers as e.g. parametric oscillators of sufficient output power are used for the measurements [52].

Acknowledgement

The support of the Bundesministerium für Forschung und Techno-
logie is gratefully acknowledged. We would like to thank also
A. Tönnißen, P. Kirchner, and W. Baumer for their help with the
measurements.

References

1. G. Mie, Annal. der Physik 25, 377 (1908)
2. W. C. Bain, M. C. W. Sandford, J. Atmosph. Terr. Phys. 28,
   543 (1966)
3. G. S. Kent, B. R. Clemesha, R. W. Wright, J. Atmosph. Terr.
   Phys. 29, 169 (1966)
4. G. Grams, G. Fiocco, J. geophys. Res. 72, 3523 (1967)
5. K. Bartusek, D. J. Gambling, W. G. Elford, J. Atmosph. Terr.
   Phys. 32, 1535 (1970)
6. G. S. Kent, R. W. H. Wright, J. Atmosph. Terr. Phys. 32,
   917 (1970)
7. P. B. Russel, W. Viezee, R. D. Hake, Jr., R. T. H. Collis,
   Paper presented at Fourth Conference on the Climatic Impact
   Assessment Program, Febr. 1975, Cambridge Mass.
8. M. R. Bowman, A. J. Gibson, M. C. W. Sandford, Nature 221,
   456 (1969)

9. M. C. Sandford, A. J. Gibson, J. Atmosph. Terr. Phys. 32, 1423 (1970)
10. A. J. Gibson, M. C. W. Sandford, Nature 239, 509 (1972)
11. F. Felix, W. Keenliside, G. Kent, M. C. W. Sandford, Nature 246, 345 (1973)
12. J. A. Gelbwachs, M. Birnbaum, A. W. Tucker, C. L. Fincher, Opto-electronics 4, 155 (1972)
13. D. A. Leonard, Nature 216, 142 (1967)
14. J. A. Cooney, Appl. Phys. Lett. 12, 40 (1968)
15. S. H. Melfi, J. D. Lawrence, Jr., M. P. McCormic, Appl. Phys. Lett. 15, (1969)
16. H. Inaba, T. Kobayashi, Nature 224, 170 (1969)
17. H. Inaba, T. Kobayashi, Opto-electronics 2, 45 (1970)
18. T. Kobayashi, H. Inaba, Appl. Phys. Lett. 17, 139 (1970)
19. H. Inaba, T. Kobayashi, Opto-electronics 4, 101 (1972)
20. T. Hirschfeld, E. R. Schildkraut, H. Tannenbaum, D. Tannenbaum, Appl. Phys. Lett. 22, 38 (1973)
21. R. G. Strauch, V. E. Derr, R. E. Cupp, Appl. Optics 10, 2665 (1971)
22. P. L. Hanst, Opt. Quant. Electron. 8, 87 (1976)
23. P. L. Kelley, R. A. McClatchey, R. K. Long, A. Snelson, Opt. Quant. Electron.
24. E. D. Hinkley, Opt. Quant. Electron. 8, 155 (1976)·
25. E. D. Hinkley, P. L. Kelley, Science 171, 635 (1971)
26. R. T. Menzies, M. S. Shumate, Science 184, 570 (1974)
27. A. E. Siegmann, Appl. Optics 5, 1588 (1966)
28. M. Subramian, J. Opt. Soc. Am. 62, 677 (1971)
29. J. Cooney, Appl Optics 11, 2374 (1972)
30. B. J. Brinkworth, Appl. Optics 12, 427 (1973)
31. V. E. Derr, C. G. Little, Appl. Optics 9, 1976 (1970)
32. R. T. H. Collis, E. E. Uthe, Opto-electronics 4, 87 (1972)
33. P. A. Davis, Appl. Optics 8, 2099 (1969)
34. Ch. S. Cook, G. W. Bethke, W. D. Conner, Appl. Optics 11, 1742 (1972)
35. R. M. Measures, G. Pilon, Opto-electronics 7, 141 (1972)
36. H. Kildal, R. Byer, Proc. IEEE 59, 1644 (1971)
37. R. L. Byer, Opt. Quant. Electron. 7, 147 (1975)
38. G. Fiocco, Plenarvorträge Bd. 9, European Physical Society, Petit lancy 1972
39. P. R. Regnier, J. P. E. Taran, Appl. Phys. Lett. 23, 240 (1973)
40. R. M. Schotland, Proc. Third Symp. on Remote Sensing of the Environment, Oct. 14-16, 1964, Univ. of Michigan, Ann Arbor, Mich.
41. R. L. Byer, M. Garbuny, Appl. Optics 12, 1496 (1973)
42. R. M. Schotland, J. Appl. Meteorology 13, 71 (1974)
43. V. L. Granatstein, M. Rhinewine, A. H. Fitch, Appl. Optics 12, 1511 (1973)
44. W. B. Grant, R. D. Hake, Jr. E. M. Liston, R. C. Robins, E. K. Proktor Jr., Appl. Phys. Lett. 24, 550 (1974)
45. K. W. Rothe, U. Brinkmann, H. Walther, The Physics of Electronic and Atomic Collisions, Invited Lectures and Progress Reports VIII ICPEAC, Ed. by B. C. Cobic, M. V. Kurepa, Beograd 1973
46. K. W. Rothe, U. Brinkmann, H. Walther, Appl. Phys. 3, 115 (1974)

47. K. W. Rothe, U. Brinkmann, H. Walther, Appl. Phys. 4, 181 (1974)
48. T. Igarashi, Fifth Conference on Laser Radar Studies of the Atmosphere, Abstracts, p 57, June 1973, Willimsburg, Va.
49. S. A. Ahmed, Appl. Optics 12, 901 (1973)
50. E. R. Murray, J. E. van der Laan, J. G. Hawley, to be published
51. K. W. Rothe, H. Walther, Conference on the Applications of Laser Spectroscopy, Optical Society of America, Anaheim, Calif. March 1975
52. R. L. Byer, R. L. Herbst, R. N. Fleming, Laser Spectroscopy ed. by S. Haroche, J. C. Pebay-Peyroula, T. W. Hänsch, S. E. Harris, Springer Verlag, Berlin, Heidelberg, New York 1975
53. E. R. Murray, R. D. Hake, Jr., J. E. van der Laan, J. G. Hawley, Appl. Phys. Lett. 28, 542 (1976)
54. T. F. Deaton, D. A. Depatie, T. W. Walker, Appl. Phys. Lett. 26, 300 (1975)
55. B. N. Edward, D. E. Burch, J. Opt. Soc. Am. 55, 174 (1965)
56. S. M. Freund, D. M. Sweger, Anal. Chem. 47, 930 (1975)
57. L. B. Kreuzer, N. D. Kenyon, C. K. N. Patel, Science 177, 347 (1972)
58. D. J. Spencer, G. C. Denault, H. H. Takimoto, Appl. Optics 13, 2855 (1974)
59. W. Schnell, G. Fischer, Appl. Optics 14, 2058 (1975)
60. R. R. Patty, G. M. Russwurm, W. A. McClenny, D. R. Morgan, Appl. Optics 13, 2850 (1974)
61. W. Schnell, G. Fischer, Z. Angew. Math. Phys. 26, 133 (1975)
62. W. R. Watkins, K. O. White, Appl. Optics 15, 1114 (1976)
63. B. D. Green, J. I. Steinfeld, Appl. Optics 15, 1688 (1976)
64. C. Young, R. H. L. Bunner, Appl. Optics 13, 1438 (1974)
65. J. Schewchun, B. K. Garside, E. A. Ballik, C. C. Y. Kwan, M. M. Elsherbiny, G. Hogenkamp, A. Kazandjian, Appl. Optics 15, 340 (1976)
66. K. Asai, T. Igarashi, Opt. Quant. Electron. 7, 211 (1975)

VII. Photobiology

RESONANCE RAMAN SPECTROSCOPY:
APPLICATION OF TUNABLE LASERS TO THE STUDY OF THE MOLECULAR
MECHANISMS AND DYNAMICS OF VISUAL EXCITATION

R. Mathies *, A.R. Oseroff, T.B. Freedman and L. Stryer
Department of Molecular Biophysics and Biochemistry,
Yale University, New Haven, Connecticut, 06520, USA

Rod cells in the retinas of vertebrates are exquisitely sensitive photodetectors. In fact, a retinal rod cell can be excited by a single photon
[1]. The photoreceptor molecule is rhodopsin, which consists of an 11-<u>cis</u>
retinal chromophore and a 38,000 dalton protein called opsin [for reviews,
see 2,3]. The 11-<u>cis</u> retinal group is bound to a specific lysine side chain
of opsin by a protonated Schiff base linkage (Fig. 1). WALD [4] showed that
light isomerizes the 11-<u>cis</u> retinal group of rhodopsin to all-<u>trans</u> retinal,
which no longer fits in the binding site for the chromophore. Several transient species in the photolysis of rhodopsin have been identified by their
distinctive spectral properties (Fig. 2).

Some fundamental questions concerning the molecular basis of vision are:

1. How does the protein tune the absorption maximum of its chromophore? The absorption maximum of rhodopsin is at 500 nm, whereas
that of the 11-<u>cis</u> protonated Schiff base in solution is at 440 nm.

2. How does the protein increase the quantum yield of photoisomerization of its chromophore? The quantum yield of photoisomerization of rhodopsin is 0.67, which is three to ten times higher than
that of protonated Schiff bases of 11-<u>cis</u> retinal in organic solvents
[5].

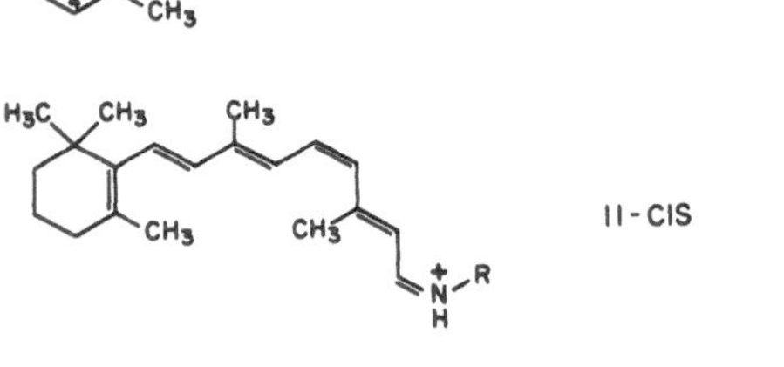

Figure 1. Protonated Schiff
bases of all-<u>trans</u> and
11-<u>cis</u> retinal.

3. How does the isomerization of 11-<u>cis</u> retinal in rhodopsin lead to visual excitation? The next defined event is a decrease in the Na$^+$ permeability and a consequent hyperpolarization of the plasma membrane of the rod outer segment [6,7]. The links between the absorption of a photon by retinal, and the subsequent decrease in the Na$^+$ permeability of the plasma membrane are not yet known. The gain of this process is at least 10^5 [7].

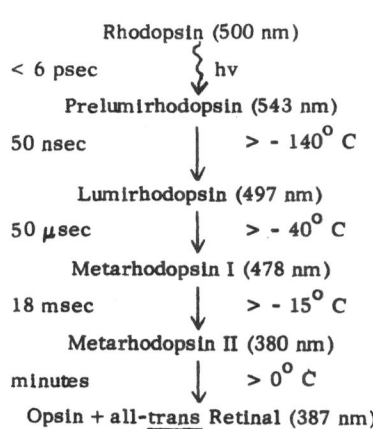

Rhodopsin (500 nm)

< 6 psec   } hv

Prelumirhodopsin (543 nm)

50 nsec   > - 140° C

Lumirhodopsin (497 nm)

50 μsec   > - 40° C

Metarhodopsin I (478 nm)

18 msec   > - 15° C

Metarhodopsin II (380 nm)

minutes   > 0° C

Opsin + all-<u>trans</u> Retinal (387 nm)

Figure 2. Sequence of intermediates in the photolysis of rhodopsin. The absorption maximum of each intermediate is shown in parentheses.

Answering these questions will require detailed information about the interactions between retinal and opsin, and about the conformational changes in the chromophore during visual excitation. Resonance Raman spectroscopy is a powerful technique for elucidating these changes [8-12]. This experimental approach takes advantage of the enhanced Raman scattering from vibrations coupled to an electronic transition when the excitation wavelength is within or near the absorption band of the chromophore [13,14]. The resonance effect makes it feasible to specifically monitor the structure of the retinal chromophore in visual pigments. However, even with resonance enhancement, Raman scattering is a very weak process. The probability that a molecule absorbs a photon is at least 10^7 greater than its probability of participating in resonance Raman scattering. If we illuminate a solution of rhodopsin with a 25 mW beam of 600 nm light focused to a diameter of 40 μm, nearly all of the rhodopsin in the beam will be isomerized within 1 msec (Fig. 3). Since a Raman spectrum is typically recorded over several minutes, it is clearly not feasible to obtain the resonance Raman spectrum of unphotolyzed rhodopsin under these conditions.

We have devised a new technique for obtaining the resonance Raman spectrum of a photolabile molecule before it is altered by light [11]. The essence of this approach is that <u>the sample is rapidly flowed through the light beam so that the fraction of isomerized molecules in the illuminated volume is very small (< 0.05).</u> For rhodopsin, the transit time must be on the order of 10 μsec. This was achieved by forming a jet stream of rhodopsin with a velocity of about 400 cm/sec and flowing it past a laser beam with a diameter of 40 μm (Fig. 4). We present here a theory for this rapid-flow technique and report the resonance Raman spectra of unphotolyzed rhodopsin and of the protonated Schiff bases of all-<u>trans</u>, 9-<u>cis</u>, 11-<u>cis</u>, and

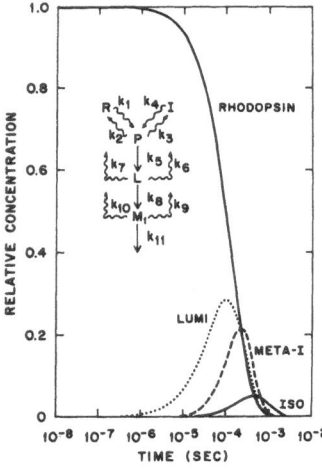

Figure 3. Calculated relative concentrations of rhodopsin and its photolytic intermediates as a function of time following the irradiation of a stationary sample of detergent-solubilized rhodopsin at 600 nm with a 25 mW beam. The relative concentrations of these species were obtained by numerical integration of the kinetic scheme shown in the figure using known rate constants and quantum yields (R = rhodopsin, I = isorhodopsin, P = prelumirhodopsin, L = lumirhodopsin, M_1 = metarhodopsin I).

13-cis retinal. These spectra are very sensitive to the conformation of retinal, and reveal that rhodopsin does not grossly distort the ground state conformation of its 11-cis retinal chromophoric group. Resonance Raman studies of the photolytic intermediates in vision and of other photolabile systems should also be highly informative. The development of tunable lasers in the ultraviolet region should permit resonance Raman studies of an even wider range of biological molecules.

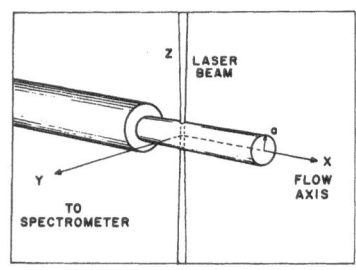

Figure 4. Geometry of the jet-stream sampling system. The flow axis x, the detection axis y, and the laser beam axis z are perpendicular to each other.

Rapid-Flow Technique

The rate of photoisomerization $k(\text{sec}^{-1})$ of a photolabile molecule such as rhodopsin is given by

$$k = I(\lambda)\sigma_A(\lambda)\phi(\lambda) \tag{1}$$

where $I(\lambda)$ is the light flux (photons $\text{cm}^{-2}\ \text{sec}^{-1}$), $\sigma_A(\lambda)$ is the absorption cross-section ($\text{cm}^2\ \text{molecule}^{-1}$), and $\phi(\lambda)$ is the quantum yield of photoisomerization. For a typical experiment on rhodopsin with a laser power of 25 mW at 600 nm and a beam diameter of 40 μm ($I = 6.01 \times 10^{21}$ photon cm^{-2} sec^{-1}, $\varepsilon = 470\ \text{cm}^{-1}\ \text{M}^{-1}$, and $\phi = 0.67$), the rate constant for photoisomerization is $7.2 \times 10^3\ \text{sec}^{-1}$. The photoisomerization time $\tau_1(\text{sec})$, which is the average time a rhodopsin molecule spends in the beam before it is isomerized, is k^{-1}. If we assume that the stream has a uniform velocity v (cm/sec) and the laser beam a uniform intensity inside a square cross-section of length ℓ (cm), then the transit time τ_t (sec) of the molecule in the beam is ℓ/v. The fraction F of rhodopsin that is isomerized while traversing the beam is then

$$F = 1 - e^{-\tau_t/\tau_i} \tag{2}$$

When the extent of isomerization is small (F << 1), this expression can be expanded to give

$$F = \tau_t/\tau_i = (I\sigma_A \phi \ell)/v \tag{3}$$

It is convenient to express F in terms of the laser power P (photons sec^{-1}), which is equal to $I\ell^2$. Also, we convert from σ_A (cm^2 molecule^{-1}) to the decadic extinction coefficient ε (cm^{-1} M^{-1}), which are related by $\sigma_A = 3.824 \times 10^{-21}\, \varepsilon$. Then, equation (3) becomes

$$F = (P\varepsilon\phi\, 3.824 \times 10^{-21})/\ell v \tag{4}$$

This expression shows that F, the photoalteration parameter, can be made suitably small (say <0.05) by having a sufficiently rapid flow.

This qualitative derivation can be carried out in a more quantitative way by considering the actual Gaussian intensity profile of the focused laser beam. The results of our original derivation [11] are in agreement with an alternative treatment by CALLENDER et al. [12]. First, the concentration of unisomerized molecules in the illuminated part of the jet stream is computed. Then, the contribution of these unisomerized molecules to the total Raman signal is obtained by weighting the composition at each point in the jet stream by the light intensity. The fraction of Raman photons that come from unisomerized molecules is then given by S'

$$S' = \frac{1}{\pi} \int_{-\infty}^{\infty}\int_{-\infty}^{\infty} [\exp - (s^2 + t^2)]\, \exp - \{\sqrt{\tfrac{2}{\pi}}\, F \int_{-\infty}^{s} \exp - (s'^2 + t^2)\, ds'\}\, ds\, dt \tag{5}$$

where s, s' and t are dummy variables. Note that this equation for S', the normalized Raman scattering, depends only on F, the photoalteration parameter. Equation (5) serves as a criterion of the feasibility of obtaining a Raman spectrum of any photolabile molecule. A plot of S' as a function of F is shown in Fig. 5. In the evaluation of F the effective length ℓ of a Gaussian laser beam is given by $\sqrt{\pi}\omega$, ω is the $1/e^2$ radius of the beam, and P, the laser power, is now given by $\pi\omega^2 I_0/2$ where I_0 is the peak light intensity.

When the extent of photoisomerization is low (F << 1), equation (5) can be simplified by expanding the exporential and evaluating the resulting integrals to give

$$S' = 1 - F/2 \tag{6}$$

When F << 1, F is simply the fraction of molecules isomerized in one transit through the beam. At a higher extent of photoisomerization, F can be interpreted as the number of absorptions that an equivalent non-photolabile molecule would experience while passing through the beam.

298

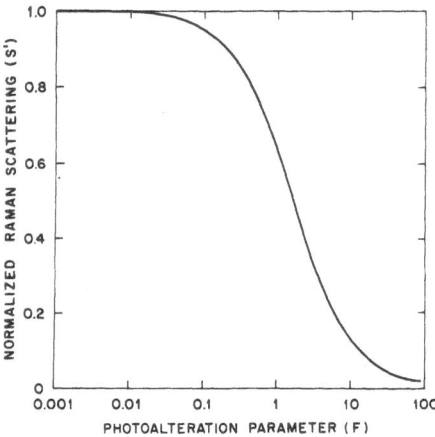

Figure 5. The normalized Raman scattering intensity S' is plotted as a function of the photoalteration parameter F, as calculated by numerical integration of equation (5).

It is important to experimentally ascertain the value of F since it is the key parameter in the rapid flow studies. When the extent of photolysis is low, F can be determined by measuring F_{bulk}, the fraction of the entire sample that has been photolyzed.

$$F_{bulk} = (2\ a\ P\ \phi\ \sigma_A\ T)/V \tag{7}$$

Here $\underline{a}$ is the radius of the jet stream, V is the total sample volume (cm^3) and T is the total irradiation time of the flowed sample. This expression may be simply derived by realizing that F_{bulk} is given by multiplying F by the number of times a molecule will, on the average, flow through the laser beam during the experiment. For a recirculating solution of visual pigments, F_{bulk} gives the fraction of molecules that have disappeared from the experiment because of bleaching. This is because the time required for bleaching is short compared with the sample recirculation time. In contrast, since the model compounds do not bleach following photoisomerization, F_{bulk} is in this case a measure of the cumulative photoisomerization in the sample.

Experimentally, the parameters in equation (4) must be carefully selected to provide a suitably small value of F. Clearly, the stream velocity $\underline{v}$ should be as high as is experimentally feasible. Also, $\underline{\omega}$ should be increased until the image of the beam waist on the monochromator slit is roughly equal to the selected slit width. It should be noted that optimizing F by varying $\underline{\omega}$ and $\underline{v}$ within the above limits results in little reduction in the ultimate signal to noise ratio of the experiment. The selection of laser wavelength and power are strongly coupled. As the laser is tuned into the absorption band, the Raman scattering is enhanced. However, to maintain a constant value of F, the laser power must be reduced. The optimum choice of excitation wavelength will depend on the ratio of the resonance Raman cross-section to the absorption cross-section. It should be noted that this treatment of the rapid-flow kinetics is also pertinent to the effect of photolysis on the operation of jet-stream dye lasers.

The photon-counting Raman spectrometer and argon ion pumped rhodamine 6G dye laser have been described previously [11]. The jet stream was formed by inserting a small variable speed pump (Micro-Pump) into a recirculating system with a 200 ml reservoir cooled in an ice bath. The solution was forced out of a teflon syringe needle having an internal diameter of 0.44 mm and the laser beam was focused on the free jet stream about 1 mm from the tip of the nozzle. The mean velocity was determined by measuring the bulk flow rate, which was about 1 ml/sec.

Results

The rapid-flow resonance Raman spectra of ethanol solutions of the protonated Schiff base derivatives of all-trans, 9-cis, 11-cis, and 13-cis retinals with n-butylamine in the 700 to 1700 cm^{-1} region are presented in Fig. 6. Previous studies of retinals and retinal derivatives suggest that the Raman lines involve contributions from C=NH$^+$ stretch (~1657 cm^{-1}), ethylenic C=C stretch (~1560 cm^{-1}), C-C stretches and CCH bends, the fingerprint region (1100-1350 cm^{-1}), C-CH$_3$ stretches (1000-1020 cm^{-1}) and CCH bend (968 cm^{-1}) [15,16,17]. The differences in the fingerprint regions of the Raman spectra of the various isomers demonstrate the sensitivity of the vibrational spectrum to the precise conformation.

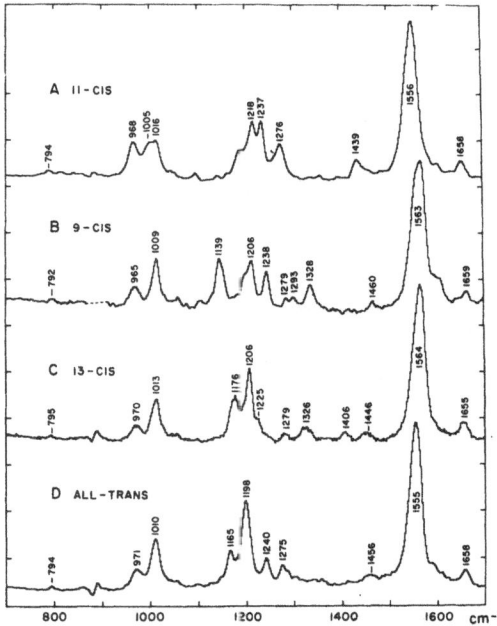

Figure 6. Rapid-flow resonance Raman spectra of protonated Schiff bases of retinals in ethanol (chloride salts of protonated n-butylamine derivatives). (a) 11-cis isomer. (b) 9-cis isomer. (c) 13-cis isomer. (d) all-trans isomer. The excitation wavelength was 514.5 nm. F was less than 0.03 and F$_{bulk}$ was less than 0.05 in all experiments.

In Fig. 7, we compare the resonance Raman spectrum of rhodopsin with that of the protonated n-butylamine Schiff base of 11-cis retinal (11-cis PSB). The close similarities in these two spectra confirm that the chromophore in rhodopsin has the 11-cis conformation. There is excellent correspondence between the 1660 cm^{-1} vibration in rhodopsin and the 1658 cm^{-1} vibration in the 11-cis PSB. The assignment of this vibration as the protonated Schiff base C=NH$^+$ stretch [10,18] demonstrates that 11-cis retinal is bound to opsin by a protonated Schiff base linkage. The prominent C=C ethylenic vibration is found at 1545 cm^{-1} in rhodopsin and at 1556 cm^{-1} in the 11-cis PSB. This frequency shift can be qualitatively correlated with the λ_{max} of these chromophores through the degree of π-electron delocalization [16,18]. In the fingerprint region, both spectra display prominent triplets with vibrations at 1216, 1240 and 1270 cm^{-1} for rhodopsin and vibrations at 1218, 1237 and 1276 cm^{-1} for the 11-cis PSB. The shoulder at 1195 cm^{-1} in the spectrum of the 11-cis PSB is due to a very intense line of the all-trans PSB which is observed as a result of the 5-10% degree of isomerization incurred as the flowed sample traversed the laser beam. In the 1010 cm^{-1} region we see two vibrations at 1000 and 1018 cm^{-1} in rhodopsin and 1005 and 1016 cm^{-1} in the 11-cis PSB. The 1000 cm^{-1} vibration has been tentatively assigned [12] as the C$_{13}$-CH$_3$ methyl stretch when retinal is in the 12-s trans configuration. The 1017 cm^{-1} region would then be due to the C$_5$- and C$_9$-methyl stretches. The last prominent vibration at 971 cm^{-1} in rhodopsin corresponds very well in frequency with the 968 cm^{-1} vibration of the 11-cis PSB. However, this vibration is much more intense in the protein.

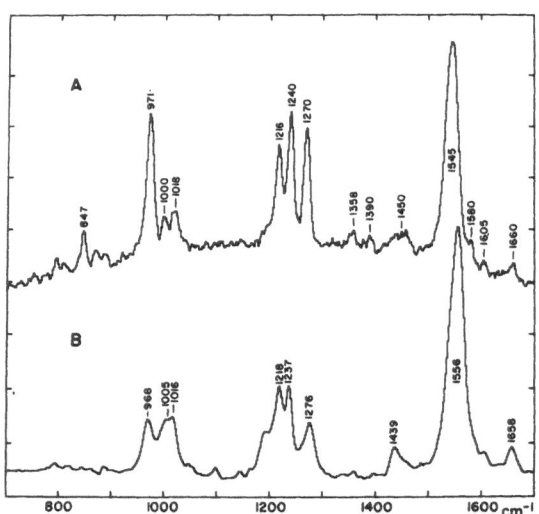

Figure 7. (A) Resonance Raman spectrum of purified rhodopsin (1.8 x 10^{-5} M) in 60 mM sodium phosphate at pH 7.0, 0.4% Ammonyx LO, 1 mM dithiothreitol, and 10 mM hydroxylamine. The laser power at 600 nm was 25 mW and the flow velocity was 350 cm/sec. The calculated value of F was 0.07. (B) Resonance Raman spectrum of the chloride salt of the protonated n-butyl amine Schiff base of 11-cis retinal in ethanol.

The comparison of these spectra leads to two important conclusions. First, the resonance Raman spectrum is very sensitive to the conformation of the retinal prosthetic group. This is illustrated by the striking differences amongst the spectra of the protonated Schiff base isomers. Second, there is remarkable agreement between the spectrum of the 11-cis PSB and of rhodopsin. This is a strong indication that the ground state conformation of retinal in rhodopsin is very similar to the conformation of the 11-cis protonated Schiff base molecule in solution. Thus, opsin does not induce gross distortions in the conformation of the 11-cis chromophore when they combine to form a visual pigment. Similarly, the resonance Raman spectrum of isorhodopsin, a synthetic pigment containing the 9-cis isomer of retinal, is very similar to that of the 9-cis protonated Schiff base.

The frequencies of the lines in a resonance Raman spectrum give the energies of the vibrational modes in the ground state of the chromophoric group. The intensities of these lines give additional information on the excited electronic states of the chromophore, since the intensities depend not only on the nature of the ground state vibrations, but also on the geometrical distortion of the molecule in the excited electronic state [13]. The distinct differences in relative intensities of some of the resonance Raman lines of rhodopsin and isorhodopsin compared to the model chromophores in solution suggest that the protein alters the geometry of the excited electronic state. This inference is consistent with the observation that the vertically excited states of retinals and of their protonated Schiff bases are highly polar [19] and hence should interact strongly with their environment.

Prospects

These experiments suggest that resonance Raman spectra of the intermediates in the photolysis of rhodopsin should reveal the sequence of conformational changes occurring in the retinal prosthetic group. One method of obtaining the Raman spectra of intermediates is to cool the rhodopsin sample to block the sequence of dark reactions following absorption of a photon (Fig. 2). Irradiation of the sample will then produce a photostationary steady-state mixture of rhodopsin and some of its photolytic intermediates. OSEROFF and CALLENDER [10] first used this method to obtain part of the Raman spectrum of prelumirhodopsin, the first intermediate. The longer-lived intermediates may be more difficult to examine with this technique because of the increasing complexity of the resultant mixtures. One possible solution would be to obtain time-resolved Raman spectra following irradiation with an intense bleaching pulse. In particular, the 1198 cm^{-1} line of the all-trans protonated Schiff base and the distinctive pattern of its fingerprint region, should unambiguously reveal when the photolyzed chromophore has attained the all-trans configuration.

The assignment of the vibrations in these resonance Raman spectra will be a very important step in the extraction of more detailed information about the conformation of retinal in the visual pigments. In addition to the spectra of the aldehyde, Schiff base and protonated Schiff base model compounds, studies of isotopically substituted retinals and of retinal analogs [20,21] will be useful in making these assignments. The correlation of these spectra with theoretical calculations [17] should then facilitate the interpretation of the spectra of rhodopsin and its photolytic intermediates.

302

We thank Mr. Jerry Johnson for expert technical assistance. This work was supported by grants from the National Eye Institute and the National Institute of General Medical Sciences. Richard Mathies is a Helen Hay Whitney Fellow (*present address: Department of Chemistry, University of California at Berkeley, 94720).

References

1. S. Hecht, S. Shlaer, and M.H. Pirenne, J. Gen. Physiol. 25, 819 (1942).
2. T. Ebrey and B. Honig, Quart. Rev. Biophys. 8, 129 (1975).
3. F.J.M. Daemen, Biochim. Biophys. Acta 300, 255 (1973).
4. G. Wald, Nature 219, 800 (1968).
5. T. Rosenfeld, B. Honig, M. Ottolenghi, and T. Ebrey, to be published.
6. T. Tomita, Quart. Rev. Biophys. 3, 179 (1970).
7. W.A. Hagins, Ann. Rev. Biophys. Bioeng. 1, 131 (1972).
8. L. Rimai, R.G. Kilponen, and D. Gill, Biochem. Biophys. Res. Commun. 41, 492 (1970).
9. A. Lewis, R.S. Fager, and E.W. Abrahamson, J. Raman Spec. 1, 465 (1973).
10. A.R. Oseroff and R.H. Callender, Biochemistry 13, 4243 (1974).
11. R. Mathies, A.R. Oseroff, and L. Stryer, Proc. Nat. Acad. Sci. USA 73, 1 (1976).
12. R.H. Callender, A. Doukas, R. Crouch, and K. Nakanishi, Biochemistry 15, 1621 (1976).
13. J. Tang and A.C. Albrecht, in Raman Spectroscopy, edited by H.A. Szymanski (Plenum Press, New York, 1970) 2, Chapter 2.
14. T.G. Spiro, Acc. Chem. Res. 7, 339 (1974).
15. L. Rimai, D. Gill, and J.L. Parsons, J. Amer. Chem. Soc. 93, 1353 (1971).
16. L. Rimai, M.E. Heyde, D. Gill, J. Amer. Chem. Soc. 95, 4493 (1973).
17. A. Warshel and M. Karplus, J. Amer. Chem. Soc. 96, 5677 (1974).
18. M.E. Heyde, D. Gill, R.G. Kilponen, and L. Rimai, J. Amer. Chem. Soc. 93, 6776 (1971).
19. R. Mathies and L. Stryer, Proc. Nat. Acad. Sci. USA 73, 000 (1976).
20. R. Crouch, V. Purvin, K. Nakanishi, and T. Ebrey, Proc. Nat. Acad. Sci. USA 72, 1538 (1975).
21. A. Kropf, B.P. Whittenberger, S.P. Goff, and A.S. Waggoner, Exp. Eye Res. 17, 591 (1973).

LASER-INDUCED FLUORESCENCE OF BIOLOGICAL MOLECULES

A. Andreoni, A. Longoni, C.A. Sacchi and O. Svelto
Centro di Studio per l'Elektronica Quantistica del C.N.R.
Instituto di Fisica del Politecnico, Milano, Italy
and
G. Bottiroli
Centro di Studio per l'Istochimica del C.N.R.,
Instituto di Anatoma Comparata, Pavia, Italy.

Abstract

A study is made of the laser-induced fluorescence properties of the complex formed by Quinacrine Mustard bound to Deoxyribonucleic Acid (DNA). Specific information is obtained on the binding mechanisms and on the base-pair sequences of the DNA. The relevance of these measurements in several fields from chemical physics to biology and medecine is discussed.

I. Introduction

The high degree of directionality and the short time duration obtainable with a laser beam are very useful properties in investigating large molecules and, in particular, molecules of biological interest [1] . Laser pulses shorter than ~ 1 nsec are indeed very useful for investigating events that occur in this, previously unexplorable, time domain. A high degree of directionality, on the other hand, is useful to focus the beam down to a spot of $\sim 1 \mu$m, a typical dimension of cells and cell organels. Note that, by contrast , a high degree of monochromaticity is not usually needed, since the absorption and fluorescence bands are generally very broad (> 50 nm) and featureless [2] . In previous works, a laser cytofluorometer, obtained by the combination of a tunable dye laser and a fluorescence microscope, was developed and successfully tested [3,4] . Since the time duration of our laser pulse is $\Delta \tau_p \cong 0.5$ nsec, events ranging from a few nanoseconds down to a fraction of nanosecond (0.1 ÷ 0.2 nsec) can be studied. The repetition rate of the laser (20 Hz) allows the use of a signal averager, and therefore makes for great improvement in signal-to-noise ratio.

In this work our instrument was used to study the fluorescence behaviour of acridine-DNA complexes. The acridines form a large family of fluorescent dyes [5,6] that easily bind to the DNA double helix. These complexes have been extensively studied during the last ten years, since the classic works of LERMAN on the Acridine-Orange-DNA complex [7] . The results that have been obtained have shown that these complexes have many interesting properties with implications of relevance in several fields from chemical physics to biophysics, pharmacology and medecine. It will be shown in this article that with our system we obtain new information on the mechanisms of the complex formation and on the base-pair sequences of the DNA. The fluorescence properties of an Acridine dye bound to the DNA of a chromosome have also been intestigated, thus obtaining a deeper knowledge of the DNA base-pair distribution in chromosomes.

II. The Acridine-DNA complex

Before proceeding it is appropriate, here to give a short description of the interesting properties of this complex.

The Acridines form a large class of fluorescent dyes, which absorb in the blue and emit in the green. For example, among the dyes directly related to our work, we mention Acridine Orange (AO), Acriflavine (AF), Proflavine (PF), 9-Amino Acridine, Quinacrine (QAC), and Quinacrine Mustard (QM). The chemical structure of some of them is shown in Fig. 1. The π-electrons

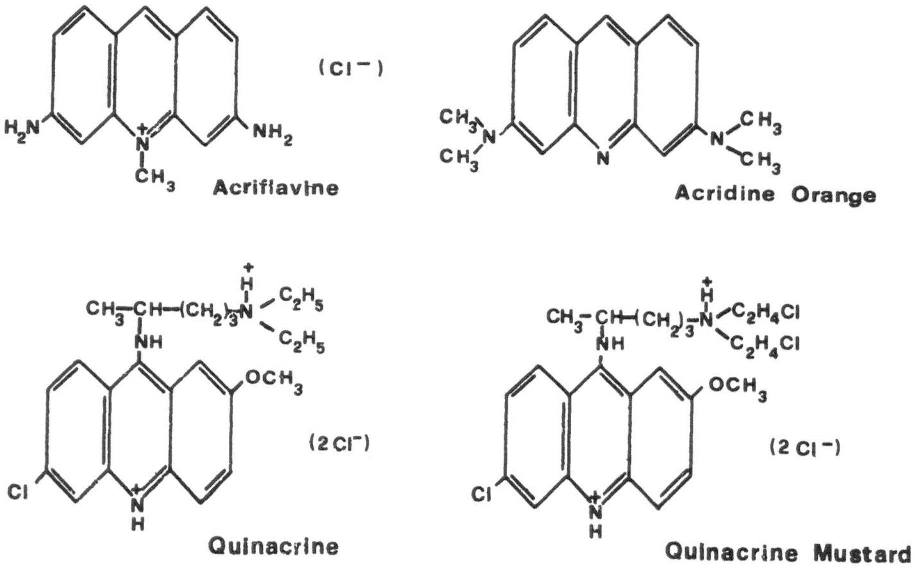

Fig.1. Chemical structure of a few dyes of the Acridine family

of the C atoms are essentially free to propagate along the acridine ring. The absorption and fluorescence spectrum of the dye of more direct interest to us (i.e. QM) are shown in Fig.2. Apart from this family of dyes, it is worth mentioning several antitumoral agents of clinical importance (Actinomycin D, Adriamycin, Daunomycin), as well as several drugs used in parasitic diseases (this class, besides Quinacrine itself includes Chloroquine, Miracil D, etc.).

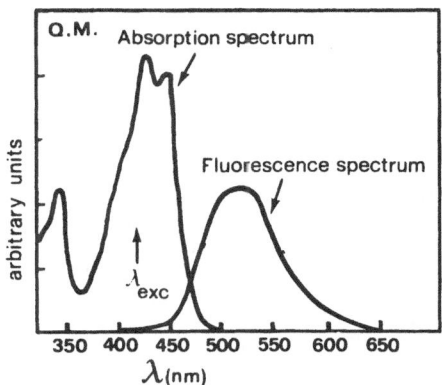

Fig. 2. Absorption and fluorescence spectra of Quinacrine Mustard (QM). The wavelength of the laser excitation used is also shown.

When a dye of the acridine family is added to a solution of DNA, an Acridine-DNA complex is formed. The binding constant of complex formation is such that, if the Acridine dye concentration is less than one tenth of the DNA phosphorus concentration, the concentration of the dye which remains free is less than 10^{-8} M. The mechanism of complex formation has been elucidated in the works of LERMAN on AO [7] . The binding mechanism is due tothe intercalation of the acridine ring between two adjacent base pairs of the DNA. The intercalation is further strengthened by an ionic binding between the N atom of the acridine ring and the O atom of the phosphate molecule. Note that the main axis of the acridine ring is oriented along the direction of the bond between the two bases of each pair.

A first very interesting property of this complex is that the fluorescence quantum yield ϕ of the bound dye molecule depends strongly on the two base-pairs in which the dye is intercalated [9,10] . When the dye intercalates two adjacent AT base pairs, ϕ increases by as much as a factor of 5 as compared to that of the same dye in a low-viscosity solution. When the dye intercalates either an AT-GC or a GC-GC sequence, the quantum yield is reduced by as much as a factor of 20, compared to the case of the dye bound to an AT-AT base pair sequence. [1] The rea-

---

[1]
This last effect does not occur with AO, where ϕ increases when the dye intercalates either an AT-AT, or an AT-GC, or a GC-GC sequence.

son for this base specificity is not clear yet. The increase in quantum yield is attributed to dye stiffening in the intercalation process (indeed, an increase in quantum yield is observed when the dye is dissolved in a high-viscosity solvent). The decrease in quantum yield in the presence of a Guanine residue is, on the other hand, attributed to the formation of a charge-transfer (CT) complex, the Guanine acting as the electron donor and the dye as the acceptor [11] . Evidence of this CT complex does not however appear conclusive [10,12] .

A second relevant property is observed when some of these dyes (notably QAC and QM, but with the exception of AO) are used to stain the eukaryotic (i.e. of higher organisms) chromosomal DNA: some parts of the chromosome are observed to fluoresce more strongly than others, giving a typical and reproducible pattern. The intensity ratio between more intensely and less intensely fluorescing regions may range between 1.5 and 2.5. These more fluorescent regions are called chromosome bands, [13], and are now widely used for chromosome analysis and characterization [14] . Figure 3, for example, shows in schematic form, the fluorescence bands observed in the M-chromosome of Vicia Faba. Also the reason for the appearance of these bands is not clear, and two interpretations have been put forward [9,15,16]: (i) The dye concentration inside the bands is higher than outside. This could be due to a larger dye intake inside the bands than outside. A different dye intake can be due to a different

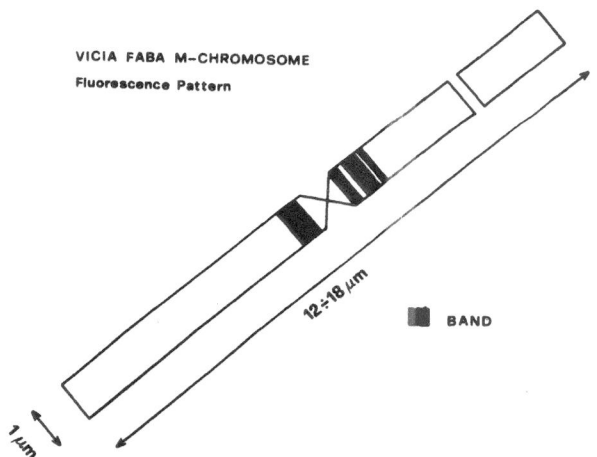

VICIA FABA M-CHROMOSOME
Fluorescence Pattern

12÷18 μm

■ BAND

1 μm

Fig. 3. Fluorescent bands of Vicia Faba M chromosome.

DNA accessibility, related either to its conformation or to a different amount of bound protein. (ii) For the second, more appealing interpretation, it is assumed that the fraction[2] of

---

[2]Note that the distance between two consecutive planes of bases in DNA is ~3.4 Å, while a typical dimension of a band is ~1μm. A large number of base-pair sequences is thus expected in the bands.

AT-AT base-pair sequence is larger inside than outside a band. We have, indeed, seen that the fluorescence quantum yield is drastically increased when the acridine dye intercalates an AT--AT base-pair sequence. To sum up, we can say that the fluorescent bands can be due either to a different dye concentration, or to a different quantum yield of the dye. The experimental data available do not permit of a clear distinction between these two possibilities.

The relevance of the two fluorescence properties of an Acridine-DNA complex discussed above (namely, base-dependent behaviour and chromosome bands) must be stressed. Indeed, a better understanding of the base-dependent behaviour, besides being a goal in itself, might help in elucidating the mutagenic properties of these dyes. An interesting possibility is that mutations depend on the change of the electronic state of Guanine induceed by the dye interaction (e.g. charge transfer formation) [10,17] . On the other hand, a better understanding of the chromosome bands might shed light on the AT-AT base-pair distribution in a chromosome. This is of obvious relevance to genetics and may also be useful in understanding anomalous banding patterns (as observed in the case of Burkitt's lymphoma [18]).

III. Plan of the experiments

To investigate some aspects of the Acridine-DNA complex formation, we decided to measure, with our apparatus, the fluorescence waveforms. We chose to study Quinacrine Mustard, since it gives chromosome bands with the highest contrast. For short-hand notation, we will call complex 1 that formed when QM intercalates an AT-AT sequence. We will call complex 2 that formed when QM intercalates either an AT-GC or a GC-GC sequence. Suppose now that the fluorescence waveform is exponential, for complexes both 1 and 2. In this case it is well known that the quantum yield is given by $\phi = \tau/\tau_{sp}$, where τ is the lifetime and τ_{sp} is the radiative lifetime of the transition. If τ_{sp} is assumed to be the same for the two types of complex [3] the corresponding lifetimes τ_1 and τ_2 will be such that $(\tau_2/\tau_1)=\phi_2/\phi_1$ (i.e. $\tau_2/\tau_1=$ =1/20) for QAC [9] }. Since τ_1 is expected to be in the range of 5 ÷ 20 nsec [10] we expect $\tau_2=0.2 ÷ 1$ nsec. Both times are thus within the reach of our apparatus. When both complexes 1 and 2 are simultaneously present with concentrations C_1 and C_2, the fluorescence waveform would be expected to consist of the sum of two exponential functions (Fig. 4), i.e. $f(t)=a_1\exp(-t/\tau_1)+$ $+a_2\exp(-t/\tau_2)$, where a_1 and a_2 are proportional to C_1 and C_2 respectively. The measurement of this waveform would then give the quantity $a_1/(a_1+a_2)$. If the dye is evenly distributed along a DNA chain containing all three possible base-pair sequences

_____
[3] This is to be expected on account of Förster's equation for the radiative lifetime (see, for instance, Ref. 19), since the shape and the absolute value of the absorption cross-section is about the same for complexes 1 and 2 [9] .

308

(i.e. AT-AT, AT-GC, GC-GC), the quantity $\alpha_1/(\alpha_1+\alpha_2)$ would then give the fraction of the AT-AT sequence of this chain. Since our beam can be concentrated into a spot smaller than the dimension of a typical fluorescent band, the fraction of the AT-AT base pair sequences can then be measured both inside and outside a band. We should thus be able to know whether fluorescent bands are due to a different fraction of the AT-AT sequences.

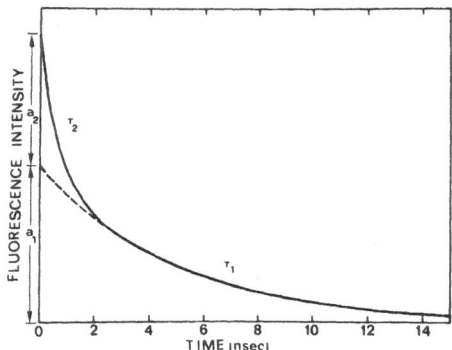

Fig. 4. The fluorescence time behaviour expected when the Acridine dye is bound to two different binding sites.

If the fluorescence is not given by the sum of two exponentials (as sometimes happens in complex molecular systems), we can still determine the characteristic decay waveforms $f_1(t)$ and $f_2(t)$ of each of the two complexes. When both complexes are present, as in natural DNA, the fluorescence waveform $f(t)$ will be expressed by $f(t)=\alpha_1 f_1(t)+\alpha_2 f_2(t)$ and the quantity $\alpha_1/(\alpha_1+\alpha_2)$, i.e. the fraction of AT-AT sequences can again be obtained. Furthermore the shape of $f_1(t)$ and $f_2(t)$ provides new information on the binding mechanisms responsible for the formation of the two complexes.

IV. Experimental apparatus and results

The experimental apparatus is described in more detail elsewhere [20]. It is sufficient here to mention that a nitrogen-pumped dye laser of special design, giving pulses of ∿0.5 nsec duration, is used as the excitation source. The laser is tuned at the wavelength of λ = 419 nm, which coincides with the peak of the absorption band of QM (see Fig.2). The laser beam is concentrated into a small spot (0.5 ÷ 1μm) by a fluorescence microscope (Fig. 5). The fluorescence waveform is monitored by an RCA type 70045 C photomultiplier with a ∿0.7 nsec time resolution (FWHM) and the electrical signal is then sent to a suitable signal averager. High signal-to-noise ratio and good accuracy (5%) are obtained, averaging over ∿40,000 laser shots. On account of the repetition rate of our laser (∿20 Hz), a measuring time of ∿35 minutes is required for each waveform.

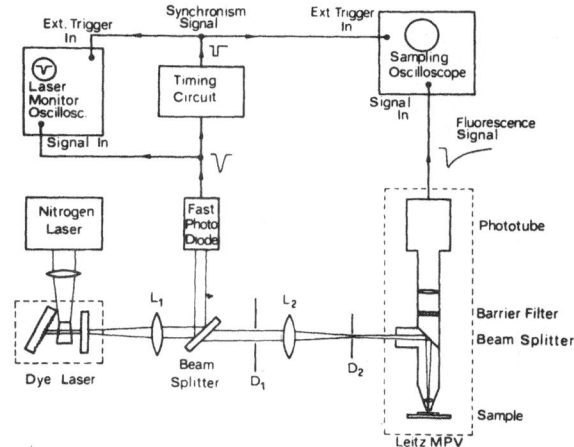

Fig. 5. Schematic apparatus of the laser fluorescence microscope pe

Two preliminary tests were performed on our apparatus for the problem to be studied. In the first one the response of the system to the laser pulse itself was tested inter alia to establish the regime of linear behaviour. A typical response curve (prompt response) in this linear regime is shown in Fig. 6. The

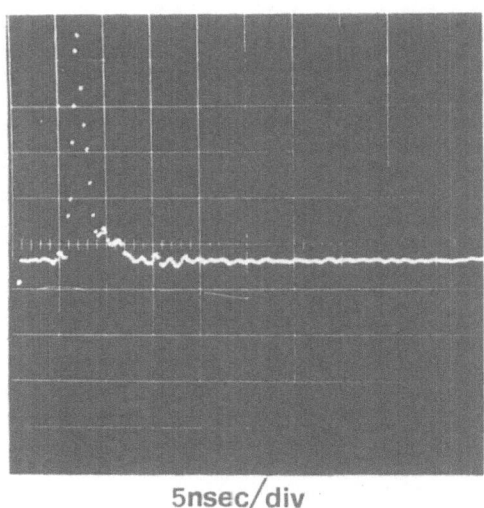

5nsec/div

Fig.6. Response of the apparatus to the exciting laser pulse

response time of~0.9 nsec (FWHM) is due to combination of the
laser pulse duration, the response time of the phototube, and
the response time of the signal averager (~150 psec). In the se
cond preliminary test the fluorescence decay of QM in an acetate buffer solution was studied. The decay curve is observed to
be a pure exponential with a decay time $\tau_0 \cong 5$ nsec. This is shown
in Fig. 7, in which a logarithmic plotting of the curve is shown.

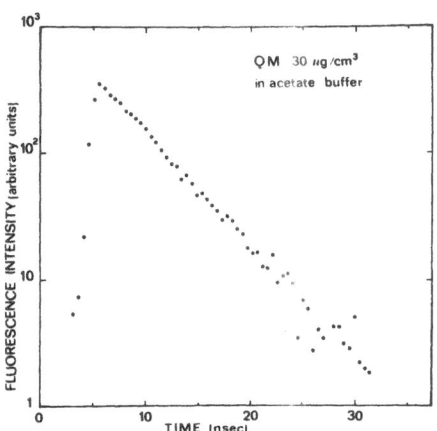

Fig. 7. Decay curve of QM in an acetate buffer solution

After these preliminary tests, the fluorescence decays of
complexes 1 and 2 were determined. For this purpose, QM was
bound to PolydA-PolydT and PolydG-PolydC, i.e. to synthetic DNA,
in which all base pairs are of the AT or GC type respectively.
For complex 1 (i.e. QM bound to PolydA-PolydT), an exponential
decay with a decay time $\tau_1 \cong 25$ nsec is observed. The increase in
quantum yield as compared to QM in the buffer solution is thus
seen to be equal to (τ_1/τ_0) i.e.~5. This is consistent with
the result achieved by PACHMAN and RIGLER for QAC. The decay
curve obtained with complex 2 (i.e. QM bound to PolydG-PolydC)
is not exponential but is, rather, given by the superposition
of a fast peak with a decay time $\tau_2' = 0.5 + 0.2$ nsec and a slower
tail (with a decay time of $\tau_2'' \cong 5$ nsec)[4] (Fig. 8). A more detail-
led analysis on this point, to be published elsewhere [21], has
led us to suggest that the decay occurs through the formation
of a non-fluorescent excimer (probably of the charge-transfer
type) between QM and the Guanine.

The next case considered is that of cell nuclei of known AT
percentage; as discussed in the previous section, the observed

[4] Note that, if a simple decay were obtained, and if the data on
the quantum yield of QAC are taken as reference [9], the expected decay time would be $\tau_2 \cong (\tau_1/20) = 1.25$ nsec.

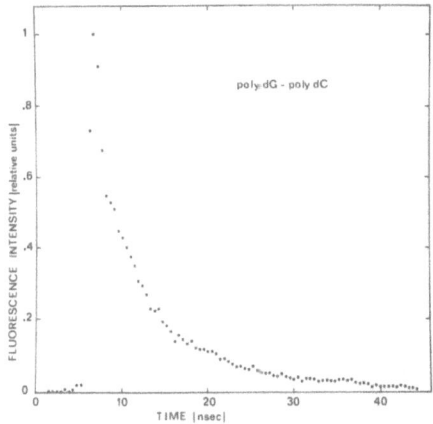

Fig. 8. Decay curve of QM bound to PolydG-PolydC

decay curve is expected to be given by a suitable superposition
of the curves of complexes 1 and 2. Indeed, as shown in Fig. 9,
the decay curve is made by the superposition of a fast peak and
a tail, which should correspond to the time behaviour of the

Triticum Vulgare:52.5%AT

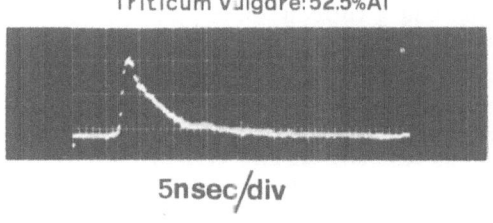

5nsec/div

Fig. 9. Decay curve of QM bound to a DNA of a given AT percenta
ge.

complex. The decay time of this tail should thus be equal to
$\tau_1 \gtrsim 25$ nsec. In fact, the decay time of the tail in Fig. 9 is seen
to be only ~ 5.5 nsec. The reduction of this time is tentatively
attributed to the presence of proteins. Further work is needed
on this point, to be able to measure the fraction of the AT-AT
sequences in a given sample.

Preliminary measurements have also been performed on the fluo
rescent bands of Vicia Faba S chromosomes. The fluorescent decay
curves obtained by focusing the laser beam inside or outside a
fluorescent band of the chromosome are shown in Fig. 10. The two
curves are seen to be different, the curve obtained from the band
containing a smaller fraction of the fast initial peak (which is

312

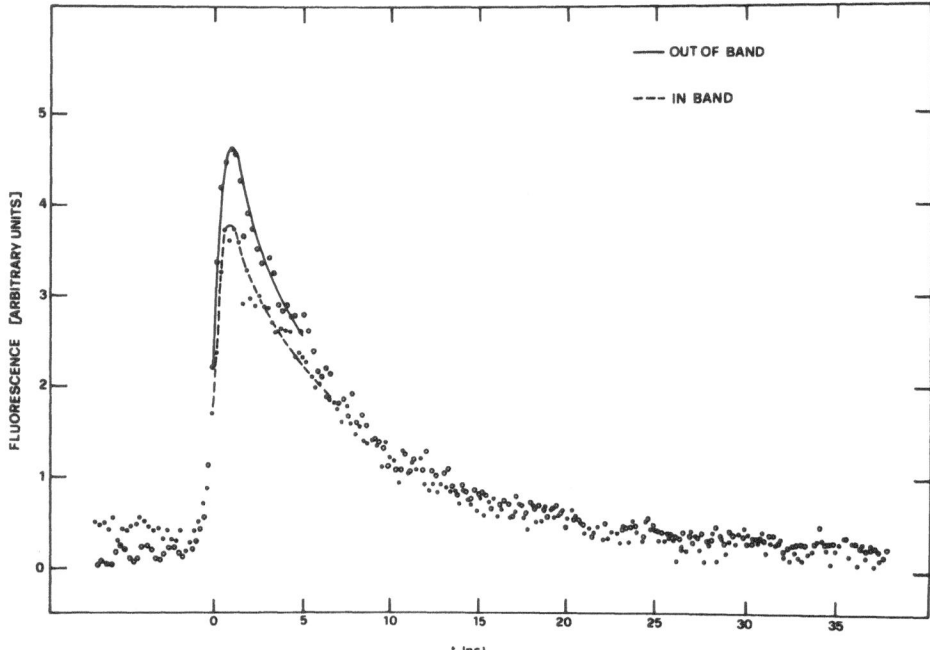

Fig. 10. Decay of the fluorescence emitted by two points, one inside and the other outside the fluorescent band of Vicia Faba S Chromosome.

indicative of the fraction of complex 2 present). Figure 10 is thus, to our knowledge, the first direct indication that a larger fraction of AT-AT sequences is present inside the band than outside it. Obviously, in this case too, more detailed work is needed to obtain a reliable measurement of these fractions.

V. Conclusions

A long-range research program on laser-induced fluorescence microscopy of the Acridine-DNA complexes is here discussed. In particular, we have addressed ourselves to the following problems: (i) mechanisms of complex formation, (ii) measurement of the fraction of AT-AT sequences of the DNA and (iii) chromosome fluorescent bands. The implications of this program for several fields, ranging from chemical physics to biology and medecine, are stressed.

The measurements given here should be considered initial mea surements in this research program. From these measurements, ho wever, we have been able to show that the complex formed when the Acridine QM is bound to PolydG-PolydC shows an interesting time behaviour of the fluorescence, which we attribute to an ex cimer formation. Preliminary measurements to show the feasibi-

lity, with our system, of measuring the AT-AT sequence percenta
ge in cells and chromosomes are also given.

The results already obtained show that our system is indeed
very useful for obtaining information previously not available
on the biological systems considered.

References

1. See, for example, Laser in Physical Chemistry and Biophysics
   ed. by J.Joussot-Dubien (Elsevier Scientific Publ.Co.,Amster
   dam, 1975)
2. A notable exception occurs when the molecule is kept at very low temperature (a few degree ^{0}K). See, for example, L.
   A. Bykovskaya, R.I. Personov and B.M.Kharlamov, Chem.Phys.
   Letters 27, 80 (1974)
3. C.A.Sacchi, O.Svelto and G.Prenna, Histochemical Journal 6,
   251 (1974)
4. S.Cova, G.Prenna, C.A.Sacchi, and O.Svelto, in Excited States of Biological Molecules ed. by J.B.Birks, pp.223-232
   (Wiley-Interscience, New York, 1975)
5. A.Albert: The Acridines, 2 nd ed. (Edward Arnold Publishers
   Ltd, London, 1966).
6. Acheson: Acridines (Interscience, New York, 1956)
7. L.S.Lerman, J.Mol.Biol. 3, 18 (1961)
8. Ritva-Kaisa Selander, Biochem. J. 131, 749 (1973)
9. U.Pachman and R.Rigler, Exp. Cell. Res. 72, pp.602-608 (1972)
10. J.P.Schreiber and M.P.Daune, J.Mol. Biol. 83, 487 (1974)
11. R.K. Tubbs, W.E.Ditmars, and Q.Van Winkle, J.Mol.Biol. 9,545
    (1964)
12. Edward J.Modest and Sisir K.Sengupta in "Fluorescence Techni-
    ques in Cell Biology" ed. by A.A.Thaer and M.Sernetz (Sprin-
    ger-Verlag Berlin-New York 1973) pp. 125-134
13. T.Casperson et al. Exp.Cell. Res. 49, 219 (1968)
14. T.Casperson et al.,Chromosoma 30, 215 (1970)
15. Bernard Weisblum and Pieter L. de Haseth, Proc.Nat.Acad.Sci
    USA 69, 629 (1972)
16. J.R.Ellison and H.J.Barr, Chromosoma, 36, 375-390 (1972)
17. B. Pullman and A. Pullman, Rev. Mod. Phys. 32, 428 (1960)
18. G.Manolov and Y.Manolova, Nature, Lond. 237, 33 (1972)
19. W.W.Ware and B.A.Baldwin, J.Chem.Phys. 40, 1703 (1964)
20. A.Andreoni et al., in Lasers in Physical Chemistry and Bio-
    physics ed. by J.Joussot-Dubien (Elsevier Scientific Publ.
    Co., Amsterdam, Oxford, New York, 1975) pp. 413-424
21. A.Andreoni, G.Bottiroli, A.Longoni, C.A.Sacchi, and O.Svelto
    to be published

FLUORESCENCE SPECTROSCOPY APPLIED TO DYNAMICS AND STRUCTURE OF BIOPOLYMERS

M. Ehrenberg and R Rigler
Department of Medical Biophysics
Karolinska Institute, Stockholm, Sweden

Abstract

Pulsed fluorescence |1,2,3| and fluorescence correlation spectroscopy |4,5,6,7| in the analysis of macromolecular conformations and motion are briefly reviewed and results from pulsed measurements on fluorescence labelled phenylalanine-transfer RNA (tRNA^{Phe}) from yeast are presented.

I. Pulsed Fluorescence Spectroscopy

A structural conformation can often be fingerprinted by the fluorescence lifetime of an inserted label or a naturally fluorescing group |3|.
Three conformational states in dynamic equilibrium

$$A_1 \rightleftharpoons A_2 \rightleftharpoons A_3$$

where each structure has a characteristic fluorescence lifetime τ_i; i=1,2,3; give in pulsed experiments birth to a relaxation curve with three exponentials

$$I(t) = \sum_{i=1}^{3} a_i e^{-t/\tau_i},$$

where the weighting factor a_i is related to the concentration of A_i, $[A_i]$:

$$a_i = \text{const}[A_i]\varepsilon_i k_i^e; \quad i=1,2,3.$$

ε_i is the extinction coefficient and k_i^e the emission rate of species i.

Identification and quantitative evaluation of lifetimes and weighting factors in pulsed fluorescence experiments give the probability $P_i = [A_i] / \sum_{j=1}^{3} [A_j]$ of each structural state. The influence of conditions changing P_i can thus be studied. This analysis demands accurate determination of the parameters in a sum of exponentials. A high experimental accuracy is necessary. The single photon counting technique offers time resolution in the subnanosecond region and can give the large dynamic range necessary for the resolution of complicated relaxation curves. Light pulses for excitation of moderate intensity and frequency ($\sim$100 kHz) and with a rapid decay over many decades are ideal.

Pulsed fluorescence gives information about structure through the rotational Brownian motion of macromolecules. The simple case of a symmetric rigid rotor is seen in Fig. 1. The exciting light pulse is linearly polarized in the z-direction. The fluorescence emission is probed with respect to its polarization in the z- and y- directions. The anisotropy $r=(I_\parallel - I_\perp)/(I_\parallel +2I_\perp)$, where $I_\parallel$ and $I_\perp$ are the fluorescence components analyzed parallel with and perpendicular to the polarization vector of excitation, consists of three exponentials. The amplitudes b_i are related to the orientation of the absorption vector $\bar{\mu}_a$ and the emission vector $\bar{\mu}_e$ in a molecular coordinate system. The exponents $1/\tau_{ri}$ are linear combinations of D_1, D_2 and D_3 which are the rotational diffusion constants in that coordinate system which diagonalizes the diffusion tensor |16|.

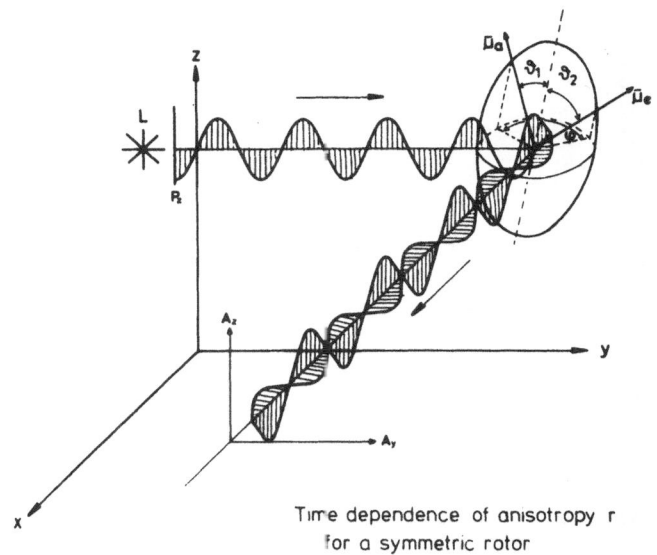

Time dependence of anisotropy r
for a symmetric rotor

Fig. 1 Rotational diffusion of molecules as detected in pulsed fluorescence experiments.

Explicitly we have for the anisotropy $r(t)$:

$$r(t) = \sum_{i=1}^{3} b_i e^{-t/\tau_{ri}},$$

with

$$b_1 = \frac{6}{5} \sin\theta_1 \cos\theta_1 \sin\theta_2 \cos\theta_2 \cos\phi,$$

$$b_2 = \frac{3}{10}(\sin\theta_1)^2(\sin\theta_2)^2[(\cos\phi)^2-(\sin\phi)^2],$$

$$b_3 = \frac{1}{10}[3(\cos\theta_1)^2-1][3(\cos\theta_2)^2-1] \quad \text{and}$$

$$1/\tau_{r1} = 5D_1 + D_3, \quad 1/\tau_{r2} = 2D_1 + 4D_3, \quad 1/\tau_{r3} = 6D_1, \quad D_1 = D_2.$$

A scheme of our single photon counting system for pulsed fluorescence spectroscopy is shown in Fig. 2.

PULSED SINGLE PHOTON-SPECTROMETER.

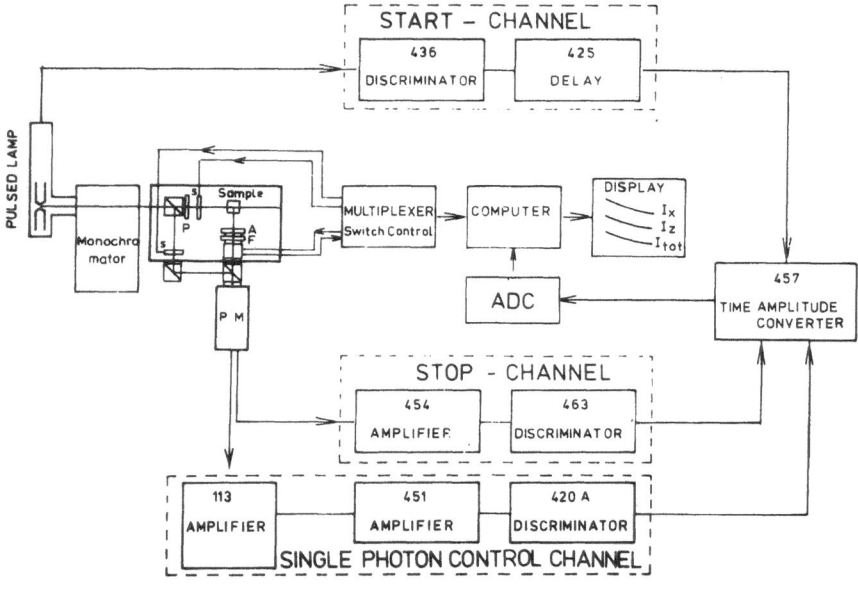

Fig. 2 Blockscheme of the pulsed single photon counting system. Light-pulses are obtained either from a free running spark gap lamp or from switching a continuous argon laser beam with a rotating prism. In the latter case pulses with a FWHM about one nanosecond can be obtained. The polarized fluorescence components and the system response are recorded in a sequence iterated many times during an experiment. This makes absolute calibration between different relaxation curves possible and minimizes experimental error.

II. Fluorescence Correlation Spectroscopy

Analysis of molecular motion in a time interval considerably longer than the fluorescence lifetime is impossible in pulsed fluorescence experiments. The application of fluorescence correlation spectroscopy |4,5|, where this limitation is avoided, to the analysis of rotational motion of large polymeric structures |6,7| is therefore a project of considerable interest. A sharply focused linearly polarized laser beam traverses a thin sample volume |4,7|, (Fig. 3):

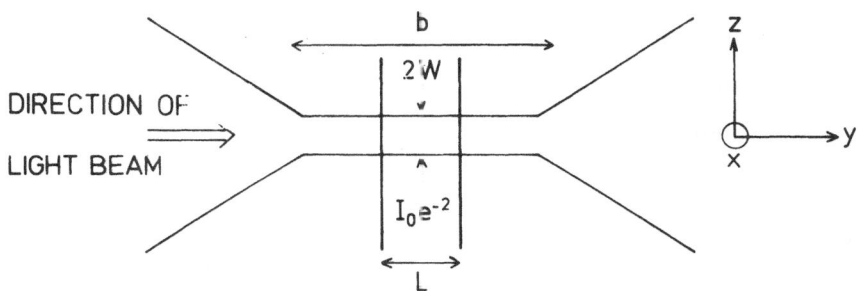

Fig. 3 W is the laser beam radius at focus, L is the thickness of the cuvette and b is the confocal length. The exciting light is linearly polarized in the z-direction.

With angular homogeneity in the detection system, i.e. the probability of photon detection is independent of the orientation of the emission vector, correlations in the direction of the absorption vector are recorded as shown in Fig. 4 and this leads to simple theoretical expressions for the rotational motion |6|. Fig. 4 shows the situation for a molecule with position $\bar{r}_o$ and orientation ω_o at time zero (where ω_o is a short notation for three Euler angles),which is excited with the probability $f_{exc}(r_o,\omega_o)=const(\bar{\mu}_a(\omega_o)\hat{z})^2 I(\bar{r}_o)$. Given this position and orientation at t=0 the particle is at $(\bar{r},\omega)$ at time t_1 with the conditional probability $P(\bar{r},\omega,t_1|\bar{r}_o,\omega_o)$ and has the excitation probability $f_{exc}(\bar{r},\omega)=const(\bar{\mu}_a(\omega)\hat{z})^2 I(\bar{r})$, where $I(\bar{r})$ is the space dependent intensity of the incoming light. The function $P(\bar{r},\omega,t_1|\bar{r}_o,\omega_o)$ depends on the molecular rotational diffusion and therefore contains information about structure. The autocorrelation g of the rate of photon emission i(t), $g(t_1)=<i(t+t_1)i(t)>$, is in the case of a symmetric rotor given by

$$g(t_1) = <i>^2(1+\frac{1}{N}\left(\sum_{j=1}^{3} 2b_j\ e^{-t_1/\tau_{rj}}+\frac{1}{1+(4Dt_1/W^2)} - \frac{9}{5}e^{-t_1/\tau}\right))$$

$$\text{for } \frac{1}{\tau} >> \frac{1}{\tau_{rj}} >> \frac{4D}{W^2}$$

The sum of exponentials is, except for the factor 2/N, identical with what is found in pulsed experiments (Fig. 1). D is the coefficient of translational diffusion, N is the average number of particles in the laser beam and τ is the lifetime of fluorescence. Since the exponential $e^{-t/\tau}$ is an additive term $\tau_{rj} \gg \tau$ can be evaluated in a correlation experiment in contrast to a pulsed experiment |6,7|.

FLUORESCENCE INTENSITY CORRELATIONS DEPENDING ON ABSORPTION PROBABILITY.

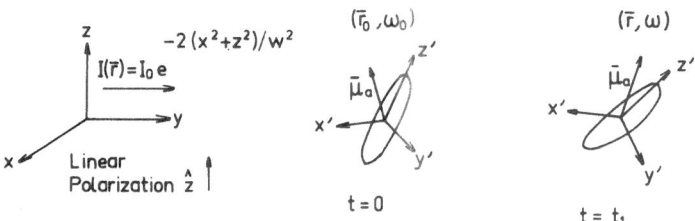

Fig. 4 Time dependent correlations in the orientation of the absorption vector.

Thus the information available using fluorescence correlations is essentially the same as that which can be obtained by pulsed fluorescence. However, in correlation spectroscopy the time interval is not limited by the fluorescence lifetime.

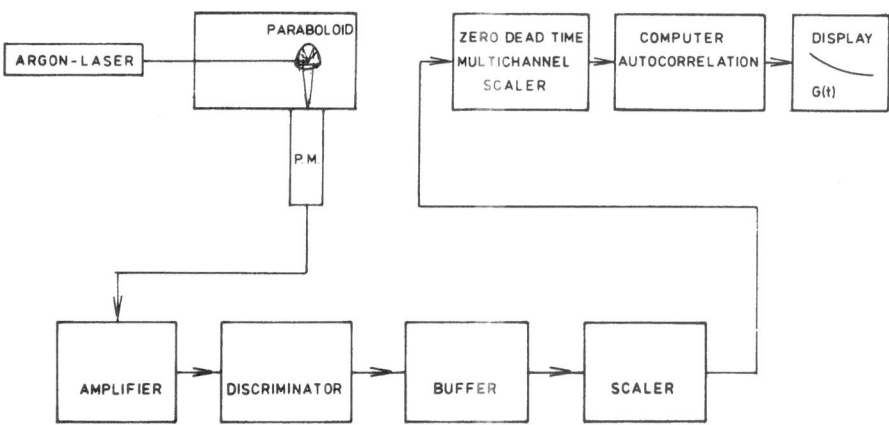

Fig. 5 Blockscheme for fluorescence correlation measurements. The thin sample cuvette is placed in a reflecting paraboloid. The fluorescence emission is collected over a solid angle of 2π.

The signal related to the correlations of rotational and translational diffusion is small in relation to the background caused by the average intensity <i>. A small number N of fluorescing particles has to be illuminated with high intensity. Photodegradation of the fluorescing dyes puts an ultimate limit to the experimental accuracy |ᴸ|.

Our experimental system is based on photon counting and computation of the autocorrelation function in a small computer as shown in Fig. 5.

The optical part of the measurement system for pulsed fluorescence and fluorescence correlation spectroscopy is shown in Fig. 6.

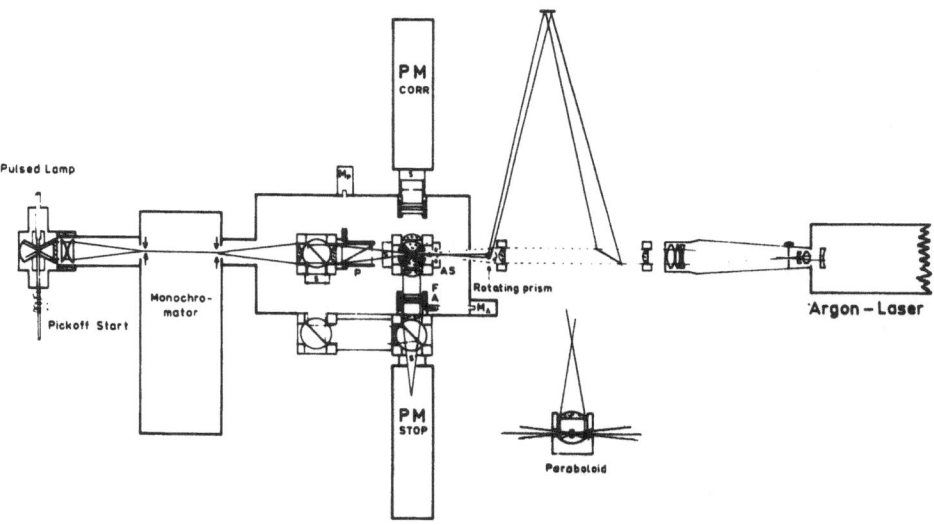

Fig. 6 Optical part of the combined system for pulsed fluorescence and fluorescence correlation spectroscopy. P is a polarizer on rotatable mount, F is a cut off filter and A is an analyser on a rotatable mount. S are electronic shutters for sequential recording of lamp pulse and sample response at the photomultiplier (PM). In correlation experiments the cavity in the measurement box is exchanged for the reflecting paraboloid to the right.

III. Structural Equilibria of tRNA^Phe in Solution Analysed by Pulsed Fluorescence Spectroscopy

We have, in pulsed experiments, investigated the fluorescence from the label ethidium bromide in position 37 in the anticodon loop or in positions 16 and 17 in the dhU-loop of tRNA^Phe |18|.

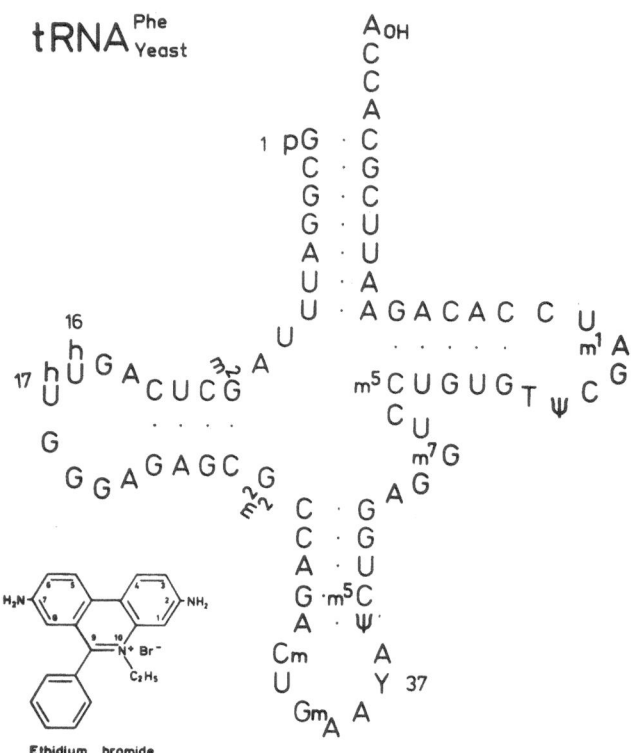

tRNA$_{\text{Yeast}}^{\text{Phe}}$

Ethidium bromide

Fig. 7 Primary and secondary structure of tRNA$_{\text{yeast}}^{\text{Phe}}$ [17] and the fluorescence label ethidium bromide.

There are twenty different transfer-RNA molecules, one for each aminoacid. The aminoacid, in our case phenylalanine, is coupled to the CCA-end of its tRNA in a reaction catalysed by phenylalanine-tRNA synthetase. The aminoacid is, in a forthcoming step of the protein synthesis, transferred from the CCA-end of its tRNA to a growing peptide chain at the ribosome. The correct aminoacid according to the genetic code is selected by an interaction between a messenger-RNA triplett and the complementary nucleotide triplett in the anticodon loop of tRNA. This interaction is highly specific in the ribosomal framework. The chemical pathways leading to the production of polypeptides according to the DNA-code are well known. However, a physical understanding of the remarkable fidelity of biological selection is still lacking.

The crystal structure of tRNA$_{\text{yeast}}^{\text{Phe}}$ determined by X-ray diffraction |9,10| did not immediately lead to an elucidation

of the molecular recognition mechanisms involved. It is clear
that the tRNA-structure is a system whose dynamics must be
taken into consideration. The structure must also be related
to its interactions with the aminoacyl-tRNA synthetases and
with the ribosomal system.

Experiments indicate that conformational changes of the
tRNA-molecule play an important role in synthetase binding |11|
upon aminoacylation |12| and in complementary anticodon-
messenger RNA codon interaction at the ribosome |13|.

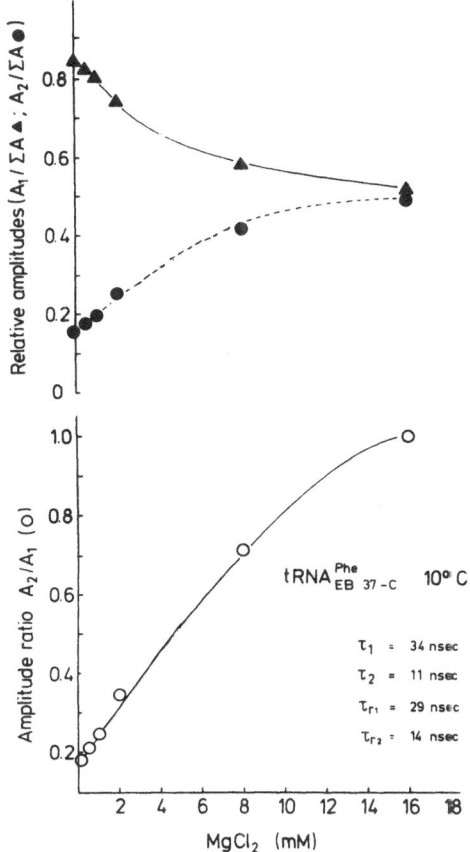

Fig. 8 Shift of equilibrium
between two conformations of
tRNA^{Phe} labelled with ethidium
at position 37 by adding Mg^{2+}
at 10°C. τ_1 and τ_2 are the
fluorescence lifetimes and τ_{r1}
and τ_{r2} are two rotational
relaxation times; one for each
lifetime.

We have found two distinct
and invariant lifetimes of the
ethidium label in the anticodon loop at 10°C which indicate the existence of two
conformations of free tRNA^{Phe}
in solution. The weighting
factors for the lifetimes
could be related to the
relative concentrations of
the two species. In Fig. 8 we
can see how Mg^{2+} ions in
increasing amounts shift the
equilibrium, measured by the
amplitude ratio A_2/A_1, towards
a conformation characterized
by a shorter rotational relaxation time and a shorter
fluorescence lifetime. tRNA^{Phe}
labelled at position 16/17
responded to increasing Mg^{2+}
concentration as tRNA^{Phe}
labelled at position 37 indicating that the observed conformational change is not only
related to the anticodon loop
of tRNA^{Phe}.

At higher temperatures a third lifetime appears which might be related to the melting of the tertiary solution structures. The Mg^{2+} dependence of the probabilities for the three structures at 35°C can be seen in Fig. 9.

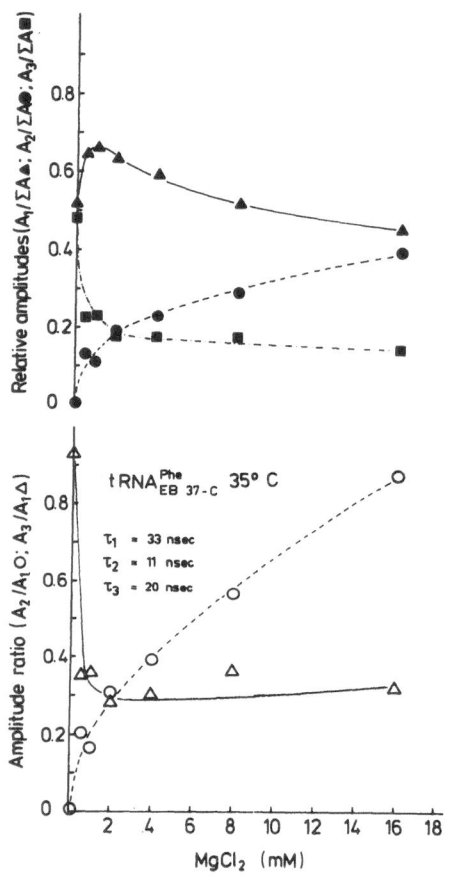

Fig. 9 Shift of equilibria between three conformations of tRNA^{Phe} labelled with ethidium at position 37 by adding Mg^{2+} at 35°C. From the three weighting factors A_i; i=1,2,3; corresponding to the three lifetimes τ_1=33 nsec, τ_2=11 nsec and τ_3=20 nsec respectively the Mg^{2+} dependent conformation probabilities

$$P_i = [A_i]/(\sum_{j=1}^{3} [A_j])$$ have been

determined.

Measurement of the chemical kinetics of the ethidium labelled tRNA^{Phe} with stopped flow and temperature jump techniques |14| shows one chemical relaxation time at 10°C and two relaxation times at 25°C (Fig. 10). At 25°C one relaxation time increases from about 10 ms to 100 ms and the other increases from about 100 ms to 600 ms when Mg^{2+} is varied from effectively zero concentration to 20 mM indicating that the transition rates between the structural states are slowed down. The results give further evidence for the notion that tRNA^{Phe}

in solution has got two well defined tertiary conformations where Mg^{2+} ions in increasing amounts function as allosteric effectors shifting the equilibrium towards a more compact structure.

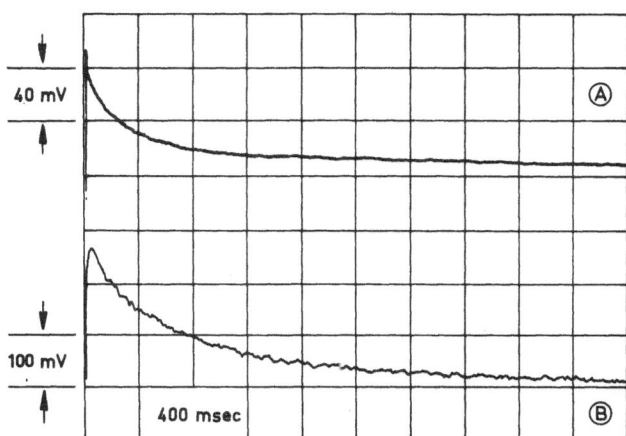

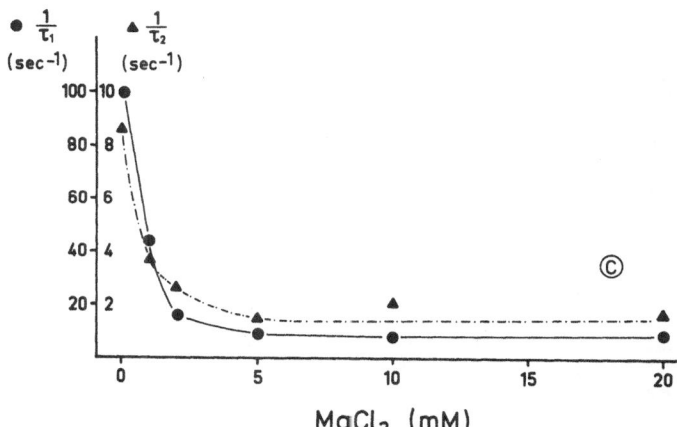

Fig. 10 Time dependent change of the fluorescence intensity of tRNA^{Phe} labelled at position 16/17 with ethidium:
   a) after rapid mixing of tRNA^{Phe} and Mg^{2+} (final concentrations: tRNA 1×10^{-6} M, Mg^{2+} 20 mM),
   b) after a temperature jump of $\Delta T = 7^{\circ}C$ at $25^{\circ}C$ of a mixture of tRNA $(1 \times 10^{-6}$ M) and Mg^{2+} (20 mM),
   c) dependence of the reciprocal relaxation time $1/\tau$ on the concentration of Mg^{2+}.

We can summarize the findings in a scheme which describes the experimental situation at 10°C but has to be enlarged by a third structure at higher temperatures.

$$
\begin{array}{ccc}
T_1 & \underset{L_o}{\rightleftharpoons} & T_2 \\[4pt]
\updownarrow & \underset{L_o C}{\rightleftharpoons} & \updownarrow \\[4pt]
T_1 M_1 & & T_2 M_2 \\[2pt]
\vdots & & \vdots \\[2pt]
T_1 M_n & \underset{L_o C^n}{\rightleftharpoons} & T_2 M_n
\end{array}
$$

T_1 and T_2 are the two tRNA conformations without Mg^{2+} characterized by the equilibrium constant

$$
L_o = [T_2]/[T_1].
$$

We have for simplicity assumed n identical binding sites for Mg^{2+} at each structure with association constants K_1 and K_2. We have introduced $C=K_2/K_1$. The equilibrium relations

$$
K_i = \frac{j}{n-j+1} \cdot \frac{[T_i M_j]}{[M][T_i M_{j-1}]} \quad , \quad j=1,2\ldots n; \; i=1,2 \quad ,
$$

are valid. $T_i M_j$ is tRNA structure i which has bound j Mg^{2+} ions and M is the Mg^{2+} concentration. The amplitude ratio a_1/a_2 found in pulsed experiments is now, since extinction coefficients and emission rates are the same in T_1 and T_2, given by

$$
a_1/a_2 = ([T_1] + \sum_{i=1}^{n} [T_1 M_i]) / ([T_2] + \sum_{i=1}^{n} [T_2 M_i]),
$$

which can be expressed as

$$
a_1/a_2 = (1+[M]K_1)^n / (L_o (1+[M]K_2)^n).
$$

With the chemical rate constant $\overrightarrow{l_o}$ for the transition $T_1 \longrightarrow T_2$ and assuming that only the "backrates" from T_2 to T_1 are effected by Mg^{2+} binding we obtain the expression

$$
1/\tau([M]) = (1 + \frac{(1+[M]K_1)^n}{L_o (1+[M]K_2)^n}) \; \overrightarrow{l_o}
$$

for the Mg^{2+} dependent relaxation time between the two structures as seen in temperature jump and stopped flow experiments at 10°C.

Knowledge of the behaviour of the free tRNA^{Phe} structure in solution is a starting point for applying our fluorescence methods to tRNA^{Phe} interacting with phenylalanine-tRNA synthetase and with the ribosomal system where Mg^{2+} ions play an important role in the mRNA-codon induced structural change of tRNA^{Phe} |13|. Here different structural states of the tRNA-molecule seem to be important for its proper selection by the mRNA-codon |15|.

Acknowledgement

This work has been supported by the Swedish Cancer Society, the Natural Science Research Council and by funds of the K. and A. Wallenberg foundation.

References

1. T.J.Chuang, K.B.Eisenthal: J.Chem.Phys. 57, 5094 (1972)
2. J.Yguerabide: Methods in Enzymology vol. 26 (ed. C.H.W.Hirs, S.N.Timasheff, Academic Press, New York 1972) part C, p. 498.
3. R.Rigler, M.Ehrenberg: Quart.Rev.Biophys. 9, 1 (1976)
4. D.Magde, E.L.Elson, W.W.Webb: Biopolymers 13, 29 (1974)
5. E.L.Elson, D.Magde: Biopolymers 13, 1 (1974)
6. M.Ehrenberg, R.Rigler: Chem.Physics 4, 390 (1974)
7. M.Ehrenberg, R.Rigler: Quart.Rev.Biophys. 9, 69 (1976)
8. M.Ehrenberg, R.Rigler: Chem.Phys.Lett. 14, 539 (1972)
9. S.H.Kim, F.L.Suddath, G.J.Quigley, A.McPherson, J.L.Sussman, A.H.J.Wang, N.C.Seeman, A.Rich: Science 185, 435 (1974)
10. J.D.Robertus, J.E.Ladner, F.T.Finch, D.Rhodes, R.S.Brown, B.F.Clark, A.Klug: Nature 250, 546 (1974)
11. R.Rigler, U.Pachmann, R.Hirsch, H.G.Zachau: Eur.J.Biochem. 65, 307 (1976)
12. M.Caron, N.Brisson, H.Dugas: J.Biol.Chem. 251, 1529 (1976)
13. U.Schwarz, H.M.Menzel, H.G.Gassen: Biochemistry 15, 2484 (1976)
14. M.Eigen, L.DeMaeyer: Technique of Organic Chemistry, vol. 8 (ed. S.L.Friess, E.S.Lewin, A.Weissberger, J.Wiley&Sons, New York 1963)
15. C.G.Kurland, R.Rigler, M.Ehrenberg, C.Blomberg: Proc.Nat. Acad.Sci. U.S.A. 72, 4248 (1975)
16. L.D.Favro: Phys.Rev. 119, 63 (1960)
17. U.L.RajBhandary, S.H.Chang, A.Stuart, R.D.Faulkner, R.M. Hoskinson, H.G.Khorana: Proc.Nat.Acad.Sci. U.S.A. 57, 751 (1967)
18. W.Wintermeyer, H.G.Zachau: FEBS Letters 18, 214 (1971)

VIII. Spectroscopic Applications of Tunable Lasers

APPLICATIONS OF HIGH RESOLUTION LASER SPECTROSCOPY*

T.W. Hänsch
Department of Physics, Stanford University
Stanford, California 94305, USA

Introduction

The advent of high monochromatic tunable laser sources has stimulated important advances in optical spectroscopy, as documented by several recent good reviews [1-3].

The highest resolution in optical spectroscopy is achieved by eliminating the Doppler broadening of atomic or molecular spectral lines. The classic approach is the transverse observation of a well collimated *molecular beam*, and quite spectacular results have been obtained by this method with the help of tunable lasers[4]. But the high intensity of laser light has also led to the development of a variety of new nonlinear spectroscopic techniques, which permit Doppler-free observations of a simple gas sample. *Saturated absorption spectroscopy* [1-3,5-9] or Lamp Dip spectroscopy is the oldest and perhaps the most widely used of these methods. Here the spread of atomic velocities along the direction of observation is effectively reduced by velocity-selective bleaching and probing with two counterpropagating monochromatic laser beams. *Saturated fluorescence spectroscopy* [10] and in particular the sensitive technique of *intermodulated fluorescence* [11] extend the potential of this method to optically very thin fluorescent samples. The nonlinear interaction of two counterpropagating laser beams in a gas can also be detected via changes in the refractive index rather than in absorption, as demonstrated by *saturated dispersion spectroscopy* [12]. Very recently we have developed a new technique, *laser polarization spectroscopy* [13], which achieves very high sensitivity by monitoring small changes in light polarization.

Other, different approaches to high resolution laser spectroscopy have been suggested, though not yet demonstrated. In particular it has been pointed out [14] that it should be possible to cool a gas sample very rapidly to a fraction of one Kelvin by resonant *radiation pressure* and hence to reduce the Doppler-width of its spectral lines. According to another

*This work was sponsored by the National Science Foundation under Grant No. MPS74-14786A01, the U.S. Office of Naval Research under Contract No. N00014-75-C-0841, and a Grant from the National Bureau of Standards.

suggestion [15] the random thermal motion of gas atoms could be eliminated by *trapping* of slow atoms in the nodes of a three-dimensional *standing wave field*. The possibility of combining radiation cooling and radiation trapping has also been discussed [16].

The need for cooling or velocity selection is circumvented in the perhaps simplest and most elegant approach to high resolution laser spectroscopy, the recently demonstrated and already widely used method of *Doppler-free two-photon spectroscopy* [17-20,1-3]. It complements the other techniques because of its different selection rules. The technique can in principle be generalized to *multi-photon transitions* [18], and Doppler-free three-photon spectroscopy has recently been demonstrated [20].

It is impossible within our limited space to discuss all the interesting applications of high resolution laser spectroscopy which have already been reported or suggested. This holds true even if we ignore such applications as chemical or isotope analysis, trace detection or pollution monitoring, where the high sensitivity of laser methods is often more important than the potential high resolution. The new techniques of Doppler-free spectroscopy are naturally of particular interest for detailed investigations of atomic and molecular energy levels, including their isotope shifts, fine and and hyperfine splittings, Zeeman and Stark effect, light shifts or collision and pressure effects [1,2]. Another very important application is the locking and stabilization of a laser wavelength to some atomic or molecular resonance line. Such lasers have already proven valuable tools for precision metrology and promise future new standards of length and time. High resolution laser spectroscopy has opened interesting new possibilities for basic physics research. The values of two fundamental constants, the speed of light [2] and the Rydberg constant [22], have already been dramatically improved using the new methods. Interesting test of special relativity are presently underway, in particular a precise study of the second order Doppler-effect due to the relativistic time dilation [23] and a new search for "ether drifts," i.e. for possible anisotropies in the vacuum velocity of light [24]. Other experiments can be interpreted as tests of quantumtum electrodynamic calculations. They include the measurement of Lamb shifts in hydrogen [22,25] and detailed studies of spontaneous emission under monochromatic excitation [26]. A particularly ambitious future project is the proposed search for a possible very small (10^{-15} eV) energy difference between the excited levels of two isomeric molecules of left handed and right handed configuration which may originate from parity-violating neutral currents [27].

The subsequent chapters will review some recent applications of high resoltuion laser spectroscopy to the study of atomic hydrogen in our laboratory at Stanford University. During the past few years we have become increasingly interested in precision spectroscopy of hydrogen, because this simple atom permits fascinating detailed comparisons between experiment and theory. Three different nonlinear spectroscopic methods have so far been successfully used: Doppler-free two-photon spectroscopy, saturated absorption spectroscopy and polarization spectroscopy. A review of our most recent experiments will not only serve to illustrate these techniques and to discuss some of their characteristics, advantages and limitations, but also to point out some possible future improvements and extensions.

Doppler-Free Two-Photon Spectroscopy of Hydrogen 1S-2S

Doppler-free two-photon spectroscopy requires in principle only a very simple experimental setup: a gas cell is placed in a standing wave field which is generated by reflecting the output of a tunable laser back into itself. Atoms are excited from the ground state to some excited state of the same parity by absorption of two counterpropagating laser photons, whose combined energies provide the required excitation energy and whose first order Doppler shifts cancel [1-3,17-20].

The line width of the resulting narrow two-photon resonance is ultimately only limited by the nautral line width, although the second order Doppler-effect, collision effects, light shifts, the finite interaction time with the laser beams or the laser band width may impose larger practical limits. A possible drawback is the requirement of a rather intense, highly monochromatic laser, since the transition probability is proportional to the square of the light intensity and can be rather small in the absence of any near-resonant intermediate state. On the other hand, all atoms contribute to the resonant signal, not just a few with a selected velocity, as in almost all other methods of Doppler-free spectroscopy, and the two-photon excitation can often be monitored with very high sensitivity via the subsequently emitted fluorescent light. A dramatic increase in signal, although at the expense of resolution, can in some cases be achieved by near-stepwise excitation through some intermediate level with two lasers of different wavelength [28].

Ever since the new technique was first suggested, numerous authors have pointed out that a particularly interesting subject would be the transition from the 1S ground state of atomic hydrogen to the metastable 2S state (Fig.1).

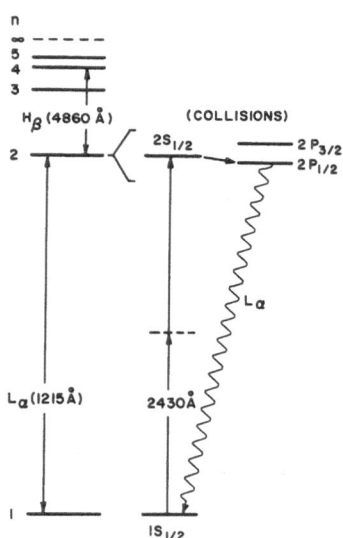

Fig. 1. Simplified term diagram of atomic hydrogen, illustrating
Doppler-free two-photon spectroscopy of the 1S-2S transition.
The Lyman-α and Balmer-β transitions are also indicated

In the absence of external perturbations, the upper 2S state decays via two-photon emission with a lifetime of about 1/7 sec, implying a natural line width of only about 1 Hz or an ultimate resolution limit of less than 1 part in 10^{15}. Even as long as the practical resolution remains far from this limit, spectroscopic studies of this transition open interesting new possibilities for precision measurements of fundamental constants and for tests of quantum electrodynamic calculations.

Unfortunately, such experiments require ultraviolet laser light at 2430Å, i.e. at twice the Lyman-α wavelength, which is somewhat difficult to generate with present tunable laser technology. For our experiments [28,25] we have employed a powerful pulsed dye laser oscillator-amplifier system, pumped by a 1 MW nitrogen laser, which provides peak powers on the order of 50 kW at the visible wavelength of 4860Å, with a line width of about 100 MHz, close to the Fourier-transform limit of its 10 nsec long pulses. Gas pressure tuning provides a continuous wavelength scan range up to several Å. The desired ultraviolet wavelength is generated by frequency doubling the visible laser output in a nonlinear crystal of lithium formate with an efficiency of 1-2%.

The scheme of our two photon spectrometer is shown in Fig. 2. The ultra-

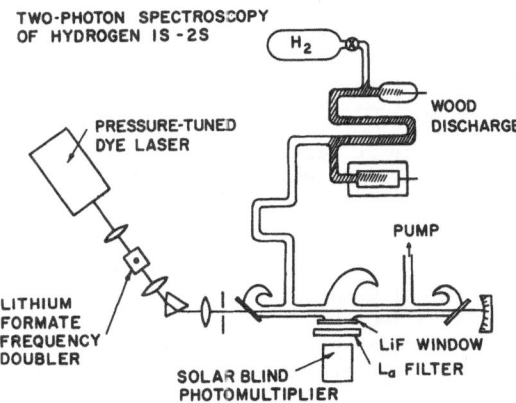

Fig. 2. Two-photon spectrometer for observation of hydrogen 1S-2S

violet coherent radiation is focused into the observation chamber and a spherical mirror reflects the beam to provide the standing wave field. The hydrogen atoms are generated by dissociation of molecular hydrogen in a low pressure gas discharge. The two-photon excitation is observed by monitoring the subsequent emission of vacuum ultraviolet Lyman-α photons from the 2P state through a LiF window and interference filter with a solar blind photomultiplier. At the operating pressure of about 0.2 torr, collisions provide sufficient mixing between the 2S and 2P states for this observation scheme. At lower pressures it would be necessary to apply a mixing static or radio-frequency field. An alternative would be the detection of charged particles which are produced by photoionization of the excited hydrogen.

Figure 3 shows some 1S-2S two-photon spectra obtained in this way.

330

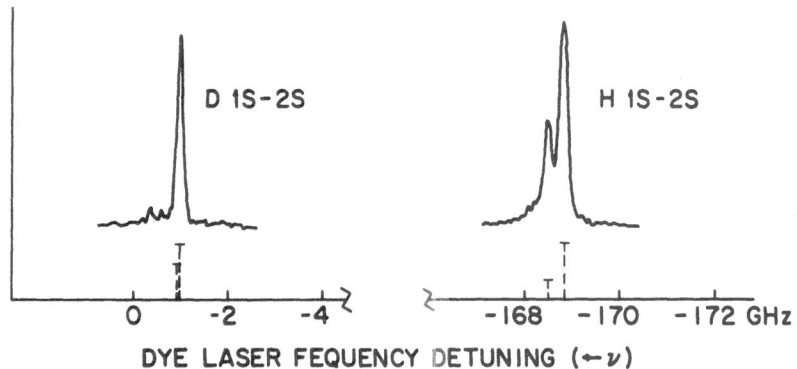

DYE LASER FEQUENCY DETUNING (←ν)

Fig. 3. Doppler-free two-photon spectra of the 1S-2S
transition in atomic deuterium and hydrogen

The spectra were recorded in a mixture of deuterium and hydrogen, and the
wide laser tuning range permitted a continuous scan over the large isotope
shift. The line width is less than 2% of the Doppler width and is limited
only by the exciting laser. Light shifts and transit time broadening are
expected to be less than 1 MHz, and the second order Doppler effect contributes a line broadening of only about 50 kHz at room temperature. The
spectrum of light hydrogen reveals the expected hyperfine doublet, the corresponding structure in deuterium is unresolved.

Depite the relatively large laser band width, these spectra provided a
new value of the Lyman-α isotope shift which exceeds the accuracy of earlier
measurements in the vacuum ultraviolet by several orders of magnitude. This
improvement is not only due to the superior resolution, but also due to the
fact that all wavelengths can now be measured conveniently in the visible.
The measured 1S-2S isotope shift, 670.933±0.056 GHz agrees within its error
limits with the theoretical prediction, 670.9949±0.008 GHz. If the experimental precision could be improved by a factor of 100, it would exceed the
theoretical accuracy. The latter is limited by our knowledge of the ratio
of electron mass to proton mass, and the envisioned more precise isotope
shift measurement would in fact confirm or improve the value of this important mass ratio. It would also be interesting to precisely measure the
absolute wavelength of the 1S-2S two-photon resonance. Even the present
limited resolution should be sufficient to determine a new, more accurate
value of the Rydberg constant.

There are several possible approaches to markedly improve the resolution
of the 1S-2S two-photon spectrum. The most straightforward one would be the
replacement of the transform-limited pulsed laser by a cw source of tunable
radiation. Although cw dye lasers can operate at 4860Å, there is unfortunately at present no known nonlinear optical material which would permit the
efficient 90°-phase-matched second harmonic generation at this short wavelength. But it is possible to generate 2430Å radiation as the sum frequency
of two different laser wavelengths in a crystal of ADP, cooled to near its
Curie temperature [29]. CARL WIEMAN and JIM ECKSTEIN, two graduate students
in our laboratory, have already succeeded in generating about .1 mW at 2430Å
in a single frequency in this way, by summing the output of a 500 mW Kr[+]

laser at 4131Å and a 200 mW rhodamine 6G dye laser in a 5 cm long ADP crystal at 160 K. This power may be sufficient for a detectable two-photon excitation, if the gas sample is placed inside an intensity-enhancing confocal resonator. A remaining problem would be the line broadening due to the short transit time of the moving atoms through the required narrow beam waist.

A rather ingenious solution to the transit time problem has recently been suggested by CHEBOTAEV [30]. In the proposed scheme an atomic beam would be sent through two consecutive transverse standing wave fields. After inter-acting with the first field, the atoms will be found in a coherent super-position of states, i.e. they will oscillate at the two-photon resonance frequency, even though there is no observable dipole moment at this frequency. Depending on the phase of this oscillation relative to the second light field, the atoms entering the second field will be either further excited or they will return to the ground state by stimulated two-photon emission. It should thus be possible to observe the optical analog of the well-known Ramsey fringes which are routinely utilized in radiofrequency spectroscopy of molecular beams. The resolution would then be limited by the travel time between the two fields rather than by the transit time through each waist. The Doppler-free two-photon excitation ensures that the unavoidable spread of transverse atomic velocities together with the short light wavelength do not lead to random relative phases which would smear out any fringe structure, if single-photon excitation was attempted.

Similar "Ramsey fringes" should also be observable with a pulsed laser source, and without need for an atomic beam, if a gas cell is irradiated by two sequential phase-coherent standing wave light pulses. The spectral resolution would then be limited by the separation between the two pulses rather than by the pulse width. Experimentally this would require only a minor modification of the previously described, two-photon spectrometer. As shown in Fig. 4a, a second, delayed laser pulse could be provided by the same

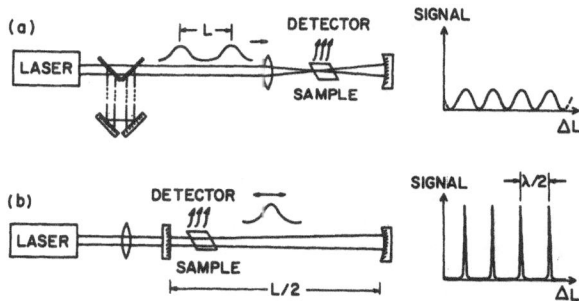

Fig. 4a. Doppler-free two-photon excitation with two phase-coherent light pulses. The excitation probability depends on the delay L between the two pulses. The pulse light is not drawn to scale. Each pulse has to be long enough to provide a standing wave field at the sample

Fig. 4b. Doppler-free two-photon excitation with a train of phase-coherent pulses. The sample is placed inside an optical resonator and a single light pulse is bouncing back and forth between the mirror. Sharp signal peaks rather than sinusoidal Ramsey fringes are expected

pulsed laser with the addition of some beamsplitters and mirrors. The phases of the observed fringes will then be determined by the exact length of the optical delay line. But the expected sinusoidal Ramsey fringes would make the observation of any more complicated spectral structure rather difficult. In a way such an experiment can be considered as the analog of the diffraction of light at a double slit.

At Stanford we are presently working at another approach which corresponds to the optical diffraction at a multiple slit or diffraction grating. We are attempting to observe Doppler-free two-photon excitation by a whole train of phase coherent pulses. Multiple interference should then lead to sharp resonant peaks, if the excitation signal is recorded versus the pulse separation. A very simple way of generating such a pulse train is illustrated in Fig. 4b. The sample is placed inside an optical resonator near one mirror, and a laser pulse is injected into this cavity, either through a partially transmitting end-mirror or better with the help of some active "inverse cavity dumper" to avoid high losses. The expected sharp "multiple-pulse Ramsey fringes" can be interpreted as originating from the modes of the optical resonator. The interpretation of the spectrum with its multiple orders can then proceed along the lines of conventional high resolution Fabry-Perot interferometry. But unlike a passive filter interferometer, our resonator with actively injected pulse does not throw away any light. On the contrary it makes much more efficient use of the light than would be possible in a single-pulse experiment. As long as relaxation can be neglected, the probability of two-photon excitation for small intensities is proportional to the square of the number of pulse roundtrips, i.e. if the pulse recurs a hundred times, the two-photon signal at resonance should increase ten thousandfold. This enhancement should permit it to operate with large beam diameters to alleviate the problem of transit time broadening. As an additional benefit, such a scheme would reduce any light shifts, in particular if the pulse is short compared to the cavity length so that the atoms oscillate during much of the time in the dark at their undisturbed eigenfrequency. We expect that a resonator length of 2m will be sufficient to approach the resolution limit of 50 kHz which is set in hydrogen by the relativistic Doppler effect.

Saturated Absorption Spectroscopy of the Balmer-β Line

The successful observation of the hydrogen 1S-2S two-photon resonance with a frequency-doubled dye laser has opened the possibility for another very interesting experiment: the simultaneous observation of the Balmer-β line, i.e. the n=2 to n=4 transition (see Fig. 1), with the fundamental dye laser output. If Bohr's formula were correct this latter interval would be exactly equal to 1/4 the Lyman-α interval, and we would find the two resonances at exactly the same dye laser frequency. The actual displacement is due to relativistic and quantum electrodynamic corrections plus some small nuclear structure effects. And the accurate comparison of the two optical energy intervals is providing for the first time an opportunity to precisely measure these level corrections. Such a measurement is in particular a sensitive test of the Lamb shift of the hydrogen ground state, which cannot be observed by radiofrequency spectroscopy, because there is no nearby P reference level. By recording a simple Doppler-broadened absorption spectrum of the Balmer-β line in a Wood-type gas discharge we were able to obtain first preliminary values of the 1S Lamb shift for hydrogen and deuterium [28]. An improvement in the resolution by some method of Doppler-free nonlinear spectroscopy was then a very desirable next step.

To obtain a high resolution spectrum of the Balmer-β line, we decided to use the same method of saturated absorption spectroscopy which we had used earlier to resolve the fine structure of the red Balmer-α line and to determine a new value of the Rydberg constant from its absolute wavelength [22]. The scheme of this method is meanwhile well known: part of the visible dye laser output is split into a weak probe beam and a stronger saturating beam which are sent in opposite directions through the absorbing gas. Each monochromatic beam is generally absorbed by a different group of atoms, those with the right axial velocity to be Doppler-shifted into resonance. At the center of a Doppler-broadened absorption line, however, the two counter-propagating beams are resonantly interacting with the same atoms, those with zero axial velocity, and the saturating beam can bleach a path for the probe. To detect small bleaching, the saturating beam is chopped, and the synchronous modulation of the probe is recorded with a phase-sensitive detector. To further reduce the noise due to laser intensity fluctuations it has proven useful to use a second "dummy" probe beam, which does not cross the bleached region, in a differential detection scheme.

Figure 5 shows a saturation spectrum of the Balmer-β line in hydrogen which was in this way recorded simultaneously with the 1S-2S two-photon spectrum [25]. If the 1S state were not shifted above its Dirac-position by the Lamb shift, the two-photon spectrum would appear displaced to the left by about two scale divisions. By measureing the relative line positions and comparing them with their theoretical values, we were able to determine a 1S Lamb shift of 8.20±0.10 GHz for hydrogen and 8.25±0.11 GHz for deuterium. Both shifts agree within their error limits with the respective theoretical values of 8.14943(8) GHz and 8.17223(12) GHz. For hydrogen our result represents the first measurement of the ground state Lamb shift; for deuterium it offers a tenfold improvement over a determination by HERZBERG [31], which was based on a difficult absolute measurement of the vacuum ultraviolet Lymann-α wavelength.

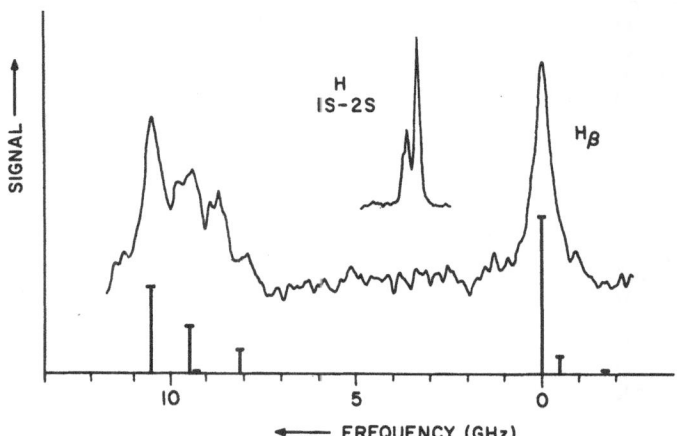

Fig. 5. Saturated absorption spectrum of the hydrogen Balmer-β line
with theoretical fine structure spectrum, and simultaneously
recorded 1S-2S two-photon spectrum

The main limitation of the quoted accuracy is the obviously poor resolution of the Balmer-β saturation spectrum. A number of causes contribute to the large observed line widths which substantially exceed the 100 MHz laser bandwidth. A residual Doppler-broadening of about 50 MHz originates from a small finite crossing angle between the probe beam and the saturating beam. In order to obtain an acceptable signal-to-noise ratio it was necessary to operate at relatively high optical densities i.e. at high discharge currents, implying excessive Stark broadening in the discharge plasma. It was also necessary to work with laser intensities near or above the saturation intensity so that substantial power broadening was unavoidable. Particularly troublesome uncertainties in the line positions result from unresolved "cross-over" resonances which are expected halfway in between any two line components which share a common upper or lower level.

All these difficulties made it desirable to find a different, more sensitive method of high resolution laser spectroscopy. Intermodulated fluorescence spectroscopy [11] unfortunately does not offer any sensitivity advantage for a study of the Balmer-β line because the weak fluorescence from the laser-excited n=4 atoms is almost completely obscured by a strong background of spontaneous emission from the gas discharge. Moreover this method requires both laser beams to be sufficiently strong to saturate and it would thus aggrevate the problem of power broadening.

Doppler-Free Laser Polarization Spectroscopy

The problems encountered in saturated absorption spectroscopy of the hydrogen Balmer-β line provided the motivation for the development of the new, sensitive technique of laser polarization spectroscopy [13] which will be discussed in this concluding chapter.

A laser polarization spectrometer, as shown in Fig. 6, requires only minor modifications of a conventional saturation spectrometer. Again the output of a tunable laser is split into a probe beam and a saturating beam. But this time, the probe is sent through a gas sample between crossed or nearly crossed linear polarizers so that only a small fraction of its light reaches a photodetector. The saturating beam is made circularly polarized by a quarter-wave plate and is sent in opposite direction through the sample.

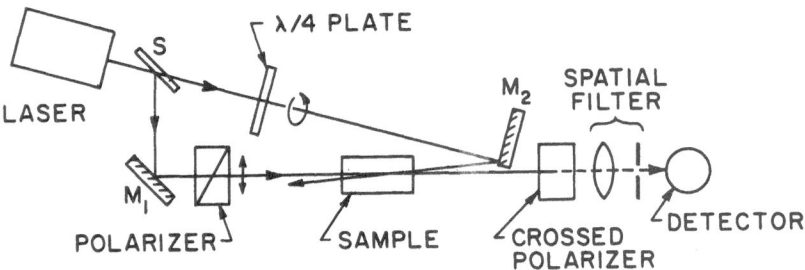

Fig. 6. Laser polarization spectrometer

Since the absorption cross section for circularly polarized light depends
generally on the orientation of the atomic angular momentum relative to the
direction of light propagation, bleaching and optical pumping by the
saturating beam will lead to a velocity-selective atomic orientation. The
sample thus becomes optically anisotropic and is able to change the polarization of the probe light. Such a polarization change can be detected with
high sensitivity via the change in light flux reaching the detector.

As in saturated absorption spectroscopy, a resonant signal is expected
only near the center of a Doppler-broadened absorption line where the two
counterpropagating beams are interacting with the same atoms. But
in this older technique, the signal appears as a small intensity
change on a much stronger background which carries obscuring noise and fluctuations. Polarization spectroscopy offers high sensitivity because it
permits a very effective supression of this background. An improvement in
signal-to-noise ratio of 100 - 1000 is quite readily achieved in practice
under otherwise identical conditions. The new technique is thus of particular interest for studies of optically thin samples or weak lines or for
measurements with weak or fluctuating laser sources.

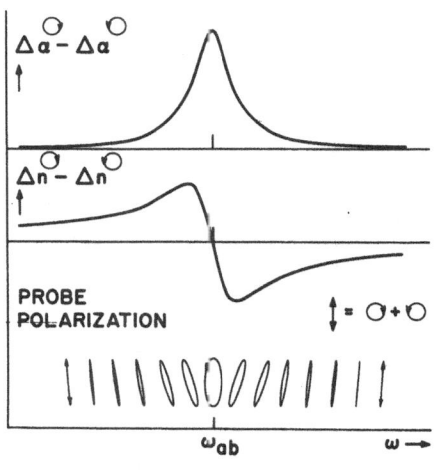

LASER FREQUENCY

Fig. 7. Optical anisotropy in Doppler-free laser polarization spectroscopy
versus laser frequency. The originally linearly polarized probe beam is
decomposed into a right hand (Ⓞ) and a left hand (Ⓞ) circularly
polarized component which experience different changes in
absorption coefficient, $\Delta\alpha$ (top), and in refractive index,
Δn (center). The polarization of the probe light after
passing through the sample is shown at the bottom

For a more quantitative description, the probe beam can be decomposed
into a right hand and a left hand circularly polarized component. At low
intensities these components can be considered separately. Each experi-
ences a light-induced change in absorption coefficient, $\Delta\alpha$, which, in the
limit of a large Doppler width, is given by a Lorentzian function of the

laser frequency with the natural line width. But the magnitude of this saturation effect is different for the two polarizations. The resulting dichroism will make the probe light elliptically polarized, as illustrated in Fig. 7. The Kramers Kronig relation requires corresponding different light-induced changes of the refractive index, Δn, for the two components. The resulting gyrotropic dichroism follows a dispersive function and will rotate the axis of the probe polarization as indicated at the bottom of Fig. 7 for some exaggerated anisotropy.

If the probe light is analyzed with a perfectly perpendicular polarizer, the transmitted light flux is given by the square of the horizontal projection of the probe field vector. It is a Lorentzian function of the laser frequency and corresponds to the natural absorption line, without Doppler broadening. An example of such a signal is shown in Fig. 8(a). If the analyzer is slightly rotated from its perpendicular position, the detector will register some finite background but the light flux is now sensitve to the sign of the polarization rotation and one obtains a dispersive resonance,

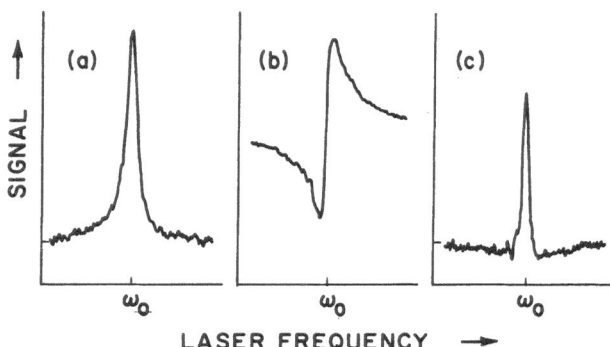

Fig. 8. Polarization spectra of the $2P_{3/2}$-$4D_{5/2}$ component of the hydrogen Balmer-β line, observed with a cw dye laser.
  (a) Lorentzian signal, recorded with perfectly crossed analyzer
  (b) Dispersive signal, observed with slightly rotated analyzer
  (c) Derivative of dispersive signal (b)

as illustrated in Fig. 8(b). This latter signal is almost entirely due to the induced birefringence. It is proportional to the light-induced anisotropy, whereas the former is a quadratic function and becomes negligible for very small absorption depths or saturating intensities. It is obvious that such a dispersive signal is ideally studied for the locking of the laser frequency to some atomic or molecular resonance line. It is also possible to electronically record the first derivative of the dispersive signal by frequency modulating the laser and using a phase sensitive detector for the probe signal. The resulting resonance is non-Lorentzian, as shown in Fig. 8(c), but, quite remarkablly, exhibits a line width less than half the natural width, i.e. it permits one to resolve line components which are separated by less than their natural width without the need for troublesome deconvolution.

To observe the hydrogen Balmer-β line in a Wood gas discharge by the method of polarization spectroscopy, we used a single-frequency cw Coumarin dye laser, pumped by a UV argon laser, with an (unstabilized) line width of about 10 MHz and a continuous scan range of 4 GHz [13]. Standard Glan Thomson prism polarizers for the probe beam reach an extinction ratio of better than 10^{-7}, if residual stress birefringence in the quartz windows of the discharge tube is compensated by gentle squeezing with adjustable clamps. The resulting high sensitivity permitted us to operate at low laser power (less than 1 mW) as well as at low discharge current and gas pressure, thus eliminating many of the problems of the earlier saturated absorption experiment. It proved even feasible to make the two counterpropagating beams truly collinear and thus to avoid any residual Doppler-broadening, by replacing the mirror M_2 in Fig. 6 by a beam splitter, despite the relatively large laser intensity fluctuations resulting from feedback of light into the resonator.

The best spectra were obtained by recording the derivative of dispersive-resonances, as in Fig. 8(c). A portion of the Balmer-β spectrum, recorded in this way, is shown in Fig. 9. The three strongest fine structure components are indicated in the level diagram on top, and the positions of the cross-over lines due to a common upper or lower level are indicated by arrows. The spectrum exhibits many more components which have to be ascribed

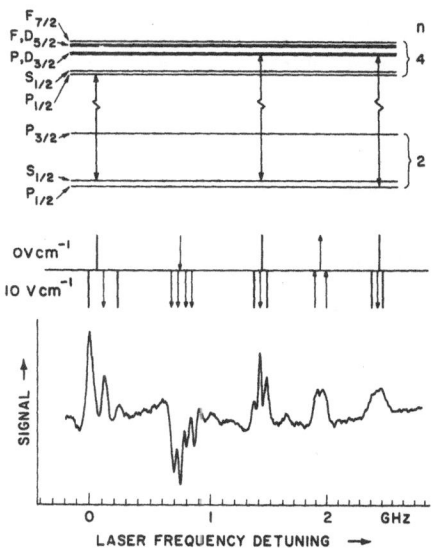

Fig. 9. Polarization spectrum of a portion of the deuterium Balmer-β line, corresponding to the line group at the right hand side of the saturation spectrum in Fig. 5. The three strongest fine structure comments and the positions of the strongest Stark components for an axial electric field of 10 V/cm are shown on top for comparison. The positions of cross-over lines due to a common upper (↑) or lower (↓) level are indicated by arrows

to the Stark effect in the axial electric field of the positive discharge column. The splitting of the closely spaced fine structure levels in the n=4 state is essentially a linear function of the field and can be calculated exactly by diagonalizing the Hamiltonian. The calculated Stark pattern for a field of 10 V/cm agrees well with the observed spectrum. The inverted lines in Fig. 9 represent cross-over lines with inverted polarization rotation.

The observed Stark pattern changes quite drastically if the laser beams are laterally displaced from the tube axis, indicating the presence of radial electric fields due to volume and surface charges. Polarization spectroscopy is thus, for the first time, providing a way to accurately map the electric field distribution inside a low pressure gas discharge. This ability is of course extremely welcome for our intended precision spectroscopy because it permits us to make any necessary corrections for Stark shifts.

The spectrum clearly reveals the different line widths for components originating in the short living 2P state and in the metastable 2S state. The narrowest observed width is about 40 MHz and is limited by the finite laser band width and by factors like residual Stark broadening. The third component from the left corresponds to a transition from the 2S state to the long-living 4S state and would be forbidden in the absence of an external electric field. The natural line width of this component at low fields should be only about 1 MHz.

Polarization spectroscopy of the Balmer-β line is thus offering an interesting alternative to Doppler-free two-photon spectroscopy of the 1S-2S transition for the determination of the electron-proton mass ratio from the H-D isotope shift and for future precise measurements of the Rydberg constant. A combination of the two experiments promises a vastly improved value for the hydrogen ground state Lamb shift and should provide a very stringent test of quantum electrodynamics.

We have seen that the powerful new methods of nonlinear high resolution laser spectroscopy have opened numerous opportunities for interesting spectroscopic research, even if we restrict our attention to the simplest of all atoms. Naturally a much richer field lies ahead if we think more generally about atomic and molecular spectroscopy, although the wealth of potential information may then be much harder to digest. In this context we should note that the new techniques do not only provide "magnifying glasses" for the study of very fine spectral details but that they can also be immensely useful tools for more qualitative survey work. The techniques of saturated absorption spectroscopy and polarization spectroscopy can for instance be applied to unravel the complexities of molecular spectra by identifying lines which originate in a common level, as demonstrated in recent studies of Na_2 by lower level labeling [31] and by polarization labeling [32].

References

1. "Laser Spectroscopy of Atoms and Molecules," *Topics in Applied Physics*, *Vol. 2*, H. Walther, ed., Springer-Verlag, Heidelberg, 1976.
2. "High Resolution Laser Spectroscopy," *Topics in Applied Physics*, K. Shimoda, ed., Springer-Verlag, Heidelberg, 1976.
3. V. S. Letokhov and V. P. Chebotaev, "Principles of Nonlinear Laser Spectroscopy," Springer-Verlag, Heidelberg, 1977.

4. L. A. Hackel, K. H. Casleton, S. G. Kukolich, and S. Ezekiel, Phys. Rev. Letters 35, 568 (1975).
5. A. Szöke and A. Javan, Phys. Rev. Letters 10, 521 (1963).
6. W. E. Lamb, Jr., Phys. Rev. A134, 1429 (1964).
7. P. H. Lee and M. L. Skolnick, Appl. Phys. Letters 10, 303 (1967).
8. C. Bordé, C. R. Aca. Sci. Paris, 271, 371 (1970).
9. T. W. Hänsch, I. S. Shahin, and A. L. Schawlow, Phys. Rev. Letters 27, 707 (1971).
10. C. Freed and A. Javan, Appl. Phys. Letters 17, 53 (1970).
11. M. S. Sorem and A. L. Schawlow, Opt. Comm. 5, 148 (1972).
12. C. Bordé, G. Camy, B. Decomps, and L. Pottier, Colloques Internationaux du C.N.R.S. No. 217, Paris (1974), pp. 231.
13. C. Wieman and T. W. Hänsch, Phys. Rev. Letters 36, 1170 (1976).
14. T. W. Hänsch and A. L. Schawlow, Opt. Comm. 13, 68 (1975).
15. V. S. Letokhov, Science 190, 344 (1975); Appl. Phys. 9, 229 (1976).
16. V. S. Letokhov, V. G. Minogin, and B. D. Pavlik, Opt. Comm., to be published.
17. L. S. Vasilenko, V. P. Chebotaev, and A. V. Shishaev, JETP Letters 12, 113 (1970).
18. B. Cagnac, G. Grynberg, and F. Biraben, J. Phys. (Paris) 34, 845 (1973).
19. F. Biraben, B. Cagnac, and G. Grynberg, Phys. Rev. Letters 32, 643 (1974); M. D. Levenson and N. Bloembergen, Phys. Rev. Letters 32, 645 (1974); T. W. Hänsch, G. Meisel, and A. L. Schawlow, Opt. Comm. 11, 50 (1974).
20. B. Cagnac, in *Atomic Physics 5, Proceedings of the Fifth International Conference*, Berekley, U.S.A. (1976), to be published.
21. K. M. Evenson, J. S. Wells, F. R. Petersen, B. L. Danielson, G. W. Day, R. L. Barger, J. L. Hall, Phys. Rev. Letters 29, 1346 (1972).
22. T. W. Hänsch, M. H. Nayfeh, S. A. Lee, S. M. Curry, and I. S. Shahin, Phys. Rev. Letters 32, 1336 (1974).
23. J. J. Snyder and J. L. Hall, in *Laser Spectroscoppy, Lecture Notes in Physics, Vol. 43*, S. Haroche et al., eds., Springer-Verlag, Heidelberg, 1975.
24. P. Franken, private communication.
25. S. A. Lee, R. Wallenstein,and T. W. Hänsch, Phys. Rev. Letters 35, 1262 (1975).
26. R. E. Grove, F. Y. Wu, and S. Ezekiel, Phys. Rev. Letters 35, 1426 (1975); Opt. Comm. 18, 61 (1976).
27. V. S. Letokhov, Science 190, 344 (1975).
28. T. W. Hänsch, S. A. Lee, R. Wallenstein, and C. Wieman, Phys. Rev. Letters 34, 307 (1975).
29. F. Zernike and J. E. Midwinter, *Applied Nonlinear Optics*, Wiley, New York, 1973.
30. Y. V. Baklanov, V. P. Chebotaev, and B. T. Dubetsky, Appl. Phys., to be published.
31. M. E. Kaminsky, R. T. Hawkins, F. V. Kowalski, and A. L. Schawlow, Phys. Rev. Letters 36, 671 (1976).
32. R. Teets, R. Feinberg, T. W. Hänsch, and A. L. Schawlow, Phys. Rev. Letters (1976), accepted for publication.

APPLICATIONS OF FAR INFRARED LASERS

B. Lax[*]
Francis Bitter National Magnet Laboratory[†]
Massachusetts Institute of Technology,
Cambridge, Massachusetts 02139, USA

Introduction

The earliest useful sources of coherent radiation in the far infrared or submillimeter region were the electron tube devices including harmonic generators and carcinotrons. These have been used for molecular spectroscopy [1] and for resonance studies of solids [2][3]. The development of the HCN laser [4] at 337 μm in 1964 soon revolutionized resonance spectroscopy in semiconductors [5] and magnetic materials. With the addition of isotopic species of HCN and H_2O, a number of additional lines from 10 to 120 cm^{-1} became available for such studies [6]. The potential for far infrared applications has been even further enhanced by the development of optically pumped molecular lasers [7] and by the nonlinear mixing of two lasers to provide tunable radiation over two decades, including the submillimeter region [8]. The cw sources provide high resolution and power levels of milliwatts and the pulse sources provide high power in the ten to hundred kilowatt range. Linear and nonlinear spectroscopic studies of molecular vibrations, transitions between excited states of atomic species, vibration spectra of suitable crystals, resonance spectroscopy of electric and magnetic dipole transitions of all sorts in solids and gases and a variety of new diagnostic and other studies of plasmas are in progress or are contemplated for the immediate future.

Far Infrared Sources

The discovery of optically pumped polar molecules by a CO_2 laser in 1970 has subsequently produced over 500 new laser lines between 12 μm and 1.8 mm [9]. Polarized cw output levels of milliwatts and pulsed power of hundreds of kilowatts in some molecules are now available. The former are very stable with linewidths as low as 10 kHz. The high power lasers can be made superradiant by pumping a long tube of gas with low reflection windows. Such a laser has produced 10 kW at 1.2 mm [10]. These can have linewidths of 500-1000 MHz and are useful for breakdown and plasma heating. In order to produce relatively narrow linewidth at high power it is necessary to build a cavity oscillator together with an amplifier [11] as shown in Fig. 1. A zig-zag pumping arrangement to accommodate a mirror and a double mesh output coupler produces a 30 MHz linewidth in CH_3F at 496 μm, which is then amplified to yield as much as 250 kW with a 200 MW CO_2 laser as the pump. Such a laser is to be used for Thomson scattering in tokamaks.

---

*Also Physics Department, Massachusetts Institute of Technology
†Supported by the National Science Foundation

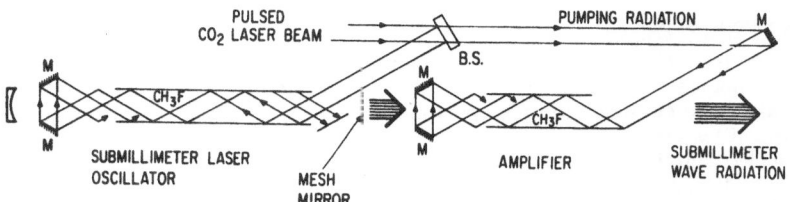

Fig. 1 Schematic diagram of high power optically pumped molecular laser
system. B.S. indicates beam splitter, M indicates mirror

Another approach for generating far infrared radiation has been that of
difference-frequency mixing. A variety of schemes have been demonstrated but
the one which we have found more versatile is the use of noncollinear phase-matched
mixing of CO_2 lasers [8]. The initial pulse technique in GaAs which generates step
tunable radiation from 70 µm to a few millimeters, has been recently improved to
produce 4 kW in the submillimeter region [12]. The new arrangement is shown in
Fig. 2, which shows two 3 MW TEA lasers tunable with a grating over the 9.6 and
10.6 µm regions directed at an angle θ_{1E} coincidentally onto the input face of the
GaAs crystal. As the lasers are tuned mirror M_1 moves laterally and is rotated

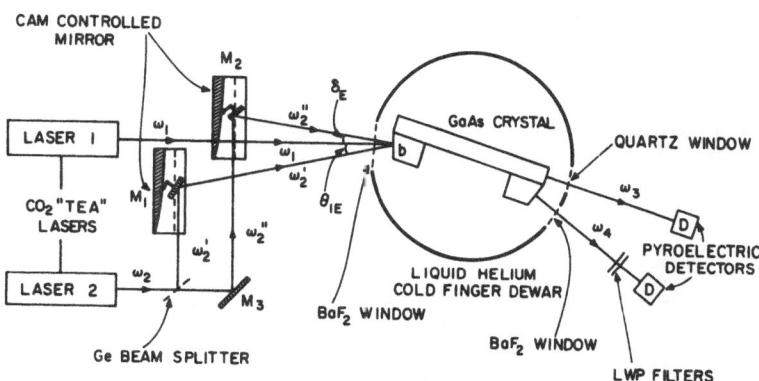

Fig. 2 A schematic of the experimental setup for the two step difference-
frequency mixing in GaAs at liquid helium temperature. For $\omega_1 =$
1045 cm^{-1} and $\omega_2 = 946$ cm^{-1}, $\theta_{1E} \approx 8^\circ$, $\delta_E \approx 1.6^\circ$

by a cam arrangement to preserve coincidence of the two beams. The difference-
frequency $\omega_3 = \omega_1 - \omega_2$ grows along the length of the crystal and is detected as
shown. If the beam of laser 2 is split and is directed onto the crystal face at angle
δ_E by another adjustable mirror M_2 then noncollinear phase-matching is achieved
with ω_3 to give $\omega_4 = \omega_2 - \omega_3 = 2\omega_2 - \omega_1$. In this way tunable radiation at the
12 µm region has been obtained with power levels of the order of 4 kW. The power
output can be scaled up with higher power lasers, larger crystals, and increased
laser intensity from 3 MW/cm^2 to 30 MW/cm^2 below the damage threshold limit.
The noncollinear mixing scheme in GaAs has also produced tunable cw radiation
in the 70 µm to the millimeter region with 100 kHz linewidth at µW power level.

Spectroscopy of Solids

One of the first experiments using the molecular lasers of HCN and H_2O and its isotopic species was cyclotron resonance in semiconductors [6]. Many new materials were studied and mass parameters were obtained. The most dramatic experiment was in CdTe [13] where the polaron effect was definitively correlated with the Froehlich theory in a magnetic field as developed by LARSEN [14]. The theory predicted an effective mass change with increasing magnetic field and, hence, with frequency. This is illustrated in Fig. 3, which compares the theoretical curve with experiment [15] for a value of the electron phonon coupling parameter $\alpha = 0.40$.

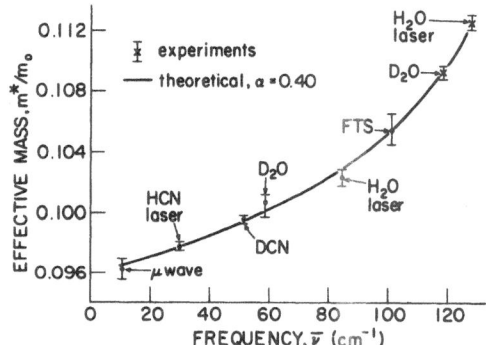

Fig. 3 Cyclotron resonance in CdTe with far infrared lasers showing polaron effect on the mass with frequency and magnetic field

This is the best value obtained as contrasted to the estimates made from optical data. Other resonant experiments with these lasers involve magnetic dipole transitions in antiferomagnets and electron spin measurements in semiconductors [16]. The latter has illustrated the anomalous spin relaxation of electrons in InSb as the temperature increases to about 40 K. This information is important for the operation of a spin-flip laser. In order to explain the anomalous behavior of the spin-flip laser at high fields, it was also necessary to examine the higher order cyclotron resonance transitions in InSb [17]. This was done with a tunable CO_2 laser which exhibited harmonic cyclotron resonance absorptions, phonon assisted, and combination resonances in the 9 and 10 μm regions where the spin-flip laser is pumped.

A different type of experiment using the tunable cw noncollinear mixing source in the far infrared was the examination of the absorption of GaAs over a broad spectral range. This information is necessary in order to predict and improve the operation of the noncollinear devices. Hence, ROSENBLUH and AGGARWAL [18] made careful measurements with their noncollinear cw tunable source. Some of their results are compared with those obtained with an interferometer as shown in Fig. 4. The circles represent the most accurate data taken in this region of the spectrum.

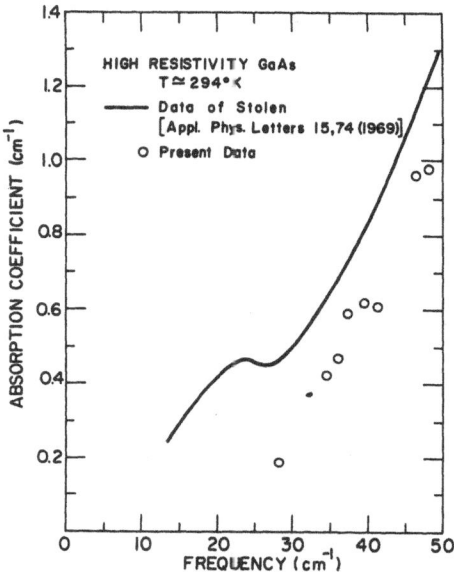

Fig. 4.   Absorption coefficient versus frequency in GaAs at room
temperature. The solid curve is after Stolen and circles
are data obtained with noncollinear tunable source.

Plasma Applications

The possibility of cyclotron resonance breakdown of gases with a high power molecular laser was theoretically considered [19]. The phenomenological theory shows
that the breakdown threshold shows a dramatic resonant behavior. The experimental study of breakdown involved the observation of the light intensity output as
a function of magnetic field [20] in a low current discharge at 3 torr, into which
a high power CH_3F laser was focussed to intensities of about 40 kW/cm^2. The
electron density was soon saturated after breakdown and, hence, the light intensity
observed was shown to be proportional to the electron energy or power gain P_g
from the laser radiation

$$P_g = \frac{e^2 E^2}{m} \frac{\nu_c}{(\omega - \omega_c^2) + \nu_c^2} \tag{1}$$

where ν_c is the electron collision frequency and ω_c the cyclotron frequency. At
higher pressures the experimental curves shown in Fig. 5 fit the Lorentzian line
shape given by Eq. (1). At lower pressures the phenomenological theory does not
apply because the electron energy gain between collisions exceeds the ionization
potential so that the assumption of equilibrium no longer holds. Nevertheless, the
pronounced resonant enhancement of light emission at 218 kG, corresponding to the
496 μm wavelength of the CH_3F laser is clearly evident.

344

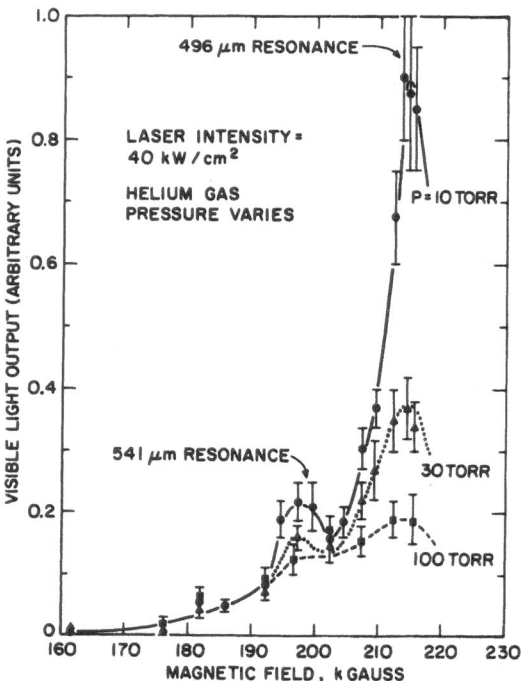

Fig. 5. Light output versus magnetic field for helium at p = 10, 30 and 100 torr, for a 496 μm laser intensity of 40 kW/cm². The curves are drawn in as a visual aid.

A second important application of high power far infrared lasers that is being implemented is the Thomson scattering from tokamak plasmas [21]. At the wavelength of the CH_3F laser $\lambda > \lambda_D$ where the latter is the Debye wavelength of the electron plasma. Under such conditions the light scatters from the Debye sphere of electrons surrounding the ions and are modulated by the Doppler frequency of the ions. The frequency spectrum then measures the ion temperature. According to the theory of Saltpeter this occurs when the following condition is satisfied, namely:

$$\alpha = \frac{\lambda}{4\pi\lambda_D\sin\theta/2} \geq 3 \tag{2}$$

θ is the scattering angle. For $\lambda = 496$ μm, $N_e = 10^{14}/cm^3$, $T_e = 2$ keV for a tokamak $\theta \simeq 30°$. With a narrow band laser oscillator-amplifier of 1 MW output and a GaAs Schotky barrier diode detector with a sensitivity of 10^{-18}W/Hz it is possible to detect the scattered signal with a heterodyne system. It is expected that the localized ion temperature and ion impurity content will be measured by such a system.

With high power superradiant optically pumped pulsed lasers, which are now scaleable into the megawatt range, it is now possible to heat electrons locally in tokamaks at the harmonics of the cyclotron frequency. By monitoring the localized electron temperature decay with electron Thomson scattering by a ruby laser, one can measure the transverse thermal diffusion coefficient of electrons [21]. Analogous experiments of ion transport with ion Thomson scattering would be of utmost importance in the understanding of the operation of tokamaks. With such high power lasers one can also observe the excitation of parametric instabilities in a plasma arc [21].

Optically pumped cw far infrared lasers are the ideal sources for interferometric measurements of electron densities in tokamaks. They are highly stable and can be brought close to the plasma machines, unlike electrically pumped lasers which are affected by the magnetic fields. The shorter wavelengths available are necessary for higher density large tokamaks to reduce reflection and refraction of the beams by the plasma. Using two methyl alcohol lasers pumped with the same CO_2 laser, operating at 118 μm, a modulation interferometer shown in Fig. 6 has been developed for use with Alcator, a high field tokamak [22]. The modulation is provided at 1 MHz by adjusting the length of one of the laser cavities accordingly.

TWO FREQUENCY BEAT– MODULATED
SUBMILLIMETER INTERFEROMETER

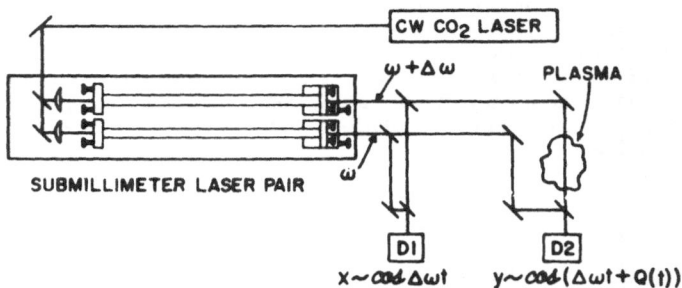

Fig. 6. Dual Beam Modulated Interferometer System. Two optically pumped lasers are operated at frequencies differing by $\Delta\omega$. Mixing the two outputs in detector D1 provides a modulated signal $x \sim \cos(\Delta\omega t)$, while detector D2 sees a similar signal $y \sim \cos(\Delta\omega t + \phi)$ with an additional phase shift due to the plasma. The magnitude of this phase shift is then obtained directly from comparison of the two signals.

A reference signal is provided by mixing the two laser outputs before and after one of the laser beams is passed through the plasma. The modulated outputs of the reference and the phase shifted output from the plasma are rectified and fed into a flip-flop, then through suitable electronics, which then measure the phase shift directly. The latter represents the integrated product of the electron density and path length through the plasma. By this technique the plasma density as a function of time is accurately measured with no disturbance due to fluctuations. Multiple channels of the interferometer intersecting the plasma can determine the plasma profile if there is cylindrical symmetry of the plasma. However, in a tokamak there are MHD modes which distort the symmetry of the plasma, especially if these become unstable. Hence, if there were several multiple channels at different angular positions, one could apply the principles of tomography to determine the plasma profile as a function of position and time.

Another application of the cw sources to tokamaks is the measurement of cyclotron harmonic absorption. Cyclotron harmonic emission has been observed by a fast scanning Fourier transformed Michelson interferometer [23]. The disadvantage to this scheme is that the inhomogeneous field broadens the scattered radiation that is reflected from the walls and is observed by the instrument. The absorption technique using multiple lines of an optically pumped laser or the tunable noncollinear mixer measures the cyclotron harmonic locally and provides a narrow linewidth. The two methods complement each other in that the absorption depends on the derivative of the electron distribution function and the emission on the distribution function itself. These measurements are sensitive to instabilities and are convenient monitors of the relative plasma temperature as a function of time.

Conclusions

The multiplicity of far infrared sources of fixed and tunable frequencies offers many new opportunities for application not only for solid state and plasma physics but other areas as well. High resolution atomic and molecular spectroscopy at long wavelengths are now possible. For example, transitions between the excited hydrogenic states of atomic helium and sodium which are pre-excited by either electrons or two photons of a dye laser can be studied with submillimeter lasers in a magnetic or electric field. Spectroscopic investigations of shallow donors and acceptors with narrow linewidths in semiconductors with and without magnetic fields are very suitable with these new sources in the far infrared. Other experiments for radio astronomy, communications, radar, biological and atmospheric studies can now be explored with the high resolution tunable sources and the high power optically pumped lasers.

Acknowledgements

The author wishes to thank R. L. Aggarwal, D. R. Cohn, K. J. Button, N. Lee, M. Rosenbluh, R. J. Temkin, S. M. Wolfe and M. P. Hacker for many useful discussions and assistance in the preparation of this manuscript.

References

1. W. Gordy, Pure and Applied Chemistry 11, 403 (1965).
2. P. Goy and B. Castsing, Int. Conf. on Submm. Waves and Their Applications, p. 169 (Conf. Digest, 1974).
3. E. G. Rudashevsky, A. S. Prokhorov and L. V. Velikov, IEEE Transactions on Microwave Theory and Techniques, MTT-22, 1064 (1974).
4. H. A. Gebbie, F. D. Findley, N. W. B. Stone and J. A. Rose, Nature 202, 169 (1964).
5. K. J. Button, H. A. Gebbie and B. Lax, IEEE J. Quant. Electr. 2, 202 (1966).
6. B. Lax, Proc. of Symp. on Submm. Waves, p. 27, Polytech Press (1970).
7. T. Y. Chang and T. J. Bridges, Optics Communication 1, 423 (1970); T. Y. Chang, IEEE Trans. on Microwave Theory and Techniques, MTT-22, 983 (1974).
8. R. L. Aggarwal, N. Lee and B. Lax, Int. Conf. on Submm. Waves and Their Applications (Conf. Digest 1974).
9. M. Rosenbluh, R. J. Temkin, K. J. Button, Applied Optics (Aug. 1976).
10. M. P. Hacker, Z. Drozdowicz, D. R. Cohn, K. Isobe and R. J. Tamkin, Phys. Letters A (to be published).
11. Z. Drozdowicz, R. J. Temkin, K. J. Button and D. R. Cohn, Appl. Phys. Lett. 28, 328 (1976).
12. N. Lee, R. L. Aggarwal and B. Lax, Appl. Phys. Lett. 29, 45 (1976).

13. J. Waldman, D. Larsen, P. E. Tannenwald, C. C. Bradley, D. R. Cohn and B. Lax, Phys. Rev. Lett. 23, 1033 (1969).
14. D. M. Larsen, Phys. Rev. 135, A419 (1964).
15. C. W. Litton, K. J. Button, J. Waldman, D. R. Cohn and B. Lax, Physical Review B (in press).
16. B. D. McCombe, Proc. Int. Conf. on the Application of High Magnetic Fields in Semicon. Phys., p. 146, Wurzburg (1974).
17. G. Favrot, R. L. Aggarwal and B. Lax, Solid State Commun. 18, 577 (1976).
18. M. Rosenbluh, Master's Thesis at M.I.T. (Aug. 1975).
19. B. Lax and D. R. Cohn, Appl. Phys. Lett. 23, 363 (1973).
20. M. P. Hacker, R. J. Temkin and B. Lax, Appl. Phys. Lett. (to be published).
21. D. L. Jassby, D. R. Cohn, B. Lax and W. Halverson, Nucl. Fusion 14, 745 (1974); B. Lax and D. R. Cohn, IEEE Transactions on Microwave Theory and Techniques, MTT-22, 1049 (1974).
22. S. M. Wolfe, K. J. Button, J. Waldman and D. R. Cohn, Appl. Opt. (to be published).
23. A. E. Costley, J. Chamberlain, K. Muraoka and D. D. Burgess, Int. Conf. on Submm. Waves and Their Applications, p. 147 (Conf. Digest 1974).

TUNABLE LASER SPECTROSCOPY IN MINERAL PROSPECTING

S.T. Eng and E. Max
Department of Electrical Measurements
Chalmers University of Technology
S-402 20 Gothenburg, Sweden

Introduction

The great global demand for minerals is leading to many new
exploration studies to locate new ores [1]. Attention has been
given recently to surface gases associated with metallic ore
deposits [2]. Since the most sensitive techniques for detection
and measurement of very low gas concentrations are optical ones,
it is very natural to apply laser spectroscopy to mineral prospecting.

Some of the more significant vapors which can be detected,
the type of deposits from which they emanate, and a few of the
optical sources useful for field detection are listed in Table I.

Table I. Some gases indicating ore deposits and optical sources
to be used in measurements

Vapor	Deposits	Optical Source (Field Use)
SO_2	Sulfide	Semic. laser
Hg	Zn-Pb-Cu sulfides	Mercury lamp Dye laser
H_2S	Sulfide	Dye laser Semic. laser
CO_2 (or CO_2/O_2)	Sulfide, gold	CO_2 laser
NO, NO_2	Nitrate deposits	Semic. laser Dye laser

As an example among ore types in which SO_2 anomalies can be measured are base metal sulfides, porphyry copper deposits, and pyrites. Mercury vapor occurs in close association with many types of ore materials such as ZnS and HgS deposits. Furthermore, hydrogen sulfide can be present in the oxidized zone of an ore deposit. Also, the oxidation of sulfide ore will result in the natural gas mixture becoming richer in CO_2 and poorer in O_2. Thus, in the soil gases above a sulfide zone, the ratio of carbon dioxide will be increased from regional background.

We have chosen in this paper to focus on SO_2 detection using a PbSe[1] diode laser [3].

Experiment and Results

The laboratory experiment performed consisted of measurement of SO_2 above rocks containing metal sulfides of which pyrite (FeS_2) is the most abundant. Other minerals in the rocks include pyrrhotite (Fe_7S_2-FeS), zinc sulfide, and lead sulfide. The rocks were kept in a glass container to which two teflon tubes were attached; one tube introduced fresh air into the container, and the other led the gases to the measurement cell.

A lead selenide (PbSe) diode injection laser was used as the spectrometer source. The diode was operated at a temperature of 77 K, a pulse width of 30 μs, a duty cycle of $1.6\cdot10^{-3}$, and a pulse current of 7 A (which is just above threshold current). In every pulse the laser frequency spectrum varied from 7.326 μm (1365 cm^{-1}) to 7.278 μm (1374 cm^{-1}). Only 13 μs of the pulse is being utilized in the experiment.

The optical setup is shown in Fig. 1 and is briefly explained as follows.

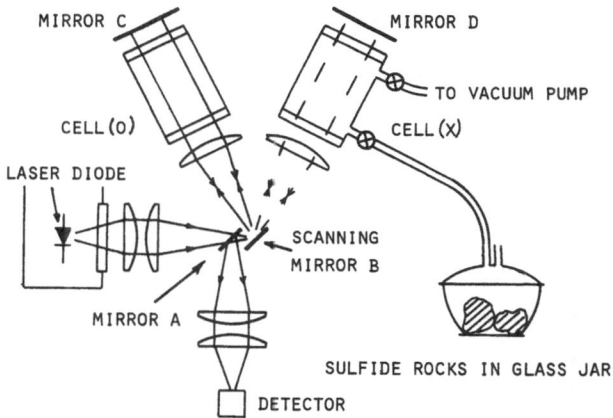

Fig. 1. The experimental optical arrangement. The lenses and windows are made of BaF_2. The length of the cells is 0.2 m

---

[1] Diode made by Laser Analytics, Inc., Mass., U.S.A.

The beam from the pulsed diode laser is passing just above the mirror (A) and is focused on the scanning mirror (B), which alternately directs the beam through the cells (X) and (O). By the two plane mirrors (C and D) the beam is reflected and then again focused on the scanning mirror (B) but this time 3 mm underneath the first focus spot. This lower orientation of the beam makes the reflected radiation first hit the mirror (A) and then the detector. Gas with an unknown concentration, C_x, of SO_2 from the glass jar is connected to the cell (X). The ratio between the signals from the two cells is detected and processed by the processing electronics. The laser frequency is tuned during the pulse, and the time derivative of the amplitude of the detected signal gives an approximation of the derivative of the transmission spectrum. By this method weak absorption lines are accentuated.

A block diagram of the electronics used to compute the derivative spectrum $D_x(t')$ in this experiment is shown in Fig. 2.

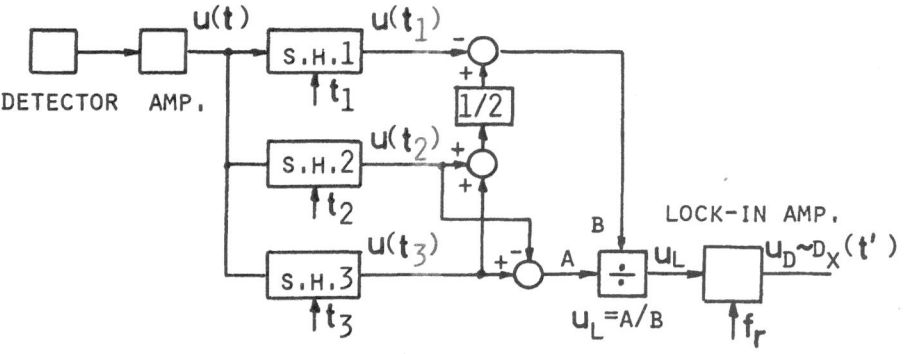

Fig. 2. Block diagram of the electronics used to compute the derivative spectra

The time elapsed since the origen of the pulse is here denoted t'. The laser output is switched between the cells with a frequency f_r, and an equal number of laser pulses passes through cell (X) in the first half of every switching period as through cell (O) in the second half of the period. There are three sample-and-hold circuits (SH 1-3), of which the first one takes a sample, $u(t_1)$, of the detector voltage at time t_1 just before the start of every laser pulse. The SH2 and SH3 circuits take the samples $u(t_2)$ and $u(t_3)$ during the pulse at times $t_2 = t_1+t'$ and $t_3 = t_2+\Delta t$. The voltage at the input of the lock-in amplifier can be approximated as

$$u_L = \frac{1}{u(t_2)-u(t_1)} \left[\frac{du(t)}{dt}\right]_{t=t_2} k_1 \Delta t \qquad (1)$$

The lock-in amplifier output can be written

$$u_D = k \, \Delta t \, D_X(t') \qquad (2)$$

where

$$D_X(t') = -\left[\frac{1}{u_X(t')} \frac{du_X(t')}{dt'} - \frac{1}{u_O(t')} \frac{du_O(t')}{dt'}\right] \qquad (3)$$

$D_X(t')$ is taken as a difference between the normalized derivatives of the output detector voltages for the two cells, and k is a scale factor. In the same fashion the spectrum $D_N(t')$ is determined by replacing the unknown concentration C_X with a known concentration C_N. The correlation between $D_X(t')$ and $D_N(t')$ is a useful measure of the product $C_X C_N$, and the autocorrelation of $D_N(t')$ is a measure of C_N^2. The ratio C_X/C_N is calculated from the correlation functions.

The first step in the measurement procedure was to take two plots (1 and 2) of the derivative spectrum with cell (X) containing clean air at a pressure of 200 torr and cell (O) containing 200 torr clean air. Then three more plots were taken, where cell (X) initially was filled with 200 torr gas from the glass container (plot 3), then with 200 torr clean air (plot 4), and finally with 1500 ppm SO_2 in 200 torr air (plot 5). Using a computer, we performed an analysis of the derivative spectrum. The result is shown in Table II.

Table II. Calculated SO_2 concentrations (C_X) from the measurements

$D_O(t')$ from plot	$C_R = \dfrac{C_X}{C_N}$	C_X
1	$0.88 \cdot 10^{-2}$	$13 \cdot 10^{-6}$
2	$0.35 \cdot 10^{-2}$	$5 \cdot 10^{-6}$
4	$0.52 \cdot 10^{-2}$	$8 \cdot 10^{-6}$

The calculations indicate a concentration of approximately 10 ppm of SO_2. The spectrum $D_O(t')$ represents the zero level taken with 200 torr clean air in the cells. Therefore, $D_O(t')$ has to be subtracted from $D_X(t')$ and $D_N(t')$ when taking the correlations.

Conclusions

The laser spectroscopy method for measuring low concentrations of SO_2 in air that has been suggested in this paper seems promising for applications in mineral prospecting. The major limitation on the sensitivity of the system is the dissimilarities between the measurement cell and the reference cell. Improvements of the laser characteristics such as lower threshold current, higher duty cycle, and higher output power are also desirable.

The experimental results indicate a concentration of around 10 ppm of SO_2 above the sulfide rocks in the glass jar.

Acknowledgment

This work is supported by the Swedish Board for Technical Development (STU). We appreciate the information provided by the Technical Staff of Boliden AB.

References

1. A.R. Barringer: Western Miner, 13 (February 1974).

2. Q. Bristow and I.R. Jonasson: Canadian Mining Journal, 93, 39 (1972).

3. E.D. Hinkley: "Development of in Situ Prototype Diode Laser System to Monitor SO_2 Across the Stack." Final Report by MIT Lincoln Laboratory for the Environmental Protection Agency (May 1973).

STUDY ON PHASE-MATCHING CHARACTERISTICS OF OPTICAL SECOND HARMONIC GENERATION IN NONLINEAR THIN-FILM WAVEGUIDES USING A TUNABLE PARAMETRIC OSCILLATOR

H. Ito and H. Inaba
Research Institute of Electrical Communication
Tohoku University, Sendai 980, Japan

Abstract

Phase-matched second harmonic generation was performed experimentally in non-linear optical waveguide structures with the configurations of Air-ZnS-Glass and Air-ZnS-LiNbO$_3$. Nonlinear interactions take place in the polycrystalline ZnS thin film for the first configuration, and in both this film and the LiNbO$_3$ substrate for the second case. The phase matching properties was examined for both cases by using a tunable optical parametric oscillator operating near 1 - 1.1 μm as the fundamental wave. The effective second-order non-linear coefficient of oriented ZnS thin film and the generated harmonic power for the waveguide with these configurations were also studied analytically.

I. Introduction

Nonlinear optical interactions such as harmonic generation, parametric oscillation, and frequency conversion in thin film optical waveguides have many advantages in integrated optical circuitry. Since most of the optical energy is concentrated within and near the thin film layer, the power density inside the optical waveguide can easily be very large to realize appreciable degree of the nonlinear coupling. Furthermore, as these thin film waveguides have modal dispersion or dimensional dispersion, the phase matching condition can be established without relying upon the conventional method based on the bire-fringence of nonlinear crystals.

There has been, however, only a limited number of studies directed towards exploring experimentally the second harmonic generation (SHG) and other non-linear optical phenomena in a guided wave structure made of thin film materials. For nonlinear optical effects in a guided wave structure, the interaction can take place in a) the thin film itself [1 - 5], b) the substrate [6 - 10], and c) both of a) and b) [11 - 14]. These three types of interaction schemes are schematically depicted in Fig. 1. Since most of the optical energy is confined customarily in the thin film layer, it is evident that the nonlinear interaction in the thin film is superior to that in the substrate, if a good nonlinear optical film is available. Nevertheless, the most effective nonlinear interactions in the waveguide structure can be expected in the form of c), because both the nonlinearities of film and substrate are substantially utilized.

We report in this paper the experimental study on phase-matching properties of SHG in a guided wave strucutre empolying the second-order nonlinearities inherent to the film and also to both the film and substrate. For phase-matched SHG in the waveguide, its thickness must be quite precise for fixed wavelength fundamental sources. In our experiment, a tunable optical parametric oscillator was used as the fundamental source to realize easily a remarkable resonance in the second-harmonic output by tuning the source wavelength through the phase-matching point.

The second-order nonlinearity can be expected to a polycrystalline thin film even if it consists of randomly oriented small single grains, since there will be nonvanishing components of effective nonlinearity averaged over the volume formed by a number of such crystalline grains. Glass substrates were utilized in the first experiment to verify the optical nonlinearities of the polycrystalline ZnS film and the capability of waveguide phase matching [3, 4]. In the second experiment, the $LiNbO_3$ single crystal was employed as the nonlinear substrate to achieve SHG incorporated with ZnS thin film waveguide under the phase-matched condition [11, 12].

II. Fabrication of ZnS Thin Film Waveguides and Their Nonlinear Properties

ZnS thin film waveguides were fabricated by the vacuum deposition on BK-7 glass substrates and polished X-cut $LiNbO_3$ substrates. From the x-ray pinhole photographs and diffractometer analyses of deposited ZnS thin films, the Debye-Sherrer ring patterns only from the (111) plane of the films were observed for both the combinations, even if the temperature of the substrates during the deposition is lower than 100 °C, and became a form of arc as heating the substrates above 100 °C. Typically at about 200 °C, the ZnS thin film is well crystallyzed and shows appreciable scattering losses mainly due to the surface irregularities. Then the ZnS waveguides used in our experiment were deposited at the temperature of the substrates less than 100 °C.

By measuring the synchronous angle of of a He-Ne laser beam in a prism-film coupler, the refractive indices and thickness of ZnS films were determined with the aid of computor calculation. From repeated measurements for a number of ZnS thin films, the average value for the refractive index of vacuum deposited ZnS thin film is revealed to be 0.3 % smaller than that of ZnS bulk crystal. However, because of lack of dispersion information of the refractive indices of the ZnS deposited film and also its quality dependence of fabrication conditions, the

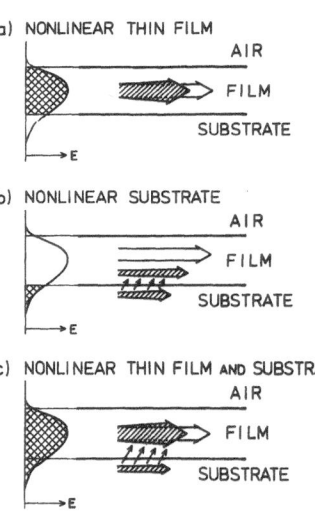

a) NONLINEAR THIN FILM

b) NONLINEAR SUBSTRATE

c) NONLINEAR THIN FILM AND SUBSTRATE

⟹ FUNDAMENTAL WAVE

▧⟫ SECOND HARMONIC WAVE

⬤ REGION OF NONLINEAR INTERACTION

Fig. 1 Schematic illustrations of three types of SHG interactions in the guided wave structure

Sellmeier's equation for the bulk crystal [15] was utilized to calculate the phase matching condition for the SHG as described in the following sections.

It is known that many polycrystalline thin films fabricated by various methods have more or less some indication of their orientations along specific crystal axis depending on the substrate temperature and circumstances during fabrication [16]. If materials of thin films are of non-centro symmetric, the second-order optical nonlinearity is expected based upon the non-vanishing components of averaged second-order nonlinear tensors. Moreover, the oriented polycrystalline nonlinear films are considered to possess somewhat larger effective second-order nonlinearity under the specific combination of interacting modes than the randomly oriented case.

In order to characterize the second-order nonlinearity of this film, the spatially averaged values of the effective second-order nonlinear coefficient $<d^2>$ was calculated depending on the point group of the crystal and on the oriented direction of the thin film. The problem of angular averages for second-order nonlinear coefficients of randomly oriented crystalline particles has been treated by Cyvin, Rauch and Decius [17], and Kurtz and Perry [18].

We consider two coordinates (X,Y,Z) and (x,y,z) corresponding to the laboratory-fixed and the crystalline-fixed systems, as shown in Fig. 2, respectively. The small single grain ZnS crystals are oriented in the [111] direction parallel to the Z-direction, but assumed to be oriented randomly in the XY-plane. By introducing the direction cosines Φ_{Fi} relating the two coordinate systems, second-order nonlinear coefficient d_{ijk} of the crystal is transformed to that of laboratory-fixed system utilizing the tensor transformation properties;

$$d_{FGH} = \sum_{ijk} \Phi_{Fi} \Phi_{Gj} \Phi_{Hk} d_{ijk} \qquad (1)$$

where each of suffices i, j and k corresponds to x, y and z of the crystalline-fixed system, and each of F, G and H corresponds to X, Y and Z of the laboratory-fixed system, respectively. For the SHG in the nonlinear thin film waveguide, the interacting modes couple through the effective nonlinear coefficient d_{FGH}. The electric fields for TE and TM modes are given by E_y and E_z, if one neglects the E_y field for TM modes. The following relation shows the correspondence between the component of d_{FGH} and the combination of interacting modes for SHG in the guided wave structure;

$$
\begin{array}{lclcclcl}
TE + TE \rightarrow TE & : & d_{XXX} & & TE + TE \rightarrow TM & : & d_{ZXX} \\
TE + TM \rightarrow TE & : & d_{XZX} & & TE + TM \rightarrow TM & : & d_{ZZX} \\
TM + TM \rightarrow TE & : & d_{XZZ} & & TM + TM \rightarrow TM & : & d_{ZZZ}
\end{array} \qquad (2)
$$

where TE + TM → TM, for instance, indicates that the combination of fundamental waves characterized by TE and TM modes generate the second harmonic wave of TM mode.

The angular averaged second-order nonlinear coefficients can be derived

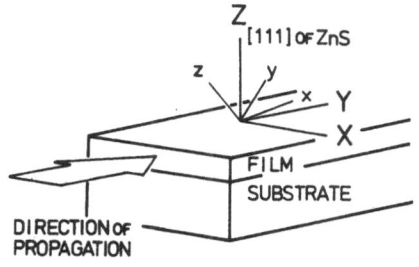

Fig. 2 Thin film waveguide geometry and the relation between two coordinates (X,Y,Z) and (x,y,z) corresponding to the laboratory-fixed and crystalline-fixed systems. The [111] direction of crystalline particles is locked to Z-direction in the laboratory-fixed coordinate system

by the integration over the all solid angle;

$$<d_{FGH}^2> = \sum_{ijki'j'k'} (\int \Phi_{Fi}\Phi_{Gj}\Phi_{HK}\Phi_{Fi'}\Phi_{Gj'}\Phi_{HK'}d\Omega \ / \int d\Omega \ ) \ d_{ijk}d_{i'j'k'} \qquad (3)$$

Cubic ZnS crystal belongs to the point group $\overline{4}3m$, and three non-zero coefficients for second-order optical nonlinearity are all equal ($d_{14}=d_{25}=d_{36}$), if Kleinman's symmetry conjecture is assumed [19]. Φ's are determined so as to orient the [111] direction of ZnS thin film normal to the substrate surface as found by the x-ray analyses. The computed results for propagating mode combinations given by (2) are summarized in Table 1. We found that TE + TE → TM type interaction gives $(7/12)d_{14}^2$ for $<d^2>$, and TE + TE → TE type yields $(1/3)d_{14}^2$, respectively.

III. Phase-matched SHG in Nonlinear Optical Waveguides

In the guided wave structure, nonlinear interactions of optical waves can be phase-matched by employing proper mode dispersion. The waveguide thickness should be selected so that the fundamental and harmonic waveguide modes take the same phase velocity. Numerical analysis was performed on the thin film thickness to realize the phase matching as a function of the fundamental wavelength [4, 11]. The overlap integrals of the fundamental mode intensity and the second harmonic field profile in the nonlinear optical waveguide were also calculated.

For the phase-matched SHG in an optical waveguide with fixed fundamental frequency sources, the larger the difference of the refractive indices between the film and substrate, the more precise control of the film thickness is necessary. When a coherent tunable source for the fundamental is available, however, the complete phase matching is actually obtained without encountering this problem.

The experiment was performed with two types of the ZnS waveguide; one is deposited on the glass substrate (Air-ZnS-Glass), and another is on the X-cut $LiNbO_3$ crystal substrate (Air-ZnS-LiNbO$_3$) as shown schematically in Fig. 1 a) and c), respectively.

A. Air-ZnS-Glass Waveguide

Figure 3 shows the computed results of the thin film thickness and the effective refractive index β/k to achieve the phase matching for the SHG as a function of the fundamental wavelength. For the calculation, the Sellmeier's equation for the bulk ZnS crystal was tentatively employed as mentioned previously. Mode combination (m m n) of TE_m wave for the fundamental and TM_n wave for the harmonic corresponding to $TE_m + TE_m → TM_n$ was considered for this interaction.

Taking into account the transverse spatial coupling coefficient and other experimental parameters of the thin film waveguide, the phase-matchable mode combination of $TE_1 + TE_1 → TM_4$ which is abbreviated simply by (114) in Fig. 3 was selected for the SHG experiment of 1.06 μm output from a Nd:YAG laser. The cal-

Table I   Spatially averaged effective second order nonlinearities $<d^2>$ for orientated ZnS polycrystalline thin films for SHG.

$\omega_1 + \omega_1 → \omega_2$	$<d^2>$
TE + TE → TE	$\frac{1}{3}d_{14}^2$
TE + TE → TM	$\frac{7}{12}d_{14}^2$
TE + TM → TE	$\frac{7}{12}d_{14}^2$
TE + TM → TM	0
TM + TM → TE	0
TM + TM → TM	$\frac{4}{3}d_{14}^2$

culated value of the phase-matchable film thickness is around 1.3 μm, and the spatial coupling coefficient for this condition is 5.5 x 10^{-2}. The thickness of the ZnS film was controlled around 1.3 μm using a calibrated thin film deposition monitor installed inside the vacuum chamber and measured by the interference thickness meter.

Phase-matched SHG was performed using the experimental arrangement shown in Fig. 4, where a Chromatix Nd:YAG laser with four oscillation wavelengths (1.0 52, 1.061, 1.064 and 1.074 μm), and an optical parametric oscillator (Chromatix Model 1020) operating in the same wavelength region were used as the fundamental. The fundamental beam was focused onto the rutile prism coupler to excite the TE_1 mode inside the ZnS thin film waveguide. The second harmonic beam with sharp directivity decoupled from the same prism was detected by a photomultiplier. The propagating mode of the harmonic wave in this nonlinear waveguide was found to be TM_4 as expected.

The complete phase-matching profile as a function of the fundamental wavelength was measured as illustrated in Fig. 5. The result obtained by the Nd:YAG laser outputs was also depicted in the figure. The generated second harmonic power was measured to be about 2 mW at 1.052 μm from the Nd:YAG laser with a peak power of 300 W. From the measured profile, the phase-matched wavelength is estimated to be 1.054 μm for this ZnS thin film and the thickness of the waveguide for the phase-matched SHG is found to be 1.28 μm. The effective interaction length in the waveguide was derived by fitting a theoretical tuning curve to the measured profile expected from mismatching behavior of nonlinear interaction between the propagating modes. The theoretical curve for 35 μm interaction length is also shown in Fig. 5, which provides the best fit around the central portion of the measured profile.

B. Air-ZnS-LiNbO$_3$ Waveguide

Figure 6 shows the phase-matchable film thickness and effective refractive index β/k for Air-ZnS-LiNbO$_3$

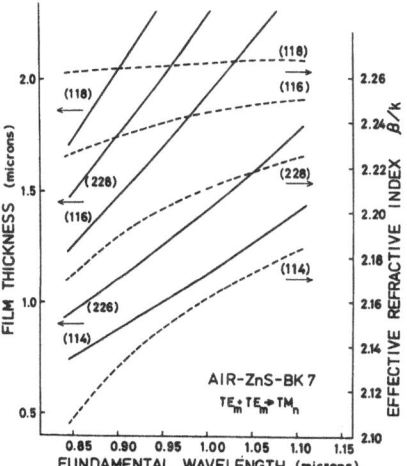

Fig. 3 Thin film waveguide thickness and effective refractive index of Air-ZnS-Glass waveguide for which the specific mode combination (m m n) is phase-matched as a function of the fundamental wavelength

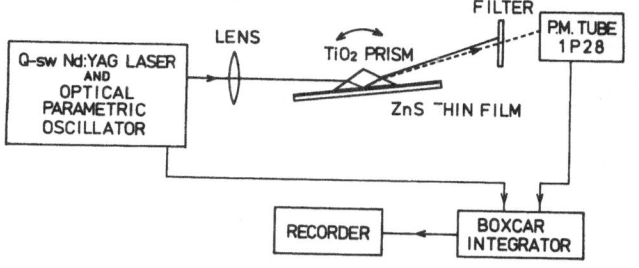

Fig. 4 Experimental setup of phase-matched SHG in a ZnS poly-crystalline thin film waveguide using outputs from a Q-switched Nd:YAG laser and an optical parametric oscillator as the fundamental wavelength around 1 - 1.1 μm

waveguide SHG as a function of the fundamental wavelength. The guided wave propagation is made along the Y-direction and the c-axis of the substrate is oriented parallel to the X-direction as shown in the insert of Fig. 6. Actually this orientation of the crystal was determined so as to utilize the largest effective second-order nonlinear coefficient $d_{eff} = -d_{33}$ of LiNbO$_3$ ($d_{33} = -(97 \pm18)$x 10^{-9} esu $\simeq 10d_{31}$) [20] which requires the nonlinear interaction only among transverse electric (TE) waves in this configuration; TE + TE → TE. The birefringence of the LiNbO$_3$ substrate is irrelevant in this case, since all waves are polarized to the direction of c-axis, that is, extraordinary rays. Taking into account the transverse spatial coupling coefficient and other experimental parameters of this Air-ZnS-LiNbO$_3$ waveguide, we chose the phase-matchable mode combination of TE$_0$ + TE$_0$ → TE$_2$; (0 0 2) for the experiment. For the fundamental wavelength of 1.1 μm, the calculated value of phase-matched film thickness is 0.83 μm and the spatial coupling coefficient for this condition is 6 x 10^{-2}, which is the largest among other mode combinations.

The experimental setup similar to Fig. 4 was employed where two rutile right angle prisms with the base angle of 60° were used to excite and decouple the waveguide modes inside the thin film [12]. It is easily confirmed that the propagating mode of the harmonic wave in the nonlinear waveguide is TE$_2$ as anticipated from Fig. 6.

The measured phase matching profile as a function of fundamental wavelength is shown in Fig. 7. It is known from the result that the phase-matched wavelength is 1.11 μm for this ZnS film which leads to the thickness of the waveguide for the SHG to be 0.85 μm, while the deposition monitor gave 0.82 μm. The input fundamental power in front of the prism coupler was approximately

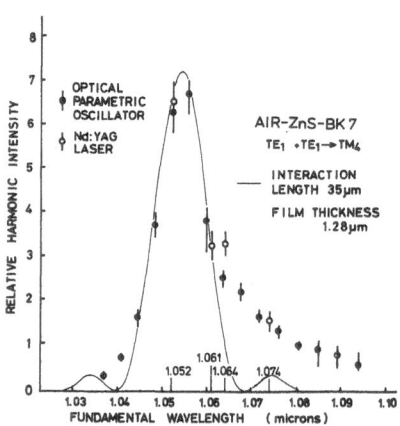

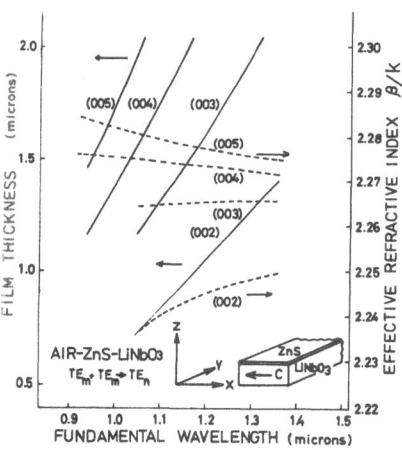

Fig. 5 Tuning curve of the phase-matched SHG in the Air-ZnS-Glass waveguide as a function of the fundamental wavelength. The nonlinear interaction was performed at the TE$_1$ mode for the fundamental and TM$_4$ for the harmonic waves. The solid curve gives the theoretical tuning characteristics assuming an interaction length of 35 μm in the ZnS thin film waveguide

Fig. 6 Thin film waveguide thickness and effective refractive index of Air-ZnS-LiNbO$_3$ waveguide for which the specific mode combination (m m n) is phase-matched as a function of the fundamental wavelength. The insert shows waveguide geometry where the direction of wave propagation is along the Y-axis, and the c-axis of the LiNbO$_3$ crystal is coincident with the X-direction

70 W, and measured decoupled second harmonic power was about 5 μW. Based on the measured tuning curve in Fig. 7, the effective interaction length was calculated to be about 40 μm by fitting a theoretical curve to explain the observed profile, which is also drawn in Fig. 7 by a solid line.

Ⅳ. Discussions and Concluding Remarks

In both the experiments described above, the spot size on the film of the fundamental beam from the parametric oscillator was approximately 45 μm. Then the estimated effective interaction lengths of 35 μm and 40 μm in Figs. 5 and 7 are quite feasible, since the nonlinear interaction was restricted by the geometrical situation of the prism couplers and almost confined in the spot size region for both the cases. It should be noted that the coherence length of the ZnS single crystal is only about 5 μm for the SHG of 1.1 μm wavelength.

The measured second harmonic powers are fairly well compared with the known theories [21]. For the case of Air-ZnS-LiNbO₃ waveguide, the second harmonic power P_{SH} generated in a guide of length ℓ for the fundamental wavelength λ_F is given by

$$P_{SH} = (512\pi^5/cn_{eff}^3)(\ell/\lambda_F)^2(P_F^2 c^2/Wt_F^2 t_{SH}) \sin^2(\Delta\ell/2)/(\Delta\ell/2)^2 \quad (esu) \quad (4)$$

and

$$c^2 = (d_{sub}I_{sub} + d_{film}I_{film})^2 \quad (5)$$

where d_{sub} and d_{film} are the effective nonlinear coefficients of the substrate and film, $\Delta = 2\beta_F - \beta_{SH}$, $n_{eff} = (\beta/k)$, W is the width of the beam in the plane of the guide. I_{sub} and I_{film} denote the transverse spatial coupling, and t_F and t_{SH} are the effective thickness of the guide defined by

$$I_{sub, film} = \sqrt{8}_{sub}, \int_{film} g_F^2(z)g_{SH}(z)dz, \quad t_{F,SH} = 2\int_{-\infty}^{+\infty} g_{F, SH}^2(z)dz, \quad (6)$$

respectively. Here $g_F(z)$ and $g_{SH}(z)$ are the transverse mode amplitudes at the fundamental and harmonic frequencies, respectively.

Calculated value for P_{SH} is about 40 μW at 70 W fundamental power input assuming a prism coupling efficiency of 25 % to the film and a decoupling efficiency of 50 %. This value appears to yield rather good agreement with the measured value of 5 μW within the accuracy of the experimental errors, since the second harmonic power depends critically on the coupling prism adjustment, and the quality and smoothness of the film as well as the substrate. For the present case of Air-ZnS-LiNbO₃ waveguide, the refractive index difference

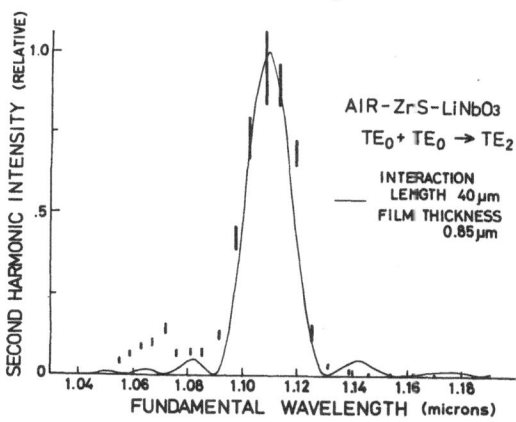

Fig. 7 Tuning curve of the phase-matched SHG in the Air-ZnS-LiNbO₃ waveguide as a function of the fundamental wavelength. The nonlinear interaction takes place both in the film amd substrate at the TE₀ mode for the fundamental and TE₂ for the harmonic waves. The solid curve indicates the theoretical tuning characteristics assuming an interaction length of 40 μm in the ZnS thin film waveguide

between the substrate and film is 0.073 at the wavelength of 1.1 μm. From the numerical analysis, 11.5 % of the fundamental power is found in the substrate, while 88.4 % is confined inside the thin film for TE_0 mode. From (4) and (5), the contribution of the substrate nonlinearity to the second harmonic power is estimated to be about 1 %, whereas the interference term between the film and substrate nonlinearities could contribute about 19 % to the total second harmonic power. The substrate nonlinearity would play more significant role, if we choose the combination of the substrate and film with less refractive index difference.

We also note that nonlinear Cerenkov radiation in the $LiNbO_3$ substrate has not been observed in our experiment. The condition for this radiation [6], $n_{sub,F} < \beta/k < n_{sub,SH}$ could not be satisfied in our experimental condition of $\beta/k \simeq 2.245$, because of $n_{sub,F} = 2.1534$ and $n_{sub,SH} = 2.2258$.

In conclusion, the capability of the phase-matched SHG in oriented polycrystalline thin film waveguides formed on linear and nonlinear substrates should enforce attractive fields of various nonlinear elements and devices for useful applications in optical integrated circuits.

References

1. D. B. Anderson, J. T. Boyd: Appl. Phys. Letters 19, 266 (1971)
2. S. Zemon, R. R. Alfano, S. L. Shapiro, E. Conwell: Ibid., 21, 327 (1972)
3. N. Uesugi, H. Ito, H. Inaba: Technical Digest of 1973 IEEE Internat. Electron Devices Meeting, Washington, D. C., U. S. A. p. 330 (1973)
4. H. Ito, N. Uesugi, H. Inaba: Appl. Phys. Letters 25, 385 (1974)
5. J. P. van der Ziel, R. M. Mikulyak, A. Y. Cho: Ibid., 27, 71 (1975)
6. P. K. Tien, R. Ulrich, R. J. Martin: Ibid., 17, 447 (1970)
7. Y. Suematsu, Y. Sasaki, K. Shibata: Ibid., 23, 137 (1973); Y. Suematsu, Y. Sasaki, K. Shibata, S. Ibukuro: IEEE J. Quantum Electron. QE-10, 222 (1974)
8. W. K. Burns, A. B. Lee: Appl. Phys. Letters 24, 222 (1974)
9. B.-U. Chen, C. L. Tang, J. M. Telle: Ibid., 25, 495 (1974)
10. B.U. Chen, C. L. Tang: IEEE J. Quantum Electron. QE-11, 177 (1975)
11. H. Ito, H. Inaba: Report of the Technical Group of Optical and Quantum Electron., The Institute of Electron. and Communica. Engineers of Japan, Tokyo, OQE 74-51, 43 (1974)
12. H. Ito, H. Inaba: Opt. Commun. 15, 104 (1975)
13. M. M. Hopkins, A. Miller: Appl. Phys. Letters 25, 47 (1974)
14. J. P. van der Ziel, R. C. Miller, R. A. Logan, W. A. Nordland Jr., R. M. Mikulyak: Ibid., 25, 238 (1974)
15. S. J. Czyzak, W. M. Baker, R. C. Crane, J. B. Howe: J. Opt. Soc. Am. 47, 240 (1957)
16. N. F. Foster: J. Vac. Sci. Technol. 6, 111 (1969)
17. S. J. Cyvin, J. E. Rauch, J. C. Decius: J. Chem. Phys. 43, 4083 (1965)
18. S. K. Kurtz, T. T. Perry: J. Appl. Phys. 39, 3798 (1968)
19. J. F. Nye: Physical Properties of Crystals (Oxford Univ. Press, London 1957)
20. Cf. J. E. Midwinter, J. Warner: Brit. J. Appl. Phys. 16, 1135 (1965); G. D. Boyd, R. C. Miller, K. Nassau, W. L. Bond, A. Savage: Appl. Phys. Letters 5, 234 (1964)
21. E.g., J. T. Boyd: IEEE J. Quantum Electron. QE-8, 788 (1972); E. M. Conwell: Ibid., QE-9, 867 (1973); A. Yariv: Ibid., QE-9, 919 (1973)

CONTROL TECHNIQUES FOR CW DYE LASERS

J.L. Hall[*] and S.A. Lee
Joint Institute for Laboratory Astrophysics
National Bureau of Standards and University of Colorado
Boulder, Colorado 80309, USA

The laser pioneers have been very farsighted, with imagination to see the many future applications and possibilities opening up with each new advance in laser techniques. On the other hand, it is notoriously easy for enthusiasm to influence the reliable estimation of difficulties still to be overcome before each new potential laser capability can really be widely applied in other research. For example, 5 years were to pass after the famous MIT stable laser experiments first suggested the possibility of an improved wavelength standard, before the demonstration in 1968 of a pure neon laser that did <u>somewhat</u> outperform the incoherent krypton standard. This was closely followed by lasers stabilized to molecular transitions: they provided advances of 2-3 orders of magnitude and completely vindicated the early optimism. Now the methane work has advanced to a state approaching parity with man's best measurement standard, the cesium beam primary frequency standard, and workers in the standards field are excited about dramatic further possible improvements. Thus experience confirms the vision and insight of the early stable laser workers, but teaches caution relative to the expected ease of application and rapidity of development.

Of course, in more recent times enthusiasm has been expressed about widely-tuneable lasers. Again one can detect a significant leading phase angle between the discussion and the wide-spread application of real lasers. The pulsed dye lasers have been perhaps the first to mature: These lasers are so powerful that, even with their vastly broader line widths, they can produce dramatic effects in atomic, molecular and ionic quantum absorbers. Numerous and impressive works already have been done with the aid of pulsed dye lasers. Even though short-pulse lasers are apparently not ideally suited to measurements of the highest ultimate precision, dramatic progress has been made in controlling these lasers and a number of beautiful fundamental measurements have been made.

---

*Staff member, National Bureau of Standards.

In some ways we believe the cw dye laser is now also near the threshold of widespread and effective application. Certainly a number of experimental groups have seen exciting applications for cw dye lasers, but have soon found themselves instead in the dye laser research business per se. It apparently has been much easier to perceive beautiful laser opportunities in physics than to teach good single-mode manners to a cw dye laser! As we also have seriously underestimated this challenge, perhaps we may be excused for presenting a partial list of the signposts along this way "where angels fear to tread." See Table I. Although we do not suggest this list of problems is complete, most of these problems will require some attention. However primarily we want to report here on the brighter side: (1) a cw jet stream laser which is comparatively pleasant to use, and (2) an automatic scanning interferometer that gives a realtime numerical wavelength display of sub-Doppler accuracy.

To us, one interesting application of a cw dye laser concerns the possibility of seeing the frequency shift (≤ 1.4 GHz) of a beam of rapidly moving atomic clocks (neon atoms) due to the Relativistic Time Dilation.[1] Since this experiment rejects the accompanying large first-order Doppler shifts (~10 Å) mainly by virtue of the planar wavefronts used in the interaction region, it was clear from the start that good wavefront quality would be one of the most important considerations in the design of our laser. An axially-pumped, prism-tuned three-mirror astigmatically-compensated cavity was chosen and extremely good laser performance was obtained.[2] Unfortunately the instability in the laser's output direction was too serious for the intended application and we were thus led to the compensated design illustrated in Fig. 1. Consider two multicolor waists: a small one fixed in the dye jet and a larger one fixed on the T = 10% plane output mirror. The two prism systems form two linearly-dispersed spectra along the folding mirror. If the prism dispersion -- pathlength products of the two systems are matched, it is clear that the output beam can remain stationary as the folding mirror is rotated to coarse-tune the

Figure 1   STABILIZED SINGLE MODE DYE LASER

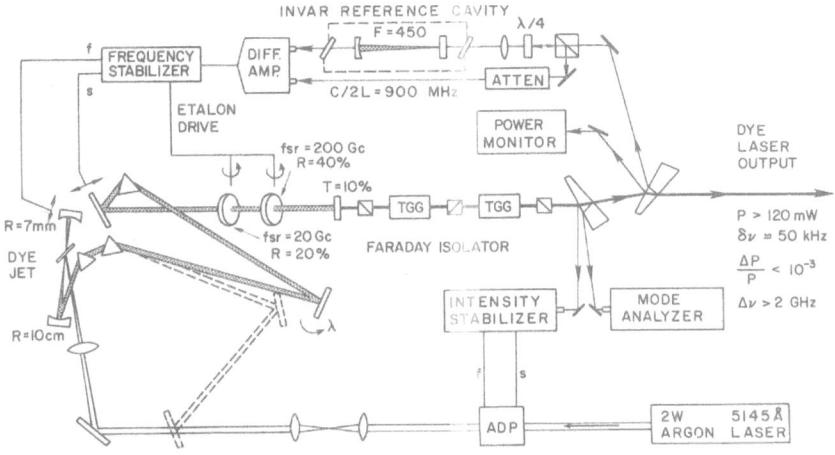

Table I. Typical Problems of Single-Frequency cw Dye Lasers for Spectroscopy

Symptom	Cause or Remark
1. Runs single mode only near threshold	Mode control inadequate a) Thick etalon resolution insufficient b) Thin etalon free spectral range too small
2. Rapid mode jumps and/or gross frequency instability	Excessive pressure variations in liquid feed to jet nozzle -- use "low pass" acoustic filter in line
3. Low single-mode power A) high threshold, poor slope efficiency near threshold	A) Excessive optical losses a) error of astigmatism - compensation b) poor etalon parallelism, reflectivity mismatch c) etalon "walk off" loss due to small beams or excessive etalon tilt angle d) insufficient focus of dye laser in jet to swamp spontaneous emission loss
B) low single-mode conversion efficiency even with good optics	B) Optical scattering from jet surface roughness: a) turbulence due to excessive velocity, or b) inadequate nozzle internal smoothness especially near exit
C) slope efficiency decreases rapidly with pump power	C) Dye flow speed too small -- interacts with items 3B, 5 and 6
4. Slow laser tuning jumps over desired frequency	Accidental etalon in cavity (thinnest birefrigent tuner plate, cell windows, outer surface of etalon plates...)
5. Laser frequency stabilizer unlocks rapidly (~seconds)	Laser goes off completely for 1-10 μsec due to mini-bubbles: a) Aggregate and remove bubbles with ultrasonics b) improve smoothness of jet catcher tube and choose angle of incidence carefully. Interacts with items 3B, C
6. Laser frequency stabilizer unlocks occasionally (minutes) or jumps to next reference cavity order	Fast frequency excursion drives over top of fringe and changes servo sign. Residual microbubbles reduce jet optical thickness, thus increase laser frequency -- be sure to lock on high frequency side of cavity fringe.
7. Small single-frequency tuning range	Inadequate etalon centering and/or tracking of cavity tuning
8. Hystersis and drift when power is changed	Etalon heating by dye laser. (Also loss is power dependent above ~1 1/2 watts incident on etalon)
9. Unstable single frequency tuning range, mode jumping	Feedback from external load back into laser: need $\gtrsim$-40dB return loss. Need Faraday-, polarization-, or acousto-optic-isolator

laser. For very large color changes (~200 Å) the Brewster incidence
condition on the prisms can be refined to decrease the losses although
of course this moves the output spot position somewhat. At very high pump
levels, say 3 W (~300 mW single frequency output), it is effective to
insert a birefringent plate to suppress oscillation on adjacent orders of
the thinner etalon (±200 GHz). In other experiments, using birefringent
elements as the primary tuner brought problems due to an etalon-effect in
the thinnest plate: to reach an atomic line of interest it was generally
necessary to override this effect with another thin etalon gang-tuned with
the birefringent tuner.[3]

The laser's broad-band spectral density is sufficient to easily find a
strong atomic line by observing fluorescence. Thus we have mounted both
etalons on precision slides which allow them to be removed from the cavity
and reinserted without further optical alignment. After finding a line
broadband, the etalons are individually inserted and the line is recovered
by electrically tuning each etalon using the respective frequency centering
control. These controls provide an input to the respective analog square-
root modules which well approximate the arccosine functions desired for
tilt-tuning the solid etalons according to the relation $(f/f_0)\cos\theta = 1$.
Appropriate power buffers drive the torque motors which tilt the etalons.
A zero-centered voltage representing the cavity length is also supplied
to these square-root modules to provide the tilt tuning to match the
cavity length tuning, and gives 15-25 cavity orders single frequency
tuning range (1.5-2.5 GHz).

The laser's output frequency may be servo-controlled to the side of a
fringe of the invar-spaced reference cavity, using the fast differencing
technique previously described.[2] A fast piezo driver of improved design
moves the small cavity end mirror (7 mm radius of curvature, 4.7 mm
diameter, 1.5 mm thick). Using shear-wave filters and a PbO-filled
polyurethane damping material, a resonance-free bandwidth beyond 700 kHz
was achieved. Unfortunately acoustic propagation delay in the mirror
thickness -- together with other delays -- limits the useful frequency
servo bandwidth to about 250 kHz. The main open-loop frequency excursions
of the dye laser are caused by residual dye pump pressure variations (±1/2
MHz excursions at 300 Hz), atmospheric pressure variations in the closed
but not sealed cover box (~ ± 3 MHz in 1 sec) and long-term thermal drift
of the laser's massive 2×36×24" Al jig plate mounting base. With the laser
operating under servo control these low frequency perturbations are
essentially totally suppressed and the loop noise of 50 kHz RMS equivalent
laser line width is dominated by high-frequency components. This high
frequency noise can be definitively associated with high frequency variations in the jet thickness. A spectral width below 5 kHz is obtained with
the faster loop response provided by an optional intracavity AD*P phase
modulator.[4]

While this laser is sufficient and effective for many interesting kinds
of spectroscopy, especially sub-Doppler spectroscopy, the single-mode
tuning range is really not adequate to search for a weak, narrow line just
guided only by the 1/2 Å pointing capability of a usual small monochromator.
Thus we have developed, and report here, a new automatic fringe-counting
dual wavelength interferometer which presents a realtime numerical wavelength display.[5]

As may be seen from the schematic diagram (Fig. 2), this device is basically an automatic-scanning Michelson interferometer which uses corner-cube retroreflectors. With the two folding mirrors, motion of the carriage lengthens one arm and shortens the other. The corner cubes insure accurate optical alignment independent of possible angular wander of the carriage. To eliminate optical feedback into the two laser sources, and to keep the beams separate, we "stack" them carefully into uncoupled sectors of the corner cubes.[6]

As the reference wavelength and dye laser wavelength differ, a different number of fringes of each will be counted per unit travel of the carriage. The fringe interpolations could be handled with the techniques of length metrology, as in the Sigma Meter.[7] It turns out to be attractive to use a constant translation velocity to map the two <u>spatial</u> fringe patterns into two fringe <u>frequencies,</u> since we then have available the powerful techniques of frequency metrology. In particular we note that 100 resolution levels (counts) per fringe can be obtained by phase-locking 1/100 the output frequency of an oscillator to match the fringe frequency. Essentially the phase-locked oscillator provides the fringe <u>interpolation</u> information via <u>digital</u> counting. Naturally the phase-lock loop must be sufficiently precise and fast, and the translation adequately uniform so that the phase error between the optical fringe and its electrical analog can be as small as desired. Peak phase errors less than 1/50 fringe were readily obtained with a primitive carriage translation mechanism and 100 μsec loop settling times. In fact, phase-locked oscillators of similar loop dynamics are used on both fringe detector signals to minimize carriage acceleration errors as well as to provide signal/noise enhancement by averaging a number of successive fringe zero crossings. As illustrated in Fig. 2, direct counter reading, C, in absolute vacuum wavelength units can be obtained by taking the preset count number, N, to be $N = \lambda_{ref,vac} [1 + (n_\lambda - n_{ref}/n_{ref})]$, where

Figure 2

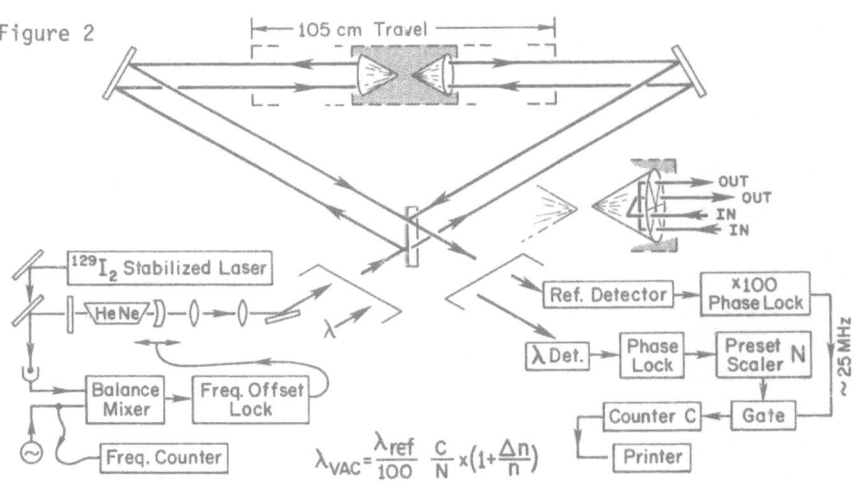

CORNER CUBE INTERFEROMETER SYSTEM
《 Lambda Meter 》

n_λ and n_{ref} are the appropriate indices of refraction. The counted reference wavelength, $\lambda_{ref,vac}$ = 632.991 045 nm, is obtained from the wavelength[8] of the $^{129}I_2$-stabilized He-Ne laser[9] by allowing for the -724 MHz frequency offset of the measured laser.

An important aspect of this frequency-space interpolation concept is that wavelength information of a given level of resolution is obtained in 100-fold less elapsed time than with simple fringe-counting. Thus with a modest carriage velocity of 4 cm/sec, the fringe rate is 250 kHz while the phase-locked oscillator produces 25×10^6 counts per second: one has thus four measurements each second of $\sim 10^{-7}$ resolution (about 1/10 of a typical Doppler relative width)! Resolution and display update rate can be exchanged conveniently by changing the preset number, N, in decade steps.

In summary, we listed a few problems confronting users of cw dye lasers. We then described a cw dye laser which coarse tunes smoothly without skipping spectrally, has a 50 kHz spectral width, is single mode tuneable over 2 GHz, at 5900 Å gives 120 mW output of almost gas laser beam quality from 1 W of 5145 Å pump light, has intensity stabilization to $\approx 10^{-4}$, and has sufficient optical isolation to allow 100% mode matched retroreflection without noticeable effect.

To tune the dye laser to an absolute wavelength, we developed an automatic-scanning, dual-wavelength interferometer which presents a real-time numerical wavelength display of sub-Doppler resolution. At a resolution of 2×10^{-7} four display updates are obtained per second. Decade resolution/update rate choices are provided, with the highest resolution being 2×10^{-9} (10 MHz out of 5×10^{14} Hz) for the present apparatus. Accuracy of at least 2×10^{-7} has been demonstrated with the first saturated absorption experiments in neon.[10]

References

1. J. J. Snyder and J. L. Hall, Laser Spectroscopy, ed. by S. Haroche, J. C. Pebay-Peyroula, T. W. Hänsch and S. E. Harris, Springer Verlag, Heidelberg, 1975, p. 6.

2. See Ref. 1, and R. L. Barger, M. S. Sorem and J. L. Hall, Appl. Phys. Lett. 22, 573 (1973).

3. R. L. Barger and J. B. West, private communication.

4. J. L. Hall and S. A. Lee, unpublished. See also R. L. Barger, J. B. West and T. C. English, Appl. Phys. Lett. 27, 31 (1975).

5. J. L. Hall and S. A. Lee, Appl. Phys. Lett., to appear in issue of Sept. 15, 1976.

6. W. R. C. Rowley, IEEE Instrum. & Meas. IM-15, 146 (1966).

7. P. Juncar and J. Pinard, Opt. Commun. 14, 438 (1975).

8. CCDM, Session of 1973, Bureau International des Poids et Mesures, Sevres, France.

9. W. G. Schweitzer, Jr., E. G. Kessler, Jr., R. D. Deslattes, H. P. Layer and J. R. Whitstone, Appl. Optics 12, 2927 (1973).

10. See Ref. 5, and to be published.

OPTICALLY PUMPED GAS LASERS*

H. Kildal and T.F. Deutsch
Lincoln Laboratory, Massachusetts Institute of Technology
Lexington, Massachusetts 02173, USA

I. Introduction

Recent advances in techniques for remote probing of trace gases in the atmosphere [1, 2], in laser induced photochemistry [3, 4], and in laser isotope separation [5, 6] have created a need for new infrared laser sources at frequencies that match molecular vibrational absorption bands. The work on optically pumped infrared lasers described here has been motivated primarily by the needs of the Los Alamos Scientific Laboratory UF_6 isotope separation program which requires lasers at 7.74, 8.62, 12.1 or 15.9 μm [7].

Optical pumping is a powerful technique for surveying new infrared laser systems. It is a more general method for obtaining laser action on vibrational transitions than conventional electric discharge excitation since it avoids molecular dissociation and since the excitation energy is deposited into a single vibrational mode, avoiding the need for differential relaxation between the upper and lower laser levels. A large number of lines can be generated starting with only a single efficient pump laser (i.e., HF or CO_2). The compact and sealed-off laser cells use only a small gas volume, making operation with isotopes practical. Finally, the efficiency of the optically pumped lasers can be high.

Although optical pumping has produced hundreds of far infrared laser lines [8], its use in the 5-20 μm region is more recent. CHANG and WOOD demonstrated laser action in CO_2 by direct pumping with a HBr laser [9] and in N_2O using collisional energy transfer from optically pumped CO_2 [10]. The latter system allowed pressures up to 42 atm to be reached. Direct pumping with a CO_2 laser has produced laser action in OCS at 19 μm [11], in SF_6 at 15.9 μm [12], and in NH_3 at

---

* This work was sponsored by the U.S. Energy Research and Development Administration under a subcontract from the Los Alamos Scientific Laboratory and by the Department of the Air Force.

12.8 μm [13]. Recently an HF laser has been used to pump the $10^{\circ}1$ level of several CO_2 isotopes producing laser action at 4.3, 10.6 and 17 μm [14]. Lasing on the CO_2 bending mode transition has also been accomplished using an HBr laser to pump low pressure HBr-CO_2 mixtures [15]. A number of optically pumped molecular lasers have also been obtained using E-V(Electronic-Vibrational) transfer from electronically excited atomic bromine produced by a flash lamp [16].

We have previously reported optically pumped laser action in a number of molecules (OCS, CO_2, N_2O, C_2H_2 and CS_2) [17]. In each case the molecules were pumped by resonant vibrational energy transfer from CO gas excited by a frequency doubled CO_2 laser. Here we briefly review that work and report on recent progress which includes laser action at 7.9 μm in the nonlinear molecule SiH_4, lasing at 6.6 μm in CS_2, grating tuning of the optically pumped OCS laser, high pressure operation of the CO_2 system, and increased output energies and efficiencies.

The CO molecule is ideal for storage of vibrational energy because of its exceedingly slow vibration to translation transfer rate of 1.9×10^{-3} Torr^{-1} sec^{-1} [18]. It is efficiently excited by the second harmonic of the CO_2 P(24) line in the 9.6 μm band which falls within 0.003 cm^{-1} of the CO 0→1 P(14) transition. With resonant energy transfer from the CO it is possible to vibrationally excite a number of molecules without relying on an exact frequency match between the pump laser and the active molecule; the mismatch can be up to 200 cm^{-1}. In addition infrared inactive modes can be excited. By contrast direct optical excitation of the lasing molecule, which we have also investigated, requires a frequency coincidence better than 0.1 cm^{-1} between the pump laser and an absorption line of the laser gas.

II. Experimental

Figure 1 shows a schematic of the experimental setup. The grating controlled CO_2 TEA laser is weakly focused by a 1.1 m f.l. mirror and the uncoated $CdGeAs_2$ doubling crystal [19, 20] is placed so that the converging beam fills most of the crystal cross-section. The crystal is kept at 77 K to reduce optical absorption losses. The LiF filter blocks the CO_2 radiation so that only the second harmonic pulse is incident on the laser cell. We have used cells with path lengths from 0.4 to 54 cm, some having provision for cooling to -80°C. Cavities with both internal and external dielectric coated mirrors and with gratings have been used. The input energy to the cell and the output energy are measured using calibrated pyroelectric detectors and the output wavelengths are measured using a 1 m grating spectrometer.

The $CdGeAs_2$ crystal technology, vital to our experiments, has improved substantially in the course of this work. In our initial experiments 10 mJ of second harmonic energy was incident on the laser cell while in the most recent experiments (OCS, high efficiency CO_2, direct pumped CS_2) we applied up to 50 mJ to the cell.

The highest second harmonic energy we have measured to date is 200 mJ from an uncoated $CdGeAs_2$ crystal. This result was obtained in a setup similar to the one shown in Fig. 1, but without a grating in the TEA laser cavity. Figure 2 shows the dependence of the second harmonic output energy on the CO_2 laser energy incident on the doubling crystal. The relatively low second harmonic conversion efficiency, 5 to 7 percent, in these experiments is due to the multimode beam and to

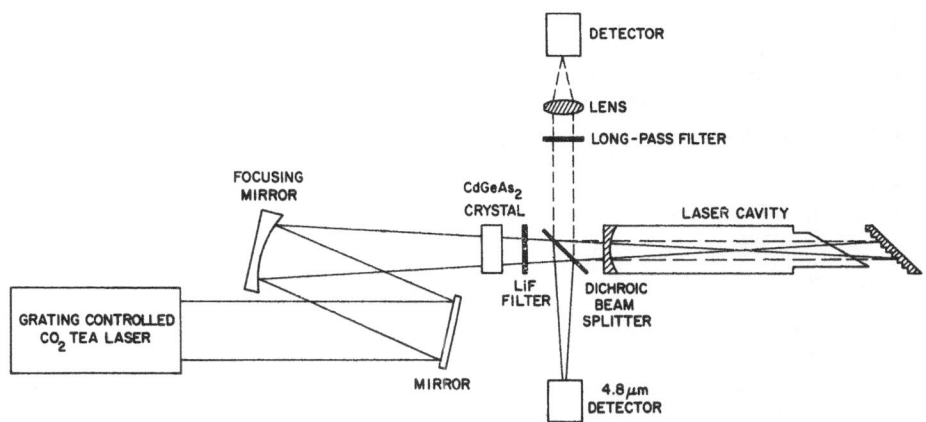

Fig. 1. Schematic diagram of optical pumping setup

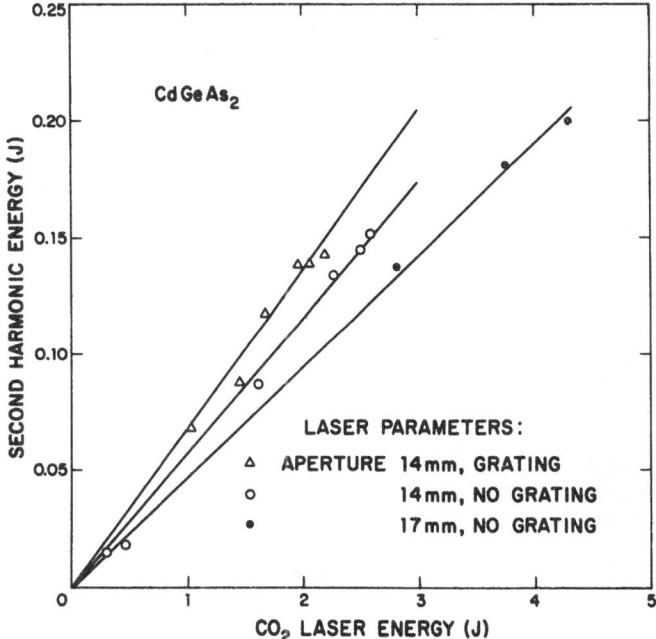

Fig. 2. Pulsed second harmonic generation of CO_2 laser radiation in an 11 mm long, 2×12 mm^2 cross-section, uncoated CdGeAs$_2$ crystal using different apertures within the CO_2 laser cavity and operating the laser with and without a grating

the fact that a portion of the pulse energy was in the low power nitrogen tail. The linear power dependence is probably due to saturation caused by linear absorption in the crystal and to the non-ideal character of the laser beam. In another experiment with a 70 nsec TEM_{00} mode CO_2 laser pulse and an A.R. coated $CdGeAs_2$ crystal we have obtained an external energy conversion efficiency of 27%. Based on these results we estimate that a CO_2 TEA laser with higher TEM_{00} mode energy and an A.R. coated state-of-the-art $CdGeAs_2$ doubling crystal it should be possible to achieve second harmonic energies approaching 1 J.

III. Results

Figure 3 shows a simplified energy level diagram for CO and the lasing molecules and gives the line centers for the laser emission. Table I summarizes our results. Not all quantities in the table were obtained under the same conditions; for example, maximum output energy is generally not obtained at maximum pressure. The quoted efficiencies are slope efficiencies, obtained from plots of output energy versus input energy to the gas.

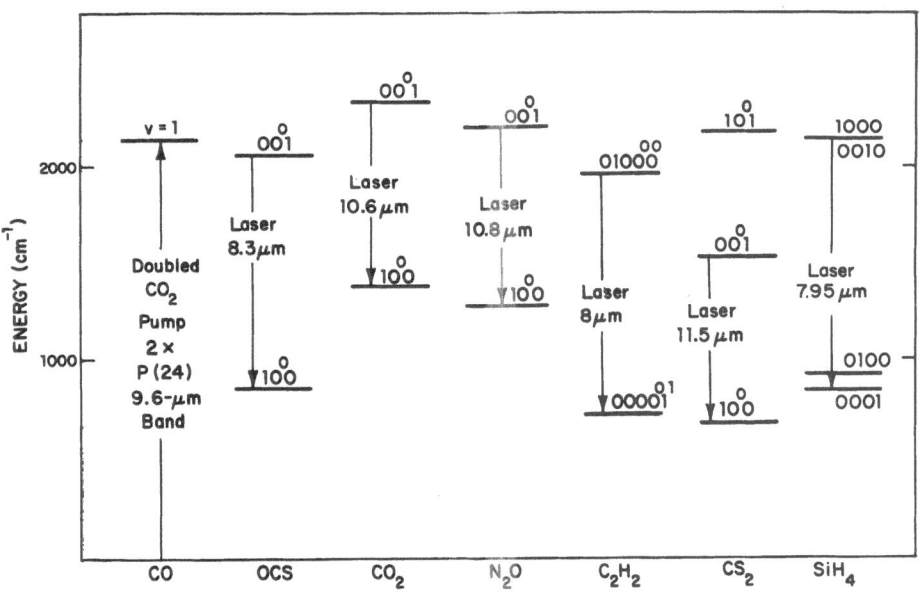

Fig. 3. Simplified energy level diagram for CO
and the molecules pumped by nearly
resonant energy transfer

Table I

Summary of performance of optically pumped lasers

| System | Maximum Pressure (Torr)* | Minimum Threshold | | Maximum Efficiency (Percent) | Maximum Output (mJ) |
		Focused (mJ)	Unfocused (mJ)		
OCS (direct-pumped)	55	0.1	0.6	19.0	5.2
CO-OCS	420	0.1	1.0	7.6	1.3
CO-CO$_2$	12,400 (16.3 atm)	-	·1.6	34.0	13.0
CO-C$_2$H$_2$	610	-	2.0	3.5	0.12
CO-CS$_2$ (11.5 μm)	20	-	2.5	0.5	0.03
CS$_2$ (6.6 μm)	25	-	6.3	0.1	0.03
CO-SiH$_4$	35	1.7	-	0.6	0.03

*With H$_2$ or He buffer

A. OCS

For applications such as isotope separation line selection is important. With a grating it is possible to tune over the whole rotational manifold. We have examined tuning techniques for the OCS laser. At a total OCS-CO pressure of 12 Torr and with an active length of 14 cm the threshold energy for a cavity with a Brewster window and a grating was 2.2 mJ, about ten times higher than for the internal mirror cavity. The grating tuned OCS laser provided over 80 lines between 8.19 and 8.46 μm and the output tuned from R(8) to R(47) and from P(6) to P(48) on the $00^\circ 1 \rightarrow 10^\circ 0$ transition.

Cooling of the OCS laser reduces the threshold. The slope efficiency, however, remains the same. For a 54 cm internal mirror cell with 1 percent output coupling and a OCS:CO ratio of 1:1 the threshold decreased by a factor of 3.5 when the laser cell was cooled from room temperature down to -80°C.

We have also obtained laser action near 8.65 μm using O^{13}CS [21]. The threshold was higher and the slope efficiency was lower by about a factor of two than with the normal isotope. This may be due to a lower V-V energy transfer rate from CO into $O^{13}CS$ since the energy mismatch is 134 cm^{-1} compared to 81 cm^{-1} for the normal isotope [22]. Furthermore the self deactivation rate for the upper laser level may be larger for $O^{13}CS$ because of a closer energy match between the $00^{\circ}1$ and the $12^{\circ}0$ and $04^{\circ}0$ levels. By tuning the $O^{13}CS$ laser with a grating, emission is expected in the 8.5 to 8.8 μm region. Since this overlaps the $SO_2(\nu_1)$ absorption band near 8.7 μm this laser could be useful for remote detection of this pollutant. The $O^{13}CS$ laser also overlaps the $UF_6(\nu_2 + \nu_3)$ combination band at 8.6 μm.

B. C_2H_2

In the case of OCS, CO_2 and N_2O the energy is transferred from CO into a vibrational mode with a dipole moment to the ground state. These molecules could also be excited directly if a suitable laser source were available. The C_2H_2 system provides an example of transfer into an infrared inactive mode.

We have previously reported wavelength measurements for the C_2H_2 laser [17]. We have now identified the laser emission as the Q(9), Q(11), Q(13) and Q(15) lines of the $01000 \to 00001^1$ transition. The observed lines, together with the measurements of SHELTON and BYRNE [23], are listed in Table II and there is good agreement with the calculated frequencies. Only odd J transitions are observed since the odd rotational levels have three times the statistical weight of the even levels. We also list the Q-branch transitions for the C_2HD and the C_2D_2 molecules which can probably also be pumped by energy transfer from CO. It may be possible to extend the laser action in C_2H_2 to the P-branch by inserting a grating in the cavity. R-branch operation is less likely because of the $(\nu_4^1 + \nu_5^1)$ combination band absorption near 7.5 μm.

The threshold of the acetylene laser was reduced by removing the traces of acetone normally present in commercial C_2H_2 by a cold trap at -80°C. Adding either He or H_2 to the C_2H_2-CO system improved the laser performance significantly, increasing the high pressure limit, the maximum output power, and the slope efficiency. A maximum output power of 0.12 mJ was obtained with a 1:3:3 C_2H_2:CO:H_2 mixture in a 14 cm cell at 30 Torr. The maximum slope efficiency of 3.1% was observed in a 1:3:30 C_2H_2:CO:He mixture at 250 Torr total pressure.

C. CS_2

In the CS_2 system the main energy transfer is probably into the $10^{\circ}1$ combination band [26]. While combination bands can be excited, their excitation will rapidly be transferred to fundamental modes by nearly resonant V-V collisions. Recent wavelength measurements have confirmed that laser action in CS_2 at 11.5 μm is on the $00^{\circ}1 \to 10^{\circ}0$ transition [27]. The origin of a number of anomalous intensity variations observed with this laser is presently not well understood [27-29].

Since the excitation technique generates population in both the upper and the lower laser levels, the efficiency of the CS_2 laser is lower than for the other linear triatomic laser systems. The efficiency can be improved some, however, by adding hydrogen which increases the deactivation rate of the lower laser level [17].

Table II

Observed C_2H_2 laser lines			Calculated frequencies for the $01000-00001^1$ Q(J) - branch transitions		
This Work	Ref. a	J	C_2H_2 (ref. b & c)	C_2HD (ref. b)	C_2D_2 (ref. b)
		1	1245.17	1176.04	1226.11
		2	1245.14	1176.02	1226.09
		3	1245.08	1175.99	1226.06
		4	1245.02	1175.94	1226.02
		5	1244.93	1175.89	1225.97
		6	1244.83	1175.83	1225.91
	1244.70	7	1244.71	1175.75	1225.84
		8	1244.57	1175.67	1225.76
1244.52	1244.46	9	1244.42	1175.57	1225.66
		10	1244.25	1175.46	1225.56
1244.12	1244.09	11	1244.06	1175.34	1225.45
		12	1243.85	1175.22	1225.33
1243.75	1243.64	13	1243.63	1175.08	1225.20
		14	1243.39	1174.93	1225.05
1243.08	1243.12	15	1243.13	1174.77	1224.90
		16	1242.86	1174.60	1224.73
		17	1242.57	1174.42	1224.56
		18	1242.26	1174.23	1224.38
		19	1241.93	1174.03	1224.18
		20	1241.59	1173.82	1223.98

a. Ref. 23
b. Ref. 24
c. Ref. 25

We have also obtained laser action at 6.6 μm on the CS_2 $10°1 \rightarrow 10°0$ transition. In this experiment we used direct pumping. The second harmonic of the 9.2 μm CO_2 R(30) line, which lies within 0.041 cm^{-1} of the CS_2 $00°0 \rightarrow 10°1$ P(60) transition [30], was used for excitation. We observed only a single line, the P(60) line, and not the corresponding R(58) line. We tried unsuccessfully to obtain more lines by adding argon, which should promote rotational equilibration without contributing significantly to the deactivation rate of the $10°1$ level. The measured threshold energy of 6.3 mJ in Table I was obtained for a 54 cm long internal mirror cell operated at room temperature with 8 Torr of CS_2. The mirrors were a 1 percent output coupler and a gold total reflector. The laser pulse length varied from 140 nsec at 1.5 Torr to less than 10 nsec at the pressure limit at 25 Torr.

D. SiH_4

The molecules discussed above are all linear molecules consisting of 3 or 4 atoms. The thresholds for laser action in nonlinear molecules are expected to be higher for several reasons. The rotational degeneracies are lifted, reducing the gain of each vibrational-rotational transition. The number of modes of vibration is also larger, increasing the number of channels for vibrational relaxation and making the production of an inversion more difficult. The limitations due to rotational and vibrational relaxation can be overcome by using direct rather than transfer pumping and operating at sub-Torr pressures so that no collisions occur during the pump pulse [12].

We have obtained laser action in SiH_4-CO mixtures at total pressures up to 35 Torr. This is the first reported laser action in SiH_4 and it is also the first time an optically pumped energy transfer laser has been demonstrated for other than a linear molecule.

A total of 6 SiH_4 laser lines were observed at 1251.25, 1254.86, 1257.10, 1258.02, 1262.31, and 1265.50 ± 0.04 cm^{-1}. SiH_4 is a tetrahedral molecule and its four modes are: 2187 cm^{-1}(ν_1), 975 cm^{-1}(ν_2), 2191 cm^{-1}(ν_3), and 914 cm^{-1} (ν_4). Only the ν_3 and the ν_4 modes are infrared active. The CO energy is transferred into the ν_1 and the ν_3 modes. There are several possible laser transitions in the 8 μm region originating from these levels: $\nu_1 \rightarrow \nu_4$ (1272.8 cm^{-1}), $\nu_3 \rightarrow \nu_4$ (1276.4 cm^{-1}), and $\nu_3 \rightarrow \nu_2$ (1216.0 cm^{-1}). The spectroscopic constants for SiH_4 are not well enough known to identify the observed laser lines. We can only state that the lasing is not on the $\nu_3 \rightarrow \nu_2$ transition because its band center is too far away from the observed laser frequencies. The wavelength measurements were performed without any frequency selecting element in the laser cavity. With a grating it might be possible to extend the tuning range in the 7.9 μm region and to obtain laser action near 8.2 μm on the $\nu_3 \rightarrow \nu_2$ transition.

Typical CO:SiH_4 ratios ranged from 1:3 to 4:1, with a 1:1 ratio being optimum. In contrast to the C_2H_2 and CS_2 systems, the addition of helium did not increase either the output energy or the operating pressure.

E. High Pressure Operation

High pressure operation of optically pumped molecular lasers offers the potential of continuous tunability. We have examined the high pressure limit for a number of systems; the results are summarized in Table I. The upper pressure limit is determined by the deactivation rate of the upper laser level. Experimentally, it is found that the pump power necessary to reach threshold increases linearly with pressure up to a critical pressure above which the threshold increases exponentially. This change in pressure dependence occurs when the deactivation time is comparable to the pump laser pulse length. Since the deactivation rate due to helium is usually slower than the self-deactivation rate, helium was added to the laser medium to increase the pressure limit.

The CO_2 laser is particularly suitable for high pressure operation because of the low self-deactivation rate of the $00^{\circ}1$ mode, $342 \text{ sec}^{-1} \text{ Torr}^{-1}$ [31], and we have operated a CO_2-CO-He system up to 16 atm, a limit imposed by the cell windows. At this pressure the rotational lines of CO_2 are sufficiently broadened to allow continuous tuning. The use of helium rather than CO_2 for pressure broadening has the advantage that it pressure broadens CO_2 about 2/3 as much as CO_2 itself, but deactivates the CO_2 upper level only 1/4 as rapidly. The three component gas system thus allows the absorption of the pump laser, the density of the active medium, and the pressure broadening to be controlled somewhat independently.

Figure 4 shows the input-output curve for the CO:CO_2:He laser for different helium partial pressures using a 4.3-cm cell with internal mirrors and a one percent output coupler. For a 9:1:57 mixture the measured threshold energy at 16 atm was 2.1 mJ, the slope efficiency was 6 percent, and the maximum output energy was 0.25 mJ. The pulse length at 16 atm was about 50 nsec.

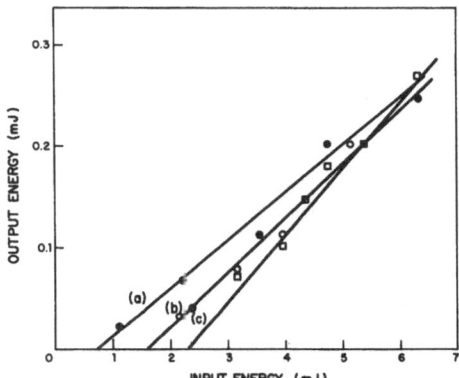

Fig. 4. CO_2 laser output energy versus input energy for a 4.3-cm cell with a one percent output coupler, filled with 2.2 atm of CO_2, 0.24 atm of CO, and He partial pressures of 0 atm (a), 8.8 atm (b), and 13.9 atm (c)

At 16 atm the rotational lines of CO are broadened into a semicontinuous band and it is therefore not necessary to rely on a specific coincidence between the CO_2 second harmonic and a CO absorption line. This was verified by producing laser action using the P(20) and P(22) lines of the CO_2 9.6 μm band as well as the usual P(24) line.

A tunable high pressure CO_2 laser will need to incorporate a grating, an etalon, and apertures for mode selection in an external mirror cavity. When the 16 atm CO_2 laser was operated with an external cavity the additional window losses and the increased build-up times raised the threshold to ~11 mJ. Grating tunable operation was estimated to require at least twice this energy and could not be obtained with the pump energy available at the time, but should now be achievable. This will make possible an efficient, high repetition rate, continuously tunable CO_2 laser suitable for spectroscopy.

F. Efficiency Considerations

As Table I shows, high efficiencies are obtained in some optically pumped systems. The measured maximum slope efficiency of 34 percent for $CO-CO_2$ is close to the maximum possible efficiency of 45 percent, which corresponds to unit quantum efficiency. By contrast, even after considerable efforts at optimization, the OCS laser shows a lower efficiency which we believe is due to a bottle-neck at the lower laser level. For the direct pumped OCS laser at 10 Torr, the laser pulse terminates within 500 nsec of the initiation of pumping, well before collisional deactivation of the upper laser level, which takes 2 μsec at 10 Torr [26]. We believe the reduced efficiency is due to the fact that the lower level deactivation time is long compared to the 200 nsec pulse length. The deactivation rate for OCS has not been measured. On the basis of energy level considerations, however, it is expected to be slower than the comparable rate in N_2O. For N_2O the deactivation time is 5 μsec at 10 Torr [32]. Based on these considerations the maximum efficiency for the direct pumped OCS laser should be half of unit quantum efficiency. The observed efficiency of 19 percent is in reasonable agreement with a theoretical efficiency of 29 percent for a quantum efficiency of 0.5.

We have demonstrated that optical pumping combined with vibrational energy transfer is a general technique for producing new infrared lasers and have obtained numerous laser lines in the 6 to 12 μm region with output energies up to 13 mJ and efficiencies ranging from 0.1 to over 30 percent.

REFERENCES

1. R. L. Byer, Optics and Quant. Electron. 7, 144 (1975).

2. E. D. Hinkley, Optical and Quant. Electron. 8, 155 (1976).

3. J. T. Knudtson and E. M. Eyring, Annual Rev. Phys. Chem. 25, 255 (1974).

4. F. Klein, F. M. Lussier, and J. I. Steinfeld, Spect. Lett. 8, 247 (1975).

5. R. V. Ambartsumian, Yu. A. Gorokhov, V. S. Letokhov, and G. N. Makarov, JETP Lett. 21, 171 (1975).

6. J. L. Lyman, R. J. Jensen, J. Rink, C. P. Robinson, and S. D. Rockwood, Appl. Phys. Lett. 27, 87 (1975).

7. R. J. Jensen, J. G. Marinuzzi, C. P. Robinson, and S. D. Rockwood, Laser Focus 12, 51 (May 1976).

8. M. Rosenbluh, R. J. Temkin, and K. J. Button, Applied Optics (to be published Nov 1976).

9. T. Y. Chang and O. R. Wood, Appl. Phys. Lett. 21, 19 (1972).

10. T. Y. Chang and O. R. Wood, Appl. Phys. Lett. 24, 182 (1974).

11. H. R. Schlossberg and H. R. Fetterman, Appl. Phys. Lett. 26, 316 (1975).

12. H. R. Fetterman, H. R. Schlossberg, and W. E. Barch, Optics Commun. 15, 358 (1975).

13. T. Y. Chang and J. D. McGee, Appl. Phys. Lett. 28, 526 (1976).

14. M. I. Buchwald, C. R. Jones, H. R. Fetterman, and H. R. Schlossberg, Appl. Phys. Lett. (to be published).

15. R. M. Osgood, Jr., Appl. Phys. Lett. 28, 342 (1976).

16. A. B. Petersen, C. Wittig, and S. R. Leone Appl. Phys. Lett. 27, 305 (1975).

17. H. Kildal and T. F. Deutsch, Appl. Phys. Lett. 27, 500 (1975).

18. M. A. Kovacs and M. E. Mack, Appl. Phys. Lett. 20, 487 (1972).

19. H. Kildal and J. C. Mikkelsen, Opt. Commun. 10, 306 (1974).

20. The $CdGeAs_2$ doubling crystals used in present experiments were grown by Dr. G. W. Iseler at Lincoln Laboratory.

21. We are grateful to Dr. M. I. Buchwald at the Los Alamos Scientific Laboratories for supplying us the $O^{13}CS$ isotope.

22. Y. Morino and T. Nakagawa, J. Mol. Spectrosc. 26, 496 (1968).

23. C. F. Shelton and F. T. Byrne, Appl. Phys. Lett. 17, 436 (1970).

24. H. Fast and H. L. Welsh, J. Mol. Spectrosc. 41, 203 (1972).

25. J. Pliva, J. Mol. Spectrosc. 44, 145 (1972).

26. J. K. Hancock, D. F. Starr, and W. H. Green, J. Chem. Phys. 61, 3017 (1974).

27. T. F. Deutsch and H. Kildal, Chem. Phys. Lett. (to be published).

28. C. K. N. Patel, Appl. Phys. Lett. 7, 273 (1965).

29. L. Y. Nelson, C. H. Fisher and S. R. Byron Appl. Phys. Lett. 25, 517 (1974).

30. A. G. Maki and R. L. Sams, J. Mol. Spectrosc. 52 233 (1974).

31. G. Inoue and S. Tsuchiya, J. Phys. Soc. Japan 38, 870 (1975).

32. R. T. V. Kung, J. Chem. Phys. 63, 5305 (1975).

CARS TECHNIQUES AND APPLICATIONS

J.-P. Taran

Office National d'Etudes et de Recherches Aérospatiales (ONERA)
92320 Châtillon, France

Coherent anti-Stokes Raman Scattering (CARS) now ranks among the most useful phenomena in nonlinear optics. A large number of applications of nonlinear optics had previously been geared to the generation of new frequencies by harmonic generation or multiple wave mixing. In CARS on the contrary, one is ultimately interested in the material which supports the optical mixing: the fluxes generated in the sidebands are merely collected and measured, so as to yield the value of the nonlinear susceptibility. Since that susceptibility depends on the invariants of the Raman tensor and exhibits the same Raman vibrational, rotational and electronic resonances, one has in CARS an alternate tool for Raman spectroscopy. But this new tool offers the advantage of very large optical signals, due to the parametric-like character of the sideband generation.

MAKER and TERHUNE [1] were the first to perform CARS spectroscopy in liquids with a ruby laser: they were able to plot the dispersion curves of benzene, bromobenzene and toluene. Several original applications of CARS were then proposed in both solids [2] and gases [3-5]. However, the possibility of doing Raman spectroscopy by CARS was not realized until the advent of tunable lasers [6-8].

Solids, liquids and gases have all been tackled by CARS. Although the most complete demonstration of the possibilities of CARS is only possible in crystals, the method definitely has a broader range of applications in liquids and gases. As a matter of fact, CARS may well become the best approach to fast, high resolution Raman spectroscopy and point, instantaneous nonintrusive analysis of biological and gaseous samples. Our purpose here is to give an update on several recent achievements of CARS in the investigation of reactive media. We shall, however, examine first some aspects of the theory which help in the understanding of the fundamental limitations. The prospects for new applications and performance amelioration will be discussed in the last section.

1 Waves and Vibrations

The problem of the generation of the anti-Stokes sideband in the course of stimulated Raman scattering was treated long ago [9-12]. An approach analogous to that of MAIER et al [13] is adequate.

Assume plane waves of frequencies ω_1 and ω_2 ($\omega_2 < \omega_1$) with parallel field vectors k_1 and k_2 aligned along the z axis. Let these waves traverse a pure gas of number density N, with an anharmonic, Raman active, vibrational mode with $0 \rightarrow 1$ transition at $\omega_v \simeq \omega_1 - \omega_2$. The nonlinear coupling of the pump waves at ω_1 and ω_2 through the resonant medium is responsible for the generation of sidebands in the form of plane waves collinear with the pump at $\omega_3 = 2\omega_1 - \omega_2$ and $\omega_4 = 2\omega_2 - \omega_1$. If we assume power densities I_i such that $I_1 \gg I_2$, then $I_4 \ll I_3$ and the influence of I_4 can be neglected. The derivation of I_3 and calculation of its reaction on I_1 and I_2 is a straightforward matter when the interaction is weak, while the perturbation caused to the vibrational population is also easily calculated. In the often achieved phase-matched case, one finds :

$$I_3 \simeq \left(\frac{4\pi^2\omega_3}{c^2}\right)^2 I_1^2 I_2 \left|\chi^2\right| z^2 \tag{1}$$

with susceptibility $\chi = \dfrac{2c^4}{\hbar\omega_1\omega_2^3} N\delta g \dfrac{d\sigma}{d\Omega} \dfrac{\omega_v}{D}$,

N number density of molecules, g weighting factor, δ population probability difference between the 0 and 1 states, $\dfrac{d\sigma}{d\Omega}$ Raman cross section, $D = \omega_v^2 - (\omega_1 - \omega_2)^2 - i\gamma(\omega_1 - \omega_2)$, $\gamma = T_1^{-1} + T_2^{-1}$. It is also interesting to examine the rates of exchange of quanta per unit volume. Using the photon fluxes $N_i = I_i / \hbar\omega_i$, one can show that [14]

$$\frac{\partial N_1}{\partial z} = a\left\{|D/D''W_2 + 2W_3\left[-D'^2\cos\frac{\Delta kz}{2} + D'D''\sin\frac{\Delta kz}{2}\right]\right\} \tag{2a}$$

$$\frac{\partial N_2}{\partial z} = a\left\{-|D|D''W_2 + W_3\left[(D'^2 - D''^2)\cos\frac{\Delta kz}{2} - 2D'D''\sin\frac{\Delta kz}{2}\right]\right\} \tag{2b}$$

$$\frac{\partial N_3}{\partial z} = a\left\{ \qquad W_3\left(D'^2 + D''^2\right)\cos\frac{\Delta kz}{2} \qquad \right\} \tag{2c}$$

$$\frac{N}{2}\frac{\partial\delta}{\partial t} = a\left\{|D|D''W_2 + W_3\left[2D''^2\cos\frac{\Delta kz}{2} - 2D'D''\sin\frac{\Delta kz}{2}\right]\right\} - \frac{N}{2T_1}(\delta - \delta^\circ) \tag{2d}$$

with $a = 8\pi^2 c^2 N\delta \dfrac{d\sigma}{d\Omega} \dfrac{W_1^2 W_2}{|D|^3} \dfrac{\omega_v}{\omega_1\omega_2^3}$, $W_i = (N_i\omega_i)^{1/2}$, $D = D' + iD''$,

$\Delta k = 2k_1 - k_2 - k_3$, δ° equilibrium value of δ prior to interaction.

The relaxation term in (2d) represents the action of collisions tending to restore BOLTZMANN equilibrium. One can recognize three distinct channels of energy flow between optical and phonon modes [9] .

(i) The dominant terms are those of stimulated Raman scattering Fig. 1a, which feeds energy, quantum for quantum, from ω_1 to ω_2, the balance being taken up by the medium. The coefficient is proportional to the imaginary part of χ (represented here by D'').

(ii) The cosine terms stand for two distinct processes contributing to CARS: a parametric process associated with the real part of χ Fig.1b, and a Raman process associated with the imaginary part of χ Fig.1c.

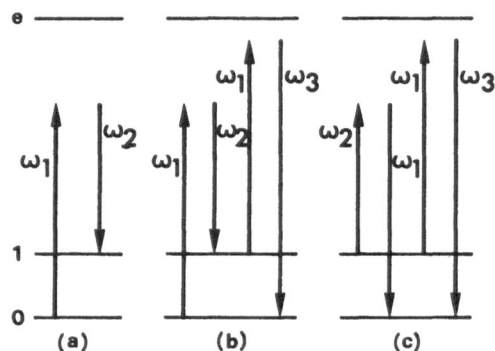

Figure 1 - Stimulated Raman scattering and CARS. Level 0 is the ground level, level 1 any Raman active level. Resonance enhancement can take place if the transition sequence brings the molecule close to an electronic level or an absorption continuum

(iii) Finally, an interesting interaction shows up when the waves are out of step (sine terms): the presence of the anti-Stokes induces an exchange between ω_1 and ω_2 through the sum of two processes of types 1b and 1c with same coefficient $D'D''$, leaving N_3 unaffected and changing the population by two quanta.

This presentation of CARS is academic. In real situations, one has different beam geometries, a much more complicated level structure of the molecules, and a relaxation of vibration energy that is not necessarily described by a simple term.

Focusing the beams gives spatial resolution because the generation of the anti-Stokes takes place only near the focus. REGNIER showed that spatial resolution is of the order of a few times the confocal parameter; in addition, the signal is nearly independent of focal length [15].

The molecular vibrational levels are split into many rotational sublevels, giving complex Raman spectral contours due to rotation-vibration coupling. The susceptibility then reads $\chi = \sum_j \chi_j^{res} + \chi'''$, where the χ_j^{res} are individual contributions from the resonances ω_{vj} close to $\omega_1 - \omega_2$, in the form given by Fig.1. Far away resonances give only small, slowly varying contributions, and they are grouped in the χ''' term which also includes a minute electronic contribution. CARS spectroscopy in gases is often confined to Q-branch analysis. Like in normal Raman, the data are used for extraction of concentrations and temperatures. The interpretation, however, is complicated somewhat by the interference between the different resonances. The case of an isolated resonance (e.g. most O, S lines, Q lines in H_2, D_2, HCL) is simple [14], the problem reducing to extraction of the magnitude of the resonance from a spectrum distorted by the flat background.

When $\omega_1 - \omega_2$ is swept across ω_{v_j}, the extremity of χ in the complex plane describes a circle of diameter d proportional to Nδ. This circle is tangent to the real axis at a point displaced by χ^{nr} from the origin Fig.2a. The modulus, which is the quantity accessible to experiment, thus goes through a maximum and a minimum whose difference is d irrespective of χ^{nr} Fig.2b. Therefore one gets Nδ from the experimental contours, and N if the BOLTZMAN equilibrium at a known temperature is established (which gives δ). Unfortunately, this treatment, which is valid for Lorentzian line shapes, produces approximate results on Q-branches when they are considered as single lines because of their asymmetry. In this case, only computer generated curves will reduce the data properly, with two parameters (concentration and temperature) left to adjust.

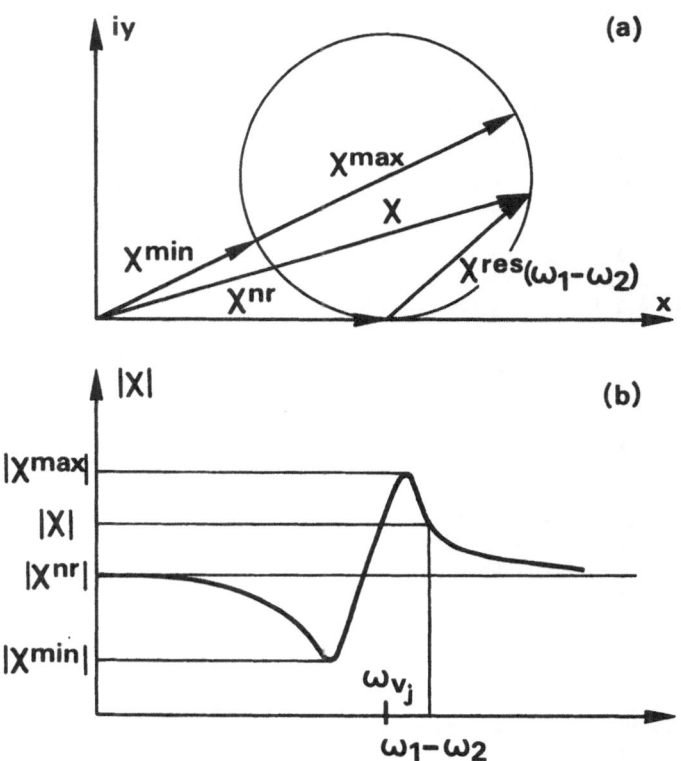

Figure 2 - Vector representation of χ in the complex plane and dispersion profile about resonance ω_{v_j}

Finally, the measurements can be considered as nonintrusive only in the limit $\frac{\partial \delta}{\partial t} = 0$. The product $I_1 I_2 t_p$, where t_p is the pulse duration, thus has to be kept small. We calculate that $I_1 = I_2 = 1$ GW/cm^2 would cause a 10% decrease in δ on resonance ($\omega_1-\omega_2 = \omega_{rj}$) in 20 ns for simple diatomic gases like N_2 or CO if rotational relaxation effects are ignored. The latter are important, the coupling between adjacent rotational levels being very fast with time constants under 1 ns. This eases somewhat the constraints on I_1 and I_2 in most gases (H_2 at room temperature is one of the few exceptions to this rule: 70% of the population is in the $J = 1$ state so that for the Q(1) line the much longer (x 10^6) vibrational relaxation time must be taken for T_1).

Finally, one may conclude that for a given f-number, there exists a maximum admissible pump power which depends on Raman linewidth, on population relaxation and on pulse duration. There results a certain CARS signal. One can only try and increase it by raising the power P_1, and decreasing P_2 as P_1^{-1} to keep P_1P_2 constant: P_3 will increase as P_1, until P_1 is large enough to cause breakdown or produce a large stimulated Raman gain for P_2.

2 Instrumental Specification and Experimental Results

The conditions under which significant temperature and concentration measurements can be done follow from the preceding discussion and from simple considerations:

(i) Diffraction limited laser pulses in the 10^5 - 10^6 W range at ω_1 and ω_2 are amply sufficient for a comfortable signal level; when larger powers are used, Δ_j, P_1 and P_2 may change, thus causing an error in the measurement.

(ii) Careful matching and alignment of the laser beams will insure optimum generation of the anti-Stokes, and signal reproduction.

(iii) For high resolution Raman analysis, laser linewidth and frequency stability should be equal to or better than the Raman linewidths. Even when these precautions are taken, we find that the performance of the set-up still leaves something to be desired. The anti-Stokes signal reproduction remains fairly poor due to the nonlinear nature of the interaction; until recently [14,16] , rms fluctuations in excess of 30% were observed after correction for laser power fluctuations by use of a reference; with improved mechanical stability, the figure was cut down to 10%. This figure is close to our error in reading the oscilloscope traces and corresponds to a 5% uncertainty on number densities.

(iiii) Discrimination between two species in a mixture is limited, even when their Raman resonances are far apart, because of their associated χ^{nr} which gives a signal throughout the spectral range and poses a problem analogous to the detection of a weak line over a uniform background of stray light in normal Raman scattering. For usual diatomic gases, $\chi^{nr}/\chi^{res}| \approx 10^{-2}$, and the minimum mole fraction that can be detected is about 10^{-3} with our present measurement accuracy.

Our set-up incorporates a single mode ruby laser and a tunable dye laser. Both are diffraction limited, with excellent stability [14,16] . Other systems, like that built at NLR [17] with a doubled yag and tunable dye laser, also meet the specifications listed above and operate satisfactorily. Nitrogen

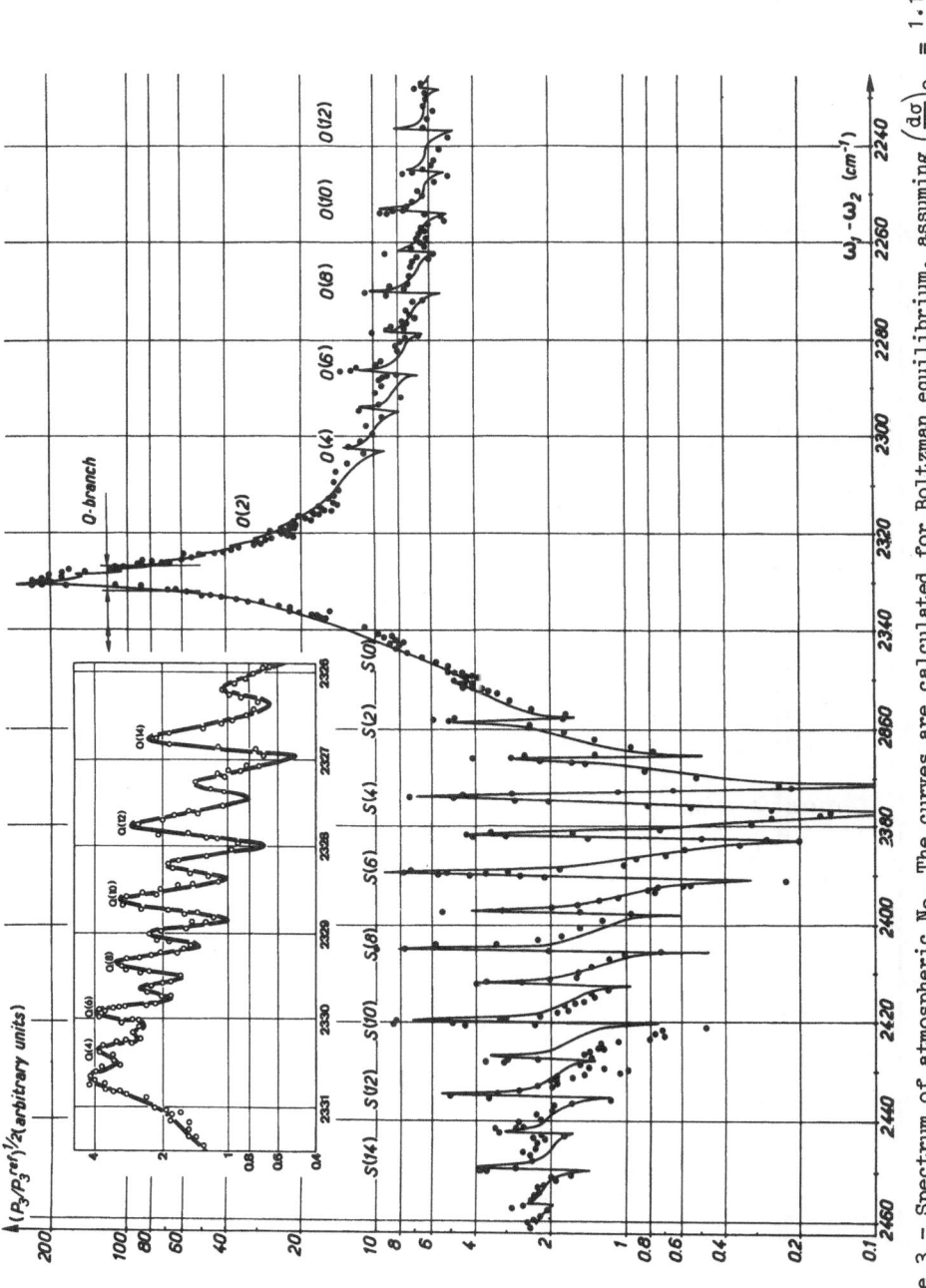

Figure 3 - Spectrum of atmospheric N_2. The curves are calculated for Boltzman equilibrium, assuming $\left(\frac{d\sigma}{d\Omega}\right)_Q = 1.1 \times 10^{-31}$ cm^2. Fitting parameters were $\left(\frac{d\sigma}{d\Omega}\right)_{0,S} = 0.55 \times 10^{-32}$ cm^2, $\chi^{nr} = 4 \times 10^{-18}$ cm^3/erg. This last value is in agreement with [3]. The vertical scale is roughly in units of 10^{-18} cm^3/erg

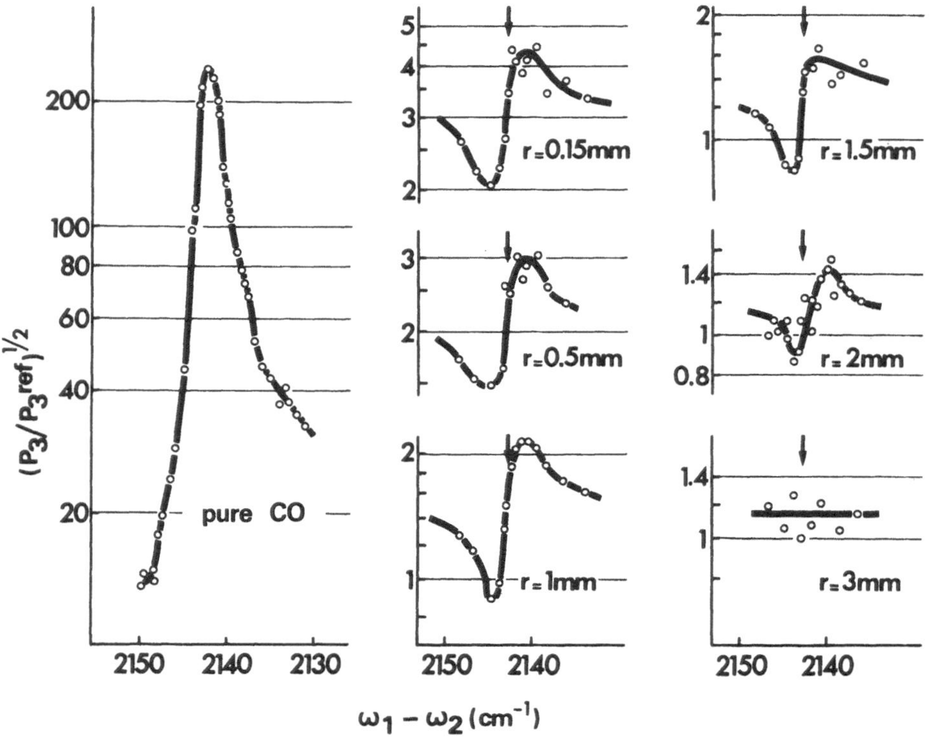

$(P_3/P_3^{ref})^{1/2}$
200
100
80
60
40
20
pure CO
2150
2140
2130
5
4
3
2
r=0.15mm
r=0.5mm
r=1mm
1
r=1.5mm
1.4
0.8
r=2mm
r=3mm
$\omega_1 - \omega_2 (cm^{-1})$

Figure 4 - Search for CO in the flame. All plots are on the same scale. The contour of pure CO at 300 K is given for calibration. The other plots are taken in the flame, 4 to 6 measurements being averaged at each point

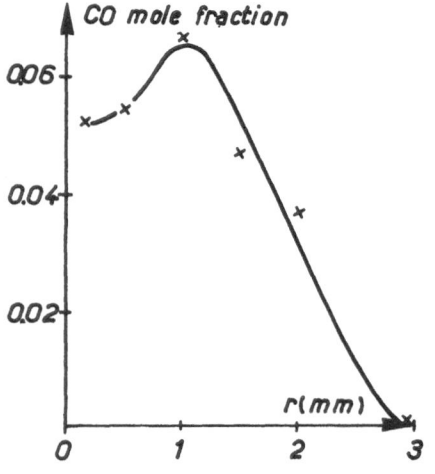

CO mole fraction
0.06
0.04
0.02
r(mm)
0
1
2
3

Figure 5 - Horizontal distribution of CO vs r

laser pumped dye lasers and flash pumped dye lasers do not offer sufficient power and beam quality at the present time for dilute gas phase studies.

Our recent efforts concentrated on the combustion of ethylene glycol on a spherical porous burner of 26 mm diameter. In order to obtain temperature data, N_2 was monitored first, then a search for other gases was undertaken.

A spectrum of atmospheric N_2 illustrates the capabilities of the set-up Fig.3. The high resolution plot (insert) was obtained with 0.15 cm^{-1} resolution. On the main O,Q,S plot, the Q-branch envelope is well resolved, and the individual lines are barely seen. A marked interference is seen on either side of the Q branch: a deep hole is "burnt" in the nonresonant background on the S branch side, and a slowly decaying wing extends over the O branch. The S lines benefit from the destructive interference and stick out strongly, while the O lines, which are just as strong, look flattened due to the log display. The agreement between the data and theory is satisfactory. The maximum discrepancy between line positions, which is about 1 cm^{-1} for S(11), is attributed to the uncertainty in the etalon free spectral range and to imperfect mechanical scanning. A gross discrepancy in signal amplitude is also seen between the S(10) and S(12) lines, which we attribute to a systematic reading error. The discrepancies in amplitude at the peaks of O and S lines are caused by the coarseness of our step size (0.5 cm^{-1}) and possible dye laser linewidth fluctuations.

Similar spectra were taken in the flame, then best fits with theoretical contours were sought, yielding N_2 temperature and concentration distributions. For CO, a 15 cm^{-1} wide spectral region about the 2143 cm^{-1} resonance was scanned at several locations across the flame. The spectra were recorded in 30 minutes by firing the laser at 3 shots per minute Fig.4. Comparing them with computer generated contours yielded the concentration profile of Fig. 5.

These results are encouraging. However, because of the presence of the nonresonant background contributed mostly by N_2, it does not seem possible at present to measure mole fractions much below 10^{-3}.(It is important to note that normal Raman scattering would offer poorer detectivities in this flame because of chemiluminescence and of incandescence of particulate matter excited by the laser [18]). There are strong hopes, though , of reducing the time needed for recording CARS spectra and of improving the detectivity.

3 Future Developments

(1) Multiplex CARS

When work on unstable combustion is done, rapid fluctuations in density and temperature occur, causing fluctuations in the Raman spectra. The mean spectrum recorded is only grossly representative of the mean concentration, and is not representative of the mean temperature. In fact, since the spectral density is a nonlinear function of temperature, it is even impossible to retrieve a meaningful mean temperature from a mean spectral contour. The problem can be solved, however, by using a broadband dye laser in conjunction with a monochromatic solid state laser, dispersing the whole anti-Stokes spectrum in a spectrograph, and recording it on a vidicon or an optical multichannel analyzer. Both temperature and density can then be obtained from a single shot. Spectra of various gases were obtained by SCHREIBER and coworkers recently [19] . That of HF is typical (Fig.6). The method also can be used for simultaneous monitoring of several species; this would be of considerable importance for turbulent combustion studies.

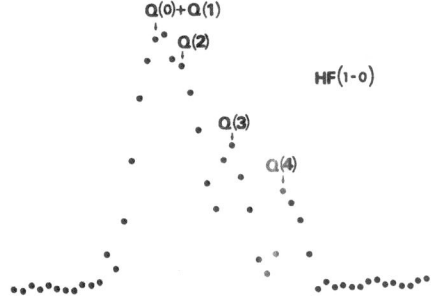

Figure 6-CARS Spectrum of HF (Q-branch of 1-0 band)

(2) Detectivity Improvement

One way of improving detectivity is to enhance χ^{res} with respect to χ^{nr} by tuning one of the field frequencies near the electronic absorption of the molecules detected, but staying in the transparent region of the buffer gas. This is evidenced in the quantum mechanical expression of χ given in terms of discrete eigenstates of the unperturbed molecules :

$$\chi \approx \chi_{buffer}^{nr} + \frac{N\delta^{(o)}}{\hbar^3 3!} \times \frac{1}{(\omega_{ba} - \omega_1 + \omega_2 - i\Gamma_{ba})} \times$$

$$(3)$$

$$\left(\sum_r \frac{\langle a|P|r\rangle\langle r|P|b\rangle}{(\omega_{ra} - \omega_3 - i\Gamma_{ra})} + \frac{\langle a|P|r\rangle\langle r|P|b\rangle}{(\omega_{ra} + \omega_1 + i\Gamma_{ra})} \right) \times \left(\sum_r \frac{\langle b|P|r\rangle\langle r|P|a\rangle}{(\omega_{ra} + \omega_2 - i\Gamma_{ra})} + \frac{\langle b|P|r\rangle\langle r|P|a\rangle}{(\omega_{ra} - \omega_1 - i\Gamma_{ra})} \right)$$

where $|r\rangle$ stands for any eigenstate of the molecules, $|a\rangle$ and $|b\rangle$ being the vibrational states coupled by the Raman transition; $\omega_{ba} \simeq \omega_v$ is the Raman shift; ω_{ra} are the absorption frequencies of the molecules with damping Γ_{ra} taking into account the homogeneous and inhomogeneous linewidths of the absorption; P is the dipole moment operator. We have assumed in (3) that the initial population of state $|b\rangle$ is negligible. This expression shows resonant enhancement if ω_1 or ω_3 approaches the frequency $\omega_{r_0 a}$ of an electronic absorption of the medium [20]. One can also increase the signal by tuning ω_1 or ω_3 into the vibrational dissociative absorption continuum of the molecules : a similar expression, but of integral form, is obtained for χ. The gain expected from these two types of excitation can be inferred from experimental results obtained in resonant Raman Scattering (RRS) where the cross-section reads :

$$\frac{d\sigma}{d\Omega} \propto \left| \sum_r \frac{\langle b|P|r\rangle\langle r|P|a\rangle}{(\omega_{ra} - \omega_1 - i\Gamma_{ra})} + \frac{\langle b|P|r\rangle\langle r|P|a\rangle}{(\omega_{ra} + \omega_2 - i\Gamma_{ra})} \right|^2$$

Sensitivities of about 1 ppm are anticipated, with minimal interference from fluorescence. However the effects of a single photon resonance on CARS cannot be predicted as easily as in RRS, and one should be very careful with interferences arising when the initial population of |b> is not negligible and when ω_2 is also in resonance.

Another alternative is to reduce χ^{nr} in a double resonance experiment as proposed by LYNCH et al. [21] . Here, a third frequency ω_0 is also exciting the medium so that an additional resonance is caused to interfere. By proper selection of ω_0, one can substantially reduce the background response. A quick look at Fig. 3 shows that tuning $\omega_0 - \omega_2$ to the dip between the S(4) and S(5) lines of N_2 would cut background down by a factor of 20.

(3) CW CARS and Density Fluctuations

BARRETT and BEGLEY recently demonstrated a cw CARS experiment [22]. The advantage of using cw lasers in CARS is the excellent spectral purity of the lasers. One can thus perform ultra-high resolution Raman spectroscopy with large signal to noise ratio. Using an intracavity arrangement, HIRTH [23] had powers in excess of 100 W on his argon ion laser and dye laser lines. With such powers, he was able to plot a spectrum like that on the insert of Fig. 3 in 10 ms. The CARS intensity is in fact so high that the time necessary to collect 10^4 photoelectrons, i.e. sufficient for a 1% measurement accuracy, is less than 100 µs. With a moderate increase in laser powers, a 1 µs response time should be achieved. This excellent sensitivity has enabled him to follow density fluctuations in a turbulent jet.

In conclusion, CARS spectroscopy in gases has an immense field of applications, ranging from analysis of reactive media to high resolution Raman spectroscopy. It is emerging as a strong competitor to normal Raman scattering in flame diagnostics. It is at present the only conceivable tool for point, non intrusive diagnostics of plamas and discharges in low pressure gases. However, CARS is a delicate technique to implement. Expensive and highly performing lasers are needed , and detectivity remains limited.

A new step will be taken by making use of electronic resonance enhancement, thus pushing the sensitivity down to the ppm level. In the process, one should not overlook the fact that resonant spectra will present a complicated structure, but there too, CARS will offer a significant advantage over normal Raman spectroscopy in that fluorescence interference will be minimized.

References

1   P.D. Maker and R.W. Terhune, Phys. Rev. 137, A801 (1965).

2   J.P. Coffinet and F. DeMartini, Phys. Rev. Letters 22, 60 (1969).

3   W.G. Rado, Appl. Phys. Letters 11, 123 (1967).

4   G. Hauchecorne, F. Kerherve and G. Mayer, J. de Physique 32, 47 (1971).

5   J. Lukasik and J. Ducuing, Phys. Rev. Letters 28, 1155 (1972).

6   J.J. Wynne, Phys. Rev. Letters 29, 650 (1972).

7   M.D. Levenson, C. Flytzanis and N. Bloembergen, Phys. Rev. 6, 3462 (1972).

8   R.F. Begley, A.B. Harvey and R.L. Byer, Appl. Phys. Letters 25, 387 (1974).

9   N. Bloembergen, "Nonlinear Optics", W.A. Benjamin, Inc, New York (1965).

10   E. Garmire, F. Pandarese and C.H. Townes, Phys. Rev. Letters 11, 160 (1973).

11   A. Yariv, "Quantum Electronics", John Wiley and Sons, Inc., New York, Second Edition (1975).

12   C.L. Tang, Phys. Rev. 134, A1166 (1964).

13   M. Maier, W. Kaiser and J.A. Giordmaine, Phys. Rev. 177, 580 (1969).

14   F. Moya, S. Druet, M. Péalat and J.P.E. Taran, "Flame Investigation by Coherent Anti-Stokes Raman Scattering", AIAA 14th Aerospace Sciences Meeting, Washington D.C., (1976), AIAA Paper 76-29.

15   P. Régnier, "Application of Coherent Anti-Stokes Raman Scattering to Gas Concentration Measurements and to Flow Visualization", ONERA, Technical Note No. 215, 92320 Châtillon, France, 1973, European Space Agency Technical Translation No. ESATT 200, Nov. 1975.

16   F. Moya, S.A. J. Druet and J-P. E. Taran, in " Laser Spectroscopy", (S. Haroche, J.C. Pebay-Peyroula, T.W. Hansch and S.E. Harris, Eds.) Springer-Verlag, New York, 1975, p. 66.

17   J.W. Nibler, J.R. Mc Donald and A.B. Harvey, "CARS Measurements of Vibrational Temperatures in Electric Discharges" Optics Com., to be published.

18   A.C. Eckbreth, AIAA Paper No 76-27.

19   P.W. Schreiber, W.B. Roh and J-P.E. Taran, 27th Pittsburgh Conference on Analytical Chemistry and Applied Spectroscopy, Cleveland, Ohio, March 1976, Paper 389.

20   S.A.J. Druet, thesis, to be published.

21   R.T. Lynch, S.D. Kramer, H. Lotem and N. Bloembergen, Optics Com., in press

22   J.J. Barrett and R.F. Begley, Appl. Phys. Letters 27, 129 (1975)

23   A. Hirth, in AGARD Conference on Applications of Non-Intrusive Instrumentation in Fluid Flow Research, Saint-Louis, France, May 1976, Paper 15.

DEVELOPMENT OF CARS FOR MEASUREMENT
OF MOLECULAR PARAMETERS

S.A. Akhmanov, A.F. Bunkin, S.G. Ivanov, N.I. Koroteev,
A.I. Kovrigir and I.L. Shumay
University of Moscow, USSR

Introduction

A new modification of Coherent Active Raman Spectroscopy (CARS) technique which gives a possibility of coherent subtraction of a nonresonant
term in the nonlinear susceptibility is proposed and experimentally investigated. The CARS method with a broadband pump is discussed for the purpose of instantaneous analysis of rarefied gases and samples with low vapor
pressure. CW CARS was used to study liquids. It was shown that CARS
using single-frequency cw tunable lasers is an excellent method for ultra-
high resolution Raman spectroscopy. CARS with tunable lasers [1,2]
has recently been extensively applied both to measurements of nonlinear
susceptibilities [3,4,5] and for analytic purposes, particularly for analyzing gas mixtures [6,7,8]. Attention will be paid here to the new
aspects of CARS for measurements of molecular parameters.

First, as known, serious difficulties arise when measuring the molecular or lattice parameters of liquids and solids due to a nondispersive
background in the CARS spectra [2,3,8]. The difficulties are caused by
the usual coherent four-wave interaction due to an electronic contribution
χ^{NR} to the third order nonlinear susceptibility $\chi^{(3)} = \chi^R + \chi^{NR}$. Usually,
the resonant contribution χ^R of non-totally symmetric vibration of the first
and, especially, higher orders is less than the electronic contribution
($\chi^R \lesssim \chi^{NR}$ or even $\chi^R \ll \chi^{NR}$). We propose the compensation of the
electronic contribution to CARS spectrum by adding specially selected admixtures, which either can absorb at one of the pump frequencies or are
transparent, but have negative non-linear susceptibility χ^E compared with
the Raman-active substance.

Second, it is demonstrated that CARS with broadband pump proposed
in [2,9] can become a unique method for remote analysis of rarefied gases
and samples with low vapor pressure, because the intense spectrum of the
Raman-active vibrations can be recorded in a broadband spectral range during a single shot of a Q-switched or mode-locked laser. In this case, the
spectrum of the coherently scattered light is a broadband pump spectrum
transformed by coherent nonlinear interaction into the anti-Stokes spectral

range and modulated by the value $|\chi_a^{(3)}(2\omega_1 - \omega_2)|^2$. In this way, the dispersion curve $|\chi^{(3)}|^2$ can be resolved by properly normalizing the CARS spectrum with respect to the broad-band pump spectrum.

Finally, the experimental results of cw CARS of about the 2326 cm^{-1} resonance in liquid nitrogen are reported. Ultra-high resolution Raman spectroscopy can be realized when a cw laser with an extremely narrow and stable spectrum is used.

CARS of mixtures: coherent subtraction of the electronic background

We observed a strong distortion of the spectrum of CARS signal from a transparent liquid when a small amount of a dye absorbing one of the pump frequencies was mixed into it. This distortion is similar to that which takes place when both Raman- and single-photon resonance are in the same molecules. The effect is due to the interference between the nonlinear susceptibilities of the different molecules of the solution and consists in coherent subtraction of the nonresonant electronic background χ^{NR} of the transparent solvent by the dye molecules.

Our experiment was carried out with liquid toluene $(C_6H_5CH_3)$ whose "active" (i.e. CARS) spectrum is known [2]. The Raman line with frequency $\Omega_R/2\pi c = 1209$ cm^{-1} was studied. We measured the intensity of the coherently scattered anti-stokes signal (frequency ω_a) of the second harmonic (SH, frequency ω_1) of a repetitively Q-switched Nd : YAG laser when tuning the pump frequency ω_2. The light from the tunable dye-laser pumped with a fraction of the SH light was used as a source of tunable pump frequency, so that $\omega_a = \omega_1 + (\omega_1 - \omega_2)$.

We were mainly interested in the connection between the dye concentration and the contrast ratio k of the toluene "active" spectrum. The contrast ratio k is the ratio of the maximum coherent anti-Stokes signal $I_{max}^{(a)}$ to the minimum signal $I_{min}^{(a)}$

$$k = \frac{I_{max}^{(a)}}{I_{min}^{(a)}} \simeq \frac{\max|\chi^{(3)}(\omega_a')|^2}{\min|\chi^{(3)}(\omega_a'')|^2}$$

The character of the spectrum distortion is shown in Fig. 1. Experimental dependence of the contrast ratio k of a solution on the absorbing molecule's density N_2 is shown in Fig. 2. It can be clearly seen from Fig. 2 that k increases by more than an order of magnitude when $N_2 = 0.4 \cdot 10^{16}$ cm^{-3}. This enhancement of k is due to the coherent subtraction of an electronic background.

The following expression can be written for the non-linear susceptibility $\chi^{(3)}$ of a medium consisting of two kinds of molecules, one of which exhibits Raman resonance $\omega_1 - \omega_2 = \Omega_R$, and the other single-photon resonance [10, 11] :

$$\chi_{1111}^{(3)}(\omega_a = 2\omega_1 - \omega_2) = 3\chi_{1111}^{NR} + \frac{\bar{\chi}_{1111}^R}{i - \Delta} + \chi_{1111}^{E'} - i\chi_{1111}^{E''} \tag{1}$$

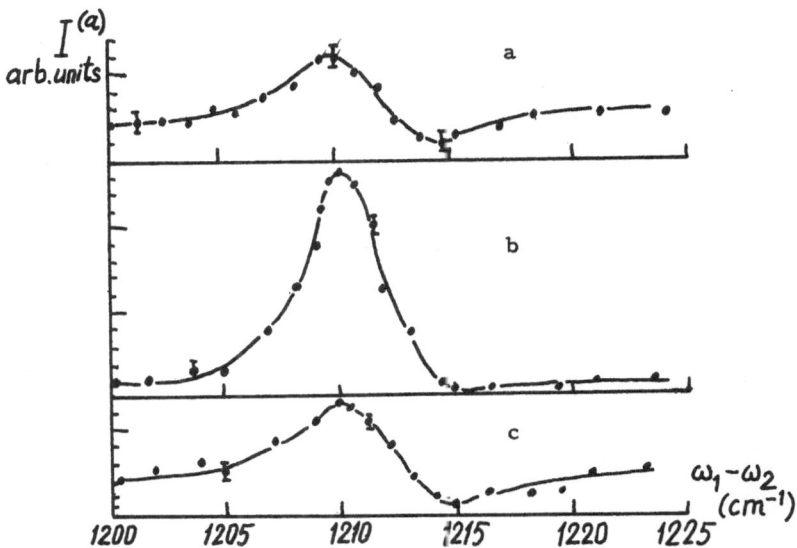

Fig. 1   Distortions of the anti-Stokes CARS signal spectrum of the
1209 cm^{-1} line of toluene with an absorbing admixture:
a - pure toluene; b - toluene with rhodamine 6G, $N_2 = 0.45 \cdot 10^{16}$
cm^{-3}; c - toluene with rhodamine 6G, $N_2 = 1.7 \cdot 10^{16}$ cm^{-3}

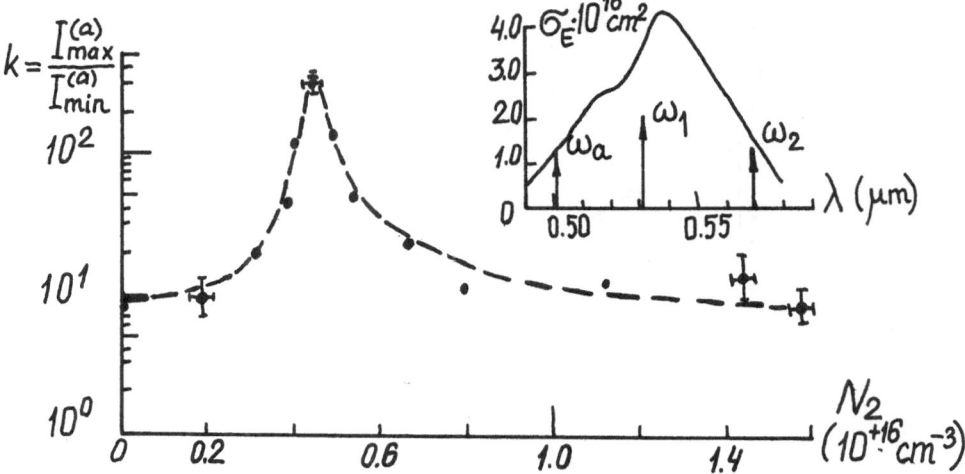

Fig. 2   Experimental dependence of the contrast ratio $k = I_{max}^{(a)}/I_{min}^{(a)}$
of an "active" spectrum $|\chi_{1111}^{(3)}(\omega_a)|^2$ near Raman resonance
$\Omega_R/2\pi c = 1209$ cm^{-1} on molecular density N_2 of rhodamine 6G.
In the insert, the absorption band of rhodamine 6G is shown.

392

Here $\Delta = (\omega_1 - \omega_2 - \Omega_R)/\Gamma_R$; Γ_R is HWHM of the Raman-resonance, $\overline{\chi}^R = (N_1 c^4/\hbar \omega_2^4 \Gamma_R)(d\sigma/d\Omega)$; N_1 is the density of the Raman-active molecules; $d\sigma/d\Omega$ is the cross-section of the spontaneous Raman scattering; $\chi^{E'}$ and $\chi^{E''}$ are proportional to the density of absorbing molecules N_2, and are the real and imaginary parts respectively of the resonant contribution of dye molecules to $\chi^{(3)}$.

Fig. 3 shows the third-order nonlinearity $\chi^{(3)}$ in the complex plane. Contributions from the various terms of eq. (1) to $\chi^{(3)}$ are shown in Fig. 3 separately. When scanning ω_2 the complex value of $\chi^{(3)}$ in a pure liquid follows the circle shown by the solid curve. When some amount of a dye is dissolved in such a liquid, the circle translates along the complex vector χ^E (the dashed line). Obviously, the direction of the translation is determined by the ratio $\chi^{E''}/\chi^{E'}$, and the length of the translation vector χ^E is proportional to the dye concentration N_2. The broken line presents the locus of

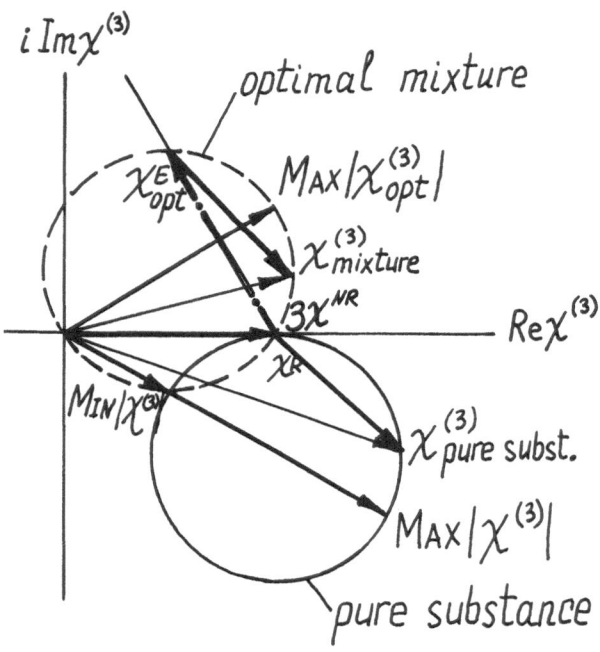

Fig. 3   Geometric representation of the third-order nonlinear susceptibility, $\chi^{(3)}$, of a mixture; all different contributions to $\chi^{(3)}$ are
shown. The broken line indicates a locus of $\chi^{(3)}$ when concentra-
tion of N_2 is optimal: $k = \max |\chi^{(3)}_{opt}|^2 / \min |\chi^{(3)}_{opt}|^2 \to \infty$.

$\chi^{(3)}$ of a solution with the optimal concentration of the dye molecules, since in this case

$$k = \max | \chi^{(3)}_{opt} |^2 / \min | \chi^{(3)}_{opt} |^2 \to \infty$$

Now it can easily be shown that the contrast ratio k of an "active" spectrum of a medium with $\chi^{(3)}_{1111}$ in the form of eq. (1) is in general expressed as

$$k = \frac{\max | \chi^{(3)}_{1111}(\omega'_a) |^2}{\min | \chi^{(3)}_{1111}(\omega''_a) |^2} = (\frac{1+\delta}{1-\delta})^2 \tag{2}$$

where

$$\delta = \frac{\bar{\chi}^R_{1111}/2}{\sqrt{(\chi^{E''}_{1111} + \bar{\chi}^R_{1111}/2)^2 + (3\chi^{NR}_{1111} + \chi^{E'}_{1111})^2}} \tag{3}$$

It can easily be seen that $k \to \infty$ when $\delta \to 1$, i.e. the contrast ratio of the $| \chi^{(3)}(\omega_a) |^2$ dispersion curve increases to infinity in the vicinity of any arbitrary weak Raman line. This is achieved when

$$-\bar{\chi}^R_{1111} \cdot \chi^{E''}_{1111} = (\chi^{E''}_{1111})^2 + (\chi^{E'}_{1111} + 3\chi^{NR}_{1111})^2 \tag{4}$$

This condition, eq. (4), can be satisfied by properly selecting the dye concentration N_2 if $\bar{\chi}^R_{1111} \cdot \chi^{E''}_{1111} \leq 0$.

If eq. (4) is not accurately satisfied for any N_2, k is maximized by maximizing δ. This is achieved by selecting N_2 properly.

The diagram presented in Fig. 3 corresponds exactly to our experimental situation since it is shown that in the pure substance $\delta = 0.53$ and $k = 10$ as was experimentally obtained for the $\Omega_R/2\pi c = 1209$ cm^{-1} line of toluene. Besides that, the experimental function $k(N_2)$ shown in Fig. 2 exhibits only a single peak.

We conclude from our experimental data and Fig. 3 that

$$| \chi^{E}_{1111}| = 3\chi^{NR}_{1111} = 4.5 \times 10^{-14} \text{ cm}^3/\text{erg}$$

when $N_2 = 0.45 \times 10^{16}$ cm^{-3} (the value of χ^{NR}_{1111} is taken from [2] with correction for the degeneracy frequency factor). The resonant hyperpolarizability of a rhodamine 6G molecule can be estimated from the expression

$$|< \gamma^{E}_{1111} >| = \frac{| \chi^{E}_{1111}|}{L^4 N_2} = \frac{4.5 \times 10^{-14}}{4 \times 0.45 \times 10^{16}} \frac{\text{cm}^6}{\text{erg}} = 2.5 \times 10^{-31} \frac{\text{cm}^6}{\text{erg}}$$

where $L = (n^2+2)/3$ is the Lorentz factor of the internal field, and

$$\text{arctg}\,(<\gamma^{E''}_{1111}>/<\gamma^{E'}_{1111}>) = 2\,\text{arctg}\,(\overline{x}^{R}_{1111}/6\chi^{NR}_{1111}) \simeq 1.14$$

i.e. $<\gamma^{E''}_{1111}>/<\gamma^{E'}_{1111}> \simeq 2.2$; both $<\gamma^{E'}_{1111}>$ and $<\gamma^{E''}_{1111}>$ being negative.

In addition to this way of compensating the nonresonant electronic susceptibility, one can mix the liquid or gas under study with a substance with $\chi^{E'} < 0$ and $\chi^{E''} = 0$.

This is now experimentally demonstrated in our laboratory. Summarizing, we should say that the CARS modification proposed here in which mixtures are used, is of interest from two aspects: first, as a method of coherent "subtraction" of a nonresonant electronic background and, second, as a way of measuring the resonant nonlinear susceptibility of an absorbing component by observing the interference quenching of the mixture nonlinearity.

CARS with a broad band pump

Here we want to emphasize that the modification of CARS with a broad pump first proposed and experimentally demonstrated in [2,9], can become a useful technique for the study of the dispersion curve of $|\chi^{(3)}(\omega_a)|^2$ in a broad spectral range during a single laser shot with duration from 10^{-8} sec to 10^{-12} sec.

Fig. 4 shows the spectrum of the CARS signal. It was obtained in the spectral range near the Raman doublet 1004 - 1031 cm^{-1} of toluene using a broad-band tunable pump. The lineshape of the pump is shown by the dashed line in Fig. 4. The result of normalizing the "active" spectrum over the broad-band pump lineshape is shown in Fig. 4 with the dotted line. As a matter of fact, this curve is the dispersion curve of $|\chi^{(3)}(\omega_a)|^2$. More detailed analysis of this CARS technique can be found in [2,9].

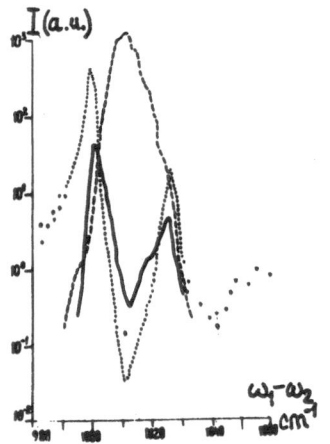

Fig. 4 Dispersion curve $|\chi^{(3)}_{1122}(\omega_a)|^2$ of pure liquid toluene (dotted line) obtained by modified CARS technique with broad-band pump. Spectrum of CARS signal (solid line). Spectrum of broad-band tunable pump (broken line).

cw CARS

cw CARS is a novel technique and only a few experiments have been published [12-15]. We report here the first experimental observations of the Raman spectrum of a liquid (condensed nitrogen) using the cw CARS technique [13]. The results obtained enable the absolute value and the sign of the ratio of the Raman and electronic contributions to the optical third order susceptibility $\chi^{(3)}_{1111}$ to be determined. It should be emphasized that the cw CARS spectroscopy with narrow-band tunable cw lasers can become a unique method for ultra-high resolution of Raman transitions.

The experimental set up is shown in Fig. 5. The output power P_1 of the single-mode argon ion laser was 2 W at λ_1 = 5145 Å. The tunable dye laser (rhodamine 6G) had an output power P_2 of 150 mW and had a linewidth of about 0.6 cm^{-1}. An achromatic lens Λ_2 focused both pump beams into a Dewar containing liquid nitrogen.

The highest signal was obtained when the angle α between the pump beams in air was 2°30', Fig. 6. The power P_a of the anti-Stokes scattering was 1.5 times that of the spontaneous Raman scattering of the same light beam. However, in the case of collinear propagation of the pump beams ω_1 and ω_2 (α = 0), the generated anti-Stokes radiation was only an order of magnitude smaller than in the case of phase matching. The anti-Stokes signal then propagated in a cone with a vertex of about 3°. Fig. 7 shows the spectrum recorded when scanning the monochromator, Fig. 5, and the output spectrum of the tunable dye laser. The spectral resolution was 2 cm^{-1}. Processing of the observed spectrum in the same manner as with a broad-band excitation [2;9] and using the Raman linewidth of liquid N$_2$ (0.067 cm^{-1}) we estimate

$$\bar{\chi}^R_{1111}/3\chi^{NR}_{1111} = +190 \pm 40$$

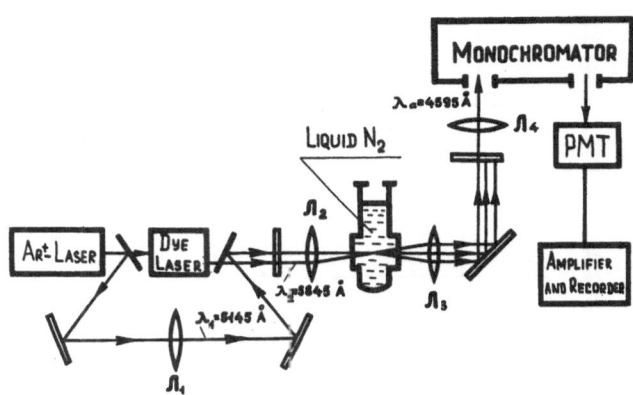

Fig. 5   Experimental set-up for cw CARS spectroscopy of liquids

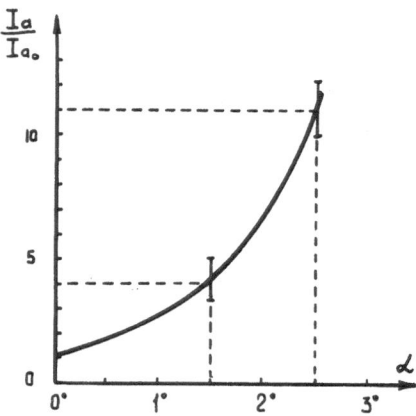

Fig. 6 Power of anti-Stokes CARS signal from the 2326 cm^{-1} Raman line of liquid nitrogen versus angle α between pump beams (in air)

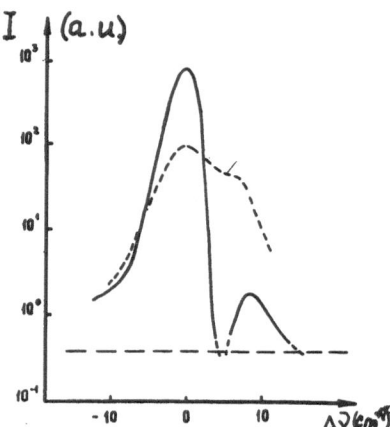

Fig. 7 Spectrum of an "active" signal from condensed nitrogen (solid line). Spectrum of the dye laser radiation used in the CARS (dotted line).

while the determination of the ratio of anti-Stokes power at resonance $(\Delta = 0)$ and far from resonance $(\Delta \neq 0)$ gives

$$| \overline{\chi}^{R}_{1111}/3\chi^{NR}_{1111} | = \left[\frac{P_a(\Delta = 0)}{P_a(\Delta \neq 0)} \right]^{\frac{1}{2}} = 200 \pm 40$$

Thus, we have experimentally shown that even using a relatively broadband cw tunable laser one can determine the molecular Raman transition parameters by taking advantage of active spectroscopy method. Much more detailed information can be obtained by using a single-frequency dye laser.

References

1. S. A. Akhmanov, V. G. Dmitriev, A. I. Kovrigen, N. I. Koroteev, V. G. Tunkin, A. I. Kholodnykh: JETP Lett. 15, 600 (1972)
2. S. A. Akhmanov, N. I. Koroteev: JETP 67, 1306 (1974)
3. M. D. Levenson, N. Bloembergen: Phys. Rev. B10, 4447 (1974)
4. R. T. Lynch, S. D. Kramer, H. Lotem, N. Bloembergen: Opt. Comm. 16, 372 (1976)
5. I. Itzkan, D. A. Leonard: Appl. Phys. Lett. 26, 106 (1975)
6. F. Moya, S. A. J. Druet, J. -P. E. Taran: Opt. Comm. 13, 169 (1975)
7. J. W. Nibler, J. R. McDonald, A. B. Harvey: Rep. on IX Intern. Conf. on Quant. Electr., Amsterdam, The Netherlands (1976)
8. J. -P. E. Taran: Tunable Lasers and Applications, Springer Verlag, Berlin - Heidelberg - New York (1976)
9. S. A. Akhmanov, N. I. Koroteev, A. I. Kholodnykh: J. Ram. Spectr. 2, 239 (1974)
10. N. I. Koroteev; Quant. Electr. 3, No. 4 (1976)
11. S. A. Akhmanov: Lectures in Fermi Summer School of Physics, Varenna, Italy (1975)
12. J. J. Barrett, R. F. Begley: Appl. Phys. Lett. 27, 129 (1975)
13. S. A. Akhmanov, N. I. Koroteev, R. Yu. Orlov, I. L. Shumay: JETP Lett. 23, 276 (1976)
14. R. L. Byer, R. L. Herbst, L. A. Kulevsky: Post Deadline paper at IX Intern. Conf. of Quant. Electr., Amsterdam, June 14-18 (1976)
15. A. Hirth, ibid

List of Participants

Mr. K.S. Aarland
Laererhogskolen
7000 Trondheim
Norway

Dr. R. Airey
Code 5540
Naval Res. Lab.
Washington DC 20375
USA

Dr. R.V. Ambartzumian
Institute of Spectroscopy
Academy of Sciences USSR
Moscow, Podolskii Rayon
Akademgordok 142092
USSR

Dr. E. Arthurs
Barr and Stroud Ltd.
Caxton St.
Anniesland-Glasgow
Scotland

Dr. A.E. Behringer
Office of Naval Research
1030 East Green St.
Pasadena
California 91106
USA

Fsj. J.O. Berg
Inst. for Atomenergi
2007 Kjeller
Norway

Dr. M. Berry
Dept. of Chem.
Univ. of Wisconsin
Madison
Wisconsin 53706
USA

Mr. A. Bjerkestrand
Norwegian Def. Res. Est.
P.O.Box 25
2007 Kjeller
Norway

Mr. A. Blenes
Norwegian Def. Res. Est.
P.O.Box 25
2007 Kjeller
Norway

Dosent K. Blotekjaer
Inst. for Teor. Elektrotekn.
NTH
Trondheim
Norway

Prof. N. Bloembergen
Div. of Applied Physics
Harvard University
Cambridge, Mass. 02138
USA

Dr. C. Bordonali
RIT - Divisione Tecnlogia, CNEN
Centro Studi Nucl. della Casaccia
Casella Postole 2400
00100 Roma
Italy

Prof. D. Bradley
Appl. Opt. Sect., Dept. of Phys.
Imp. Coll. of Science and Tech.
Prince Consort Road
London SW7 2BZ
England

Dr. R.G. Brewer
IBM Research Laboratories
Monterey & Cottle Roads
San José, Cal. 95114
USA

Dr. S.R.J. Brueck
MIT Lincoln Laboratory
P.O.Box 73
Lexington, Mass. 02173
USA

Dr. P. Burlamacchi
Lab. di Electronica
Quantistica
Via Panciatichi 56/30
Florence
Italy

Prof. R. Byer
Microwave Laboratory
Stanford Univ.
Stanford, Cal. 94305
USA

Dr. C.D. Cantrell
Los Alamos Sci. Lab.
Los Alamos
New Mexico 87545
USA

Dr. P.O. Clark
DARPA
1400 Wilson Blvd.
Arlington, VA 22209
USA

Cand Real A.J. Dale
Norsk Hydro
Oslo
Norway

Dr. T.F. Deutsch
MIT Lincoln Laboratory
P.O.Box 73
Lexington, Mass. 02173
USA

Prof. J. Ducuing
Lab. d'Optique Quant. du CNRS
Lab. de l'Ecole Polytechnique
Route de Saclay
91120 Palaiseau
France

Prof. H. Edgerton
Dept. of Electrical Eng.
P.O.Box 73
Lexington, Mass. 02173
USA

Dr. M. Ehrenberg
Karolinska Inst.
Stockholm
Sweden

Prof. S. Eng
Dept. Electr. Measurements
Chalmers Univ. Technology
Gothenburg 5
Sweden

Dr. G.J. Ernst
Twenze Univ. of Techn.
P.O.Box 217
Enschede
Holland

Dr. J. Ewing
AVCO Everett Res. Lab.
2385 Revere Beach Blvd.
Everett, Mass. 02149
USA

Mr. J. Falnes
Inst. for Eksperimental Fys.
NTH
7000 Trondheim
Norway

Prof. B.J. Finlayson Pitts
Dept. of Chemistry
California State Univ.
Fullerton, Cal. 92634
USA

Dr. P. Forrester
Royal Radar Est.
Gt. Malvern
England

Dr. S. Freund
Los Alamos Scientific Laboratory
Los Alamos, New Mexico 87544
USA

Dr. R. Frey
Lab. d'Optique Quant. du CNRS
Route de Saclay
91120 Palaiseau
France

Dr. A. Gibson
Physics Dept.
Univ. of Essex
Wivenhoe Park
Colchester, Essex
England

Dr. G. Girard
Laboratoires de Marcoussis
Centre de Recherche C.G.E.
Route de Nozay
91460 Marcoussis
France

Dr. A. Graziuk
Lebedev Physical Institute
Leninsky Prospekt 53
Moscow
USSR

Dr. J. Hall
JILA
Nat. Bureau of Standards
Bounlder, Colorado
USA

Prof. T. Hansch
Physics Department
Stanford Univ.
Stanford, Cal. 94305
USA

Prof. S.E. Harris
Microwave Laboratory
Stanford Univ.
Stanford, Cal. 94305
USA

Dr. A.B. Harvey
Naval Research Lab.
Code 6110
Washington DC 20375
USA

Prof. S.C. Haydon
The Clarendon Laboratory
Parks Road
Oxford
England

Dr. J.M. Hobart
Coherent Radiation
3210 Porter Drive
Palo Alto, Cal. 94304
USA

Dr. P.W. Hoff
U.S. ERDA
Germantown, MD
USA

Prof. H. Inaba
Research Institute of
Electrical Communication
Tohoku University
Sendai
Japan

Dr. I. Itzkin
AVCO Everett Res. Lab.
Everett
Mass. 02149
USA

Dr. P. Jacquinot
Labor. Aime-Cotton Batim. 505
Université Paris-Sud
91405 Orsay
France

Dr. T. Jaeger
Norwegian Def. Res. Est.
P.O.Box 25
Kjeller
Norway

Dr. S. Jarrett
Spectra-Physics Corporation
Mountain View
California
USA

Prof. H.S. Johnston
Dept. of Chem.
Univ. of Calif. at Berkeley
Berkeley, Cal. 94720
USA

Prof. W. Kaiser
Physik Department E 11
Techn. Univ. München
Arcisstr. 21
8000 München 2
West Germany

Dr. A. Kaldor
EXXON Corporate Research Lab.
P.O.Box 45
Linden, N.J. 07036
USA

Mr. P. Kaspersen
SIMRAD
Ensjoveien 18
Oslo 6
Norway

Dr. P.L. Kelley
MIT Lincoln Laboratory
P.O.Box 73
Lexington, Mass. 02173
USA

Prof. R.V. Khokhlov, Rector
Moscow State University
Moscow
USSR

Dr. H. Kildal
MIT Lincoln Laboratory
P.O.Box 73
Lexington, Mass. 02173
USA

Prof. B. Kleman
Res. Inst. of Nat. Def.
Linnegatan
10450 Stockholm 80
Sweden

Siv. Ing. J.O. Klepsvik
Fysisk Institutt
Univ. I Bergen
Bergen
Norway

Dr. L. Klynning
Fysik Institutionen
Stockholm Univ.
11346 Stockholm
Sweden

Prof. K. Kompa
Max-Planck-Institut für
Plasmaphysik
8046 Garching
West Germany

Dr. A.N. Kovrigin
Moscow State University
Moscow
USSR

Dr. W. Krupke
Lawrence Livermore Lab.
P.O.Box 808
Livermore, Cal. 94550
USA

Fsj. B. Landmark
Norwegian Def. Res. Est.
P.O.Box 25
2007 Kjeller
Norway

Dr. A. Laubereau
Physik Department E 11
Techn. Univ. München
Arcisstr. 21
8000 München 2
West Germany

Prof. B. Lax
Magnet Lab.
Mass. Institute of Technology
Cambridge, Mass. 02139
USA

Prof. F. Legay
Lab. de Photophysique Mol.
Univ. de Paris-Sud
91 Orsay
France

Prof. V.S. Letokhov
Institute of Spectroscopy
Academy of Sciences USSR
Moscow, Podolskii Rayon
Akademgordok 142092
USSR

Prof. R. Levine
Dept. of Physical Chemistry
Hebrew Univ.
Jerusalem
Israel

Dr. A. Levin
Nuclear Research Center
Negev
Beer Sheva
Israel

Dir. F. Lied
Norwegian Def. Res. Est.
P.O.Box 25
2007 Kjeller
Norway

Dr. J. Lindskog
Fysikum
Uppsala Univ.
Uppsala
Sweden

Dr. C.T. Lin
Univ. Estadval de Camp.
Caixa Postal 1170
13100 Campinas
Sao Paulo
Brasil

Prof. A. Lofthus
Fysisk Inst.
Oslo Univ.
Oslo
Norway

Dr. H. Lotsch
Springer-Verlag
Postfach 105280
Neuenheimer Landstr. 28-30
6900 Heidelberg
West Germany

Forsker T. Lund
Norwegian Def. Res. Est.
P.O.Box 25
2007 Kjeller
Norway

Prof. A.L. McWhorter
MIT Lincoln Laboratory
P.O.Box 73
Lexington, Mass. 02173
USA

Dr. M. Michon
Laboratoires de Marcoussis
Centre de Recherche C.G.E.
Route de Nozay
91460 Marcoussis
France

Dr. A. Mooradian
MIT Lincoln Laboratory
P.O.Box 73
Lexington, Mass. 02173
USA

Prof. C.B. Moore
Dept. of Chemistry
Univ. of California at Berkeley
Berkeley, Cal. 94720
USA

Dr. T. Morrow
Dept. of Pure and Appl. Phys.
Queen University
Belfast, BT7 1NN
Northern Ireland

Forsker P.E. Nordal
Norwegian Def. Res. Est.
P.O.Box 25
2007 Kjeller
Norway

Forsker S.O. Olsen
Norwegian Def. Res. Est.
P.O.Box 25
2007 Kjeller
Norway

Dr. C. Pidgeon
Dept. of Physics
Heriot-Watt Univ.
Edinburgh EH1 1HY
Scotland

Dr. H. Pilloff
Office of Naval Research
Washington DC
USA

Dr. J. Pinard
Labor. Aime-Cotton Batim. 505
Université Paris-Sud
91405 Orsay
France

Prof. J.N. Pitts
Statewide Air Poll. Res. Center
Univ. of California at Riverside
Riverside, Cal. 92502
USA

Dr. M. Poliakoff
Dept. of Inorganic Chem.
Univ. of Newcastle
Newcastle NE1 7RU
England

Dr. O. Poulsen
Fysisk Institut
Aarhus Univ.
8000 Aarhus C
Denmark

Prof. S.P.S. Porto
Coord Geral Dos Insti.
Univ. Estadval de Camp.
Caixa Postal 1170
13100 Campinas
Sao Paulo
Brasil

Dr. M.F. Pradere
Labor. d'Optique Quantique
Ecole Polytechnique
Route de Saclay
91120 Palaiseau
France

Dr. H. Rausch
Laser Focus
385 Eliot St.
Newton, MA 02164
USA

Dr. C. Rhodes
Stanford Res. Inst.
Menlo Park
California 92025
USA

Dr. C.P. Robinson
Los Alamos Scientific Laboratory
Los Alamos
New Mexico 87544
USA

Dr. S. Rockwood
Los Alamos Scientific Laboratory
Los Alamos
New Mexico 87544
USA

Dr. H.N. Rutt
Culham Lab.
Abingdon
Oxon
England

Dr. C.A. Sacchi
Inst. di Fisica del Pol.
Piazza Leonardo da Vinci 32
20133 Milano
Italy

Dr. V.V. Savransky
Lebedev Physical Inst.
Leninsky Prospect 53
Moscow
USSR

Prof. F.P. Schafer
Max-Planck-Institut für
Biophysikal. Chemie
3400 Göttingen
West Germany

Prof. Dr. W. Schmidt
Fachbereich 7 (Physik)
Universität Essen
Gesamthochschule
Unionstr. 2
4300 Essen
West Germany

Prof. K. Shimoda
Dept. of Physics
Univ. of Tokyo
Bunkyo-Ku
Tokyo
Japan

Prof. A.E. Siegman
Microwave Laboratory
Stanford Univ.
Stanford, Cal. 94305
USA

Prof. T. Sikkeland
Inst. for Eksperimental Fysikk
NTH
Trondheim
Norway

Dr. A. Siklosi
Tillempad Fysik
Kungl. Tekn. Hogskolan
Stockholm
Sweden

Dr. R. Smith
Dept. of Electronics
Univ. of Highfield
Southampton SO9 5NH
England

Dr. A. Sona
C.I.S.E.
P.O.Box 3986
20100 Milano
Italy

Prof. J. Steinfeld
Chemistry Department
Room 2-221
Mass. Institute of Technology
Cambridge, Mass. 02139
USA

Dr. S. Stenholm
Dept. of Techn. Physics
Helsinki Univ. of Techn.
Otaniemi
Finland

Dr. J.C. Stephenson
National Bureau of Standards
Div. 110
Washington DC 20234
USA

Dr. R. Stern
Lawrence Livermore Lab.
P.O.Box 808
Livermore
California 94550
USA

Prof. B.P. Stoicheff
Dept. of Physics
Univ. of Toronto
Toronto, Ontario M55 IA7
Canada

Forsker P. Stokseth
Norwegian Def. Res. Est.
P.O.Box 25
2007 Kjeller
Norway

Prof. L. Stryer
Yale Univ.
Box 1937, Yale Station
New Haven, Conn. 06520
USA

Dr. S. Svanberg
Dept. of Physics
Chalmers Univ. of Tech.
Fack S-40220 Göteborg
Sweden

Prof. O. Svelto
Instituto di Fisica del
Politecnico
Piazza Leonardo da Vinci, 32
20133 Milano
Italy

Prof. H. Takuma
Dept. of Engineering Physics
Univ. of Electro-Communication
1-5-1 Lhofugaoka Chofu-Shi
Tokyo
Japan

Dr. J. Taran
O.N.E.R.A.
29 Av. de la Division Leclerc
92320 Chatillon
France

Prof. M.L. Termikaelyan
Physical Res. Institute
378410 Ashtarak-2
Armenia SSR
USSR

Dr. E.L. Thomas
Dept. of Appl. Physics
Univ. of Hull
Hull
North Humberside HU6 7RX
England

Prof. A. Tonning
Inst. for Teor. Elektrotekn.
NTH
Trondheim
Norway

Dr. S. Wallace
Dept. of Physics
Univ. of Toronto
Toronto
Ontario M55 1A7
Canada

Prof. H. Walther
Sekt. Physik der Univ. München
Am Coulombwall 1
8046 Garching
West Germany

Dr. G. Wang
Norwegian Def. Res. Est.
P.O.Box 25
2007 Kjeller
Norway

Prof. K. Welge
Dept. of Physics
Univ. of Bielefeld
4800 Bielefeld
West Germany

Dr. P. Williams
Rutherford Laboratory
Chilton, Didcot
Oxon
England

Dr. O. de Witte
Laboratoires de la C.G.E.
Route de Nozay
91460 Marcoussis
France

Dr. J. Wolfrum
Max-Planck-Institut für
Strömungsforschung
Bunzenstrasse
Göttingen
West Germany

Prof. T. Yajima
Univ. of Tokyo
Roppongi, Minato-Ku
Tokyo 106
Japan

Prof. J.T. Yardley
Dept. of Chemistry
Univ. of Illinois
Urbana, Ill. 61801
USA

Prof. A. Yariv
California Inst. of Tech.
Pasadena, Cal. 91125
USA

TITLES OF RELATED INTEREST

DYE LASERS
Topics in Applied Physics, Vol.1
ed. by *F.P. Schäfer*

F.P. Schäfer: Principles of Dye
Laser Operation

B.B. Snavely: Continuous-Wave Dye
Lasers

C.V. Shank, E.P. Ippen: Mode-Locking
of Dye Lasers

K.H. Drexhage: Structure and Properties of Laser Dyes

T.W. Hänsch: Applications of Dye
Lasers

**LASER SPECTROSCOPY of Atoms
and Molecules**
Topics in Applied Physics, Vol.2
ed. by *H. Walther*

H. Walther: Atomic and Molecular
Spectroscopy with Lasers

E.D. Hinkley, K.W. Nill, F.A. Blum:
Infrared Spectroscopy of Molecules
by Means of Lasers

K. Shimoda: Double-Resonance Spectroscopy of Molecules by Means of
Lasers

J.M. Cherlow, S.P.S. Porto: Laser
Raman Spectroscopy of Gases

B. Decomps, M. Dumont: Linear and
Nonlinear Phenomena in Laser Optical
Pumping

K.M. Evenson, F.R. Petersen: Laser
Frequency Measurements, the Speed of
Light, and the Meter

LIGHT SCATTERING IN SOLIDS
Topics in Applied Physics, Vol.8
ed. by *M. Cardona*

M. Cardona: Introduction

A. Pinczuk, E. Burstein: Fundamentals of Inelastic Light Scattering
in Semiconductors and Insulators

R.M. Martin, L.M. Falicov: Resonant
Raman Scattering

M.V. Klein: Electronic Raman Scattering

M.H. Brodsky: Raman Scattering in
Amorphous Semiconductors

A.S. Pine: Brillouin Scattering in
Semiconductors

Y.-R. Shen: Stimulated Raman Scattering

HIGH-RESOLUTION LASER SPECTROSCOPY
Topics in Applied Physics, Vol.13
ed. by *K. Shimoda*

K. Shimoda: Introduction

K. Shimoda: Line Broadening and Narrowing Effects

P. Jacquinot: Atomic Beam Spectroscopy

V.S. Letokhov: Saturation Spectroscopy

J.L. Hall, J.A. Magyar: High Resolution Saturated Absorption Studies
of Methane and Some Methyl-Halides

V.P. Chebotayev: Three-Level Laser
Spectroscopy

S. Haroche: Quantum Beats and Time-Resolved Fluoresence Spectroscopy

N. Bloembergen, M.D. Levenson: Doppler-Free Two-Photon Absorption
Spectroscopy

LASER MONITORING OF THE ATMOSPHERE
Topics in Applied Physics, Vol.14
ed. by *E.D. Hinkley*

E.D. Hinkley: Introduction

S.H. Melfi: Remote Sensing for Air
Quality Management

V.E. Zuev: Laser-Light Transmission
Through the Atmosphere

R.T.H. Collis, P.B. Russell: Lidar
Measurement of Particles and Gases
by Elastic Backscattering and Differential Absorption

H. Inaba: Detection of Atoms and
Molecules by Raman Scattering and
Resonance Fluorescence

E.D. Hinkley, R.T. Ku, P.L. Kelley: Techniques for Detection of Molecular Pollutants by Absorption of Laser Radiation

R.T. Menzies: Laser Heterodyne Detection Techniques

RADIATIONLESS PROCESSES in Molecules and Crystals
Topics in Applied Physics, Vol.15
ed. by F.K. Fong

F.K. Fong: Introduction

K.F. Freed: Energy Dependence of Electronic Relaxation Processes in Polyatomic Molecules

D.J. Diestler: Vibrational Relaxation of Molecules in Condensed Media

J.C. Wright: Up-Conversion and Excited State Energy Transfer in Rare-Earth Doped Materials

R. Kopelman: Exciton Percolation in Molecular Alloys and Aggregates

INFRARED GENERATION BY NONLINEARITIES
Topics in Applied Physics, Vol.16
ed. by Y.R. Shen

Y.R. Shen   Introduction

R.L. Aggarwal, B. Lax: Mixing of CO_2 Lasers in the Far-Infrared

R.L. Byer, R.L. Herbst: Parametric Oscillation and Mixing

V.T. Nguyen, T.J. Bridges: Difference Frequency Mixing via Spin Nonlinearities in the Far-Infrared

J.J. Wynne, P.P. Sorokin: Optical Mixing in Atomic Vapors

T.Y. Chang: Optical Pumping in Gases

ULTRASHORT LIGHT PULSES
Topics in Applied Physics, Vol.18
ed. by S.L. Shapiro
With contributions by:
S.L. Shapiro; D.J. Bradley;
E.P. Ippen, C.V. Shank; D.H. Auston;
D. von der Linde; K.B. Eisenthal;
A.J. Campillo, S.L. Shapiro

LASER SPECTROSCOPY, Proceedings of the 2nd International Conference, Mégève, France, June 23-27, 1975
Lecture Notes in Physics, Vol.43
ed. by S. Haroche, J.C. Pebay-Peyroula, T.W. Hänsch, S.E. Harris

BEAM-FOIL SPECTROSCOPY
Topics in Current Physics, Vol.1
ed. by S. Bashkin

S. Bashkin: Experimental Methods

I. Martinson: Studies of Atomic Spectra by the Beam-Foil Method

L.J. Curtis: Lifetime Measurements

O. Sinanoğlu: Theoretical Oscillator Strengths of Neutral, Singly-Ionized, and Multiply-Ionized Atoms: The Theory, Comparisons with Experiment, and Critically-Evaluated Tables with New Results

W. Wiese: Regularities of Atomic Oscillator Strengths in Isoelectronic Sequences

W. Whaling: Applications to Astrophysics: Absorption Spectra

L.J. Heroux: Applications of Beam-Foil Spectroscopy to the Solar Ultraviolet Emission Spectrum

R. Marrus: Studies of Hydrogen-Like and Helium-Like Ions of High Z

J. Macek, D. Burns: Coherence, Alignment, and Orientation Phenomena in the Beam-Foil Light Source

I. Sellin: The Measurement of Autoionizing Ion Levels and Lifetimes By Fast Projectile Electron Spectroscopy

SPRINGER-VERLAG
BERLIN HEIDELBERG NEW YORK